THE EFFECTS OF TAURINE ON EXCITABLE TISSUES

W0106260

Monographs of the Physiological Society of Philadelphia

THE EFFECTS OF TAURINE ON EXCITABLE TISSUES

Proceedings of the 21st Annual A. N. Richards Symposium
of the Physiological Society of Philadelphia,
Valley Forge, Pennsylvania, April 23-24, 1979

Edited by

S. W. Schaffer, Ph.D.
Department of Physiology and Biophysics
Hahnemann Medical College, Philadelphia

Steven I. Baskin, Ph.D.
Department of Pharmacology
Medical College of Pennsylvania, Philadelphia

James J. Kocsis, Ph.D.
Department of Pharmacology
Jefferson Medical College, Philadelphia

MTP PRESS LIMITED
International Medical Publishers

Published in the UK and Europe by
MTP Press Limited
Falcon House
Lancaster, England

Published in the US by
SPECTRUM PUBLICATIONS, INC
175-20 Wexford Terrace
Jamaica, N.Y. 11432

Copyright © 1981 Spectrum Publications, Inc.
Softcover reprint of the hardcover 1st edition 1981

All rights reserved. No part of this book may be reproduced in any form, by
photostat, microform, retrieval system, or any other means without prior written
permission of the copyright holder or his licensee.

ISBN-13:978-94-009-8095-2   e-ISBN-13:978-94-009-8093-8
DOI: 10.1007/978-94-009-8093-8

Contributors

R. M. ADEME
Centro de Investigaciones en Fisiologia
  Celular
Universidad Nacional Autónoma de México
México

L. E. ALTO
Departments of Medicine and Physiology
University of Manitoba
Winnipeg, Canada

A. BABA
Department of Pharmacology
Osaka University
Osaka, Japan

J. BAHL
Department of Internal Medicine
University of Arizona
Tucson, Arizona

S. I. BASKIN
The Medical College of Pennsylvania
Department of Pharmacology
Philadelphia, Pennsylvania

H. F. BRADFORD
Department of Biochemistry
Imperial College of Science and Technology
London, England

N. BRESOLIN
Department of Pharmacology
College of Medicine
University of Arizona Health Sciences Center
Tucson, Arizona

R. BRESSLER
Department of Internal Medicine
University of Arizona
Tucson, Arizona

S. CHANG
Department of Internal Medicine
University of Arizona
Tucson, Arizona

J. CHOVAN
Department of Physiology and Biophysics
Hahnemann Medical College
Philadelphia, Pennsylvania

M. F. CRASS, III
Department of Physiology
Texas Tech University School of Medicine
Lubbock, Texas

A. N. DAVISON
Department of Neurochemistry
Institute of Neurology
London, England

N. S. DHALLA
Department of Physiology
University of Manitoba
Winnipeg, Canada

L. DURELLI
Clinica Neurologica
Università di Sassari
Sassari, Italy

L. FEUER
Chinoin Pharmaceutical and Chemical Works
  Ltd.
Budapest, Hungary

C. M. FINNEY
College of Marine Studies
University of Delaware
Newark, Delaware

C. J. FRANGAKIS
Department of Internal Medicine
University of Arizona
Tucson, Arizona

K. FURUKAWA
Department of Pharmacology
Toho University School of Medicine
Tokyo, Japan

G. GAULL
Department of Human Development and
  Nutrition
New York State Institute for Basic Research
  in Mental Retardation
Staten Island, New York

D. GROSSO
Department of Internal Medicine
University of Arizona
Tucson, Arizona

R. J. HOESCHEN
Departments of Medicine and Physiology
University of Manitoba
Winnipeg, Canada

N. HOMMA
Department of Pharmacology
Toho University School of Medicine
Tokyo, Japan

R. HORIE
Japan Stroke Prevention Center and
  Department of Pathology
Shimane Medical University
Izumo, Japan

F. C. G. HOSKIN
Department of Biology
Illinois Institute of Technology
Chicago, Illinois

R. E. HRUSKA
Department of Pharmacology
College of Medicine
University of Arizona Health Sciences Center
Tucson, Arizona

R. J. HUXTABLE
Department of Pharmacology
University of Arizona Health Sciences Center
Tucson, Arizona

R. ITO
Department of Pharmacology
Toho University School of Medicine
Tokyo, Japan

H. IWATA
Department of Pharmacology
Osaka University
Osaka, Japan

A. KARPPINEN
Department of Biochemistry
University of Oulu
Oulu, Finland

Y. KATAYAMA
Institute for Neurobiology
Okayama University Medical School
Okayama, Japan

J. C. KHATTER
Departments of Medicine and Physiology
University of Manitoba
Winnipeg, Canada

D. C. KLEIN
Laboratory of Developmental Neurobiology
National Institute of Child Health and
  Human Development
National Institutes of Health
Bethesda, Maryland

J. J. KOCSIS
Department of Pharmacology
Jefferson Medical College of Thomas
  Jefferson University
Philadelphia, Pennsylvania

P. KONTRO
Department of Biomedical Sciences
University of Tampere
Tampere, Finland

J. KRAMER
Department of Physiology and Biophysics
Hahnemann Medical College
Philadelphia, Pennsylvania

E. KULAKOWSKI
Department of Physiology and Biophysics
Hahnemann Medical College
Philadelphia, Pennsylvania

E. KUMPULAINEN
Department of Biochemistry
University of Oulu
Oulu, Finland

K. KURIYAMA
Department of Pharmacology
Kyoto Prefectural University of Medicine
Kyoto, Japan

P. LÄHDESMÄKI
Department of Biochemistry
University of Oulu
Oulu, Finland

H. E. LAIRD
Department of Pharmacology and Toxicology
University of Arizona
Tucson, Arizona

N. LAKE
Department of Visual Science
Institute of Ophthalmology
University of London
London, England

B. LARSEN
Department of Internal Medicine
University of Arizona
Tucson, Arizona

S. LIPPINCOTT
Department of Pharmacology
University of Arizona Health Sciences Center
Tucson, Arizona

J. B. LOMBARDINI
Department of Pharmacology and
  Therapeutics
Texas Tech University School of Medicine
Lubbock, Texas

A. M. LÓPEZ-COLOMÉ
Centro de Investigaciones en Fisiologia
  Celular
Universidad Nacional Autónoma de México
Oéxico

W. LOVENBERG
Section on Biochemical Pharmacology
National Heart, Lung and Blood Institute
National Institutes of Health
Bethesda, Maryland

K.-M. MARNELA
Institute of Biomedical Sciences
University of Tampere
Tampere, Finland

J. MARSHALL
Department of Visual Science
Institute of Ophthalmology
University of London
London, England

M. MATSUMOTO
Institute for Neurobiology
Okayama University Medical School
Okayama, Japan

B. A. MEINERS
Department of Pharmacology
College of Medicine
University of Arizona Health Sciences Center
Tucson, Arizona

A. MORI
Institute for Neurobiology
Okayama University Medical School
Okayama, Japan

R. MUTANI
Clinica Neurologica
Università di Sassari
Sassari, Italy

M. MURAMATSU
Department of Pharmacology
Kyoto Prefectural University of Medicine
Kyoto, Japan

Y. NARA
Japan Stroke Prevention Center and
  Department of Pathology
Shimane Medical University
Izumo, Japan

C. NAUSS-KAROL
Department of Biochemical Nutrition
Hoffmann-LaRoche Inc.
Nutley, New Jersey

S. OHKUMA
Department of Pharmacology
Kyoto Prefectural University of Medicine
Kyoto, Japan

S. S. OJA
Department of Biomedical Sciences
University of Tampere
Tampere, Finland

A. OOSHIMA
Japan Stroke Prevention Center and
    Department of Pathology
Shimane Medical University
Izumo, Japan

A.C.I. ORAEDU
Department of Visual Science
Institute of Ophthalmology
University of London
London, England

H. PASANTES-MORALES
Centro de Investigaciones en Fisiologia
    Celular
Universidad Nacional Autónoma de México
México

D. K. RASSIN
Department of Human Development and
    Nutrition
New York State Institute for Basic Research
    in Mental Retardation
Staten Island, New York

S. W. SCHAFFER
Department of Physiology and Biophysics
Hahnemann Medical College
Philadelphia, Pennsylvania

S. Y. SCHMIDT
Berman-Gund Laboratory for the Study of
    Retinal Degenerations
Harvard Medical School, Mass Eye and Ear
Boston, Massachusetts

J. SHAFFER
Department of Pharmacology
Jefferson Medical College of Thomas
    Jefferson University
Philadelphia, Pennsylvania

P. L. SONI
Departments of Medicine and Physiology
University of Manitoba
Winnipeg, Canada

R. C. SPETH
Department of Pharmacology
College of Medicine
University of Arizona Health Sciences Center
Tucson, Arizona

J. A. STURMAN
Developmental Neurochemistry Laboratory
Department of Pathological Neurobiology
Institute for Basic Research in Mental
    Retardation
Staten Island, New York

E. J. THOMPSON
Department of Neurochemistry
Institute of Neurology
London, England

T. UCHIYAMA
Department of Pharmacology
Toho University School of Medicine
Tokyo, Japan

C. VANDERWENDE
College of Pharmacy
Rutgers — The State University
Piscataway, New Jersey

M. J. VOADEN
Department of Visual Science
Institute of Ophthalmology
University of London
London, England

J. D. WELTY
Division of Biochemistry, Physiology and
    Pharmacology
The University of South Dakota School of
    Medicine
Vermillion, South Dakota

M. C. WELTY
Division of Biochemistry, Physiology and
    Pharmacology
The University of South Dakota School of
    Medicine
Vermillion, South Dakota

G.H.T. WHELER
Laboratory of Developmental Neurobiology
National Institute of Child Health and
  Human Development
National Institutes of Health
Bethesda, Maryland

S. YAMAGAMI
Department of Pharmacology
Osaka University
Osaka, Japan

K. YAMAGUCHI
Department of Medical Chemistry
Osaka Medical College
Osaka, Japan

H. I. YAMAMURA
Department of Pharmacology
College of Medicine
University of Arizona Health Sciences Center
Tucson, Arizona

Y. YAMORI
Japan Stroke Prevention Center and
  Department of Pathology
Shimane Medical University
Izumo, Japan

S. YODA
Department of Pharmacology
Toho University School of Medicine
Tokyo, Japan

I. YOKOI
Institute for Neurobiology
Okayama University Medical School
Okayama, Japan

Preface

It has become an annual custom for the Physiological Society of Philadelphia to sponsor a spring symposium in honor of A. N. Richards (1876–1966), a research pharmacologist who developed the classical micropuncture technique for studying kidney function. The A. N. Richards Symposium for 1979 was held on April 23–24 in Valley Forge, Pennsylvania. The theme of this symposium was "The Actions of Taurine on Excitable Tissues." Although taurine was discovered as a constituent of bile salts in 1857 by a chemist and an anatomist (Gmelin and Tiedemann), interest today centers chiefly on the extrahepatic actions of taurine, especially in brain, heart, and other excitable tissues. Research on taurine is clearly in a period of exponential growth.

We can be sure that the research reports presented and described herein as the "Proceedings of the Symposium" will provide impetus for further growth. Thus the report describing macromolecular receptors for taurine in myocardial sarcolemma may provide a model for exploring the molecular mechanisms that underlie the action(s) of taurine. Stabilization of membranes and modulation of ion fluxes are two fundamental actions of taurine dealt with in many of these reports. It is just these actions of taurine that have been reported by several investigators as being involved in human myotonia, diabetes, and heart failure.

Other presentations at the symposium will add to the growing consensus that taurine acts as a modulator rather than a transmitter in the CNS. Hypertension, epilepsy, and feline retinal degeneration have been described here by other investigators as diseases in which either the biosynthesis, uptake, or release of tissue taurine may be critical.

The biological effects of the newly discovered taurine analogue γ-glutamyl taurine, along with those of hypotaurine and taurocyamine, again raise the question as to whether it is taurine or its analogues that mediate "the effects of taurine." Reports describing methods for reducing tissue taurine levels provide another approach to answering this same question. Altogether the proceedings of this symposium should provide a useful summary of the rapidly growing knowledge about taurine and its biological effects.

We wish to thank the Physiological Society of Philadelphia for their sponsorship and financial support, and the officers of the society for their personal assistance in making the symposium a success. We also gratefully acknowledge the support of the many foreign participants who attended the symposium (especially our Japanese colleagues) as well as the contributions made by our respective educational institutions, especially our graduate students and technicians. We also thank the following companies for additional financial support: Merck & Co.; Arnar-Stone Laboratories; Smith, Kline & French; ICI Americas, Inc.; Upjohn Co.; Fujisawa Pharmaceutical Corporation; and Wyeth, Inc.

S. W. Schaffer
S. I. Baskin
J. J. Kocsis

THE EFFECTS OF TAURINE ON EXCITABLE TISSUES

Contents

Part IV—Clinical Implications of Taurine

PART I

Metabolism and Function of Taurine Analogues

Introduction

The most widely studied pathway in the biosynthesis of taurine involves the conversion of cysteine to taurine. Although the existence of this pathway has been known for several years, investigators have only recently begun to delineate the factors that regulate this pathway. The first two chapters in this section focus on several such factors that appear to control the rate of taurine biosynthesis. The first reaction in this pathway is catalyzed by the enzyme cysteine dioxygenase and involves the oxidation of cysteine to cysteine sulfinic acid. Yamaguchi describes the purification and partial characterization of this enzyme from rat liver. His data indicate that the activity of the enzyme is regulated by both tissue cysteine content and the presence of a protein factor. Since this reaction may be the rate-limiting step in the synthesis of taurine, he suggests that these factors play a central role in the biosynthesis of taurine in the liver. Iwata et al. have examined the uptake and metabolism of cysteine in rat brain. They find that cysteine uptake, and hence taurine biosynthesis, is regulated by both neural activity and the level of two metabolic intermediates, cysteine sulfinic acid and cysteic acid.

It has been proposed by several investigators that some, if not all, of the effects of taurine may be mediated by one of its derivatives. The next three chapters are concerned with several taurine analogues that exhibit specific biological activity. Feuer reports a newly discovered taurine-containing peptide from the parathyroid gland. He finds that this dipeptide, γ-glutamyl taurine, exhibits both vitamin A-like and taurine-like activity. Based on its biological activity and the observation that the parathyroid gland appeared only with the emergence of land vertebrates, he proposes that the peptide plays an important role in the adaptation of aquatic vertebrates to terrestrial conditions. Mori et al. have examined the distribution and biological activity of taurocyamine in bovine brain. While taurine inhibits seizure activity, they find that taurocyamine produces seizures. Kontro and Oja have characterized the hypotaurine transport system of mouse brain. They show the existence of two uptake processes, both of which are sodium dependent.

The last chapter in this section is concerned with the biosynthesis of isethionate in cephalopod nerve. Hoskin provides some evidence that the carbon for isethionate is derived from glucose while the sulfur comes from thiosulfate.

Copyright © 1981, Spectrum Publications, Inc.
The Effects of Taurine on Excitable Tissues

CHAPTER 1

Cysteine Dioxygenase and Its Possible Role on Taurine Formation in Rats

Kenji Yamaguchi

PURIFICATION AND NOVEL PROPERTIES OF CYSTEINE DIOXYGENASE IN RAT LIVER

Cysteine is recognized as a precursor of taurine in mammals. Chapeville and Fromageot (1955) showed the presence of radioactive cysteine sulfinate (CSA) and hypotaurine, when labeled cysteine was incubated with liver preparations; and Tabachnick and Tarver (1955) demonstrated the appearance of labeled cysteic acid, cystathionine, and taurine from ^{35}S methionine. More recently, rat liver supernatant has been shown to contain an enzyme catalyzing the oxidation of cysteine to CSA (Sörbo and Ewetz, 1965; Wainer, 1965; De Marco et al., 1966; Ewetz and Sörbo, 1966; Lombardini et al., 1969a; Yamaguchi et al., 1971). The enzyme responsible for the catalysis was shown to be a dioxygenase (Lombardini et al., 1969b) and was named cysteine dioxygenase (EC. 1.13.11.20) (Sakakibara et al., 1976). Most attempts to purify the enzyme, however, were unsuccessful because of the great instability of this enzyme. Furthermore, participation by the enzyme in taurine formation in mammals has been questioned because of its extremely low activity (119 nmol/hr/mg protein, Yamaguchi et al., 1971; 240 nmol/hr/g protein, Lombardini et al., 1969a). Recently, two important findings in this connection have been presented from our laboratory (Yamaguchi et al., 1973; Sakakibara et al., 1973; Sakakibara et al., 1976). The first finding was that the enzyme activity in the cytoplasmic preparation from rat liver homogenate was enhanced around 10–20-fold by preincubation with 10 mM L-cysteine under nitrogen gas at 37°. In addition, the enzyme preparation when chromatographed on DEAE-cellulose (DEAE-enzyme) was obtained as an inactive form, which was again activated by anaerobic incubation with L-cysteine. The second finding was that the activated DEAE-enzyme was rapidly and irreversibly inactivated during assay and that this inactivation is completely prevented by a distinct cytoplasmic protein in rat liver, called protein-A. Protein-A was easily separated from the catalytic protein by DEAE-cellulose column chromatography. Based on these findings, the enzyme found in the cytosol fraction of rat liver was purified to homogeneity as shown in Fig. 1.1 (Yamaguchi et al., 1978). Purification of the enzyme is summarized in Table 1.1. Its molecular weight, obtained by gel filtration on Sephadex G-75 and by electrophoresis in SDS-polyacrylamide gel, was calculated as 23,000 and 22,500, respectively. These findings suggest that the enzyme is a single peptide chain.

In previous papers, it has been reported that ferrous iron, and/or copper, is an integral part of cysteine dioxygenase (Sakakibara et al., 1973; Sakakibara et al., 1976). As shown in Table 1.2, the chelating agents having high affinity for Fe^{2+} or Fe^{3+} such as α, α'-dipyridyl, o-phenanthroline, bathophenanthroline, and 8-hydroxyquinoline markedly diminished its enzyme activity. EDTA and EGTA also caused a remarkable decrease in enzyme activity. On the other hand, neither bathocuproine sulfonate nor neucuproine, which have relatively high affinities for Cu^+, nor diethyldithiocarbamate, which is Cu^{2+} specific, exhibited any significant effect on the enzyme activity, either in the presence or absence of protein-A; however, bathophenanthroline sulfonate at a concentration of 1×10^{-4}M, did slightly depress enzymatic activity. It is of interest that the chelating agent inhibited activity only when it was added before preactivation of the enzyme.

The iron and copper contents in various preparations during the purification procedures are shown in Table 1.3. No metal other than iron was detected in significant quantities in purified cysteine dioxygenase. The calculation based on atomic absorption analysis of the highly purified enzyme indicated the presence of 0.8 g atom per 22,000 g of protein. The absorption spectrum of the native enzyme exhibited a typical protein maximum at 278 nm, but the absence of absorption at longer wave lengths suggests that iron is not present as a heme prosthetic group.

Requirement of Cofactor

It has been reported that the enzyme activity in crude cytosol preparation of rat liver was enhanced by exogenous NADH and NADPH (Sörbo and

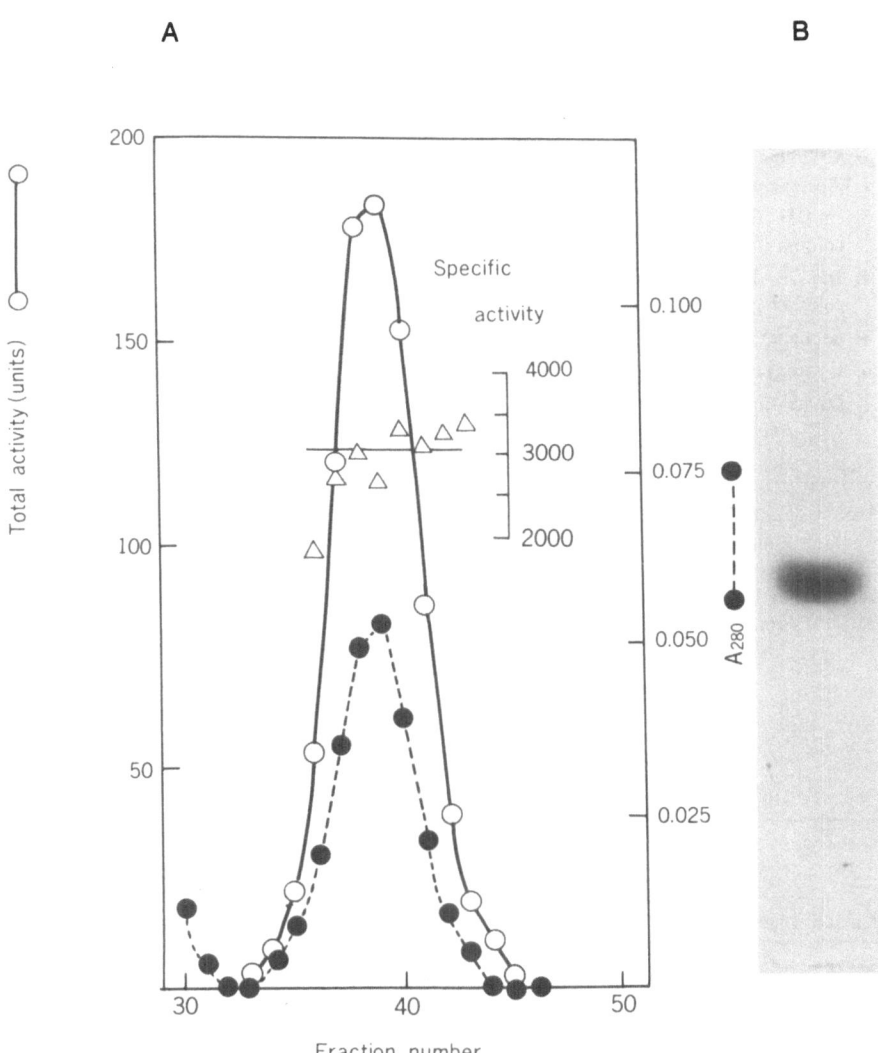

Fig. 1.1. (*a*) Sephadex G-75 gel filtration of cysteine dioxygenase. The pooled, concentrated enzyme fractions from Step 7 in the purification procedure were chromatographed and 2-ml fractions were collected. Plotted are total enzyme activity in units (○), protein concentration in A 280 (●), and specific activity in units per milligram (△). (*b*) Photograph of polyacrylamide gel obtained by electrophoresis of Fraction 39 from the Sephadex column in the presence of sodium dodecyl sulfate. Twenty micrograms of protein were applied to the gel. After electrophoresis, the gel was stained with Coomassie Brilliant Blue. (From data of Yamaguchi et al., 1978.)

Table 1.1   Summary of Purification of Rat Liver Cysteine Dioxygenase[a]

Step	Volume (ml)	Total units	Total protein	Specific activity	Yield (%)
1. Homogenate supernatant	1,414	43,187	39,261	1.100	100
2. Acetone precipitation	934	44,923	14,369	3.126	104
3. 1st DEAE-cellulose column	533	22,493	773	29.10	52
4. 2nd DEAE-cellulose column	175	9,286	175	53.06	22
5. Sephadex G-100 column	95	5,014	4.41	1,138	12
6. Hydroxyapatite column	26	2,551	1.67	1,528	6
7. DEAE-Sephadex A-25 column	25	1,598	0.76	2,103	4
8. Sephadex G-75 column	7	710	0.25	2,840	1.6

[a]Assay of enzyme activity was performed under standard assay conditions. Protein-A, 0.5 mg, was added into assay mixture using the preparation after Step 3. Units are in micromoles of CSA produced per hour. Protein is in milligrams, as determined by the method of Lowry et al., compared to bovine serum albumin as standard. Specific activity is in units per milligram. (From data of Yamaguchi et al., 1978.)

Table 1.2   Effect of Chelating Agents on Cysteine Dioxygenase[a]

Chelating agents (1×10^{-4}M)	Relative activity (%)	
	+Protein-A	−Protein-A
None	100	100
o-Phenanthroline	0	4
Bathophenanthroline sulfonate	12	0
8-Hydroxyquinoline	0	1
α, α'-Dipyridyl	0	42
EDTA	3	1
EGTA	0	5
Neocuproine	82	104
Bathocuproine sulfonate	95	62
Diethyldithiocarbamate	100	130

[a]Assay of enzyme activity was performed under standard assay conditions with or without protein-A. Chelating agents were added into reaction mixture before anaerobic activation procedure. (From data of Yamaguchi et al., 1978.)

Table 1.3   Iron and Copper Contents of Cysteine Dioxygenase[a]

Step	Specific activity (units/mg protein)	Metal contents (µg/mg protein)	
		Iron	Copper
1. Cytoplasmic supernatant	1.078	0.317	0.034
2. Acetone precipitation	2.547	0.595	0.074
3. 1st DEAE-cellulose column	17.12	0.952	N.D.[b]
4. 2nd DEAE-cellulose column	56.62	1.994	N.D.[b]
5. Sephadex G-100	1,138	1.090	N.D.[b]
6. Sephadex G-75	2,166	1.912	N.D.[b]

[a] Assay of enzyme activity was performed under standard assay conditions and 0.5 mg of protein-A was added into assay mixture using the preparation after Step 3. Iron and copper contents were determined by atomic absorption spectrophotometer. (From data of Yamaguchi et al., 1978.)

[b] No significant amount was determined.

Ewetz, 1965; Lombardini et al., 1969a) or NAD (Yamaguchi et al., 1971). However, the activating effects of these nucleotides on enzyme activity were not observed in the preparation after acetone treatment or DEAE-cellulose column chromatography. Furthermore, CSA-forming enzyme activity was not detected in this preparation without anaerobic preactivation with L-cysteine. Purified cysteine dioxygenase did not require any cofactor such as exogenous metals or hydrogen carriers for full enzyme activity.

Distribution of Cysteine Dioxygenase

It has been reported that NAD-dependent cysteine oxidase activity was found not only in liver but also in some extrahepatic tissues such as brain (Misra and Olney, 1975; Yamaguchi et al., 1971), spleen, intestine, heart (Yamaguchi et al., 1971), and retina (Di Giorgio et al., 1975). However, the enzyme activity in these tissues of the rat differed in properties from that of hepatic cysteine dioxygenase. Cysteine dioxygenase activity was found in the livers of various species in the following order of decreasing activity: mouse, rat, pig, dog, rabbit.

It is of interest that no activity was observed in rat hepatoma cells (AH 2440, AH 109A), mouse Ehrlich ascites tumor cells, or in fetal rat liver.

Properties of Partially Purified Protein-A

Protein-A was partially purified from rat liver cytosol by acetone fractionation, DEAE-cellulose, CM-cellulose, and Sephadex G-200 column chromatography (Hosokawa et al., 1978b).

Protein-A has a molecular weight around 80,000, with a pI of 7.8. The stabilizing activity of protein-A could not be duplicated by bovine serum albumin, bovine serum globulin, hemoglobin, ovalbumin, histone, or protamine. A value of 0.2 μg of partially purified protein-A was enough to stabilize 0.5 μg (10 units) of purified cysteine dioxygenase. It appears unlikely that the stabilizing effect of protein-A is associated with any metals since the stabilization was not affected by metal-chelating agents such as EDTA (Yamaguchi et al., 1978). In contrast to cysteine dioxygenase, protein-A is widely distributed in rat tissues such as liver, brain, heart, kidney, and spleen (Hosokawa et al., 1978b).

POSSIBLE ROLE OF HEPATIC CYSTEINE DIOXYGENASE IN TAURINE BIOSYNTHESIS

Dietary Control of Cysteine Dioxygenase Activity in Rat Liver

We have observed that the cysteine dioxygenase activity in rat liver was markedly responsive to changes in dietary protein. Figure 1.2 shows that although cysteine dioxygenase activity increased only slightly with dietary protein levels up to 20%, it increased dramatically at protein levels higher than 20%, reaching a plateau at levels of over 40% protein. On the other hand, cysteine desulfhydrase activity in liver was much lower than that of cysteine dioxygenase and did not significantly respond to dietary protein changes, as shown in Fig. 1.2. We have found that cysteine dioxygenase in rat liver can be induced by either cysteine or methionine (Yamaguchi et al., 1971; Hosokawa et al., 1978a).

An experiment was performed to determine whether the response of cysteine dioxygenase to increased protein intake was due to a specific increase in cysteine and its congeners or to a simple increase in the amino acid pool. The hepatic cysteine dioxygenase activity in rats fed the low-protein diet supplemented with various amino acids was examined. As shown in Table 1.4, inclusion of either cysteine or methionine in low-protein diets resulted in a marked increase of hepatic cysteine dioxygenase activity. It is noteworthy that adding 5% methionine to the diet increased hepatic cysteine dioxygenase activity more than adding 5% cysteine, but 1% added cysteine seemed to be

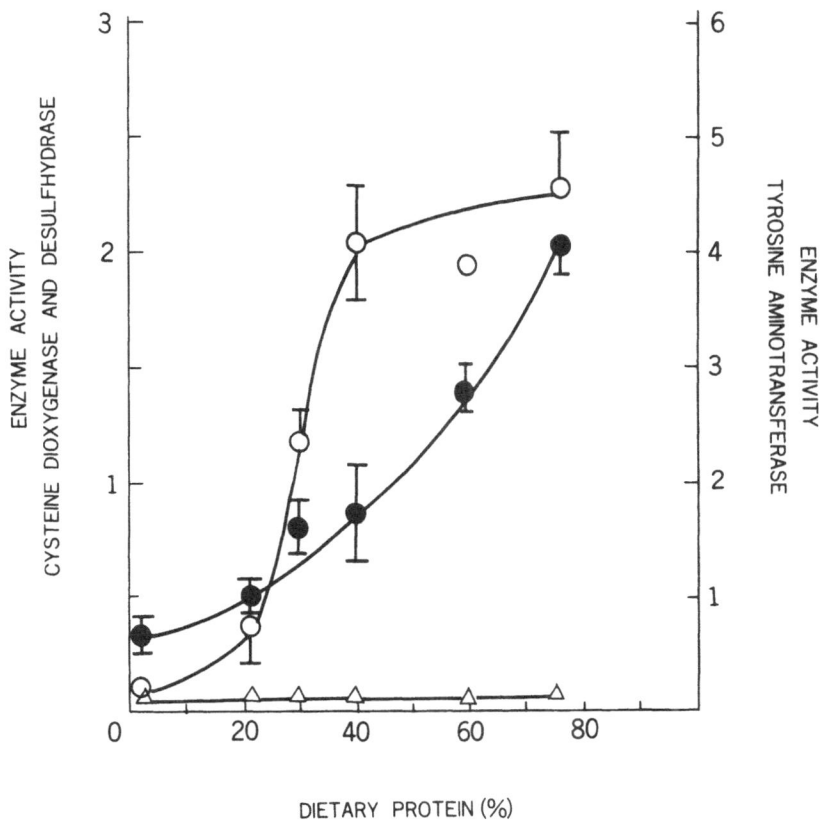

DIETARY PROTEIN (%)

Enzyme activity : μmol of product/h/mg protein

Fig. 1.2. The effects of dietary protein on the activities of cysteine dioxygenase, cysteine desulfhydrase, and tyrosine aminotransferase in rat liver. Plotted are cysteine dioxygenase activity (O), cysteine desulfhydrase activity (△), and tyrosine aminotransferase activity (●). Enzyme activities are expressed as the mean ± S.D. (represented by vertical line) of six and three animals fed on the basal (21% protein) and the other diet, respectively. Animals were fed on each experimental diet for 2 days prior to sacrifice. (From data of Kohashi et al., 1978.)

just as effective as 5% cysteine and was considerably more effective than 2.5% methionine. On the other hand, amino acids such as tryptophan, histidine, tyrosine, alanine, glycine, and leucine did not increase cysteine dioxygenase activity. However, the hepatic cysteine desulfhydrase activity was slightly induced by the inclusion of 10% cysteine but not by 5% cysteine or 2.5% methionine.

Table 1.4 The Effect of the Inclusion of Various Amino Acids in the Low-protein Diet on the Activities of Cysteine Dioxygenase, Tyrosine Aminotransferase, and Cysteine Desulfhydrase, and the Contents of Taurine and Cysteine in Rat Liver[a]

	Body weight	Food consumption (g/3 rats/2 days)	Enzyme activity (units/mg protein)			Contents (µmol/g liver)	
			Cysteine dioxygenase	Cysteine desulfhydrase	Tyrosine aminotransferase	Taurine	Cysteine
None (6)	178 ± 6	85	0.06 ± 0.01	0.05 ± 0.006	1.11 ± 0.65	4.0 ± 0.05	0.09 ± 0.01
1% Cysteine (3)	193 ± 9	55	1.39 ± 0.05	0.14 ± 0.010	0.28 ± 0.07	15.0 ± 1.56	0.25 ± 0.04
5% Cysteine (3)	167 ± 24	50	1.34 ± 0.03	0.04 ± 0.006	0.26 ± 0.02	2.5 ± 0.58	0.12 ± 0.02
2.5% Methionine (3)	173 ± 3	80	0.32 ± 0.06	0.04 ± 0.019	0.19 ± 0.05	7.8 ± 0.84	0.20 ± 0.09
5% Methionine (3)	185 ± 11	85	1.82 ± 0.06	0.05 ± 0.010	0.58 ± 0.03	6.5 ± 0.74	0.22 ± 0.01
1% S-Methyl L-cysteine (3)	193 ± 3	60	0.16 ± 0.07	0.05 ± 0.003	0.78 ± 0.13	2.1 ± 0.64	0.15 ± 0.02
5% Histidine (3)	202 ± 13	85	0.07 ± 0.01	0.05 ± 0.003	0.51 ± 0.22	2.1 ± 0.64	0.15 ± 0.02
5% Tryptophan (3)	160 ± 6	65	0.08 ± 0.01	0.08 ± 0.009	1.07 ± 0.12	6.6 ± 0.54	0.10 ± 0.01
5% Alanine (3)	190 ± 10	89	0.05 ± 0.01	0.04 ± 0.004	2.36 ± 0.64	3.8 ± 1.02	0.08 ± 0.01
5% Tyrosine (3)	207 ± 7	85	0.08 ± 0.01	0.03 ± 0.001	3.00 ± 0.89	1.5 ± 0.36	0.06 ± 0.01

[a]Rats were fed on the basal diet ad libitum for 5 days, then on the low-protein diet or the low-protein diet supplemented with various amino acids for 2 days. Results were expressed as the mean ± S.D. for three animals. Food consumption for the 2 days before sacrifice are given. Number of animals is shown in parentheses. (From data of Kohashi et al., 1978.)

Effect of Dietary Protein on the Hepatic Cysteine and Taurine Contents and on the Urinary Excretion of Taurine

The hepatic cysteine and taurine contents of rats were significantly increased with the increase of dietary protein contents up to 30% and 40%, respectively, but no further increases were observed on higher-protein diets, as shown in Fig. 1.3. On the other hand, urinary taurine excretion was increased markedly by increasing dietary protein. Figure 1.4 shows that rats fed on a 75%-protein diet excreted ten times more urinary taurine than rats fed on a 2.1%-protein diet. From these findings, it appears likely that the maximum levels of cysteine and taurine in the intact adult rat liver are 0.2–0.3 and 6–8 μmol/g liver weight, respectively.

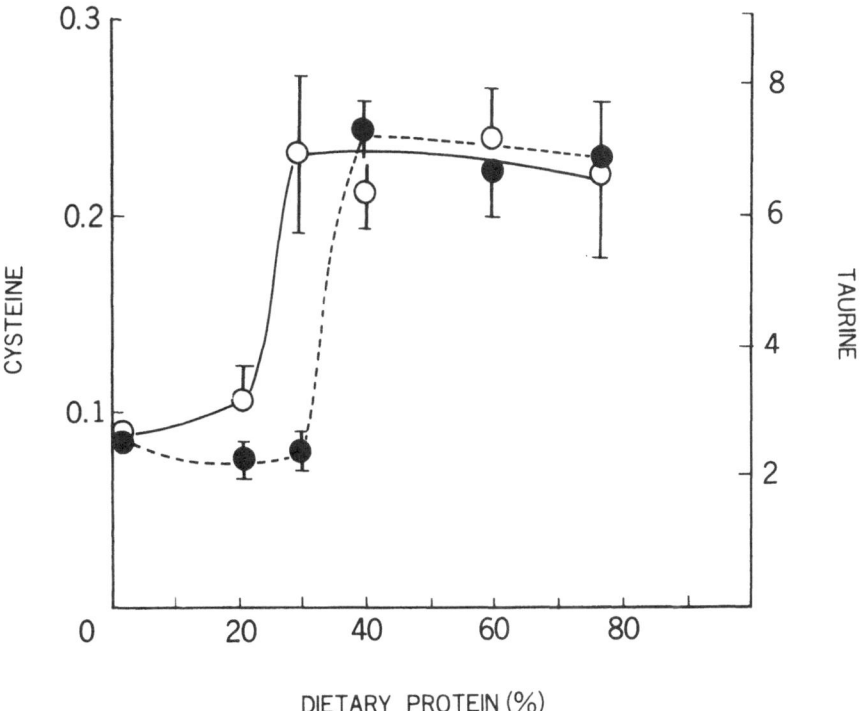

Fig. 1.3. Effects of dietary protein on the hepatic cysteine and taurine contents. Plotted are cysteine content in μmol/g liver (O) and taurine contents in μmol/g liver (●). Results were obtained from the same animals as in Fig. 1.2. (From data of Kohashi et al., 1978.)

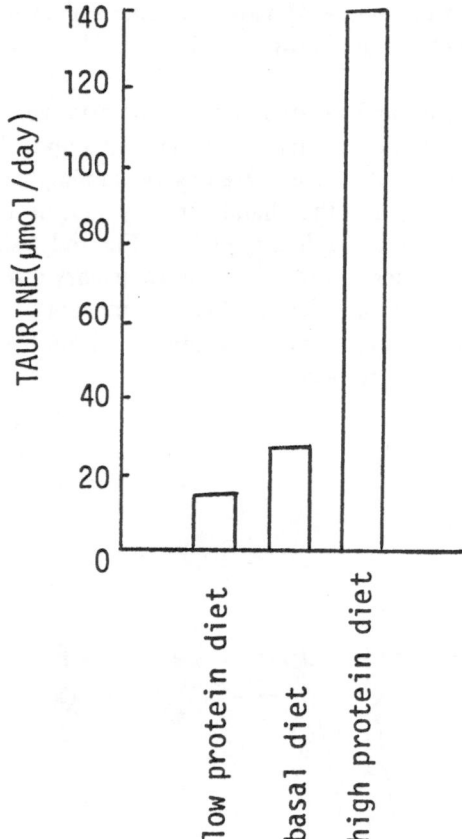

Fig. 1.4.   Effects of dietary protein on urinary taurine level. Two animals in each group were housed in one cage. Animals were fed on the diets and water ad libitum for 3 days. Urine for 24 hr was collected on the third day and used for a quantitative analysis of taurine. (From data of Kohashi et al., 1978.)

The Effect of Dietary Protein on the Metabolic Activities of Cysteine and Cysteine Sulfinate in Rat Tissues

To examine the relationship between cysteine and CSA metabolism in various tissues, tissue slices were incubated with L-[U-14C] cysteine and L-[U-14C] CSA in Eagle's minimum-essential medium containing 10% fetal calf serum for 1 hr using a Gilson respirometer, and the 14CO$_2$ eliminated from these tracers was used to measure the metabolic activity of L-cysteine or L-

CSA. L-[U-^{14}C] CSA was enzymatically synthesized from L-[U-^{14}C] cysteine using purified rat liver cysteine dioxygenase and protein-A prepared in our laboratory. As shown in Fig. 1.5, L-cysteine had significant metabolic activity in liver and kidney but not in other tissues. On the other hand, L-CSA showed metabolic activity in all tissues tested so far except intestinal mucosa, the highest metabolic activity being found in brain. In brain, the $^{14}CO_2$ elimination from L-[U-^{14}C] CSA was lowered to 6% by the 10 mM loading of L-CSA in the medium, while that in liver was lowered only to 60%. This finding suggests that brain may utilize CSA effectively in lesser concentration of CSA than liver does. In general, the metabolic activities of L-CSA in rat tissues were much higher than those of L-cysteine. Figures 1.6 and 1.7 show the effect of dietary protein on the metabolic activity of L-cysteine and L-CSA in liver, kidney, and brain. The $^{14}CO_2$ elimination from L-[U-^{14}C] cysteine in

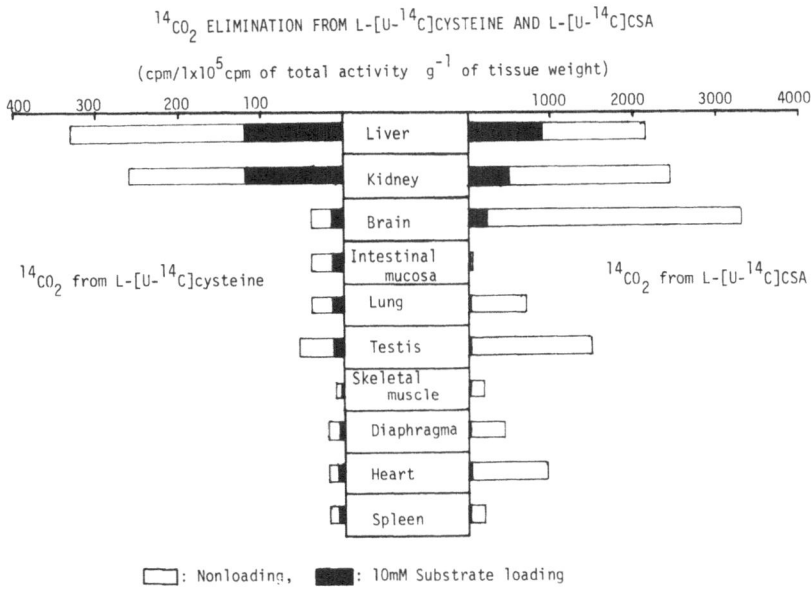

Fig. 1.5. Organ interrelationships of the metabolic activities of L-[U-^{14}C]cysteine and L-[U-^{14}C]CSA in rat. Tissue slices were prepared with Stadie-Riggs slicer. The reaction was carried out in Eagle minimum essential medium containing 10% fetal calf serum using a Gilson respirometer and its flask. A small filter paper soaked with 0.2 ml of 30% KOH was set in center well of a flask to trap $^{14}CO_2$ eliminated [^{14}C]tracer. The reaction was carried out 1 hr at 37° in concentrations of L-cysteine and L-CSA of 0.14mM and 0.1mM, respectively, with or without the loading of excess substrate. (Unpublished data of Kori et al. Preliminary communication appeared in the Abstracts of 33rd Meeting of Japanese Society of Food and Nutrition in May, 1979 at Tokyo, p. 63.)

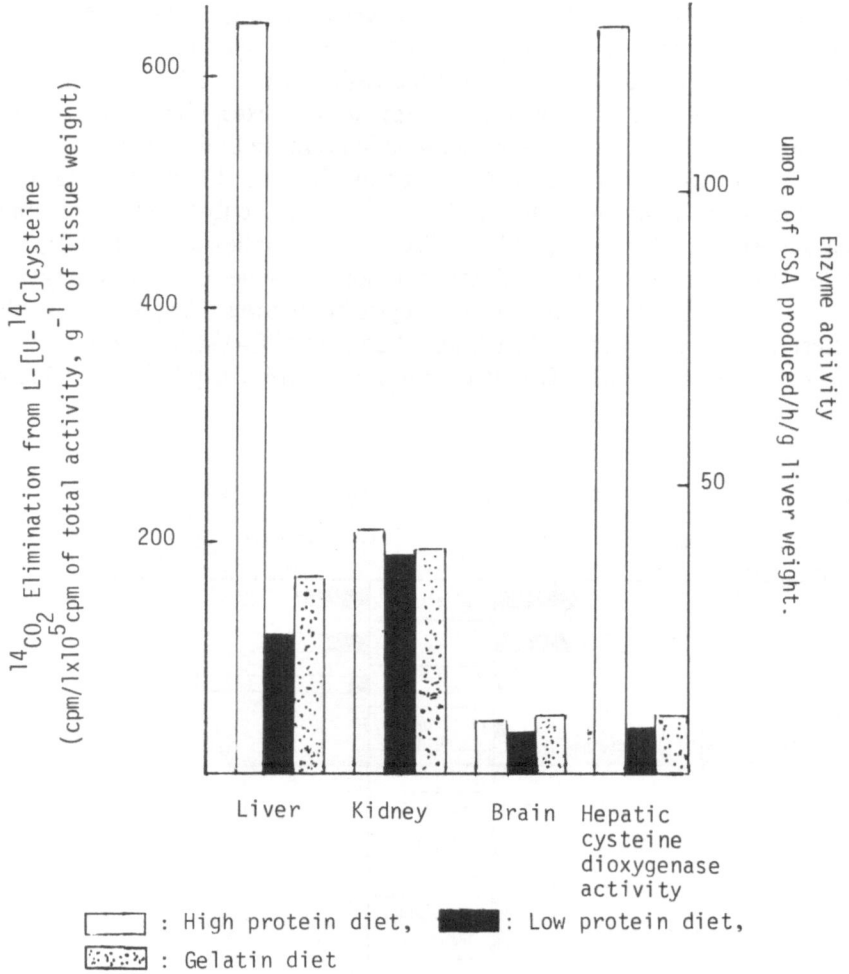

Fig. 1.6. The effect of dietary protein on the $^{14}CO_2$ elimination from L-[U-^{14}C]cysteine in liver, kidney, and brain of rat. The experiments were carried out under same conditions as described in Fig. 1.5. The casein content in high-protein diet and low-protein diet was 90% and 2.5%, respectively. The gelatin content of the diet was 18%. The cysteine dioxygenase activity was expressed as μmol of CSA produced/hr/mg protein. Animals were fed on the experimental diet for 2 days prior to sacrifice. The $^{14}CO_2$ elimination was determined without the loading of excess L-cysteine. (Unpublished data of Kori et al.; the preliminary communication appeared in the Abstracts of 33rd Meeting of Japanese Society of Food and Nutrition, May 1979, Tokyo, p. 63.)

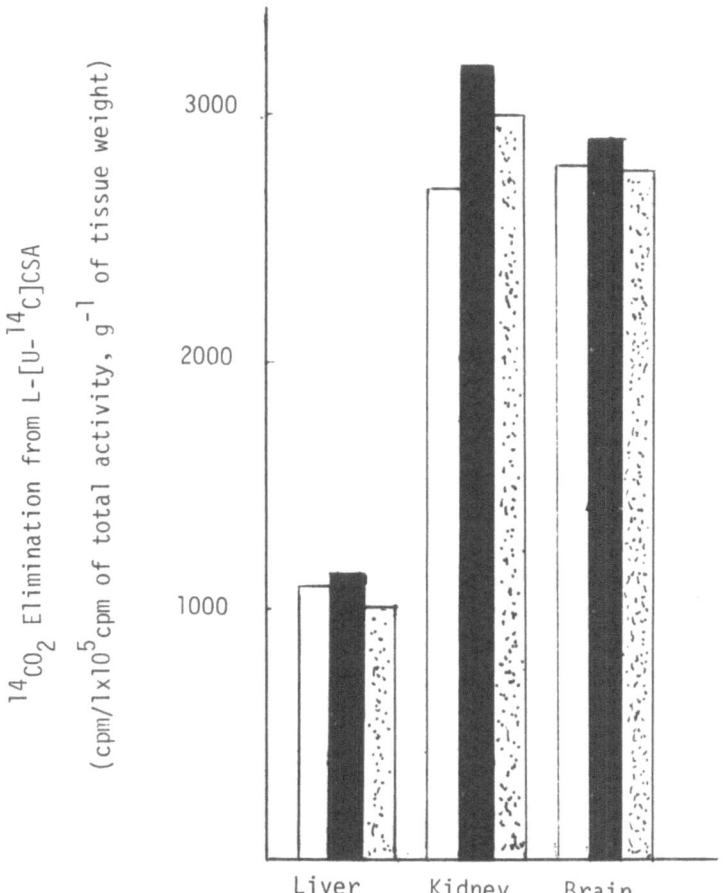

Fig. 1.7. The effect of dietary protein on the $^{14}CO_2$ elimination from L-[U-^{14}C]CSA in liver, kidney, and brain of rat. The experiments were carried out under same conditions as described in Figs. 1.5 and 1.6. (Unpublished data of Kori et al.; the preliminary communication appeared in the Abstracts of 33rd Meeting of Japanese Society of Food and Nutrition, May 1979, Tokyo, p. 63.)

the liver of rats fed on either low-protein diet or 18%-gelatin diet, in which sulfur-containing amino acids and tryptophan are almost absent, was much lower than that in liver of rats fed on a high-protein diet. Thus, the change in $^{14}CO_2$ elimination from L-[U-^{14}C] cysteine in liver was well correlated with the hepatic cysteine dioxygenase activity in rats fed on these diets. On the other hand, the metabolic activities of L-CSA in liver and kidney, and those of L-cysteine in liver, kidney, and brain did not respond to dietary protein, as shown in Figs. 1.5 and 1.6.

DISCUSSION

Cysteine dioxygenase from rat liver was purified to homogeneity. Since purified enzyme was obtained in an inactive form, the question may be raised whether a distinct enzyme catalyzing the CSA formation exists or whether an active enzyme is converted to an inactive form during purification procedures. The latter possibility seems more likely, since the enzyme preparation after acetone treatment (Step 2 in Table 1.2) did not exhibit any enzyme activity without L-cysteine treatment, but total enzyme activity was recovered after the preactivation procedure, as shown in Table 1.1. Another question is whether the NAD(H)-dependent CSA-forming enzyme observed in various tissues of the rat, as mentioned above, is the same enzyme as liver cysteine dioxygenase. Although a precise answer to this question must await purification of the NAD(H)-dependent enzyme, the enzyme in liver appears unlikely to play an important role in L-cysteine metabolism in liver, since the enzyme activity is much lower than that of cysteine dioxygenase. We have observed that the stimulating effect of NAD on the CSA-forming enzyme is enhanced by the presence of a microsomal fraction, and Misra and Olney (1975) reported that the NAD-dependent enzyme in brain was mostly associated with the microsomal fraction. These findings suggest that CSA-forming enzyme may be a protein distinct from hepatic cysteine dioxygenase. In addition, the CSA-forming enzyme of extrahepatic tissue appears unlikely to play a regulatory role in L-cysteine or taurine metabolism at the whole animal level because of its extremely low enzymatic activity; however, the enzyme may be associated with specific physiological functions in a given tissue.

Hepatic cysteine dioxygenase activity and urinary taurine excretion were both markedly increased by feeding a diet containing more than 20% protein. These results suggest that cysteine dioxygenase may control possible toxic effects of excess cysteine resulting from the high protein intake by converting the excess cysteine to more taurine, which is largely excreted into the urine. Taurine may be considered a detoxication product of cysteine, since the sulfonate group of taurine has fewer biological effects than the sulfhydryl group of cysteine.

The experiments showing $^{14}CO_2$ elimination from L-[U-^{14}C] cysteine indicate that liver and kidney have special roles in the metabolism of L-cysteine. However, the metabolic activity of L-cysteine in kidney did not respond to alterations in dietary protein contents, and cysteine dioxygenase activity was not detected in kidney. From these findings, it appears likely that a metabolic pathway other than the cysteine dioxygenase pathway may exist for L-cysteine in kidney. On the other hand, a considerable production of $^{14}CO_2$ from L-[U-^{14}C]CSA was also observed in brain, heart, testes, and lung; however, $^{14}CO_2$ production from L-[U-^{14}C] cysteine in these extrahepatic tissues was extremely low. Thus these tissues may utilize CSA as a precursor of hypotaurine and taurine. In addition, the dilution of $^{14}CO_2$ from L-[U-^{14}C] CSA from the 10 mM L-CSA loading was much higher in these tissues than in liver. This fact also supports the possibility that these extrahepatic tissues may utilize CSA more effectively than liver.

In liver, the $^{14}CO_2$ elimination from L-[U-^{14}C] CSA was much higher than that from L-[U-^{14}C] cysteine. Furthermore, the $^{14}CO_2$ elimination from L-[U-^{14}C] CSA did not respond to dietary protein alterations, while the $^{14}CO_2$ elimination from L-[U-^{14}C]cysteine did respond and the response was correlated with the change in cysteine dioxygenase activity. These facts support the hypothesis that the rate-limiting step in the formation of taurine from cysteine is the degradation of cysteine by cysteine dioxygenase, and therefore that the enzyme cysteine dioxygenase plays an important role in the regulation of cysteine metabolism.

ACKNOWLEDGMENTS

This study was supported in part by grants from the Scientific Research Fund of the Ministry of Education and Culture of Japan, and from the TANABE Amino Acid Research Fund. I am grateful to my collaborators, in particular Drs. Y. Hosokawa, N. Kohashi, Y. Kori, and S. Sakakibara, with whom most of this work was carried out. My thanks also go to Professor I. Ueda, Osaka Medical College, for his continuous encouragement and useful discussion. I also thank Professor Y. Sakamoto, Institute of Cancer Research, Osaka University Medical School, for his useful advice and encouragement. The early part of this work was carried out in Professor Sakamoto's laboratory.

REFERENCES

Chapeville, F.; and Fromageot, P. La formation de l'acide cystéinesulfinique à partir la cysteine chez le rat. *Biochim. Biophys. Acta,* 17, 275 (1955).

De Marco, C.; Mosti, R.; and Cavallini, D. Sulla ossidazione della cisteina e della cisteamina a dirivati solfinice, catalizata dal fegato di ratto. *Boll. Soc. Ital. Biol. Sper.,* 42, 94–96 (1966).

Di Giorgio, R. M.; Tucci, G.; and Macaione, S. Cysteine oxidase activity in rat retina during development. *Life Sci.,* 16, 429–436 (1975).

Ewetz, L.; and Sörbo, B. Characteristics of the cysteinesulfinate-forming enzyme system in rat liver. *Biochim. Biophys. Acta,* 128, 296–305 (1966).

Hosokawa, Y.; Yamaguchi, K.; Kohashi, N.; Kori, Y.; and Ueda, I. Decrease of rat liver cysteine dioxygenase (cysteine oxidase) activity mediated by glucagon. *J. Biochem.,* 84, 419–424 (1978a).

Hosokawa, Y.; Kohashi, N.; and Yamaguchi, K. Study on rat liver cysteine dioxygenase. *Sulfur-containing Amino Acids* [in Japanese], 1, 251–263 (1978b).

Kohashi, N.; Yamaguchi, K.; Hosokawa, Y.; Kori, Y.; Fujii, O.; and Ueda, I. Dietary control of cysteine dioxygenase in rat liver. *J. Biochem.,* 84, 159–168 (1978).

Lombardini, J. B.; Turini, P.; Biggs, D. R.; and Singer, T. P. Cysteine oxygenase: 1. General properties. *Physiol. Chem. & Phys.,* 1, 1–23 (1969a).

Lombardini, J. B.; Singer, T. P.; and Boyer, P. D. Cysteine oxygenase: II. Studies on the mechanism of the reaction with [18]oxygen. *J. Biol. Chem.,* 244, 1172–1175 (1969b).

Misra, G. H.; and Olney, J. W. Cysteine oxidase in brain. *Brain Res.,* 97, 117–126 (1975).

Sakakibara, S.; Yamaguchi, K.; Hosokawa, Y.; Kohashi, N.; Ueda, I.; and Sakamoto, Y. Two components of cysteine oxidase in rat liver. *Biochem. Biophys. Res. Commun.,* 52, 1093–1099 (1973).

Sakakibara, S.; Yamaguchi, K.; Hosokawa, Y.; Kohashi, N.; Ueda, I.; and Sakamoto, Y. Purification and some properties of rat liver cysteine oxidase (cysteine dioxygenase). *Biochim. Biophys. Acta,* 422, 273–279 (1976).

Sörbo, B.; and Ewetz, L. Enzymatic oxidation of cysteine to cysteine-sulfinate in rat liver. *Biochem. Biophys. Res. Commun.,* 18, 359–363 (1965).

Tabachnik, M.; and Tarver, H. The conversion of methionine-^{35}S to cystathionine-^{35}S and taurine-^{35}S in the rat. *Arch. Biochem. Biophys.,* 56, 115–121 (1955).

Wainer, A. The production of cysteine-sulfinic acid from cysteine in vitro. *Biochim. Biophys. Acta,* 104, 405–412 (1965).

Yamaguchi, K.; Sakakibara, S.; Hosokawa, Y.; and Ueda, I. Induction and activation of cysteine oxidase of rat liver: I. The effects of cysteine, hydrocortisone and nicotinamide on hepatic cysteine oxidase and tyrosine transaminase activities of intact and adrenalectomized rats. *Biochim. Biophys. Acta,* 237, 502–512 (1971).

Yamaguchi, K.; Sakakibara, S.; Asamizu, J.; and Ueda, I. Induction and activation of cysteine oxidase of rat liver: II. The measurement of cysteine metabolism in vivo and the activation of in vivo activity of cysteine oxidase. *Biochim. Biophys. Acta,* 297, 48–59 (1973a).

Yamaguchi, K.; Sakakibara, S.; Hosokawa, Y.; and Ueda, I. The physiological significance of cysteine oxidase in vivo. Abstract of 9th International Congress of Biochemistry (Stockholm), p. 334 (1973b).

Yamaguchi, K.; Hosokawa, Y.; Kohashi, N.; Kori, Y.; Sakakibara, S.; and Ueda, I. Rat liver cysteine dioxygenase (cysteine oxidase). *J. Biochem.,* 83, 479–491 (1978).

Copyright © 1981, Spectrum Publications, Inc.
The Effects of Taurine on Excitable Tissues

CHAPTER 2

Uptake and Metabolism of Cysteine in Rat Brain

Heitaroh Iwata
Akemichi Baba
Satoru Yamagami

During the last decade it has become apparent that certain amino acids function as synaptic transmitters in the central nervous system (CNS) (see the review by Curtis and Johnston, 1974). As a result, increasing attention has been paid to the synaptic biochemistry of these amino acids (Snyder et al., 1973). Emphasis has recently focused on the role of the amino acid neurotransmitter candidate taurine in the CNS. It is thought that taurine may function as a neurotransmitter or neuromodulator in the CNS (see the reviews by Baskin et al., 1976; and by Mandel and Pasantes-Morales, 1976).

While much evidence has accumulated concerning the uptake and release of taurine in CNS preparations (see the review by Mandel and Pasantes-Morales, 1976), neuronal regulation of taurine biosynthesis is not well documented. In order to determine the role of taurine in the CNS, the need to learn more about the physiological regulation of taurine biosynthesis became apparent. In the present study, we investigated the uptake and metabolism of cysteine in rat brain preparations, since taurine is synthesized from cysteine in rat brain slices (Collins, 1974).

METHODS

The cerebral cortex of male Sprague-Dawley rats, weighing 180–250 g, was used for the preparation of synaptosomal enriched fractions. The preparation of the synaptosomal fraction (P2) and purified synaptosomes was carried out by the method of Cotman (1974). Both the P2 and purified synaptosome fractions were suspended in Krebs-Ringer bicarbonate buffer (KR). The KR contained 138 mM NaCl, 5.6 mM KCl, 1 mM $CaCl_2$ 1 mM $MgCl_2$, 11 mM $NaHCO_3$, 1 mM NaH_2PO_4, and 10 mM glucose, and was gassed with 95% O_2–5% CO_2.

Uptake of Cysteine

The P2 suspension in KR was preincubated at 37° for 30 min. After centrifugation at 10,000 g for 10 min, the pellets were suspended in new KR (P2 from 1 g of cortex/3 ml). After 5 min of preincubation, the reaction was started by adding 200 μl of P2 into 0.8 ml of KR-uptake medium and was allowed to proceed for 5 min. Unless otherwise indicated, the KR-uptake medium contained 1 mM L-cysteine and 0.1 μCi/tube of L-[^{35}S]-cysteine hydrochloride (specific activity: 130 mCi/mmol, the Radiochemical Centre Amersham). L-[^{14}C]-cysteine hydrochloride (specific activity: 27 mCi/mmol, the Radiochemical Centre Amersham) was also used in some experiments. Sodium-independent uptake of cysteine was carried out in Na^+-free KR medium (Na^+ was replaced with sucrose). The uptake was stopped by rapid filtration through Whatman GF/C filter followed by washing four times with 4 ml ice-cold washing solution (138 mM NaCl–5.6 mM KCl). After drying the filter paper, the radioactivity on the filter was counted in a liquid scintillation spectrometer.

Separation of Cysteine Metabolites

Separation of cysteine metabolites was carried out by thin-layer high-voltage electrophoresis. Samples were spotted on a cellulose plate (Merck, thickness: 0.1 mm, width: 20 cm, length: 40 cm), and electrophoresis was carried out at 3,500 V for 40 min using acetic acid–formic acid–water (120:26:1,000, pH 1.9). After drying the plate, each metabolite was monitored fluorimetrically using the fluorescamine reaction. Each spot was collected in a vial and its radioactivity was determined.

RESULTS

Uptake of Cysteine

The time course of cysteine uptake by the P2 fraction is depicted in Fig. 2.1. The uptake was markedly lowered in Na^+-free KR (about 20–30% of the

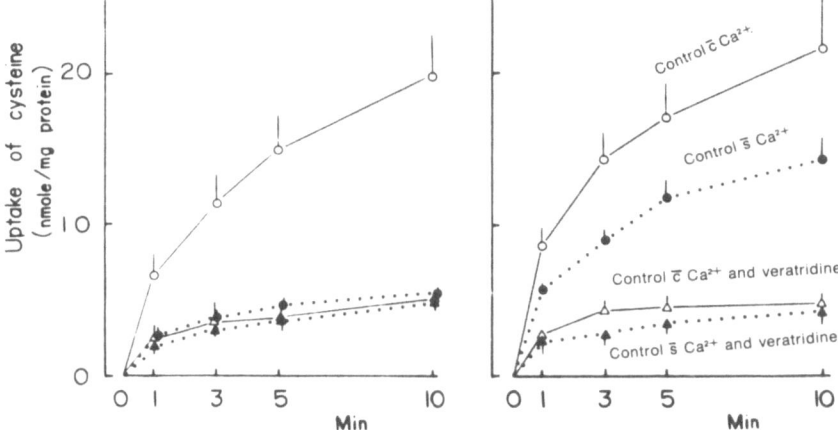

Fig. 2.1 (*a*) Effect of veratridine on the uptake of cysteine in the synaptosomal fraction (P_2) control in the presence ($\bigcirc$) and absence ($\bullet$) of Na^+. Control plus veratridine ($\triangle$), Na free ($\bullet$), Na free plus veratridine ($\blacktriangle$). (*b*) Effect of veratridine on the uptake of cysteine in the P_2 fraction. Control in the presence ($\bigcirc$) and absence ($\bullet$) of Ca^{2+}. Control plus veratridine ($\triangle$), Ca^{2+} free ($\bullet$), Ca^{2+} free plus veratridine ($\blacktriangle$). Veratridine (5×10^{-5} M) was added at 0 time.

control; Fig. 2.1a) and was also significantly reduced in Ca^{2+}-free medium (Fig. 2.1b). Addition of 5×10^{-5} M veratridine completely blocked the Na^+-dependent uptake process without altering the Na^+-independent one (Fig. 2.1a). In addition, veratridine also strongly inhibited cysteine uptake even in the absence of Ca^{2+} (Fig. 2.1b). Similar inhibition of cysteine uptake was achieved by the addition of 56 mM KCl (data not shown). A kinetic study showed that the cysteine uptake in the P2 fraction exhibited two Km values; 1.3 mM in the presence of Na^+ and 8.8 mM in the absence of Na^+. Veratridine or high K^+ competitively inhibited cysteine uptake, as noted by the increase in the Na^+-dependent Km value to 9.4 mM. Relative inhibitory potency of various amino acids and cysteine metabolites on the cysteine uptake is shown in Table 2.1. At an equal molar concentration of 1 mM, homocysteine, phenylalanine, methionine, glutamic acid, aspartic acid, and serine slightly inhibited the uptake of cysteine in the P2 fraction. Among the metabolites of cysteine, cysteine sulfinic acid (CSA) and cysteic acid (CA) had potent inhibitory effects, while hypotaurine and taurine had no effect. Inhibition of cysteine uptake by CSA was also observed at low concentrations of CSA; 50 μM of CSA caused a 27% inhibition of uptake (Fig. 2.2). In contrast, the Na^+-independent uptake of cysteine was not affected by CSA. CA showed the same dose-dependent inhibition of cysteine uptake as CSA (data not shown). Na^+-dependent uptake of cysteine in the purified synaptosomes

Table 2.1   Relative inhibitory potency of various amino acids on cysteine uptake[a]

Amino acids (1 mM)	Percent of control
Cysteine sulfinic acid	57.0
Cysteic acid	57.5
Taurine	108.0
Hypotaurine	91.9
Homocysteine	67.9
Homocysteic acid	94.1
L-Phenylalanine	67.3
γ-Amino butyric acid	93.3
L-Methionine	75.7
L-Glutamic acid	74.6
L-Aspartic acid	73.3
DL-α-Alanine	88.1
β-Alanine	93.8

[a]Cysteine was at a concentration of 1.0 mM.

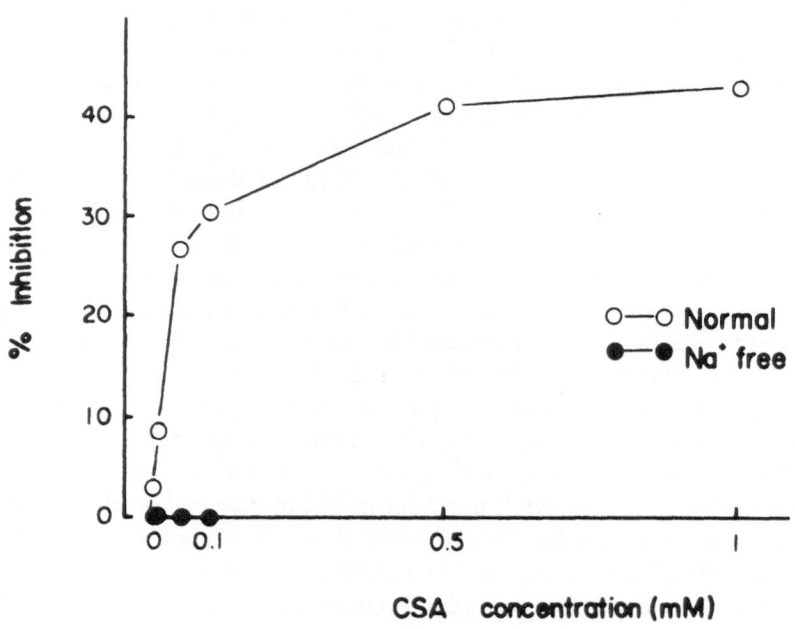

Fig. 2.2.   Effect of cysteine sulfinic acid (CSA) on the uptake of cysteine in the P_2 fraction.

was also inhibited by CSA; about 50% inhibition was observed at a concentration of 0.1 mM. From the kinetic study, it was shown that CSA decreased the Vmax of cysteine uptake without altering the Km value.

Since low-molecular-weight substances can easily bind to the membrane components either specifically or nonspecifically, it was necessary to eliminate the contribution of cysteine binding. This was accomplished by comparing cysteine uptake in both P2 and lysed P2 (P2') fractions. The P2' fraction was obtained by homogenizing the P2 fraction with a Polytron (Kinematica, Switzerland) and subsequently freeze-thawing the preparation three times (Fig. 2.3). The apparent uptake of cysteine by the P2' fraction represented the binding component and was about 20% of that in the P2 fraction. The addition of 0.1 mM CSA did not affect the binding by the P2' fraction.

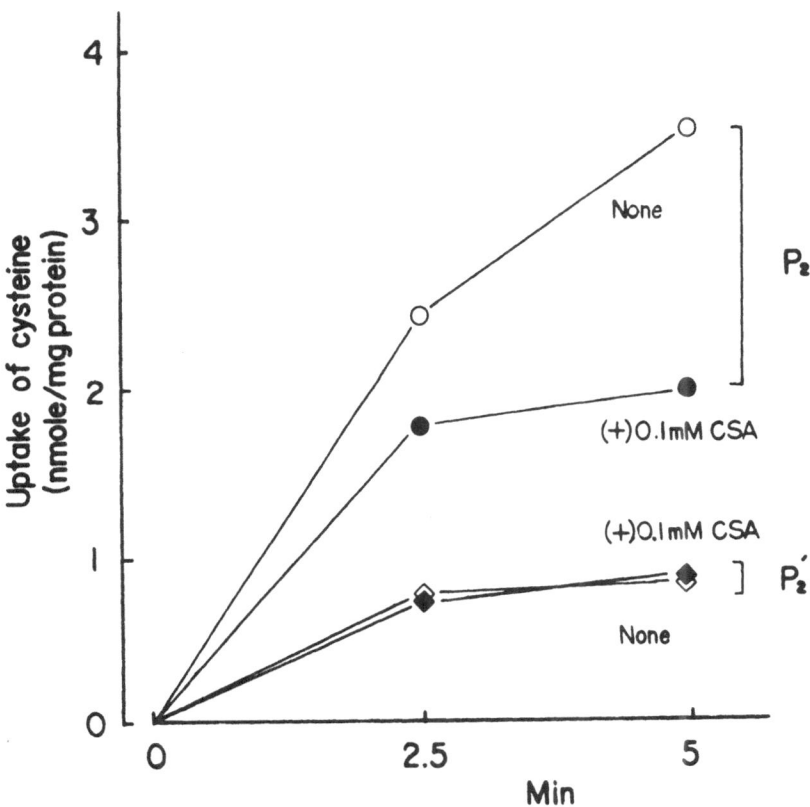

Fig. 2.3. Cysteine uptake in the P_2 and lysed P_2 fractions.

Metabolism of Cysteine

Under the 10-min incubation conditions of the cysteine uptake study, more than 80% of the total radioactivity in the preparation was recovered as cysteine, the remaining 20% being associated with its metabolic products. Figure 2.4 shows the complete separation of these cysteine metabolites in the P2 fraction following incubation with radioactive cysteine (final concentration 0.1 mM) for 30 min at 37°. When [^{35}S]-cysteine was used as a substrate, the hypotaurine, taurine, CSA, and CA spots all contained radioactivity. In addition, a new spot was detected near the cathode, which by amino acid analysis was found to be alanine.

DISCUSSION

Amino acids that accumulate at the nerve terminal by high affinity Na$^+$-dependent transport are believed to be candidates for neurotransmitters. Recent studies have demonstrated that synaptosomal particles from the cerebral cortex or the spinal cord possess high affinity Na$^+$-dependent transport systems

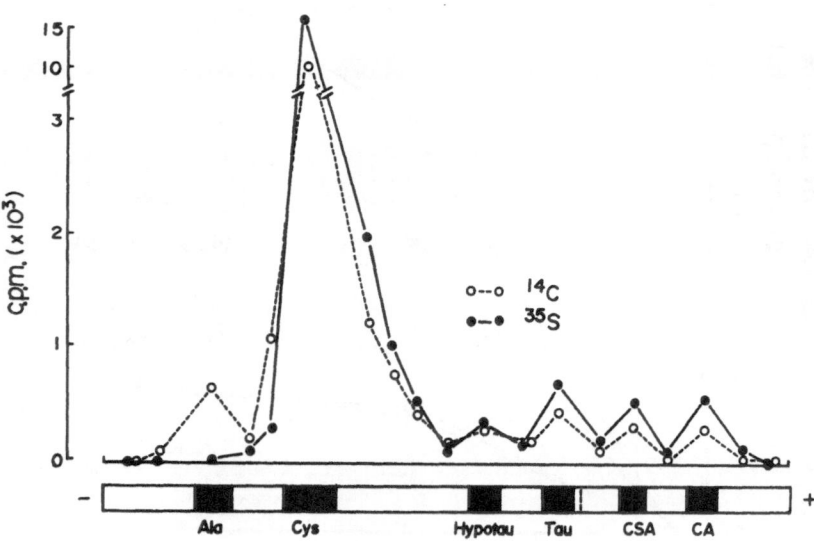

Fig. 2.4.   Separation of the metabolites of the cysteine accumulated in the P$_2$ fraction after incubation with radioactive cysteine. Reaction was stopped by the addition of trichloroacetic acid (TCA). After centrifugation, TCA in the resulting supernatant was removed by ether extractions. Electrophoresis was carried out as described in the Methods section.

for various amino acids (see the review by Snyder et al., 1973). Among the sulfur-containing amino acids, taurine is known to be transported into the nerve cells by at least two mechanisms; one is saturable and exhibits low capacity and high affinity for taurine (Km 2 to 8 $\times$ 10^{-5}), while the second transport system is nonsaturable and has a high capacity but low affinity (Km in the millimolar range) (Lähdesmäki et al., 1975; Schmidt et al., 1975; Hruska et al., 1974; Sieghart and Karobath, 1976; Lombardini, 1977; Kontro and Oja, 1978). Nothing is known about the transport system and the metabolism of cysteine in the synaptosomal fraction. Previously, Collins (1974) reported that intracisternal injection of radioactive cysteine was followed by subsequent formation of taurine, demonstrating that cysteine was transported into the nerve cells and metabolized to taurine. Here, we undertook the characterization of the transport and metabolism of cysteine by rat brain synaptosomal fractions.

As shown in Fig. 2.1, the uptake of cysteine into the synaptosomal fraction was strictly Na^+-dependent and partly Ca^{2+}-dependent. However, while the high affinity Na^+-dependent transport system for various amino acids is characterized by a low Km value, usually less than 100 μM, the Na^+-dependent transport system of cysteine in the P2 fraction had a Km value in the millimolar range. Even in the purified synaptosomes, the Km value for cysteine was about 400 μM. Thus, the transport system for cysteine in the synaptosomes may not be high affinity.

Two interesting findings emerged from the present study. First, depolarizing agents such as veratridine or high K^+ markedly inhibit Na^+-dependent uptake of cysteine; and, second, low concentrations of cysteine sulfinic acid and cysteic acid, both metabolites of cysteine, also inhibit this uptake process. During the past few years there has been an increasing awareness of the significant role of the membrane potential in Na^+-coupled transport of several substances. It has been reported that depolarization affects the high affinity uptake of choline in synaptosomes. If synaptosomes were initially depolarized and then reincubated under normal conditions, the high affinity uptake of choline was increased; whereas uptake in the presence of depolarizing agents reduced choline uptake (Barker, 1976; Murrin and Kuhar, 1976; Simon and Kuhar, 1976). By comparison, a preliminary depolarization increased the transport of glutamic acid, aspartic acid, glycine, and choline, all of which are putative neurotransmitters or neurotransmitter precursors (Murrin et al., 1978). The effect of depolarization by high K^+ on the uptake of cysteine was therefore examined, but no activation was observed (data not shown), as in the case of taurine (Murrin et al., 1978).

The inhibition of cysteine uptake by CSA in the P2 fraction seems functionally interesting. Two points should be mentioned: First, CSA caused noncompetitive inhibition of the uptake; and, second, it was an effective inhibitor even at low concentrations. In a separate experiment, using a newly developed enzyme cycling method, we determined the CSA concentration in the rat brain to be about 0.2 μmol/g wet weight, which is well within the concentration range of CSA necessary to inhibit cysteine uptake. However, the addition of 0.1 mM CSA did not affect the spontaneous, slow release of cysteine from preloaded synaptosomes or slices (data not shown), although high K^+ enhanced this Ca^{2+}-dependent process. Thus, the question arises whether the decreased synaptosomal level of cysteine following depolarization or CSA addition was actually due to an accelerated efflux of this compound from the synaptosomes rather than a decreased uptake. Several lines of evidence argue against this possibility. First, veratridine completely blocked the uptake of cysteine even in the absence of Ca^{2+}, while depolarization-induced release of cysteine was only inhibited 50%. Second CSA did not stimulate the release of cysteine. Finally, the amount of spontaneous and depolarization-induced release of cysteine was less than 2% of the total cysteine that initially accumulated in the preparation (data not shown), while depolarization completely blocked Na^+-dependent uptake of cysteine.

Following a 10-min incubation of the P2 fraction with 0.1 mM [^{35}S]-cysteine, about 80% of the radioactivity in the preparation was found to be unmetabolized cysteine. Examination of the remaining radioactivity revealed that some cysteine was metabolized to CSA, CA, hypotaurine, and taurine. In addition, when P2 was preloaded with [^{14}C]-cysteine, [^{14}C]-alanine was also detected in the preparation, indicating that the synaptosomal fraction had cysteine sulfinic desulfinase activity.

In the present study, we found that both depolarizing agents and CSA had inhibitory effects on the uptake of cysteine in the synaptosomal fraction. In addition, exogeneously applied cysteine was metabolized to CSA, CA, hypotaurine, taurine, and alanine. These results suggest that cysteine metabolism may be regulated by both nervous activity and by the level of the two metabolic intermediates, CSA and CA.

REFERENCES

Barker, L. A. Modulation of synaptosomal high affinity choline transport. *Life Sci.*, 18, 725–730 (1976).

Baskin, S. I.; Leibman, A. J.; and Cohn, E. M. Possible functions of taurine in the central nervous system. In *Advan. Biochem. Psychopharmac.*, Costa, E.; Giacobini, E.; and Paoletti, R., eds. Raven Press, New York, Vol. 15, pp. 153–164 (1976).

Collins, G. G. S. The rate of synthesis, uptake and disappearance of [^{14}C]-taurine in eight areas of the rat central nervous system. *Brain Res.*, 76, 447–459 (1974).

Cotman, C. W. Isolation of synaptosomal and synaptic plasma membrane fractions. In *Methods in Enzymol.*, Fleischer, S.; and Packer, L. eds. Academic Press, New York, Vol. 31, pp. 445–452 (1974).

Curtis, D. R.; and Johnston, G. A. R. Amino acid transmitters in the mammalian central nervous system. *Ergebn. Physiol.*, 69, 97–188 (1974).

Hruska, R.; Huxtable, R.; Bressler, R.; and Yamamura, H. Sodium-dependent high affinity transport of taurine into rat brain synaptosomes. *Proc. West. Pharmac. Soc.*, 19, 152–156 (1974).

Kontro, P.; and Oja, S. S. Taurine uptake by rat brain synaptosomes. *J. Neurochem.*, 30, 1297–1304 (1978).

Lähdesmäki, P.; Pasula, M.; and Oja, S. S. Effect of electrical stimulation and chlorpromazine on the uptake and release of taurine, -aminobutyric acid and glutamic acid in mouse brain synaptosomes. *J. Neurochem.*, 25, 675–680 (1975).

Lombardini, J. B. High affinity uptake systems for taurine in tissue slices and synaptosomal fractions prepared from various regions of the rat central nervous system: Correction of transport data by different experimental procedures. *J. Neurochem.*, 29, 305–312 (1977).

Mandel, P.; and Pasantes-Morales, H. Taurine: A putative neurotransmitter. In *Advan. Biochem. Psychopharmac.*, Costa, E.; Giacobini, E.; and Paoletti, R., eds. Raven Press, New York, Vol. 15, pp. 153–164 (1976).

Murrin, L. C.; and Kuhar, M. J. Activation of high affinity choline uptake *in vitro* by depolarizing agents. *Molec. Pharmac.*, 12, 1082–1090 (1976).

Murrin, L. C.; Lewis, M. S.; and Kuhar, M. J. Amino acid transport: Alterations due to synaptosomal depolarization. *Life Sci.*, 22, 2009–2016 (1978).

Schmid, R.; Sieghart, W.; and Karobath, M. Taurine uptake in synaptosomal fractions of rat cerebral cortex. *J. Neurochem.*, 25, 5–9 (1975).

Sieghart, W.; and Karobath, M. Uptake of taurine into subcellular fractions of C-6 glioma cells. *J. Neurochem.*, 26, 981–986 (1976).

Simon, J. R.; and Kuhar, M. J. High-affinity choline uptake: Ionic and energy requirements. *J. Neurochem.*, 27, 93–99 (1976).

Snyder, S. H.; Young, A. B.; Bennett, J. P., and Mulder, A. H. Synaptic biochemistry of amino acids. *Fed. Proc.*, 32, 2039–2047 (1973).

Copyright © 1981, Spectrum Publications, Inc.
The Effects of Taurine on Excitable Tissues

Biological Effects of Gamma-Γ Glutamyl Taurine (Glutaurine): A New Parathyroid Hormone

László Feuer

While it is well established that parathyroid hormone is produced in the parathyroid gland, it has recently become apparent that the parathyroid gland may also have another endocrine function. This hypothesis is based in part upon the observation that hypoparathyroidism is accompanied by symptoms that resemble vitamin A deficiency (Varró, 1964; Brodehl et al., 1967; Henkin, 1968; Gardner, 1969; Pastinszky and Rácz, 1974). Moreover, the parathyroid gland contains oxyphil cells, as well as the parathyroid hormone-producing chief cells. Although the role of the oxyphil cells has not been established, it has been suggested that they have an endocrine function (Tremblay and Peärse, 1959; Tremblay and Cartier, 1961; Halver, 1973).

In 1977 it was discovered that oral administration of a protein-free, parathyroid hormone-free extract derived from bovine parathyroid gland significantly increased serum vitamin A levels of test animals (Feuer et al., 1977a). The bioactive substance isolated from these extracts was subsequently found to be the dipeptide γ-L-glutamyl taurine, also known by the generic name glutaurine or by the trade name Litoralon (Feuer et al., 1977b). Identification of glutamic acid and taurine as constituents of the peptide is of interest since both amino acids are putative neurotransmitters and/or neuromodulators.

Glutaurine is known to be synthesized in the parathyroid gland, although its presence has also been demonstrated in other tissues (Furka et al., 1977). The biosynthesis of glutaurine from taurine and a glutamyl donor appears to be catalyzed by the enzyme γ-glutamyl transpeptidase (Feuer et al., 1979c). It is assumed that once the peptide is produced it is secreted into the blood and carried to its target organ. Autoradiographic studies performed on rats administered radioactive glutaurine revealed high concentrations of the substance in epithelial tissue; even after 24 hr measurable quantities were detected in small intestinal and glandular epithelia as well as in the kidneys, skin, and retina. However, very little radioactivity was detected in the brain, indicating that the peptide does not readily penetrate the blood-brain barrier (Feuer, unpublished data).

ROLE IN PHYLOGENESIS

The parathyroid gland appeared in evolutionary development simultaneously with the emergence of land vertebrates. During the metamorphosis of urodeles, the gland developed with the disappearance of gills; the amphibians, which retained their gills after metamorphosis, failed to develop the endocrine gland. Histological and embryological studies have shown that the parathyroid gland is derived from branchial epithelium, which during aquatic life has a respiratory function (Greep, 1963).

It has been proposed that the parathyroid gland plays an essential role in the adaptation of aquatic vertebrates to terrestrial conditions (Feuer, 1977). In addition to its well-accepted physiological role in the regulation of blood calcium levels, the theory suggests that the parathyroid gland also functions in the development of: (1) airways and lungs, (2) enhanced immune surveillance, (3) protective mechanisms against infection and irradiation, (4) organs of locomotor function, and (5) the capacity to regulate ion balance.

EFFECT OF GLUTAURINE ON METAMORPHOSIS

The phylogenetic hypothesis predicts that glutaurine should affect the metamorphosis of amphibians. In agreement with this hypothesis, glutaurine has been found to accelerate the early stages, and retard the later stages, of this transformation (Feuer et al., 1978a; Feuer et al., 1978b). In the early stages, the peptide accelerated tail resorption, enhanced body shortening, facilitated the emergence of forelimbs, and promoted maturation of intestinal mucosa. These effects appear to result from the vitamin A-like activity of this substance, which includes stimulation of lysosomal enzyme activity, an increase in the number of macrophages, and the promotion of mucopolysaccharide biosynthesis (Feuer et al., 1978a; Feuer et al., 1979b).

The later stages of metamorphosis are dominated by triiodothyronine. As seen in Table 3.1, triiodothyronine greatly enhanced body shortening, indicating that it accelerated the rate of metamorphosis. On the other hand, glu-

Table 3.1   Antagonism of the effect of glutaurine by prednisolone and triiodothyronine[a]

Treatment	Body length (mm)
I. Control	34.0 ± 0.6
Triiodothyronine (0.01 μg/ml)	12.5 ± 0.5
Triiodothyronine (0.01 μg/ml) plus Glutaurine (0.5 μg/ml)	28.1 ± 0.4
II. Control	33.4 ± 0.6
Prednisolone (10 μg/ml)	28.4 ± 0.3
Prednisolone (10 μg/ml) plus Glutaurine (0.5 μg/ml)	32.0 ± 0.6

[a]Larve of *Rana arvalis* were kept in chlorine-free tap water (changed daily) at room temperature of 20–25°C. The larvae were removed form this water for a period of 2 hr daily and placed in water containing the appropriate concentration of triiodothyronine, glutaurine, and/or prednisolone. The period of treatment in both studies was 10 days. Development of the larvae was observed daily, and body measurements were made photometrically (Feuer et al., 1978b).

taurine antagonized these triiodothyronine effects. This modulating activity of glutaurine is consistent with its inhibitory action on the later stages of metamorphosis.

The data in Table 3.1 show that glutaurine also inhibited the effects of glucocorticoids on metamorphosis. Prednisolone was found to potentiate the effects of triiodothyronine while antagonizing the effects of vitamin A on metamorphosis (Török et al., 1979; Feuer et al., 1979d); however, glutaurine inhibited or modulated the actions of both the glucocorticoids and triiodothyronine. This antagonism between glutaurine and the other two hormones appears to be a general property of the peptide since it is also observed in mammalian systems (Feuer et al., 1978c) and may be analogous to the neuromodulator actions of taurine in the central nervous system.

Probably as a result of its control of metamorphosis, glutaurine also improved the survival rate of animals during their transformation. As seen in Table 3.2, the mortality rate of the untreated animals was approximately twofold greater than that of the glutaurine-treated group (Feuer et al., 1978b).

OTHER VITAMIN A-LIKE EFFECTS OF GLUTAURINE

Many of the glutaurine-induced effects on metamorphosis reflect its vitamin A-like activity. Glutaurine also mediated vitamin A-like actions in mammalian systems. In rats, glutaurine reduced the signs of osteolarthyrism in-

Table 3.2   Effect of glutaurine on the mortality rate of animals undergoing metamorphosis[a]

Age of animal (days)	Percent mortality	
	Control	Glutaurine treated
65	3.0	1.6
69	43.0	15.0
73	75.0	33.0
77	92.0	52.0

[a]Thirty-day-old tadpoles (*Rana dalmatina*) were subjected daily to a 2-hr treatment of glutaurine (0.5 μg/ml), as described in Table 3.1. The mortality rate from a group of 60 animals was registered 35–47 days after initiation of treatment.

duced by β-aminopropionitryl fumarate (Feuer et al., 1980). β-Aminopropionitryl fumarate is a potent inhibitor of lysine oxidase, a metalloenzyme that plays a central role in bone matrix formation. It has been suggested that glutaurine may mediate these effects on osteolathyrism by providing sufficient levels of cofactor for the metalloenzyme (Feuer et al., 1980). Although tissue trace metal content was not measured in the above studies, an earlier report revealed that glutaurine influences serum trace metal content of rabbits receiving daily doses of the peptide (Feuer et al., 1977b). This property of glutaurine also appears to be responsible for its beneficial effect on experimental fluorosis and cadmium toxicity (Feuer et al., 1977b).

Glutaurine has also been found to stimulate epithelial proliferation and increase the number of macrophage cells in rat thymus cultures (Feuer et al., 1978c; Feuer et al., 1978d). The mononuclear cells of the treated cultures exhibited enlarged nuclei and nucleoli, suggesting an effect of glutaurine at the genetic level (Feuer et al., 1978d). Studies supporting an effect of glutaurine on transcription have been reported (Csaba et al., 1979; Feuer et al., 1979a).

TAURINELIKE ACTIVITY OF GLUTAURINE

In addition to its vitamin A-like activity, glutaurine also exhibited taurinelike activity. These taurinelike actions are observed at glutaurine concentrations ranging from two to three orders of magnitude lower than the dosage of taurine required to elicit the same response. The radioprotective effect of glutaurine is an example; daily administration of only 100 μg/kg of glutaurine

Table 3.3   Protection by glutaurine and its analogues against irradiation damage[a]

Condition	Percent survival	
	6.30 Gy[b]	9.00 Gy
Control	0	0
Before irradiation		
Taurine (10 mg/kg)	50	0
Taurine (10 mg/kg) plus AET	—	58
Glutaurine (100 μg/kg)	60	0
Glutaurine (100 μg/kg) plus AET	—	56
γ-Aminobutyryltaurine (100 μg/kg) plus AET	—	63
γ-Aminobutyryl cholamine phosphate (100 μg/kg) plus AET	—	58
After irradiation		
Taurine (10 mg/kg)	37	—
Glutaurine (100 μg/kg)	75	—

[a] Male mice were subjected to the indicated dosage of either X-ray or ^{60}Co-gamma whole body irradiation. The dose rate was 0.44 and 0.67 Gy/min for X-ray and ^{60}Co-gamma irradiation, respectively. All animals, with the exception of the glutaurine-treated animals, received a single dose i.p. of the appropriate radioprotective agent; the glutaurine group received a daily dose of 100 μg/kg for a period of 4 days. The data represent the survival rate 30 days after irradiation. β-aminoethylisothiuronium (AET) was used at a dose of 140 mg/kg.

[b] A Grey (gy) unit is equivalent to an *absorbed* dose of 100 rads.

for a period of 4 days provides as much protection against 630 rads (0.63 Gy) of irradiation as 10 mg/kg of taurine (Table 3.3).

The nature of the radioprotective effect illustrates several characteristics of glutaurine. First, better effect is usually observed after prolonged administration (Benkö and Feuer, 1979). Although some actions of glutaurine become manifest after the initial administration, others become apparent only after repeated treatment. Second, the intensity of the response is weak. The fact that glutaurine is generally not a potent effector may be because glutaurine modulates the actions of other agonists, rather than itself acting as an agonist. Glutaurine is very useful nevertheless because it can act synergistically with other radioprotectants such as β-aminoethylisothiouronium (AET) (Table 3.3). Since the toxicity of AET is much greater than that of glutaurine, combination of the two yields an improved radioprotective effect. Glutaurine is also unique as a radioprotective agent because it may be more effective after irradiation than before (Table 3.3), in contrast to most radioprotective agents such as AET. Finally, several glutaurine analogues, such as γ-aminobutyryl taurine and γ-glutamyl cholamine phosphate, mimic the radioprotective effect of the dipeptide glutaurine (Table 3.3). γ-Amino-butyryl taurine also produced many of the same effects as glutaurine, while the analogue γ-glutamyl homotaurine seemed to be a glutaurine antagonist (Feuer and Gaál, 1979).

Taurine has been shown to decrease the incidence of stroke and to reduce blood pressure in the stroke-prone, spontaneously hypertensive rat (Yamori et al., Chap. 27). There is some evidence that glutaurine may also alter blood pressure (Feuer and Gaál, 1979). Figure 3.1 shows the effect of increasing doses of glutaurine on the concentration of plasma renin. The concentration of plasma renin was maximal at a dose of 1 μg/kg i.v.; it may be noted that 10 μg/kg was less effective than 1 μg/kg. The mechanism of this response remains to be elucidated.

CONCLUSIONS

Glutaurine (γ-L-glutamyl taurine or Litoralon) is a newly discovered hormonelike substance that was isolated originally from the parathyroid gland but is now synthetically prepared. Glutaurine has biological effects that fulfill many of the predictions of the phylogenetic hypothesis. This hypothesis postulates that the parathyroid gland was essential for the evolutionary development of aquatic organisms to land forms. It may be noted that fish do not have parathyroid glands. Glutaurine has been found to function in metamorphosis and immune surveillance, and to protect organisms against infection and damage due to X-irradiation. Glutaurine, in much lower doses, produces many of the effects of taurine, but whether all of the effects of glutaurine can be reproduced by taurine, or vice versa, is not known at this time.

ACKNOWLEDGMENT

I wish to thank S. W. Schaffer and J. J. Kocsis for their aid and advice in the preparation of this manuscript.

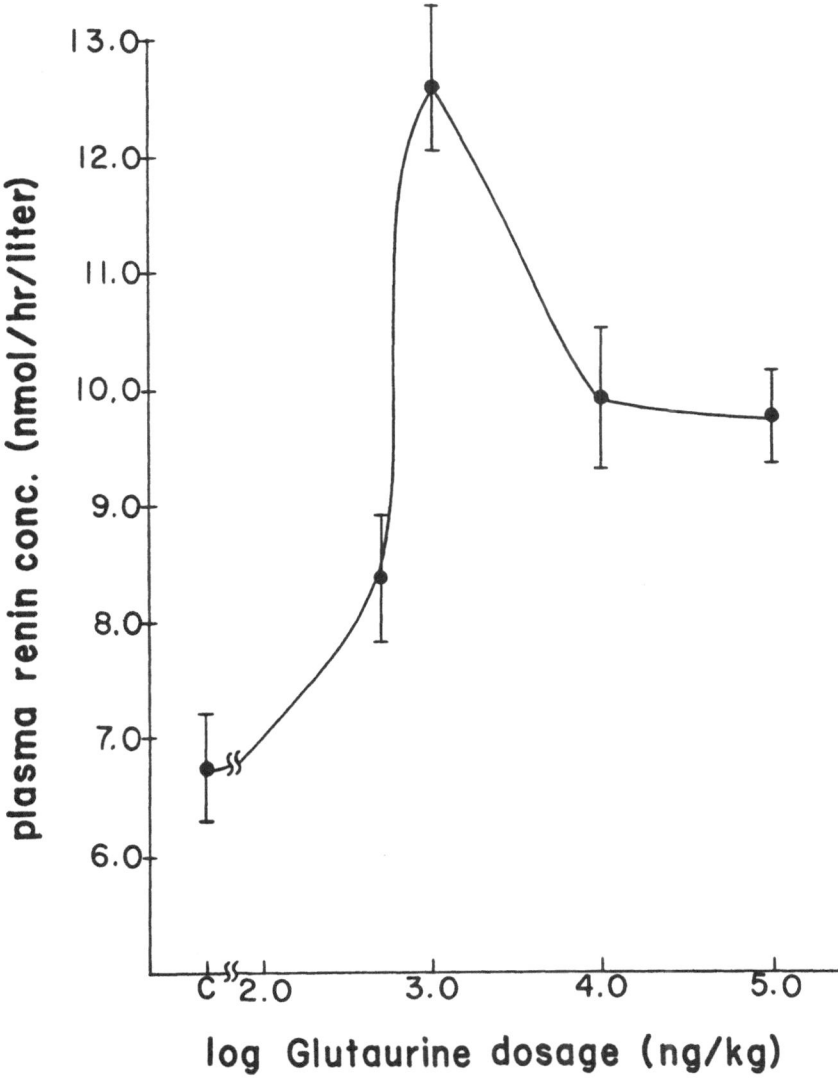

Fig. 3.1. Effect of glutaurine on plasma renin levels. Rats were injected intravenously with 0.5 ml physiological saline containing glutaurine levels varying from 0 to 100 μg/kg body weight. Fifteen minutes after the injection, blood samples were obtained following treatment with heparin (500 U i.v.). Plasma renin concentration was determined according to the method of Dauda et al. (1969); all values are expressed as nanomoles per hour per liter of angiotensin II.

REFERENCES

Benkö, G.; and Feuer, L. Investigation of the radiation protective effect of Litoralon (gamma-L-glutamyl taurine) in experimental animals. *Sixth International Congress of Radiation Research*, Tokyo, May 13–19, 1979.

Brodehl, J.; Gillissen, K.; and Kowalewski, S. Isolierter Defekt der tubulares Cystin-Ruckresorption in einer Familie mit idiopatischem Hypoparathyroidismus. *Klin. Wschr.*, 45, 38–40 (1967).

Csaba, G.; Feuer, L.; Török, L. J.; Dobozy, O. and Kovács, P. Analysis of the antagonistic effect of γ-L-glutamyl taurine on glucocorticoid and triiodothyronine. *Acta Anat.*, 104, 427–430 (1979).

Dauda, G.; Dévényi, I.; and Szokoly, V. Converting enzyme activity in experimental cirrhosis of the rat. *Acta Physiol. Acad. Sci. Hung.*, 35, 365–368 (1969).

Feuer, L. Theoretical background of the recognition of a new bioactive substance, Litoralon, isolated from the parathyroid gland. Further theoretical considerations. *Biologia* (Budapest), 25, 3–33 (1977).

Feuer, L.; and Gaál, K. Effect of glutaurine on plasma renin activity in the rat and the dog. *Gen. Comp. Endocrin.*, 39, 330–335 (1979).

Feuer, L.; Bányai, B.; and Hercsel, J. Influence of protein free aqueous extract of parathyroid powder on serum vitamin A levels in rats. *Experientia*, 33, 1005–1006 (1977a).

Feuer, L.; Török, G.; Nagy, L.G.; and Kollár, J. Effect of glutaurine (gamma-L-glutamyl taurine) on trace element levels in blood of rabbits. *Proceedings of the Third International Conference on Nuclear Methods in Environmental and Energy Research*. Missouri University, Columbia, Mo., October 10–13 (1977b).

Feuer, L.; Török, L. J.; and Csaba, G. Effect of Litoralon on the development of globlet cells of the intestinal mucosa during metamorphosis of the frog. *Acta Morph. Acad. Sci. Hung.*, 26, 319–324 (1978a).

Feuer, L.; Török, L. J.; Kapa, E.; and Csaba, G. The effect of gamma-L-glutamyl taurine (Litoralon) on the amphibian metamorphosis. *Comp. Biochem. Physiol.*, 61, 67–71 (1978b).

Feuer, L.; Török, O.; and Csaba, G. Effect of glutaurine on vitamin A and prednisolone treated thymus cultures. *Acta Morph. Acad. Sci. Hung.*, 26, 75–85 (1978c).

Feuer, L.; Török, O.; and Csaba, G. Effect of glutaurine, a newly discovered parathyroid hormone, on rat thymus cultures. *Acta Morph. Acad. Sci. Hung.*, 26, 87–94 (1978d).

Feuer, L.; Bányai, B.; and Farkas, K. Effect of parathyroid extract and of glutaurine on the manifestations of osteolathyrism. *Endokrinologie*, 75, 350–356 (1980).

Feuer, L.; Cserhalmi, M.; and Csaba, G. The effect of glutaurine on the development of amphibian larvae with inhibited RNA and protein synthesis. *Endokrinologie*, 74, 363–368 (1979a).

Feuer, L.; Kovács, P.; and Csaba, G. The effect of Litoralon (gamma-L-glutamyl taurine) on the lysosomal activity of mesenchymal cells and macrophages. *Comp. Biochem. Physiol.*, 64A, 299–303 (1979b).

Feuer, L.; Török, L. J.; and Csaba, G. Effect of taurine and taurine + triiodothyronine combination on the development of anuran larvae. *Comp. Biochem. Physiol.*, 62A, 995–997 (1979c).

Feuer, L.; Török, L. J.; and Csaba, G. The triiodothyronine antagonistic effect of gamma-L-glutamyl-taurine (Litoralon). *Endokrinologie*, 73(3), 367–369 (1979d).

Furka, A., Sebestyén, F., Feuer, L., Horváth, A., Bányai, B. and Gonda, Á. Isolation of γ-glutamyl-taurine from bovine parathyroid gland and muscle tissue. *Proceedings of The 1977 Chemistry Conference*, Debrecen, Hungary, 1977.

Gardner, L. I. *Endocrine and Genetic Diseases of Childhood.* Saunders, Philadelphia, 1969, p. 375.

Greep, R. O. Parathyroid glands. In *Comparative Endocrinology,* U. S. Euler and H. Heller, eds. Academic Press, New York, 1963, Vol. 1, p. 325.

Halver, B. Is there more than one parathyroid hormone? *New Engl. J. Med.,* 288, 321 (1973).

Henkin, R. I. Impairment of olfaction and of the tastes sour and bitter in pseudo-hypoparathyroidism. *Endocrin.,* 28, 624–628 (1968).

Pastinszky, I.; and Rácz, I. *Hautveranderungen bei inneren Krankheiten.* VEB Verlag Volk und Gesundheit, Berlin, 1974, Vol. 1, pp. 432–435.

Török, L. J.; Feuer, L.; and Csaba, G. Effect of prednisolone on the development of amphibian larvae and the antagonism of Litoralon and prednisolone. *Gen. and Comp. Endocrin.,* 38, 285–289 (1979).

Tremblay, G.; and Cartier, G. E. Histochemical study of oxidative enzymes in the human parathyroid. *Endocrinology,* 69, 658–661 (1961).

Tremblay, G.; and Peärse, A. G. E. A cytochemical study of oxidative enzymes in the parathyroid oxyphil cell and their functional significance. *Exp. Path.,* 40, 66–70 (1959).

Varró, V. *Gastroenterologia.* Medicina, Budapest, 1964, p. 243.

Williams, R. H. *Textbook of Endocrinology,* Saunders, Philadelphia, 1968, p. 883.

Yamori, Y.; Nara, Y.; Horie, R.; Ooshima, A.; and Lovenberg, W. Pathophysiological role of taurine in blood pressure regulation in stroke-prone spontaneously hypertensive rats (SHR), this volume, Chap. 27.

Copyright © 1981, Spectrum Publications, Inc.
The Effects of Taurine on Excitable Tissues

CHAPTER 4

Inhibition of Taurocyamine (Guanidinotaurine)-induced Seizures by Taurine

Akitane Mori
Yasuto Katayama
Isao Yokoi
Michiko Matsumoto

Taurocyamine, the guanidine analogue of taurine, has been found to occur naturally in various animal tissues. We have analyzed the guanidino compounds in animal brain by a liquid chromatographic method (Mori et al., 1974; Matsumoto et al., 1976a). Table 4.1 shows the distribution of arginine, guanidinosuccinic acid, γ-guanidinobutyric acid, taurocyamine, and guanidinoacetic acid in bovine brain. While arginine was the major guanidino compound found in bovine brain tissue, measureable quantities of the other guanidine derivatives were also found. The highest concentrations of taurocyamine were found in the cortical gray matter and the lowest in the cerebellum and the medulla oblongata.

A comparison of these data with that published for amino acids in the same areas of rat brain (Kandera et al., 1968; Shaw and Heine, 1965) shows that the distribution of taurocyamine, guanidinoacetic acid and γ-guanidinobutyric acid corresponds closely with that of taurine, glycine, and γ-aminobutyric acid, respectively. This suggests that ω-amino acids may be the precursors of the corresponding guanidino compounds, and that they are produced by transamidination reactions in the brain. Increased levels of taurocyamine have also been demonstrated in the taurine-loaded rat (Thoai et al., 1954) and rabbit (Mori et al., 1978). These facts suggest that taurocyamine in the brain may be derived from taurine.

Table 4.1   Levels of guanidino compounds in bovine brain (nmol/g)[a,b]

	n	G · Tau	G · SA	G · Ac	γ-G · BA	Arg
Cortex: gray matter	7	25 ± 20	45 ± 19	19 ± 7	35 ± 6	165 ± 64
white matter	7	15 ± 7	33 ± 8	16 ± 5	25 ± 14	352 ± 97
Cerebellum	7	5 ± 4	34 ± 6	6 ± 5	18 ± 3	210 ± 33
Thalamus	8	7 ± 1	41 ± 10	16 ± 5	36 ± 6	288 ± 47
Hypothalamus	7	10 ± 8	42 ± 14	14 ± 6	33 ± 13	275 ± 47
Hippocampus	8	13 ± 12	44 ± 9	19 ± 12	28 ± 6	306 ± 39
Caudate nucleus	8	7 ± 2	42 ± 16	20 ± 5	23 ± 16	220 ± 60
Pons	7	14 ± 10	32 ± 15	21 ± 7	17 ± 3	316 ± 62
Medulla oblongata	7	5 ± 3	41 ± 22	17 ± 14	5 ± 1	380 ± 155
Amygdala	7	11 ± 8	34 ± 6	15 ± 10	27 ± 10	177 ± 57

[a]Data given are mean values ± S.D.

[b]G · Tau: taurocyamine; G · SA: guanidinosuccinic acid; G · Ac: guanidinoacetic acid; γG · BA: γ-guanidinobutyric acid; Arg: arginine.

We recently performed a preliminary expriment investigating this problem using ^{14}C-labeled arginine. One milligram of ^{14}C-guanidino-labeled arginine containing 5 μCi and the same molar equivalent of nonlabeled taurine (0.825 mg) were administered orally to a mouse weighing about 15 g. Urine samples, collected at 24 and 48 hr after isotope administration, were deproteinized using a millipore filter and then analyzed using a paper chromatographic system. Table 4.2 shows that the taurocyamine spot was labeled by

Table 4.2   Determination of labeled taurocyamine in urine after administration of guanidino-^{14}C-arginine[a]

Compound	(Rf)	0–24 hrs after (dpm)	24–48 hrs after (dpm)
Arginine	(0.15)	345	142
Taurocyamine	(0.31)	269	176
Unknown	(0.38)	7,611	2,521

[a]5 μCi (1 mg) of ^{14}C-arginine was orally administered to mice. 10 μl of urine was analyzed paper chromatographically using TOYO filter paper No. 51 and a solvent (pyridine: isoamylalcholol:acetic acid:water = 8:4:1:4). Arginine and taurocyamine spots detected by Sakaguchi reagent, along with the unknown radioactive spot, were cut out, divided and extracted each by 1 ml of Soluene®-350 (Packard, Ill.) in a vial. To the extract was added 5 ml of liquid scintillator (0.4% PPO and 0.1% POPOP in toluene). Radioactivity was estimated by a liquid scintillation spectrometer (Aloka LSC 653).

the isotope derived from ^{14}C-guanidino-labeled arginine, although most of the radioactivity was found in an area on the chromatogram that was neither arginine nor taurocyamine. These data indicate that the guanidine label in taurocyamine comes from arginine, and suggest the possibility that tauro-cyamine is formed from taurine by a transamidination.

We also studied the CNS excitatory action of taurocyamine. We had previously observed that γ-guanidinobutyric acid could induce tonic and clonic seizures in rabbits and cats when injected intracisternally (Jinnai et al., 1966), suggesting that guanidino compounds may play an excitatory role in the central nervous system. We later demonstrated that taurocyamine (Mizuno et al., 1975), glycocyamine (Jinnai et al., 1969), N-acetylarginine (Mori and Okusu, 1971), and methylguanidine (Matsumoto et al., 1976) each acted as a convulsant when injected intracerebrally.

Table 4.3 shows an example of the seizures induced by intracisternal injection of taurocyamine in the rabbit. The taurocyamine used here was synthesized from taurine and S-methylisothiourea sulfate; it was recrystallized from water and ethanol and melted sharply at 267°C (uncorr.). In this experiment, 0.7 ml of 0.1 M taurocyamine was injected into the cisterna magna of a rabbit weighing 1.7 kg. Short bursts of clonic convulsions started 10 min after the injection, then repeated tonic convulsions occurred without any inter-

Table 4.3   An example of convulsions induced by intracisternal injection of 0.7 ml of 0.1 M taurocyamine in the rabbit

Time after injection (min)	Type of convulsion	Duration of convulsion(s)
10	Clonic	2–3
13	Clonic	3
40	Tonic	30
75	Tonic	17
98	Tonic	30
105	Tonic	30
110	Tonic	40
115	Tonic	45
125	Tonic	45
130	Tonic	88
140	Tonic	30
147	Tonic	52
157	Tonic	36
222	Tonic	30[a]

[a]No more convulsions followed this, and the animal regained the ability to walk.

vening clonic convulsions between 40 and 222 min after the injection. The average duration of these tonic convulsions was 40 sec, after which the animal recovered completely. In this series of 12 rabbit experiments, the threshold dosage for taurocyamine was 6–7 mg/kg, the latent time varied from 0–40 min, and the duration of the entire train of convulsive episodes was 1–3 hr.

In another series of experiments, 0.8 ml of 0.2 M taurocyamine was injected into the cisterna magna of cats, and the EEG was recorded throughout the control period and until recovery from the convulsions occurred. Cats were anesthetized with ether and fixed in a stereotaxic apparatus. Concentric needle electrodes were inserted into the motor cortex, globus pallidus, amygdala, dorsal hippocampus, and mesencephalic reticular formation for recording, the recording points being histologically verified after each experiment. One of the EEG records is illustrated in Fig. 4.1. The high voltage spike that appeared first in the dorsal hippocampus 270 sec after the taurocyamine injection was accompanied by a relatively low voltage spike in the amygdala. Continuous multiple low voltage spike discharges followed a few seconds later in both structures, continuing for 12sec (Fig. 4.1a). The same pattern of discharges reappeared after an interval of 8sec. The discharges developed later into characteristic high voltage spike discharges in the hippocampus, accompanied by medium high voltage spikes in the amygdala and also high voltage slow waves mixed with some spike activities in the motor cortex. Sporadic single high voltage spikes reappeared 150 sec later and developed progressively into multiple spike discharges (Fig. 4.1b). These repeated electroencephalographic processes continued for 2 hours.

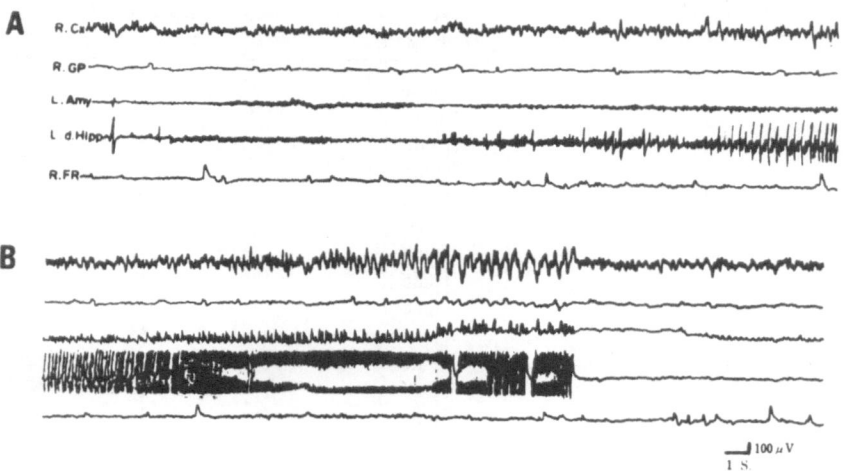

Fig. 4.1.   An example of taurocyamine-induced seizures.

The EEG observations suggest that taurocyamine-induced seizures originate in the limbic system, probably in the hippocampus.

This hypothesis was supported by a biochemical study of the distribution of taurocyamine in the rabbit brain after the intracisternal injection of ^{3}H-taurocyamine. Table 4.4 shows that taurocyamine was concentrated in the hippocampus.

We then studied the effect of taurine on taurocyamine-induced seizures in the cat. In this experiment, 0.8 ml of 0.1 M taurocyamine was injected intracisternally to induce increased EEG activity. While such a dose of taurocyamine did not induce a frank convulsion, high voltage spikes appeared in hippocampus, amygdala, and central median nucleus at a frequency of 5–6/min (Fig. 4.2).

If taurine (0.8 ml, 0.1 M in saline solution) was administered at this stage, it was found that the spike counts were reduced to almost zero for several minutes after the injection, although the spikes reappeared 20–30 min after taurocyamine administration. Figure 4.3 shows an example of these procedures. We have also administered taurine in the ictal stage in the same way, but observed no inhibitory effect of taurine on spike activity. These observations suggest that taurine affects propagation of seizure discharges.

We have also studied the effect of guanidino compounds on electrical activity of the giant neurons in the subesophageal ganglia of the African giant snail (Matsumoto et al., 1976c). No excitatory or inhibitory effects were produced by taurocyamine and the other guanidino compounds, except for an in-

Table 4.4   Distribution of ^{3}H-taurocyamine in brain (cpm/mg)[a]

| | Experiment number | | |
	1	2	3
	Time after i.c. injection		
	15 min	100 min	120 min
Frontal lobe	157.5	40.0	8.3
Amygdala	154.0	287.4	104.8
Hippocampus	152.9	1,034.0	1.653.0
Thalamus	139.7	339.0	174.2
Brain stem	198.7	308.1	619.7
Cerebellum	128.4	331.4	172.6

[a]^{3}H-Taurocyamine (12.7 μCi, 16.7 mg) was injected into the cisterna magna of rabbits. Each sample of brain tissue, (4 mg), was extracted, dissolved in 0.1 ml of 2N KOH-methanol solution in a vial, and 10 ml of liquid scintillator was added. Radioactivities were estimated by a liquid scintillation spectrometer.

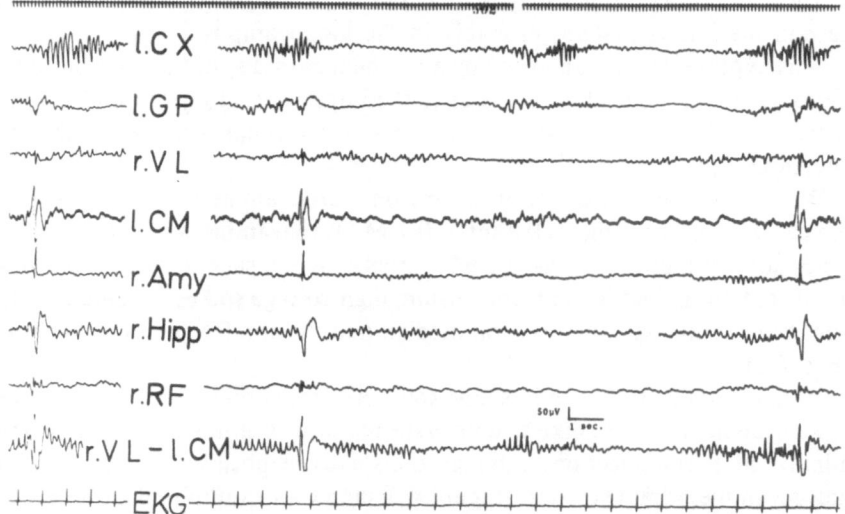

Fig. 4.2. An example of taurocyamine-induced spike activities.

hibitory tendency of guanidinoacetic acid. Takeuchi and Takeuchi (1975) reported that γ-guanidinobutyric acid, guanidinoacetic acid, and β-guanidino-propionic acid competively inhibited the action of GABA on the postsynaptic membrane of crayfish muscle.

The observation that taurocyamine produces seizures which are antago-

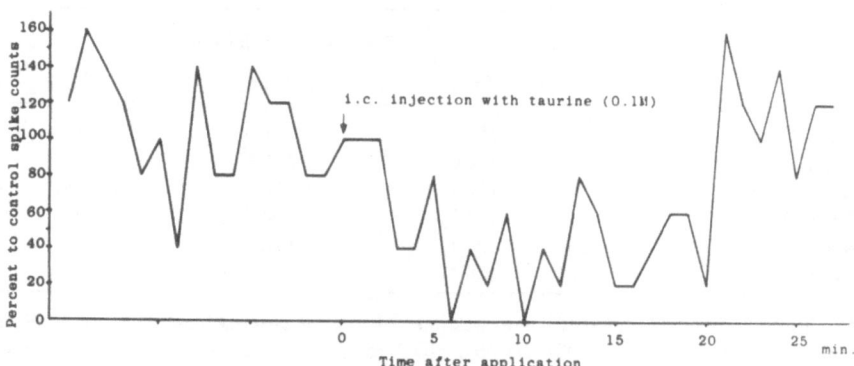

Fig. 4.3. Effect of taurine on taurocyamine-induced spike activity of cat brain.

nized by taurine suggests that these two compounds could play mutually antagonistic roles in excitation and inhibition in the central nervous system. The physiological and biochemical mechanisms, however, are still obscure.

In conclusion, taurocyamine is a naturally occurring substance widely distributed in the brain, and it may be produced from taurine and arginine by a transamidination reaction. When taurocyamine was administered intracisternally into the experimental animal, seizures were observed clinically as well as by EEG recordings. In the cat taurine was found to inhibit the spike activities induced by taurocyamine.

REFERENCES

Barker, L. A. Modulation of synaptosomal high affinity choline transport. *Life Sci.*, 18, 725–730 (1976).

Baskin, S. I.; Leibman, A. J.; and Cohn, E. M. Possible functions of taurine in the central nervous system. In *Advan. Biochem. Psychopharmac.*, Costa, E.; Giacobini, E.; and Paoletti, R., eds. Raven Press, New York, Vol. 15, pp. 153–164 (1976).

Collins, G. G. S. The rate of synthesis, uptake and disappearance of [14C]-taurine in eight areas of the rat central nervous system. *Brain Res.*, 76, 447–459 (1974).

Cotman, C. W. Isolation of synaptosomal and synaptic plasma membrane fractions. In *Methods in Enzymol.*, Fleischer, S.; and Packer, L. eds. Academic Press, New York, Vol. 31, pp. 445–452 (1974).

Curtis, D. R.; and Johnston, G. A. R. Amino acid transmitters in the mammalian central nervous system. *Ergebn. Physiol.*, 69, 97–188 (1974).

Hruska, R.; Huxtable, R.; Bressler, R.; and Yamamura, H. Sodium-dependent high affinity transport of taurine into rat brain synaptosomes. *Proc. West. Pharmac. Soc.*, 19, 152–156 (1974).

Kontro, P.; and Oja, S. S. Taurine uptake by rat brain synaptosomes. *J. Neurochem.*, 30, 1297–1304 (1978).

Lähdesmäki, P.; Pasula, M.; and Oja, S. S. Effect of electrical stimulation and chlorpromazine on the uptake and release of taurine, γ-aminobutyric acid and glutamic acid in mouse brain synaptosomes. *J. Neurochem.*, 25, 675–680 (1975).

Lombardini, J. B. High affinity uptake systems for taurine in tissue slices and synaptosomal fractions prepared from various regions of the rat central nervous system: Correction of transport data by different experimental procedures. *J. Neurochem.*, 29, 305–312 (1977).

Mandel, P.; and Pasantes-Morales, H. Taurine: A putative neurotransmitter. In *Advan. Biochem. Psychopharmac.*, Costa, E.; Giacobini, E.; and Paoletti, R., eds. Raven Press, New York, Vol. 15, pp. 153–164 (1976).

Murrin, L. C.; and Kuhar, M. J. Activation of high affinity choline uptake *in vitro* by depolarizing agents. *Molec. Pharmac.*, 12, 1082–1090 (1976).

Murrin, L. C.; Lewis, M. S.; and Kuhar, M. J. Amino acid transport: Alterations due to synaptosomal depolarization. *Life Sci.*, 22, 2009–2016 (1978).

Schmid, R.; Sieghart, W.; and Karobath, M. Taurine uptake in synaptosomal fractions of rat cerebral cortex. *J. Neurochem.*, 25, 5–9 (1975).

Sieghart, W.; and Karobath, M. Uptake of taurine into subcellular fractions of C-6 glioma cells. *J. Neurochem.*, 26, 981–986 (1976).

Simon, J. R.; and Kuhar, M. J. High-affinity choline uptake: Ionic and energy requirements. *J. Neurochem.*, 27, 93–99 (1976).

Synder, S. H.; Young, A. B.; Bennett, J. P., and Mulder, A. H. Synaptic biochemistry of amino acids. *Fed. Proc.*, 32, 2039–2047 (1973).

Copyright © 1981, Spectrum Publications, Inc.
The Effects of Taurine on Excitable Tissues

CHAPTER 5

Properties of Hypotaurine Uptake in Mouse Brain Slices

P. Kontro
S. S. Oja

Hypotaurine, 2-aminoethane sulfinic acid, is thought to be the precursor of taurine in the mammalian brain. Taurine biosynthesis has been assumed to proceed from cysteine via cysteine sulfinic acid and hypotaurine intermediates (Jacobsen and Smith, 1968), even though the final step in biosynthesis, the oxidation of hypotaurine to taurine, has never been satisfactorily demonstrated (Oja et al., 1977). Only quite recently have we been able to characterize some properties of that reaction and to confirm its enzymatic nature (Kontro and Oja, 1978a). Hypotaurine is the most potent inhibitor of taurine transport among its structural analogues, as shown first with brain slices (Lähdesmäki and Oja, 1973) and then with different CNS preparations (Schmid et al., 1975; Kennedy and Voaden, 1976; Sieghart and Karobath, 1976). The transport of taurine has been studied to some extent by several investigators (see Oja and Kontro, 1978), but no work has been done on the transport of hypotaurine itself.

METHODS

[^{35}S]Hypotaurine was prepared from [^{35}S]cystamine (118 Ci/mol) using a modified method of Scandurra et al. (1969). The radiochemical purity of the preparation was checked by thin-layer chromatography. Slices 0.4 mm thick were cut in the frontal plane from mouse cerebral hemispheres with a McIlwain tissue chopper. About 100 mg of slices were incubated under O_2 at 37°C with shaking in 5 ml of Krebs-Ringer-HEPES solution (pH 7.4): NaCl,

120; KCl, 5; CaCl$_2$, 2.5; NaH$_2$PO$_4$, 1.3; MgSO$_4$, 1.2; HEPES, 10 mmol/liter. The medium contained 10 mmol/liter of glucose, 10 μCi/liter of [^{35}S]hypotaurine, and varying amounts of unlabeled hypotaurine. Preincubation with Krebs-Ringer-HEPES solution was 30 min, and the final incubation with labeled hypotaurine lasted generally 15 min, after which the slices and the incubation media were separated and counted. The inulin space of the slices was determined according to Laakso and Oja (1976). It was 30.5 $\pm$ 1.3% (mean $\pm$ S.E.M., $n = 9$) of the weight of incubated slices. The concentration of hypotaurine in the slices was then calculated as described by Vahvelainen and Oja (1972).

In some experiments the slices were exposed to metabolic poisons, nitrogen atmosphere, or lowered temperatures, both during the preincubation and final incubation periods. In a number of experiments, sodium, potassium, calcium, or magnesium ions in the medium were replaced by equimolar amounts of choline. In the absence of exogenous calcium, the endogenous calcium was bound with 0.5 mmol/liter EDTA. At elevated sodium ion concentrations the results were corrected for the effects of hyperosmolarity with the aid of corresponding control samples in which the surplus NaCl was replaced by equimolar choline chloride. Only about 10% of the original [^{35}S]hypotaurine was oxidized to [^{35}S]taurine by the slices even when the incubation was prolonged up to 4 hr. The [^{35}S]hypotaurine preparation used contained less than 4% [^{35}S]taurine as a radiochemical impurity.

RESULTS

Hypotaurine uptake was linear up to 40 min, after which time uptake gradually slowed down as equilibrium between influx and efflux was approached (Fig. 5.1). The highest tissue water/medium ratio reached was about 80. An incubation time of 15 min was chosen for further experiments. Uptake was at least partly energy dependent and was strongly reduced under anaerobic conditions and by metabolic inhibitors such as 2,4-dinitrophenol (Fig. 5.2). Ouabain was likewise a potent inhibitor. On the other hand, cyanide and iodoacetate had no great effect. Uptake was very temperature sensitive, being completely abolished at 5°C.

Uptake was saturable in nature with no apparent signs of simultaneous nonsaturable penetration (Fig. 5.3). The rate of uptake, v, at varying hypotaurine concentrations, s, was also plotted against v/s (Fig. 5.4). This particular linear transformation of the Michaelis equation was chosen for its sensitivity to any departure in the data from linearity. Uptake seems to consist of a low-affinity and a high-affinity component. These were separated from each other as follows. The kinetic constants (V and K_m) of the low-affinity,

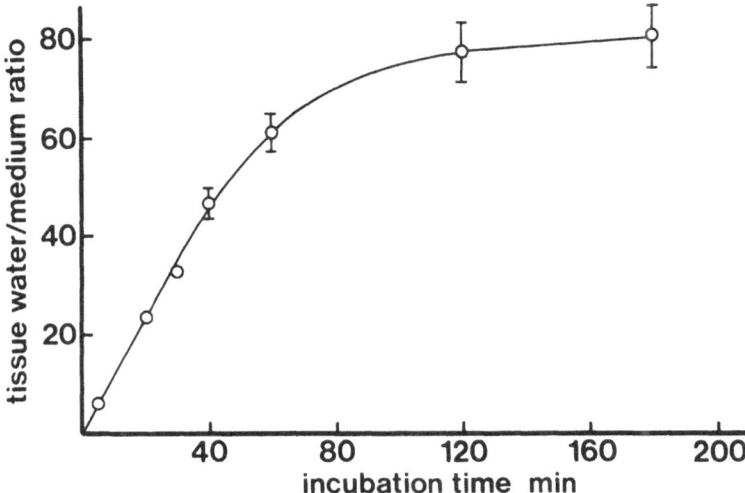

Fig. 5.1. Time course of hypotaurine (0.1 mmol/liter) uptake by mouse brain slices in Krebs-Ringer-HEPES-glucose medium at 37°C under O_2. Each point is a mean ($\pm$ S.E.M.) of three experiments.

high-capacity transport were first evaluated from the plot by drawing a best-fit straight line by the method of least squares through the unweighted experimental points corresponding to higher concentrations of hypotaurine in the medium (Fig. 5.4A). The contribution of this transport system was then

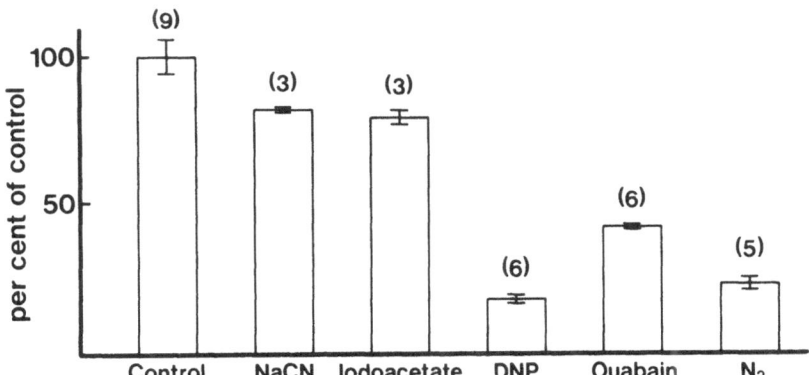

Fig. 5.2. Effects of some metabolic inhibitors and nitrogen atmosphere on hypotaurine (0.1 mmol/liter) uptake by mouse brain slices. Concentrations of sodium cyanide, sodium iodoacetate, and 2,4-dinitrophenol were 1.0 mmol/liter, and that of ouabain was 0.01 mmol/liter. Means ($\pm$ S.E.M.) are indicated with number of experiments in brackets.

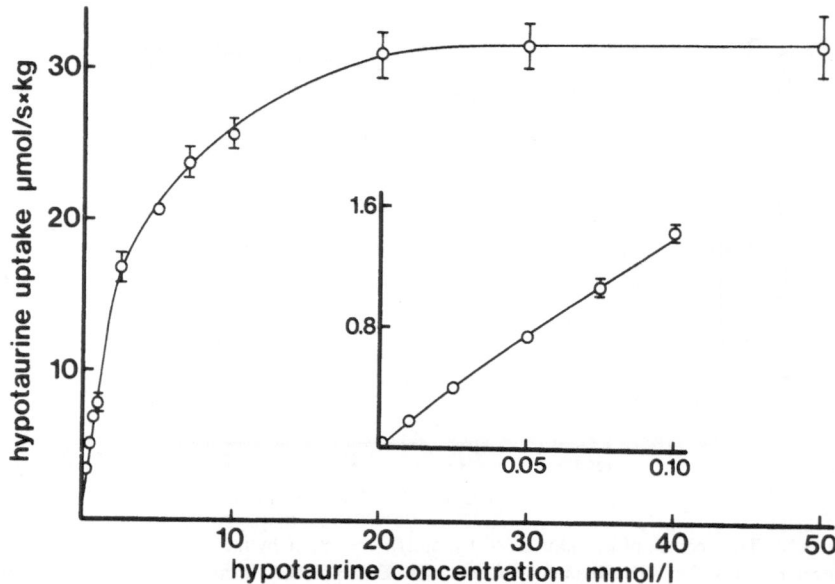

Fig. 5.3. Hypotaurine uptake by mouse brain slices incubated for 15 min at 37°C in Krebs-Ringer-HEPES-glucose medium under O_2 with varying hypotaurine concentrations. Each point is a mean (± S.E.M.) of six experiments.

subtracted from the total uptake. Residual uptake still remaining after this subtraction yielded the kinetic constants for the high-affinity, low-capacity transport system (Fig. 5.4B). The transport constants, K_m, were about 30 μmol/liter and 3 mmol/liter for the high-affinity and low-affinity components, respectively, while the maximal velocities, V, of the transport components were about 0.3 and 30 μmol/sec × kg incubated slices.

Uptake of hypotaurine was completely abolished when sodium ions in the medium were replaced by equimolar amounts of choline (Fig. 5.5). It was considerably reduced in the absence of potassium and also significantly reduced ($p < 0.01$), but to a smaller extent, in the absence of calcium or magnesium ions. Interactions of sodium with hypotaurine uptake were further examined at varying sodium and hypotaurine concentrations. In a low-sodium medium (60 mmol/liter) the uptake was still saturable and consisted of the two transport components (Fig. 5.6). Transport constants for both components were nearly fourfold higher than in the normal-sodium medium, whereas the maximal velocities of transport were about the same. The uptake of hypotaurine did not show a linear sodium dependence but rather a sigmoidal one (Fig. 5.7). Optimal sodium concentration was about 240

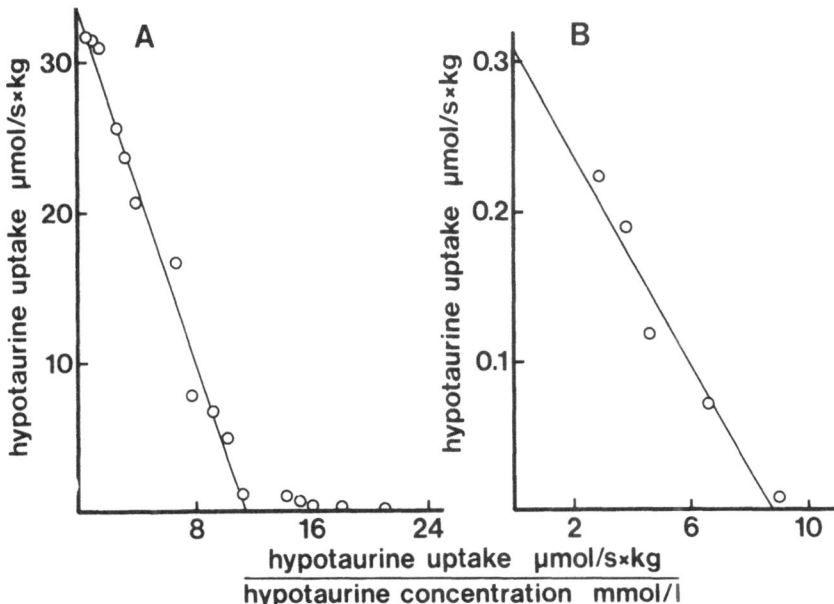

Fig. 5.4. (A) Hypotaurine uptake by mouse brain slices depicted in a v vs. v/s plot. The results cannot be fitted by only one saturable transport component. The straight line indicates the estimation of kinetic constants for the low-affinity transport. (B) The share of the high-affinity transport after subtraction of the low-affinity transport component from the total uptake. The data in this figure are derived from the experiments shown in Fig. 5.3.

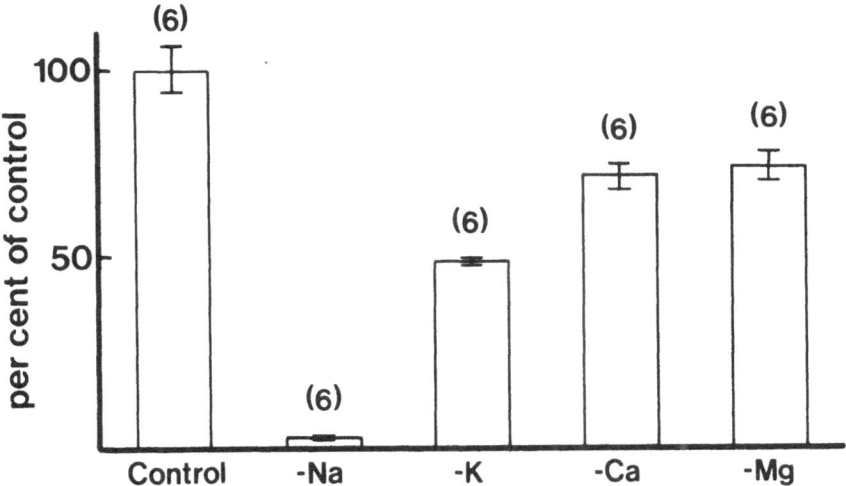

Fig. 5.5. Effect of omission of cations on hypotaurine (0.1 mmol/liter) uptake by mouse brain slices. Means (± S.E.M.) are indicated with number of experiments in brackets.

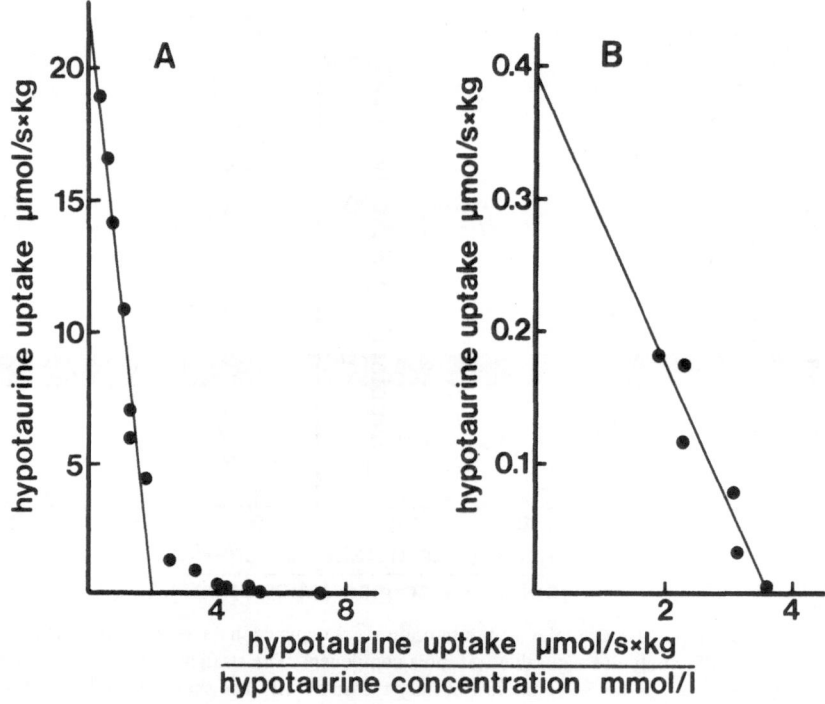

Fig. 5.6. Hypotaurine uptake by mouse brain slices in a low-sodium (60 mmol/liter) medium. The results at varying hypotaurine concentrations are depicted in v vs. v/s plots. The estimation of kinetic parameters for the low-affinity (a) and high-affinity (b) transport components was carried out as in Fig. 5.4. Each point represents a mean of three experiments.

mmol/liter for the uptake from solutions containing 0.01 mmol/liter hypotaurine. The sodium-dependence curve was very similar when the uptake was studied at a 1 mmol/liter hypotaurine concentration.

DISCUSSION

Hypotaurine uptake seems to be much faster than taurine uptake (Kontro and Oja, 1978b). The sulfinic acid group renders the molecule more amenable for transport across neural plasma membranes. The tissue water/medium ratios of hypotaurine were several fold greater than those obtained for taurine in brain slices (Oja, 1971). They are quite comparable to the highest ratios reported for GABA in rat cerebral cortex slices (Bond, 1973) and for

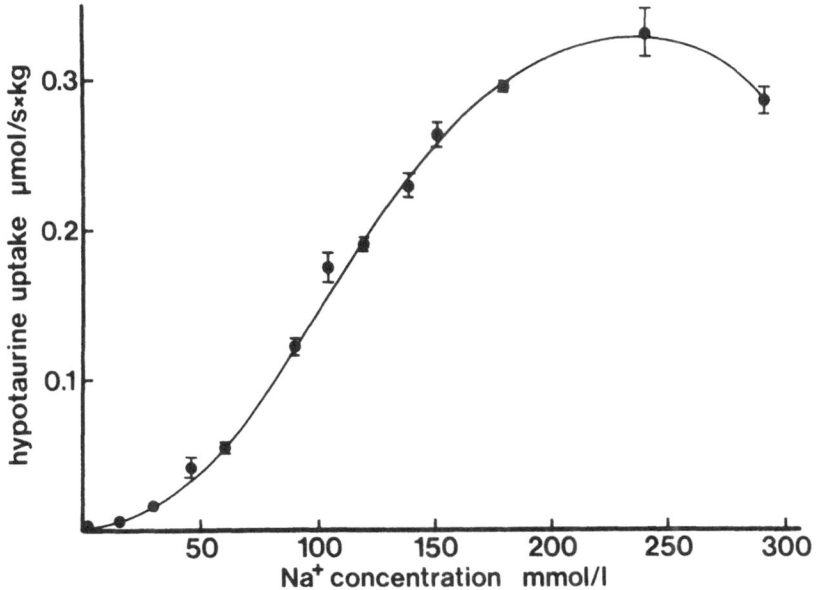

Fig. 5.7.   Sodium dependence of hypotaurine (0.01 mmol/liter) uptake by mouse brain slices. Each point is a mean (± S.E.M.) of three to nine experiments.

glycine in the spinal cord (Johnston and Iversen, 1971). Hypotaurine uptake, like the uptake of GABA, glycine, and taurine, consisted of two saturable transport components. The transport constant for the high-affinity uptake of hypotaurine was very similar to the corresponding constants generally reported for these three neurotransmitter candidates (Oja et al., 1977). The estimates for the maximal velocities of the high-affinity transport reported by different authors are hardly comparable, but they may be highest for GABA, lowest for taurine, and intermediate for both glycine and hypotaurine. On the other hand, the maximal velocity of the low-affinity uptake may be highest for hypotaurine (see Oja et al., 1977).

A strictly sodium-dependent, high-affinity transport has generally been postulated as characteristic for neurotransmitters. The high-affinity transport of taurine is also completely blocked by the absence of external sodium ions, whereas the low-affinity transport component is only moderately inhibited (Kontro and Oja, 1978c). Both uptake components of hypotaurine are also abolished in the absence of sodium. The increase of transport constants in the low-sodium medium suggests that sodium ions are required for the interaction of hypotaurine with its putative carrier sites in neural plasma membranes. Since the maximal velocities of transport remained about the same, it is unlikely that sodium ions could also enhance the translation rate of either unloaded or hypotaurine-loaded carriers across cell membranes. On the other hand, both the transport constant and the maximal velocity of GABA uptake by synaptosomes have been reported to be affected by sodium ions (Martin, 1973).

Optimal sodium ion concentration for hypotaurine uptake was found to be of the same order of magnitude as that reported for GABA uptake by the retina (Starr, 1973). With regard to its other characteristics, the sodium-dependence curve of hypotaurine uptake was similar to that of GABA and taurine in synaptosomes (Martin and Smith, 1972; Kontro and Oja, 1978c). Sodium ions thus mimic allosteric effectors in hypotaurine transport, showing positive cooperativity. This indicates that the hypotaurine carriers in the cell membrane may also contain several binding sites for sodium. This statement applies to both high-affinity and low-affinity transport components of hypotaurine.

The close resemblance of hypotaurine uptake to that of GABA and glycine has two alternative consequences. Either an energy-dependent and strictly sodium-dependent high-affinity transport system is not an essential and specific property of a neurotransmitter, or hypotaurine indeed possesses some synaptic role of its own. This latter possibility has never been considered, to the best of our knowledge.

SUMMARY

The uptake of [^{35}S]hypotaurine was studied with mouse brain slices incubated in Krebs-Ringer-HEPES-glucose medium. Hypotaurine uptake closely resembled the transport of the amino acid neurotransmitter candidates. It consisted of two transport systems. The high-affinity component was concentrative, energy dependent, temperature sensitive, and strictly sodium dependent. When compared to taurine, hypotaurine is taken up much faster owing to its larger transport capacity.

REFERENCES

Bond, P. A. The uptake of gamma-[^{3}H]-aminobutyric acid by slices from various regions of rat brain and the effect of lithium. *J. Neurochem.*, 20, 511–517 (1973).

Jacobsen, J. G.; and Smith, L. H., Jr. Biochemistry and physiology of taurine and taurine derivatives. *Physiol. Rev.*, 48, 424–511 (1968).

Johnston, G. A. R.; and Iversen, L. L. Glycine uptake in cat central nervous system slices and homogenates: Evidence for different uptake systems in spinal cord and cortex. *J. Neurochem.*, 18, 1951–1961 (1971).

Kennedy, A. J.; and Voaden, M. J. Studies on the uptake and release of radioactive taurine by the frog retina. *J. Neurochem.*, 27, 131–137 (1976).

Kontro, P.; and Oja, S. S. Source of taurine in developing brain. In *Abstracts of the International Symposium on Developmental Neurobiology*, Tehran (1978a), pp. 60–61.

Kontro, P.; and Oja, S. S. Taurine uptake by rat brain synaptosomes. *J. Neurochem.*, 30, 1297–1304 (1978b).

Kontro, P.; and Oja, S. S. Sodium dependence of taurine uptake in rat brain synaptosomes. *Neuroscience*, 3, 761–765 (1978c).

Laakso, M.-L.; and Oja, S. S. Factors influencing the inulin space in cerebral cortex slices from adult and 7-day-old rats. *Acta Physiol. Scand.*, 97, 486–494 (1976).

Lähdesmäki, P.; and Oja, S. S. On the mechanism of taurine transport at brain cell membranes. *J. Neurochem.*, 20, 1411–1417 (1973).

Martin, D. L. Kinetics of the sodium-dependent transport of gamma-aminobutyric acid by synaptosomes. *J. Neurochem.*, 21, 345–356 (1973).

Martin, D. L.; and Smith, A. A., III. Ions and the transport of gamma-aminobutyric acid by synaptosomes. *J. Neurochem.*, 19, 841–855 (1972).

Oja, S. S. Exchange of taurine in brain slices of adult and 7-day-old rats. *J. Neurochem.*, 18, 1847–1852 (1971).

Oja, S. S.; and Kontro, P. Neurotransmitter actions of taurine in the central nervous system. In *Taurine and Neurological Disorders*, A. Barbeau and R.J. Huxtable, eds. Raven Press, New York (1978), pp. 181–200.

Oja, S. S.; Kontro, P.; and Lähdesmäki, P. Amino acids as inhibitory neurotransmitters. *Progr. Pharmacol.*, 1(3), 1–119 (1977).

Scandurra, R.; Fiori, A.; and Cannella, C. Preparation of ³⁵S-labeled hypotaurine. *Ital. J. Biochem.*, 18, 19–24 (1969).

Schmid, R.; Sieghart, W.; and Karobath, M. Taurine uptake in synaptosomal fractions of rat cerebral cortex. *J. Neurochem.*, 25, 5–9 (1975).

Sieghart, W.; and Karobath, M. Uptake of taurine into subcellular fractions of C-6 glioma cells. *J. Neurochem.*, 26, 981–986 (1976).

Starr, M. S. Effects of changes in the ionic composition of the incubation medium on the accumulation and metabolism of [³H]-γ-aminobutyric acid and ¹⁴C-taurine in isolated rat retina. *Biochem. Pharmacol.*, 22, 1693–1700 (1973).

Vahvelainen, M.-L.; and Oja, S. S. Kinetics of transport of phenylalanine, tyrosine, tryptophan, histidine and leucine in brain cortex slices prepared from adult and 7-day-old rats. *Brain Res.*, 40, 477–488 (1972).

Copyright © 1981, Spectrum Publications, Inc.
The Effects of Taurine on Excitable Tissues

CHAPTER 6

Cephalopod Nerve as a Model System for the Study of the Metabolism of Taurine and Related Sulfur Compounds

Francis C. G. Hoskin

A. N. Richards's work was mainly concerned with kidney function and nitrogen metabolism and excretion. It is not certain that Richards would have been half so interested in a compound with a hydroxyl in place of an amine at one end, but that is where interest in the squid giant axon has taken this author, and along the way he has become acquainted with taurine.

In the phylum *Mollusca* there are three major and quite diverse classes—the gastropods, the bivalves, and the cephalopods. While they are related anatomically, the cephalopods show marked physiological differences from the other mollusks. There are ways in which their nervous systems and behavior are more like the vertebrates than like the invertebrates. The most visible and striking example is the cephalopod simple eye, not encountered anywhere else in the invertebrate world. In addition, the squid giant axon has served as the definitive model system for the movement of sodium and potassium ions (Hodgkin and Huxley, 1952).

The interest in these cations has overshadowed a long-known but generally overlooked anionic anomaly of the squid giant axon, namely that isethionate (2-hydroxyethanesulfonate) is present in the axoplasm at a concentration of about 150 mM (Deffner and Hafter, 1960; Hoskin and Brande, 1973), while taurine is present at about 75 mM (Deffner and Hafter, 1960). Thus, the inside of the squid giant axon is 0.2 to 0.3 molar in organosulfur compounds, a truly remarkable concentration that, one suspects, must reflect an equally remarkable level of related enzyme activities. These observations have been extended sufficiently (Hoskin, 1974; Hoskin et al., 1975; Garden et al., 1975) to permit the conclusion that cephalopod nerve in general has a similarly high intracellular concentration of sulfur compounds, especially isethionate, and further that other organs of cephalopods contain little or no isethionate.

Cephalopod nerve does indeed contain a remarkable level of a particular enzyme, DFPase, capable of hydrolyzing diisopropylphosphorofluoridate (DFP), although the relationship to isethionate, while suggested (Hoskin and Long, 1972; Hoskin, 1974), remains highly speculative. Squid-type DFPase has at least one marked difference from the other DFPases. The squid enzyme hydrolyzes DFP several times faster than it hydrolyzes ethyl N, N-dimethylphosphoramidocyanidate (Tabun) (Hoskin, 1971), whereas this order is reversed in the other DFPase (Hoskin, 1971; Mounter, 1963). A parallel between the levels of isethionate in various tissues and the levels of squid DFPase, as defined by the preferential hydrolysis of DFP over Tabun, is illustrated in Fig. 6.1. One might speculate that isethionate or a sulfonic or sulfinic precursor of isethionate might be held in a nonionic acid anhydride linkage, perhaps with phosphate or even sulfate. During brief periods when an intracellular anion deficit could not be overcome rapidly enough by inward chloride diffusion, the acid anhydride-splitting enzyme DFPase could be activated to provide isethionate.

Regarding the question of sulfonic or sulfinic precursors of isethionate, it has generally been assumed that these include taurine, hypotaurine, cysteine-sulfinate, and cysteate. The obvious final step in this pathway from taurine to isethionate does not exist in the squid (Hoskin and Brande, 1973; Hoskin et al., 1975), and now it is questionable whether it exists to any appreciable degree in other organisms (Fellman et al., 1978). Such negative findings are now strengthened by evidence of an alternate pathway to isethionate (Hoskin and Kordik, 1977), namely that ^{35}S-sulfide is metabolized to isethionate. The sulfur of ^{35}S-cysteine, but not the carbons, is also metabolized to isethionate. This suggests that the sulfur portion of the cysteine passes through a sulfide stage or its equivalent on its way to isethionate. Since our finding that sulfide is incorporated into isethionate (Hoskin and Kordik, 1977), our aim has been to find a source of the carbon atoms. This clearly pointed to the need for a method of assaying ^{14}C and ^{35}S in a double-labeled sample, a measurement not normally possible due to the virtually identical decay energies of the two isotopes. We have developed such an assay.

The basis for the ^{14}C assay is the fact that the complete oxidation of carbon yields acid-volatile CO_2 capable of passing a dry ice-acetone trap, and condensable into one of the organic base-containing scintillants (for example New England Nuclear's Oxyfluor) at liquid nitrogen temperature. The oxida-

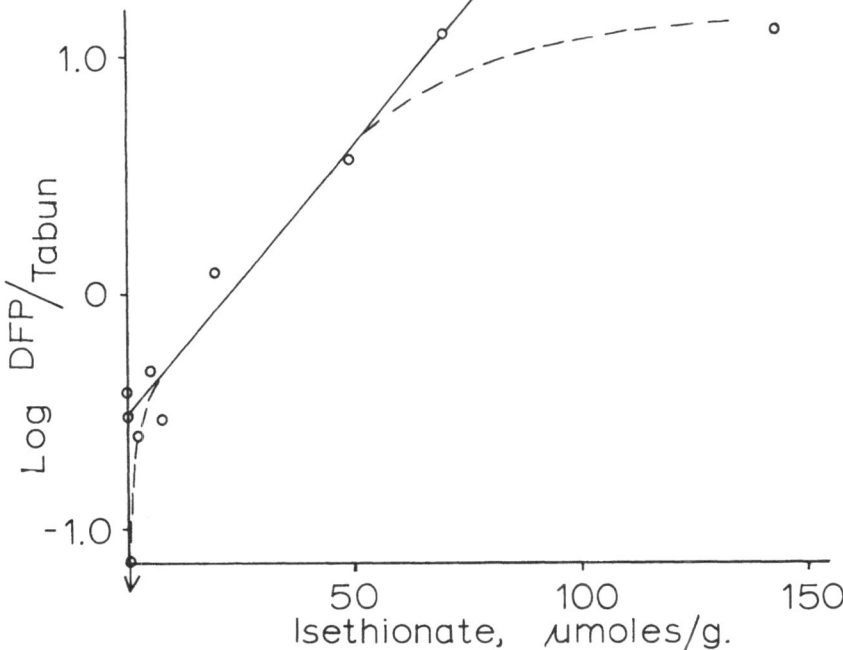

Fig. 6.1. Evidence for a parallel between the level of isethionate in various tissues and species and the presence of DFPase of the kind that hydrolyzes DFP more readily than Tabun (so-called squid-type DFPase). The two points markedly deviating from the linear relationship are for squid axoplasm (upper right), for which there is no extracellular space; and squid blood, which does not hydrolyze DFP at all, thus making "log DFP/Tabun" appear to be an infinitely large negative value for purely arithmetic reasons, and off the graph to the lower left. Reprinted from Hoskin and Brande (1973) with permission of Pergamon Press, Ltd.

tion apparatus and the associated gas train are modifications of the Van Slyke apparatus. The detailed use of the apparatus has been described in the original publication of Van Slyke and associates (1951) as modified in my own laboratory (Hoskin et al., 1975). The method permits the quantitative and exclusive determination of ^{14}C in a sample containing ^{14}C and ^{35}S. It is, of course, an absolutely essential feature of the assay that the CPM due to ^{14}C are not contaminated with CPM originating from ^{35}S. This is shown in Table 6.1; a "double-labeled"[1] sample made up to contain no ^{14}C and a high level of ^{35}S was assayed for ^{14}C by the Van Slyke method.

---

[1]The quotation marks denote a sample "double-labeled" by combining two single-labeled samples of the same compound. The analytical method is illustrated by the biosynthesis and analysis of double-labeled isethionate in which a significant number of the two different isotopic atoms may be in the same molecule.

The basis for the ^{35}S assay is the fact that the complete oxidation of sulfur yields nonvolatile barium-precipitable sulfate. The precipitate, which includes nonradioactive carrier sulfate (Hoskin and Kordik, 1977), is filtered on a 0.22 μm pore-size Millipore filter and counted in one of the millipore-compatible scintillants, in this case New England Nuclear's Aquasol. The details of the oxidation mixture, temperature, time, and so on, have been reported from this laboratory in another context (Hoskin and Brande, 1973). The method is less quantitative than for the ^{14}C assay. On the average, 75 ± 11% (SD) of a known ^{35}S content can be determined. The filtrate is found to contain, on the average, 30 ± 11% (SD) of the original content. Thus, essentially all of the ^{35}S is accounted for. As with the ^{14}C method, it is important to note that there is no contamination of the ^{35}S counts with [14]C. This is also shown in Table 6.1; data are given for a "double-labeled" sample made up to contain no ^{35}S and a high level of ^{14}C, and assayed for ^{35}S by the Hoskin-Brande method.

The biosynthesis and subsequent analysis of double-labeled isethionate was carried out primarily to illustrate an application of the [14]C–[35]S analytical procedure, but the results also address the question of the origin of the carbons of isethionate. Briefly, the method is the following. Glucose-U-^{14}C and Na_2[35]S, both at 10^{-4} M, were incubated with homogenized frozen squid optic ganglia in 10^{-3} M Hepes buffer, pH 7.5, for 5 hr at 30°C. Unused glucose and sulfide were separated from the sulfate-plus-isethionate peak on an Aminex A-4 column, sulfate and carrier sulfate were precipitated as $BaSO_4$, and hexose phosphate esters were hydrolyzed by heating with 1 N NaOH. These procedures were interspersed with additional Aminex A-4 purifications, sulfate precipitations, AG50 (H) removal of cations, and finally a paper chromatographic isolation of a radioactive product migrating with authentic 1,2-^{14}C-isethionate, and not with [35]SO_4^{2-}. This material has been assayed for total radioactivity by conventional counting, for ^{14}C by the Van Slyke procedure, and for ^{35}S by the Hoskin-Brande procedure.

The results of the two experiments performed to date are given in Table

Table 6.1 Assay for one isotope (^{14}C or ^{35}S) in the presence of the other

Composition, CPM[a]		Found, CPM[a]	
[14]C	[35]S	Van Slyke	Hoskin-Brande
6,746 ± 124	0	—	2 ± 1
0	8,230 ± 34	2 ± 2	—

[a]Average of 4 ± SD.

6.2. All samples constituting one complete experiment were counted in one 24-hr period, thus correcting for ^{35}S radioactive decay. The different radioactive compositions of the two incubation mixtures are due only to matters of laboratory convenience and do not reflect changes in chemical composition.

So far as an illustration of the application of the ^{14}C-^{35}S counting method, it can be seen that 88–89% of the radioactivity found in isethionate is accounted for. If the ^{14}C determination is quantitative, then the ^{35}S determinations are 77% and 86%, both within the range of the Hoskin-Brande method.

Turning now to the metabolic significance of the results, it may appear that the isethionate yield is very small, 0.1% to 0.4% without regard to which isotopic label is considered. In the course of purifying the isethionate there have probably been large losses of all of the metabolic products—indeed, the first Aminex A-4 separation (results not shown) suggested that about 4% of the total starting radioactivity became isethionate-plus-sulfate; the first $BaSO_4$ precipitation suggested that about half of this was isethionate. This is similar to the yields we reported (Hoskin and Kordik, 1977) for isethionate as a percentage of substrate taken up by intact squid axons. But results on freshly dissected intact single cells (as the giant axon) cannot be compared with results obtained from homogenates of frozen tissues.

Table 6.2 Synthesis of ^{14}C-^{35}S double-labeled isethionate by squid nerve tissue

Operation	Radioactivity[a]		
	CPM	%	$^{14}C/^{35}S$
Experiment I			
Incubation mixture[a]	29,769,000	100	
Glucose-U-^{14}C	19,584,000	66 }	1.94
Na_2-^{35}S	10,185,000	34 }	
Isethionate isolated	23,180	0.078	
Van Slyke for ^{14}C	12,100	0.041 }	1.46
Hoskin-Brande for ^{35}S	8,360	0.028 }	
Experiment II			
Incubation mixture[a]	15,115,000	100	
Glucose-U-^{14}C	6,754,000	45 }	0.82
Na_2-^{35}S	8,361,000	55 }	
Isethionate isolated	62,550	0.41	
Van Slyke for ^{14}C	9,180	0.061 }	0.20
Hoskin-Brande for ^{35}S	45,450	0.30 }	

[a]10^{-4} M glucose-U-^{14}C, 10^{-4} M Na_2-^{35}S, 80 mg homogenized squid optic ganglion, and 10^{-3} M Hepes buffer to give a final volume of 1.0 ml, pH 7.5, 30°C, 5 hr.

While the yields of ^{14}C and ^{35}S in isethionate differ rather markedly between the two experiments, in neither case has there been a mere trickle of radioactivity into the isethionate from one or the other isotopic source. In fact, the proportions of ^{14}C and ^{35}S in the isethionate are not unreasonably different from the proportions in the incubation mixture. Thus in the one experiment the starting $^{14}C/^{35}S$ ratio is 1.94 and the product $^{14}C/^{35}S$ ratio is 1.46; in the other, the same two ratios are 0.82 and 0.20, respectively. On the whole, these results suggests suggest a relatively direct pathway from glucose to a 2-carbon fragment, and the transfer of sulfide or a sulfide intermediate to this 2-carbon acceptor. A possibly analogous transfer is already known, namely that of the outer (sulfane) sulfur of thiosulfate to cyanide to produce thiocyanate, catalyzed by the enzyme rhodanese (Lang, 1933; Westley, 1973).

Rhodanese is known to be present in cephalopod nerves (Schievelbein et al., 1969; Hoskin and Kordik, 1977) but is also widely distributed. Recently, evidence has been presented that a physiological role of rhodanese is the incorporation of the outer sulfur of thiosulfate into the protein structure of certain iron-sulfur enzymes (FinazziArgo et al., 1971; Tanaguchi and Kimura, 1974). Another physiological role for rhodanese could be the formation of the sulfur-carbon bond of isethionate. Figure 6.2 summarizes these observations. In the squid nerve the sulfur of isethionate appears to be derived from sulfide, while the carbon backbone comes from a carbon fragment of glucose. A possible role for rhodanese, the existence of an acid anhydride $(X \sim S)$ sulfinic intermediate, and especially a role for squid-type DFPase on that pathway are all highly speculative. One of the messages of this presentation is that taurine and isethionate are only indirectly related as contrasted to the usual view. While isethionate is not synthesized from taurine, both isethionate and taurine are synthesized from sulfide.

The squid giant axon may be an ideal model system for the study of sulfur metabolism, in particular the pathways related to isethionate and taurine. A vast amount is already known about the physiology of this system. In the one species, Loligo, there is available a single cell from which the intracellular content can be removed if desired. This feature may be especially important in bypassing permeability barriers and for enzyme purification (Hoskin and Long, 1972; Garden et al., 1975). Cephalopod nerve is known to have a high intracellular content of taurine and isethionate, and it is reasonable to assume that there will be a high level of the associated synthetic, degradative, and transporting enzymes. It may also be assumed that these systems are under strict regulatory control, and that the system and its controls will be subject to suitable perturbations. The results of such studies may help to provide additional clues to the actions of taurine and related sulfur compounds on excitable tissue.

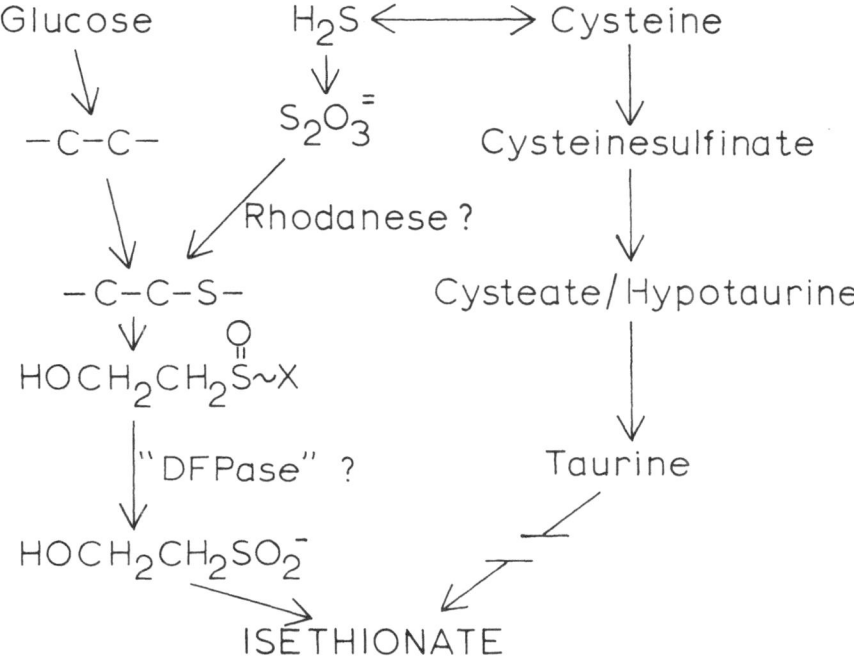

Fig. 6.2. Pathways to account for the observations that sulfide and the sulfur of cysteine are incorporated into isethionate, that the carbons of cysteine are not, and that there appears to be a fairly direct route from glucose to isethionate. The involvement is speculative for thiosulfate, and S ∼X compound, a sulfinate precursor of isethionate (isethiinate), and rhodanese and DFLase.

ACKNOWLEDGMENT

This research was supported by NIH grants NS-09090 and ES-02116.

REFERENCES

Deffner, G. G. J.; and Hafter, R. E. Chemical investigations of the giant nerve fibers of the squid. IV: Acid-base balance in axoplasm. *Biochim. Biophys. Acta,* 42, 200–205 (1960).

Fellman, J. H.; Roth, E. S.; and Fujita, T. S. Taurine is not metabolized to isethionate in mammalian tissue. In *Taurine and Neurological Disorders,* A. Barbeau and R. J. Huxtable, eds. Raven Press, New York (1978), pp. 19–24.

FinazziArgo, A.; Cannella, C.; Graziani, M. T.; and Cavallini, D. Possible role for rhodanese: Formation of "labile" sulfur from thiosulfate. *FEBS Lett.,* 16, 172–174 (1971).

Garden, J. M.; Hause, S. K.; Hoskin, F. C. G.; and Roush, A. H. Comparison of DFP-hydrolyzing enzyme purified from head ganglion and hepatopancreas of squid *(Loligo pealei)* by means of isoelectric focusing. *Comp. Biochem. Physiol.,* 52C, 95–98 (1975).

Hodgkin, A. L.; and Huxley, A. F. A quantitative description of membrane currents and its application to conduction and excitation in nerve. *J. Physiol.* (London), 117, 500–544 (1952).

Hoskin, F. C. G. Diisopropylphosphorofluoridate and Tabun: Enzymatic hydrolysis and nerve function. *Science,* 172, 1243–1245 (1971).

Hoskin, F. C. G. Squid nerve type DFPase: A consideration of molecular structures. In *Jerusalem Symposium on Molecular and Quantum Pharmacology,* E. D. Bergman and B. Pullman, eds. Reidel, Dordrecht, Holland (1974), pp. 205–207.

Hoskin, F. C. G.; and Brande, M. An improved sulfur assay applied to a problem of isethionate metabolism in squid axon and other nerves. *J. Neurochem.,* 20, 1317–1327 (1973).

Hoskin, F. C. G.; and Kordik, E. R. Hydrogen sulfide as a precursor for the synthesis of isethionate in the squid giant axon. *Arch. Biochem. Biophys.,* 180, 583–586 (1977).

Hoskin, F. C. G.; and Long, R. Purification of a DFP-hydrolyzing enzyme from squid head ganglion. *Arch. Biochem. Biophys.,* 150, 548–555 (1972).

Hoskin, F. C. G.; Pollock, M. L.; and Prusch, R. D. An improved method for the measurement of $^{14}CO_2$ applied to a problem of cysteine metabolism in squid nerve. *J. Neurochem.,* 25, 445–449 (1975).

Lang, K. Die Rhodanbildung in Thierkörper. *Biochem. Z.,* 259, 243–256 (1933).

Mounter, L. A. Metabolism of organophosphorus anticholinesterase agents. In *Handbuch der Experimentellen Pharamakologie, Cholinesterases and Anticholinesterase Agents,* G. B. Koelle, ed. Springer-Verlag, Berlin (1963), Chap. 10.

Schievelbein, H.; Baumeister, R.; and Vogel, R. Comparative investigations on the activity of thiosulfate transferase. *Naturwissenschaften,* 56, 416–417 (1969).

Taniguchi, T.; and Kimura, T. Role of 3-mercaptopyruvate sulfurtransferase in the formation of the iron-sulfur chromophore of adrenal ferredoxin. *Biochim. Biophys. Acta,* 364, 284–295 (1974).

Van Slyke, D. D.; Steele, R.; and Plazin, J. Determination of total carbon and its radioactivity. *J. Biol. Chem.,* 192, 769–805 (1951).

Westley, J. Rhodanese. *Advan. Enzymol. Relat. Areas Mol. Biol.,* 39, 327–368 (1973).

Discussion

DR. RYAN HUXTABLE (University of Arizona): Dr. Yamaguchi, your figure showing the distribution of the binding protein-A included data on the heart. Does this mean that cysteine dioxygenase is present in the heart?

DR. KENJI YAMAGUCHI (Osaka Medical College): No, I don't believe that there is any cysteine dioxygenase activity in the heart.

DR. DAVID RASSIN (Institute for Basic Research in Mental Retardation, New York): Dr. Iwata, do you know whether the cysteine you used in the uptake studies was always in the reduced form?

DR. HEITAROH IWATA (Osaka University): I don't know. We decided not to pursue that line of research since our metabolic data were disappointing.

DR. DAVID RASSIN (Institute for Basic Research in Mental Retardation, New York): It is possible that if you maintained your cysteine in its reduced form, both the transport and metabolism of the amino acid might improve.

DR. STEVEN BASKIN (Medical College of Pennsylvania): Dr. Hoskin, could the membrane in your preparation be somewhat permeable to taurine and isethionate and thus affect cation flux in the preparation?

DR. FRANCIS HOSKIN (Illinois Institute of Technology): Isethionate and taurine both pass through the membrane, although not nearly as well as cysteine. There is no question that isethionate contributes to the anion pool, but we don't know if it is required for long-term maintenance of anion balance in the cell. On a short-term basis, other anions can substitute for isethionate. In fact, the best anion for maintenance of the perfused squid axon preparation is fluoride and not isethionate.

PART II

Actions of Taurine in the Central Nervous System

Introduction

It has been proposed that taurine may function as a neurotransmitter or neuromodulator in the central nervous system. However, before taurine can be classified as a neurotransmitter, several criterial must be satisfied. First, all transmitters must be present in the central nervous system, preferably within the synaptosomes. Second, a rapid inactivating mechanism for the transmitter (either catabolism or uptake) must be present in the synaptic region. Third, the transmitter must alter the ion conductance of the postsynaptic membrane. Fourth, postsynaptic transmitter receptors must be present. Fifth, the transmitter must be released from presynaptic nerve terminals during excitation.

In the first part of this section the role of taurine in the central nervous system is discussed. Yamamura et al. describe some of the properties of a high affinity, sodium-dependent taurine uptake system of rat brain. Consistent with the suggestion that taurine is a neurotransmitter or neuromodulator, they find that kainic acid-induced lesions inhibit the taurine uptake system. Nauss-Karol and VanderWende report on a similar high affinity uptake system in human platelets. Wheler et al. have investigated the subcellular distribution of radioactive taurine following incubation of cortex slices with ^{35}S-taurine. This radioactivity is released from the slices following electrical stimulation. Most of the taurine mobilized during stimulation appears to be released from the synaptic region, implying an important role for taurine in nerve endings. Wheler and Klein show that taurine can also be released from the pineal gland, a process that is mediated by the neurotransmitter norepinephrine. The significance of this release is discussed. Oja et al. find that the synaptic vesicles of bovine brain have a high taurine content. Enrichment of these vesicles with taurine throughout various regions of the brain implies a neuromodulator role for taurine. Based on similar regional distribution studies, Kumpulainen et al. also suggest that taurine has a neuromodulator function.

The physiological role of taurine in the retina is more clearly defined than its function in other regions of the central nervous system. The evidence supporting a transmitter role for taurine in the retina is reviewed by Voaden et al. They discuss the distribution and uptake, biosynthesis, and specific actions of taurine in the retina. Pasantes-Morales et al. present evidence that taurine alters calcium transport by retinal subcellular fractions. They also review the importance of calcium in the process of photoexcitation and discuss the significance of the taurine effect.

Cats fed a taurine-free diet develop photoreceptor cell lesions. Schmidt reports that a correlation exists between the loss in tissue taurine levels in these animals and changes in electrical activity of the retina. This evidence suggests that taurine deficiency may lead to abnormal ionic conductance in the retina. She also examines the high affinity taurine uptake system, which becomes inhibited in photoreceptorless retinas. Sturman finds that taurine is transported in the axon of the optic nerve. He speculates that this process may facilitate the development of axons and the formation of synaptic connections.

Copyright © 1981, Spectrum Publications, Inc.
The Effects of Taurine on Excitable Tissues

CHAPTER 7

Effect of Kainic Acid Lesions on Taurine Transport into Rat Brain Synaptosomes

Henry I. Yamamura
Robert C. Speth
Robert E. Hruska
Nereo Bresolin
Brad A. Meiners
Ryan J. Huxtable

Recently, emphasis has been focused on the relationship between taurine and neurological diseases (Barbeau and Huxtable, 1978). Because of the high concentration of taurine in mammalian brain (Jacobsen and Smith, 1968), taurine may have a neurotransmitter or neuromodulator role in the central nervous system.

Biochemically, taurine appears to fulfil the majority of criteria that a substance must fulfill before it can be considered a neurotransmitter (Oja et al., 1976). In addition, its strong inhibitory actions on neuronal activity in cortical areas and spinal cord (Crawford and Curtis, 1964; Curtis and Crawford, 1969; Curtis and Watkins, 1960; Haas and Hösli, 1973), as well as its antagonistic effects on seizures (Izumi et al., 1974; Kaczmarek and Adey, 1974; Mutani et al., 1974), provide further indications of its neurotransmitter or neuromodulatory role.

In this chapter, we report on the characterization and the effect of kainic acid lesions on taurine transport into rat brain synaptosomes. Since taurine metabolism occurs rather slowly (Peck and Awapara, 1967; Gaitonde, 1970;

O'Keeffe and Smith, 1973; Federici et al., 1974), it seems reasonable that released taurine must be removed by an efficient high-affinity transport system.
To this end, several people have examined taurine transport in brain slices
(Kaczmarek and Davison, 1972; Lähdesmäki and Oja, 1973; Honegger et al.,
1973) and synaptosomal fractions (Schmid et al., 1975; Lähdesmäki et al.,
1975; Lombardini, 1976; Collins, 1974; Oja and Lähdesmäki, 1974; Oja et
al., 1976). However, a specific high-affinity taurine transport system was not
always demonstrated (Schmid et al., 1975), and others have reported variations in transport kinetics with Km values ranging from 20 to 60 μM (Schmid
et al., 1975; Lähdesmäki and Oja, 1973).

In order to facilitate the detection of a high-affinity taurine transport site
into rat brain synaptosomes, we used ^{3}H-taurine of high specific activity. Initially taurine was custom-tritiated for us by Dr. Richard Young at New England Nuclear to a specific activity of 2.8 Ci/mmol (Hruska et al., 1977); however, ^{3}H-taurine can now be purchased commercially at a specific activity of
23 Ci/mmol.

We examined ^{3}H-taurine transport over a wide range of concentrations
(0.05–600 μM) and found that the data best fitted a two-transport system. As
shown in Fig. 7.1, the sodium-dependent high-affinity transport component
had an apparent Km value of 4.6 μM and an apparent Vmax value of 5.3
nmole/g protein/min, whereas the apparent sodium-dependent low-affinity
Km and Vmax values were 424 μM and 132 nmole/g protein/min, respectively. At 3 μM ^{3}H-taurine, we calculated that about 65% of the total transport of taurine was via the high-affinity system (Hruska et al., 1978a,b),
which agrees nicely with other investigators (Lähdesmäki and Oja, 1973;
Lähdesmäki et al., 1975; Schmid et al., 1975).

In order to obtain the true individual high- and low-affinity transport constants, we subtracted the contributions made by the two transport systems by
calculating the transport constants using the method of Spear et al. (1971).
After performing 10 iterations, we obtained the true kinetic constants for
each of the two transport systems. The corrected Km and Vmax values are
noted in Fig. 7.1.

The effects of various changes in incubation medium are depicted in Table
7.1. The influx of ^{3}H-taurine is temperature dependent. The transport of ^{3}H-
taurine is also sodium dependent, as replacement of sodium with lithium,
choline, or sucrose inhibits taurine transport by 95%.

Figure 7.2 illustrates an Arrhenius plot of taurine transport. The calculated energy of activation (Ea) for taurine transport is 15.6 Kcal/mol, and
the energy quotient (Q_{10}) is 2.34. Both values indicate a high dependence on
energy for the transport of ^{3}H-taurine into rat brain synaptosomes.

The transport site is specific for taurine since only close analogues of taurine (hypotaurine and β-alanine) produce significant inhibition of ^{3}H-taurine

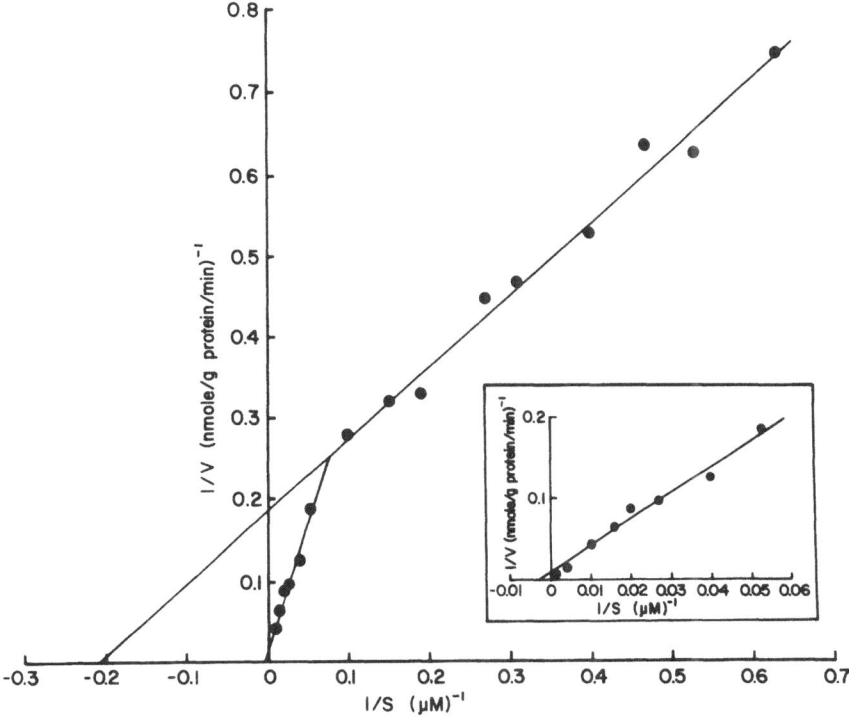

Fig. 7.1. Lineweaver-Burk plot of [³H]taurine transport. The sodium-dependent transport was best fitted by two lines. The high-affinity transport system has an apparent Km value of 4.76 μM and Vmax value of 5.35 nmol/g protein/min, while the low-affinity transport system (see insert) has respective values of 384 μM and 116 nmol/g protein/min. The true kinetic constants (see text) are: for the high affinity system, 3.20 μM and 2.96 nmol/g protein/min; for the low-affinity system, 3,340 μM and 699 nmol/g protein/min. Each point is the mean of eight experiments performed in triplicate.

transport (Table 7.2). The homologs of taurine (aminomethane sulfonic acid and 3-aminopropane sulfonic acid) were ineffective inhibitors of ³H-taurine transport. The proposed metabolic precursors of taurine (cysteine sulfonic acid, cysteamine, and L-cysteine) were also poor inhibitors. Two distant analogues of taurine (L-alanine and α-aminobutyric acid), as well as two close analogues (2-aminoethyl phosphonic acid and 2-aminoethyl hydrogen sulfate), were equally ineffective inhibitors. Putative amino acid neurotransmitters (glycine and γ-aminobutyric acid) and their antagonists (strychnine and bicuculline) were all weak inhibitors of taurine transport. These data illustrate the marked specificity of ³H-taurine for its transport site.

Table 7.1   Effect of various incubation conditions on high-affinity taurine uptake

Incubation condition	Percentage of control	
	Temperature dependent[a]	Sodium dependent[b]
Normal	100	100
2°C	0	3
NaCl replaced with LiCl[c]	3	0
NaCl replaced with choline Cl[c]	5	0
NaCl replaced with sucrose[c]	11	0

[a]Concentration of ^{3}H-taurine = 0.4 μM. Incubation temperature 30°C. Transport at 2°C is defined as 0% transport.

[b]Concentration of ^{3}H-taurine = 3.0 μM. Incubation temperature 37°C. Transport with LiCl replacement of NaCl is defined as 0% transport.

[c]The Na_2HPO_4 buffer is also replaced with a Tris-PO_4 buffer.

Table 7.3 shows the effect of intrastriatal kainic acid injections on ^{3}H-taurine transport in rat striatal synaptosomes. Kainic acid has been shown to cause destruction only of neurons with their perikarya in the injected area, and to spare axons of passage and afferent nerve terminals (Coyle and Schwarcz, 1976). We found a 40% reduction in Na^+-dependent ^{3}H-taurine transport in 5-day-lesioned rats, and this level of reduction in taurine transport was maintained in 31–40-day-lesioned animals. Comparable reductions in temperature-dependent ^{3}H-taurine transport were observed in 5- and 31–40-day-lesioned animals. These data indicate that ^{3}H-taurine uptake occurs into striatal neurons. The residual high-affinity ^{3}H-taurine transport after kainate lesions may be localized to afferent neurons or to a glial population.

Taurine uptake was also examined in rats injected with kainic acid into the cerebellum (Table 7.4). The selective destruction of cerebellar neurons by kainic acid enabled us to determine whether ^{3}H-taurine uptake into cerebellar synaptosomes was localized to neurons. Significantly, we found a 78% reduction in Na^+-dependent ^{3}H-taurine transport in 52–55-day, cerebellar kainate-lesioned rats. A 40% reduction in temperature-dependent ^{3}H-taurine transport was also seen. From the difference between the temperature-dependent and sodium-dependent, one can calculate the sodium-independent ^{3}H-taurine transport. This latter uptake system was only reduced 20%. These

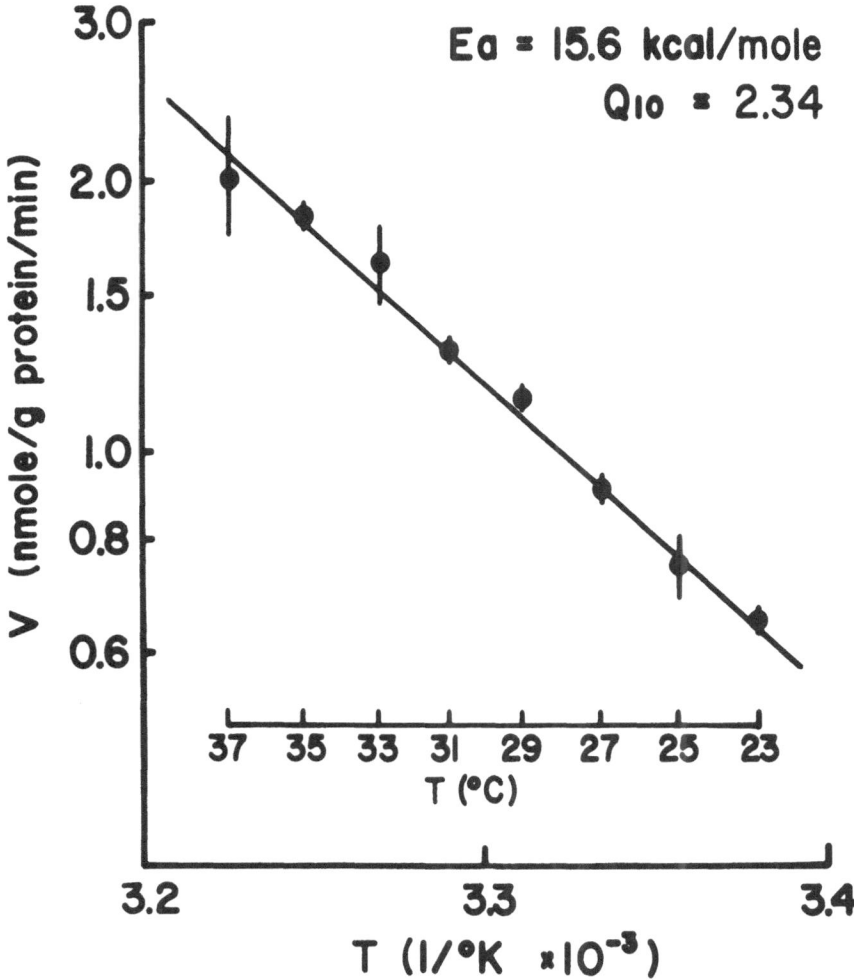

Fig. 7.2. Arrhenius plot of the transport of [³H]taurine. On the ordinate is the velocity (V) of [³H]taurine transport. The concentration of [³H]taurine was 2.8 μM. The temperature quotient (Q_{10}) and energy of activation (Ea) were determined from the linear regression line. Each point is the mean ± SEM of four experiments performed in triplicate.

data are consistent with the hypothesis that a sodium-dependent taurine up-take system is present in neurons in the cerebellum.

In summary, the results of our studies are consistent with the hypothesis that taurine is a neurotransmitter or neuromodulator in the central nervous

Table 7.2   Inhibition of high-affinity taurine uptake

	Percentage of control	
Inhibitor	Temperature dependent[a]	Sodium dependent[b]
None	100	100
Aminomethane sulfonic acid	91	65
3-Aminopropane sulfonic acid	77	100
L-Cysteine	74	100
Cysteamine	78	85
Cysteine sulfinic acid	83	75
L-Alanine	77	100
α-Aminoisobutyric acid	72	85
2-Aminoethyl phosphonic acid	77	100
2-Aminoethyl hydrogen sulfate	75	100
Glycine	83	100
GABA	56	62
Strychnine	74	85
Bicuculline	65	61
β-Alanine	48	31
Hypotaurine	22	19

[a]Concentration of ^{3}H-taurine = 4.0 μM. Concentration of inhibitor = 100 μM.
[b]Concentration of ^{3}H-taurine = 1.0 μM. Concentration of inhibitor = 100 μM.

Table 7.3   Taurine uptake into striatal synaptosomes after kainic acid lesions[a]

Days after KA	Uptake conditions	^{3}H-Taurine uptake (pmol/mg protein/4 min)		Percent change
		Control	Lesion	
5	Na$^+$dependent	15.4	9.4[b]	−40
	Temp. dependent	21.1	14.6	−40
31–40	Na$^+$dependent	18.3	11.8	−36
	Temp. dependent	27.0	19.1	−29

[a]Sodium pentobarbital-anesthetized rats were lesioned unilaterally with 2 μg/μl of kainic acid (Sigma). For striatal lesions, the sterotaxic coordinates were A-P = at bregma, L: 2.6 mm and V: 4.8 mm from the surface of the brain. One microliter of a 2-μg/μl solution of kainic acid in saline was injected over a 3-min period. ^{3}H-Taurine concentraiton was 3 μM. Na$^+$-dependent uptake is the amount taken up at 37°C in the absence of Na$^+$. Temperature-dependent uptake is the amount taken up at 37°C in the presence of Na$^+$ minus the amount taken up at 0°C in the presence of Na$^+$.
[b]$P < 0.05$.

Table 7.4   Taurine Uptake into Cerebeller Synaptosomes 52–55 Days after Kainic Acid Injections[a]

Uptake conditions	^{3}H-Taurine uptake (pmol/mg protein/4 min)		Percent change
	Controls	Lesions	
Na dependent	3.38	0.76[b]	−78
Tempature dependent	9.34	5.56[c]	−40

[a]For cerebellar lesions, two injections were made at coordinates A-P: 3.0 mm and 4.0 mm posterior to lambda, L: 2.5 and V: 3.5 mm from the surface of the brain.
[b]$P < 0.05$.
[c]$P < 0.02$.

system. There is a high-affinity, temperature-sensitive and sodium-dependent, energy-requiring, neuronally localized transport system for taurine in rat brain; and this transport system appears uniquely specific for ^{3}H-taurine.

ACKNOWLEDGMENTS

This investigation was supported by USPHS grants from the NIMH and from The Huntington Chorea Foundation and Hereditary Disease Foundation. H. I. Yamamura is a recipient of a USPHS Research Scientist Development Award from the NIMH (MH-00095). We thank Thomas McManus, David Chapman, Marla Bliss, and Andrew Chen for technical assistance. Cathy Kousen deserves special thanks for the typing of this manuscript.

REFERENCES

Barbeau, A.; and Huxtable, R. J. In *Taurine and Neurological Disorders,* ed. by A. Barbeau and R. J. Huxtable. Raven Press, New York, (1978).
Collins, G. G. S. *Brain Res.,* 76, 447–459 (1974).
Coyle, J. T.; and Schwarcz, R. *Nature,* 263, 244–246 (1976).
Crawford, J. M.; and Curtis, D. R. *Brit. J. Pharmacol.,* 23, 313–329 (1964).
Curtis, D. R.; and Crawford, J. M. *Ann. Rev. Pharmacol.,* 9, 209–240 (1969).
Curtis, D. R.; and Watkins, J. C. *J. Neurochem.,* 6, 117–141 (1960).
Federici, G.; Duprè, S.; Rosei, M. A., Granata, F.; and Orlando, M. *Physiol. Chem. Phys.,* 6, 411–416 (1974).

Gaitonde, M. K. In *Handbook of Neurochemistry*, Vol. 3, ed. by A. Lajtha. Plenum Press, New York, pp. 253–287 (1970).

Haas, H. L., and Hösli, L. *Brain Res.*, 52, 399–402 (1973).

Honegger, C. G.; Krepelka, L. M.; Steiner, M.; and von Hahn, H. P. *Experienta*, 20, 1235–1237 (1973).

Hruska, R. E.; Huxtable, R. J.; and Yamamura, H. I. *Anal. Biochem.*, 79, 568–570 (1977).

Hruska, R. E.; Huxtable, R. J.; and Yamamura, H. I. In *Taurine and Neurological Diseases*, ed. by A. Barbeau and R. Huxtable. Raven Press, New York, (1978a).

Hruska, R. E.; Padjen, A.; Bressler, R.; and Yamamura, H. I. *Molec. Pharmacol.*, 14, 77–85 (1978b).

Izumi, K.; Igisu, H.; and Fukuda, T. *Brain Res.*, 76, 171–173 (1974).

Jacobsen, J. G.; and Smith, L. H., Jr. *Physiol. Rev.*, 48, 424–511 (1968).

Kaczmarek, L. K.; and Adey, W. R. *Brain Res.*, 76, 83–94 (1974).

Kaczmarek, L. K.; and Davison, A. N. *N. Neurochem.*, 19, 2355–2362 (1972).

Lähdesmäki, P.; and Oja, S. S. *J. Neurochem.*, 20, 1411–1417 (1973).

Lähdesmäki, P.; Pasula, M.; and Oja, S. S. *J. Neurochem.*, 25, 675–680 (1975).

Lombardini, J. B. In *Taurine*, ed. by R. Huxtable and A. Barbeau. Raven Press, New York, pp. 311–326 (1976).

Mutani, R.; Bergamini, L.; Fariello, R.; and Delsedine, M. *Brain Res.*, 70, 170–173 (1974).

Oja, S. S.; Kontro, P.; and Lähdesmäki, P. In *Transport Phenomena in the Nervous System*, ed by G. Levi, L. Battistin, and A. Lajtha. Plenum Press, New York, pp. 237–252 (1976).

Oja, S. S.; and Lähdesmäki, P. *Med. Biol.*, 62, 138–143 (1974).

O'Keeffe, C. M.; and Smith, L. H., Jr. *Res. Commun. Chem. Pathol. Pharmacol.*, 6, 755–758 (1973).

Peck, E. J., Jr.; and Awapara, J. *Biochim. Biophys. Acta*, 141, 499–506 (1967).

Schmid, R.; Sieghart, W.; and Karobath, M. *J. Neurochem.*, 25, 5–9 (1975).

Spears, G.; Sneyd, J. G. T.; and Loten, E. G. *Biochem. J.*, 125, 1149–1151 (1971).

Copyright © 1981, Spectrum Publications, Inc.
The Effects of Taurine on Excitable Tissues

CHAPTER 8

High Affinity Taurine Uptake in Human Blood Platelets

Cheryl Nauss-Karol
Christina VanderWende

INTRODUCTION

Taurine is nearly ubiquitous in the body, with the highest concentration in excitable tissue. Although it has been implicated in diseases so diverse as epilepsy (Barbeau and Donaldson, 1974; Barbeau et al., 1975) and congestive heart failure (Huxtable and Bressler, 1974), little is actually known of its physiological function. Since it has a potent depressant action on neurons (Curtis and Watkins, 1965; Curtis et al., 1968; Curtis et al., 1971a; Krnjevic, 1964; Hosli and Tebecis, 1970; Haas and Hösli, 1973; Kaczmarek and Adey, 1974), it has been postulated to function as an inhibitory neurotransmitter in the central nervous system. In support of this postulation, taurine also appears to meet the other criteria necessary for a compound to qualify as a neurotransmitter (Agrawal et al., 1971; Haas and Hösli, 1973; Curtis et al., 1968; Curtis et al., 1971a,b; Hammerstad et al., 1971; Guidotti et al., 1972; Snyder et al., 1973; Starr, 1973; Schmid et al., 1975).

Evidence suggests the existence of three separate transport systems for taurine in nervous tissue. A low affinity uptake, probably corresponding to general amino acid transport that is operative in most mammalian tissues (Snyder et al., 1973), has been reported by many authors (Starr and Voaden, 1972; Lähdesmäki and Oja, 1973; Honegger et al., 1973; Lähdesmäki et al., 1975; Hruska et al., 1976). This is a relatively nonspecific form of transport and requires the presence of high taurine concentrations. Amino acid uptake into transmitter pools must be more highly specific. In fact, taurine uptake into brain slices, synaptosomes, astrocytes, glial cells, and retinae of several species does show high affinity kinetics similar to those that have been found for other putative neurotransmitters (Kaczmarek and Davison, 1972; Snyder et al., 1973; Lähdesmäki and Oja, 1973; Ehinger, 1973; Kaczmarek and Adey, 1974; Schmid et al., 1975; Lähdesmäki, et al., 1975; Hruska, et al., 1976; Sieghart and Karobath, 1976; Schousboe et al., 1976). Such uptake could efficiently terminate the synaptic action of taurine if it serves as a neurotransmitter (Kaczmarek and Davison, 1972; Schousboe et al., 1976). Recently, a unique high affinity transport system was delineated for rat brain synaptosomes, utilizing ^{3}H-taurine of a very high specific activity. In this study the affinity was approximately tenfold greater than had been previously reported (Hruska et al., 1976). A distinguishing characteristic of these high affinity transport systems is their nearly absolute sodium dependence. The low affinity uptake does not show nearly as marked a sodium requirement (Starr, 1973; Bennett et al., 1973; Oja et al., 1976).

Striking similarities in the transport, storage, and metabolism of putative neurotransmitter substances in platelets and nervous tissue have suggested the use of the human blood platelet as a model system of metabolic activity within brain synaptosomes (Paasonen, 1968; Page, 1968; Pletscher, 1968; Solomon et al., 1969; Boullin and O'Brien, 1971; Green et al., 1972; Murphy and Wyatt, 1972; Tuomisto, 1974; Gaut and Nauss, 1976). This simple model could provide a powerful tool for the screening of psychoactive drugs and other possible neurotransmitter candidates, as well as being useful for the study of chronic nervous disorders.

We have previously reported the existence of three distinct taurine transport systems in the human blood platelet (Gaut and Nauss, 1976). The high affinity uptake system correlates very closely with the unique high affinity synaptosomal system (Hruska et al., 1976). However, in those previous studies a detailed kinetic analysis was performed only on the medium affinity transport system. The present investigation was undertaken to further characterize the high affinity taurine transport system in human blood platelets, and to study the relative sodium dependence of the three different uptake systems.

METHODS

Blood was collected from healthy human volunteers in 2% ethylenedinitrilotetraacetic acid (EDTA). Platelet rich plasma was prepared by centrifugation at approximately 100 g for 30 min at room temperature. Aliquots of the plasma were placed into small, tared, nalgene tubes, and the platelets were isolated and washed twice with 0.9% NaCl containing 1.3 mM EDTA (pH 7.4), as previously described (Zieve et al., 1966; Zieve and Solomon, 1966).

Platelet pellets of approximately 30 mg wet weight were then resuspended in Ca^{2+}- and Mg^{2+}-free Krebs-Ringer bicarbonate buffer (KRB), pH 7.4. Appropriate aliquots of ^{14}C- or ^{35}S-taurine and inhibitor were added to each suspension, and the tubes were mixed on a vortex. Samples were incubated in an Eberbach metabolic shaker under O_2–CO_2 (95:5, V/V) at 37°C for either 15 or 30 min. The platelets were immediately sedimented by centrifugation at approximately 2,595 g for 5 min at 4°C. Supernatants were decanted, and the tubes were swabbed with cotton-tipped applicators to remove excess fluid adhering to the walls. The platelet pellets were then weighed and lysed in 1 ml of distilled water. Aliquots of 0.5 ml of the supernatant or the lysate were placed into 20-ml glass scintillation counting vials with 10 ml of a scintillation fluid containing 50 g naphthalene, 7 g 25-diphenyloxazole (PPO), and 50 mg 1,4-bis-[2-(4-methyl-5-phenyloxanyolyl]-benzene (POPOP) per liter of P-dioxane (Zieve and Solomon, 1966). The radioactivity was determined in a Packard TriCarb liquid scintillation spectrometer. Uptake was expressed as a gradient reflecting the ratio of the concentration of radioactivity in the intraplatelet water [I] to the concentration of radioactivity in the incubation medium [O], [I]/[O].

To examine the energy dependency of the uptake system, 1 mM of p-chloromercuribenzoic acid (PCMB), dinitrophenol (DNP), or sodium cyanide (NaCN) was added to the incubation medium to inhibit energy metabolism. The effect of incubating at 0°C was also examined. Since most transport systems exhibit sodium dependency, whether or not the taurine system depends on sodium was determined by substituting LiCl for NaCl in the KRB. In order to gain further evidence that taurine is transported in platelets by a specialized system, three structural analogues (hypotaurine, β-alanine, and diaminopropionic acid) were studied as potential competitive inhibitors, as compared to ouabain and strychnine as potential noncompetitive inhibitors.

RESULTS

The effects of metabolic inhibitors and 0°C on high affinity taurine transport (low taurine concentrations) are shown in Table 8.1. Uptake was inhibited 76% by PCMB (1 mM), 82% by DNP (1 mM), 46% by NaCN (1 mM), and 89% by incubation at 0°C. These results suggest an energy dependency of this high affinity system.

The sodium dependency of the system was demonstrated by substituting equimolar amounts of LiCl for NaCl in the KRB. At taurine concentrations of 0.5 or 5 μM, which represent high and medium affinities, respectively, the platelet uptake system exhibited a nearly absolute, linear, de-

Table 8.1   The effect of metabolic inhibitors on high affinity taurine uptake in human platelets[a]

Inhibitor	Concentration (mM)	No. of samples	Percent inhibition ± SE
0°	—	5	89 ± 0.8
PCMB	1.0	5	76 ± 2.3
DNP	1.0	5	82 ± 1.0
NaCN	1.0	5	46 ± 4.3

[a]Incubation time: 15 min.

pendence on sodium ions. This dependence was much less pronounced as the level of taurine was increased to 500 μM (Table 8.2). To determine whether the lithium itself affected taurine transport, LiCl was added to platelets suspended in buffer containing normal amounts of NaCl. These results are shown in Table 8.3. LiCl itself, even at very high concentrations, had no effect on taurine uptake.

The three structural analogues hypotaurine, β-alanine, and diaminopropionic acid, which were reported as the most potent competitive inhibitors of medium affinity uptake (Gaut and Nauss, 1976), were tested at much lower taurine concentrations, ranging from 0.1 to 2.0 μM, in order to measure the effect on high affinity uptake. Table 8.4 records the inhibitor constants resulting from these studies. As seen in Fig. 8.1, the inhibition was competi-

Table 8.2   Sodium dependence of the three taurine uptake systems in human platelets[a]

Buffer content		Gradient, % difference from control		
LiCl	Na	500 μM taurine	5 μM taurine	0.5 μM taurine
mM	mM	± SE	± SE	± SE
0	144	—	—	—
107	37	−36 ± 3	−84 ± 0.9	−81 ± 1
84	61	−30 ± 3	−71 ± 2	−75 ± 2
60	85	−15 ± 2	−55 ± 4	−58 ± 2
36	109	+7 ± 3	−23 ± 5	−32 ± 3
12	132	+5 ± 4	−9 ± 6	−11 ± 5

[a]An equimolar amount of LiCl was substituted for NaCl in KRB. Sodium in the form of NaHCO$_3$ remained unchanged at 25 mM. Platelets were incubated with 500 or 5 μM ^{14}C-taurine for 30 min, or 0.5 μM ^{35}S-taurine for 15 min. n = 10 for each data point.

Table 8.3   The effect of LiCl on high affinity taurine uptake in human platelets

Number of samples	LiCl concentration (mM)	Gradient, % difference from control ± SE
10	200	+21 ± 6
10	100	+13 ± 6
10	50	+0.5 ± 8
8	25	−1 ± 4
10	10	+5 ± 10
10	5	−1 ± 9
10	1	−16 ± 5

tive. Strychnine and ouabain were also examined for their effect on the high affinity uptake. At a concentration of 0.5 μM taurine, strychnine inhibited uptake by 50% at 310 μM, whereas ouabain showed on average inhibition of 40% over a concentration range of 1–1,000 μM (Table 8.4). This inhibition was not competitive. The results with the structural analogues and with strychnine and ouabain are similar to those with the medium affinity uptake (Gaut and Nauss, 1976).

DISCUSSION

The existence of three separate taurine transport systems in the human

Table 8.4   Inhibitory effects of structural analogues, strychnine, and ouabain on high affinity taurine uptake in human platelets[a]

Inhibitor	Ki (μM)	ID$_{50}$ (μM)
Structural analogues		
Hypotaurine	56	—
β-Alanine	280	—
Diaminopropionic acid	769	—
Others		
Strychnine	—	310
Oubain	—	b

[a]Incubation time: 15 min. ^{35}S or ^{14}C taurine concentration range yielding a Km of 2.9 μM was used. Data analyzed by method of Lineweaver and Burk (1934). $n = 5$ for each data point.
[b]Maximum inhibition obtainable was 40% regardless of concentration employed in the range of 1–1,000 μM.

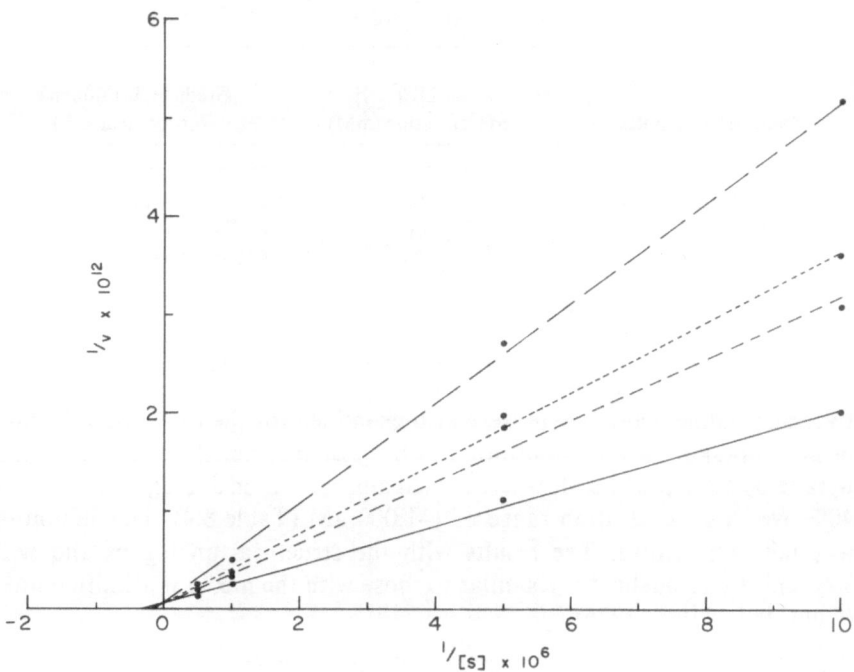

Fig. 8.1. Double reciprocal plots of high affinity taurine uptake by the platelet. Platelets were incubated for 15 min with ^{35}S- or ^{14}C-taurine at concentrations of 0.1, 0.2, 1.0, and 2.0 μM. Means are shown for five experiments per data point for inhibitors, and for 15 per data point for the control curve. Velocity is expressed as mol/mg of platelet/15 min and substrate concentration is expressed as mol/liter. Key: —— —— —— α-alanine, – – – hypotaurine,— — — diaminopropionic acid, ———————— control.

blood platelet was previously demonstrated, but only the medium affinity transport system was studied in detail at that time (Gaut and Nauss, 1976). In the present studies, the high affinity system was more closely examined so that it could be compared to the medium affinity system. A detailed study of the sodium dependency was done for all three systems.

The low affinity uptake showed a much weaker dependency on sodium than the other two, suggesting that this represents the function of a more generalized amino acid transport system. Hruska et al. (1976) demonstrated the presence of a low affinity system in rat brain synaptosomes, which exhibited a Km of 400 μM. This is strikingly similar to the system of the blood platelets, where the Km is 360 μM.

High affinity transport processes appear to be of special importance relative to nervous function. The medium and high affinity systems of the platelet were identical in their dependency on sodium, with the inhibition of taurine transport being about double that of the low affinity system at reduced sodium concentrations. Hruska et al. (1976) also reported a unique high affinity uptake system for taurine into rat brain synaptosomes, with a Km value of 3 μM. This, again, is strikingly similar to high affinity uptake seen in the platelet, which had a Km of 2.9 μM (Gaut and Nauss, 1976). Both of these studies utilized radioactive taurine of high specific activity. Based on the present studies, there also appears to be good correlation between the systems of the platelet and synaptosome in their sodium dependency and inhibition by structural analogues or metabolic inhibitors.

High affinity transport systems for taurine in neural tissue of several species have previously been reported by a number of other groups (Kaczmarek and Davison, 1972; Neal et al., 1973; Lähdesmäki and Oja, 1973; Ehinger, 1973; Schmid et al., 1975; Lähdesmäki et al., 1975; Schousboe et al., 1976). These showed high affinity kinetics similar to those that have been found for other putative neurotransmitters (Snyder et al., 1973) and had an average Km value of 47 μM, with considerable variability. However, none of these investigators used radioactive taurine of sufficient specific activity to allow them to test concentrations as low as those yielding the Km values of 3 or 2.9 μM reported by Hruska et al. (1976) and Gaut and Nauss (1976) for rat synaptosomes and human platelets, respectively. Therefore, these so-called high affinity uptake systems previously reported for synaptosomes are probably analogous to the medium affinity carrier of the platelet reported by Gaut and Nauss (1976), which has a Km value of 100 μM. There is good correlation between the brain systems and that of the platelet so far as specificity of inhibition by structural analogues and metabolic inhibitors is concerned. Temperature and sodium ion dependency are also similar. However, these previously reported high affinity systems of brain appear to be less specific than the truly high affinity systems.

With a medium containing either 5 μM or 0.5 μM taurine, uptake by the platelet was almost completely abolished by the metabolic inhibitors PCMB and DNP, or by incubation at 0°C. Sodium cyanide inhibited uptake at either taurine concentration, by about 50%. Strychnine, previously shown to be a noncompetitive inhibitor of taurine uptake in platelets (Gaut and Nauss, 1976), exhibited a 50% inhibition at 310 μM no matter which concentration of taurine was utilized. It appears that the high affinity taurine uptake system of the platelet is four to five times more specific than medium affinity taurine uptake, since the inbititor constants of the three structural analogues hypotaurine, β-alanine, and diaminopropionic acid shift from the values of 11, 71, and 150 μM previously determined for medium affinity transport, to

56, 280, 769 μM, respectively, shown in the present studies for high affinity transport. Ouabain, another noncompetitive inhibitor, reduced uptake a maximum of 60% at medium taurine concentrations, but only 40% at low taurine concentrations. Thus, the high affinity transport system appears to be somewhat less sensitive to ouabain.

Suppression of uptake at either taurine concentration by metabolic inhibitors or incubation in a O°C bath, indicates that this is an active, energy-requiring process. The absolute sodium dependency and ouabain sensitivity suggest that this transport is dependent for its metabolic energy on a functioning Na^+–K^+ ATPase system. However, the limited sensitivity to ouabain suggests that the Na gradient is probably not solely responsible for taurine accumulation. Sodium ion may also influence substrate binding, which may involve Na^+ as a cosubstrate in the transport process. This situation is analogous to those proposed for the transport of several putative neurotransmitters in neural tissue (Starr, 1973; Bennett et al., 1973). Synaptosomal uptake of putative neurotransmitters is Na^+ dependent and ouabain sensitive, whereas accumulation of other compounds is much less sensitive to low sodium ion cencentrations (Bennett et al., 1973). Glutamic acid, gamma-aminobutyric acid, and taurine have been reported to be accumulated by unique populations of synaptosomes in the presence of Na^+ (Bennett et al., 1973; Sieghart and Karobath, 1974; Bennett et al., 1972), while at low Na^+ concentration there is no difference in localization of glutamic acid (Bennett et al., 1972) from the general synaptosomal population. Thus, the absolute Na^+ dependence of high affinity transport systems for putative neurotransmitters could prove useful in differentiating populations of synaptosomes (Bennett et al., 1973).

The three taurine transport systems apparent in the human blood platelet seem to have their counterparts in neural tissue. Both the medium and high affinity uptake systems are operative at relatively low taurine concentrations and have a high affinity for substrate, a high degree of specificity, and an absolute dependence on Na^+. Therefore, both of these may be relevant to nervous tissue, and they possibly represent models of two distinct populations of neural cells.

SUMMARY

Low, medium, and high affinity uptake of taurine has been reported to occur in the human blood platelet, but only the medium uptake system was studied for its specific characteristics. In this study, the high affinity system was examined in detail so that it could be characterized and compared to both the medium and low affinity systems. High affinity uptake was inhibited by the metabolic inhibitors PCMB, DNP, and NaCN as well as by reducing the incubation temperature to 0°C, demonstrating an energy dependency similar to that of the medium affinity system. The high affinity system appears to be more specific than the medium affinity system since the Ki values for the structural analogues are four to five times higher for inhibition of high affinity uptake as compared to medium affinity uptake. Both the high and medium affinity systems showed considerable sodium dependency. This was contrasted by a low sodium requirement for the low affinity system. High affinity was inhibited by strychnine and showed sensitivity to the effects of ouabain. The high and medium transport systems would appear to be representative of the specialized neurotransmitter uptake of the neuron.

REFERENCES

Agrawal, H. C.; Davidson, A. N.; and Kaczmarek, L. K. Subcellular distribution of taurine and cysteinesulphinate decarboxylase in developing rat brain. *Biochem. J.*, 122, 759–763 (1971).

Barbeau, A.; and Donaldson, J. Zinc, taurine and epilepsy. *Arch. Neurol.*, 30, 52–58 (1974).

Barbeau, A.; Inoue, N.; Tsukada, Y.; and Butterworth, R. F. Minireview: The neuropharmacology of taurine. *Life Sci.*, 17, 669–678 (1975).

Bennett, J. P.; Logan, W. J.; and Snyder, S. H. Amino acid neurotransmitter candidates: Sodium dependent high-affinity uptake by unique synaptosomal fractions. *Science*, 178, 997–998 (1972).

Bennett, J. P.; Logan, W. J.; and Snyder, S. H. Amino acids as central nervous system transmitters: The influence of ions, amino acid analogues, and ontogeny on transport systems for L-glutamic and L-aspartic acids and glycine into central nervous synaptosomes of the rat. *J. Neurochem.*, 21, 1533–1550 (1973).

Boullin, D. J.; and O'Brien, R. A. Abnormalities of 5-hydroxytryptamine uptake and binding by blood platelets from children with Down's syndrome. *J. Physiol.* (London), 212, 287–297 (1971).

Curtis, D. R.; Hösli, L.; and Johnston, G. A. R. A pharmacological study of the depression of spinal neurones by glycine and related amino acids. *Exp. Brain Res.*, 6, 1–18 (1968).

Curtis, D. R.; Duggan, A. W.; Felix, D.; and Johnston, G. A. R. Bicuculline, an antagonist of GABA and synaptic inhibition in the spinal cord of the cat. *Brain. Res.*, 32, 69–96 (1971a).

Curtis, D. R.; Duggan, A. W.; Felix, D.; Johnston, G. A. R.; and McLennan, H. Antagonism between bicuculline and GABA in the cat brain. *Brain Res.*, 33, 57–73 (1971b).

Curtis, D. R.; and Watkins, J. C. The pharmacology of amino acids related to gamma-aminobutyric acid. *Pharmacol. Rev.*, 17, 347–391 (1965).

Ehinger, B. Glial uptake of taurine in the rabbit retina. *Brain Res.*, 60, 512–516 (1973).

Gaut, Z. N.; and Nauss, C. B. Uptake of taurine by human blood platelets: A possible model for brain. In *Taurine*, R. Huxtable and A. Barbeau, eds. Raven Press, New York (1976), pp. 91–98.

Green, A. R.; Boullin, D. J.; Masarelli, R.; and Hanin, I. Can the human blood platelet be used as a model for the cholinergic nerve ending? *Life Sci.*, 11, 1049–1058 (1972).

Guidotti, A.; Badiani, G.; and Pepeu, G. Taurine distribution in cat brain. *J. Neurochem.*, 19, 431–435 (1972).

Haas, H. L.; and Hösli, L. The depression of brain stem neurones by taurine and its interaction with strychnine and bicuculline. *Brain Res.*, 52, 399–402 (1973).

Hammerstad, J. P.; Murray, J. E.; and Cutler, R. W. P. Efflux of amino acid neurotransmitters from rat spinal cord slices. II: Factors influencing the electrically induced efflux of ^{14}C-glycine and ^{3}H-GABA. *Brain Res.*, 35, 357–367 (1971).

Honegger, C. G.; Krepelka, L. M.; Steiner, M.; and Von Hahn, H. P. Kinetics and subcellular distribution of S^{35}-taurine uptake in rat cerebral cortex slices. *Experientia* (Basel), 29, 1235–1237 (1973).

Hosli, L.; and Tebecis, A. K. Actions of amino acids and convulsants on bulbar reticular neurones. *Exp. Brain Res.*, 11, 111–127 (1970).

Hruska, R. E.; Huxtable, R. J.; Bressler, R.; and Yamamura, H. I. Sodium dependent high affinity transport of taurine into rat brain synaptosomes. *Proc. West. Pharmacol. Soc.*, 19, 152–156 (1976).

Huxtable, R.; and Bressler, R. Taurine concentrations in congestive heart failure. *Science*, 184, 1187–1188 (1974).

Kaczmarek, L. K.; and Adey, W. R. Factors affecting the release of [^{14}C] taurine from cat brain: The electrical effects of taurine on normal and seizure prone cortex. *Brain Res.*, 76, 83–94 (1974).

Kaczmarek, L. K.; and Davison, A. N. Uptake and release of taurine from rat brain slices. *J. Neurochem.*, 19, 2355–2362 (1972).

Krnjevic, K. Micro-iontophoretic studies on cortical neurons. *Int. Rev. Neurobiol.*, 7, 41–98 (1964).

Lähdesmäki, P.; and Oja, S. S. Effect of Electrical stimulation on the influx and efflux of taurine in brain slices of newborn and adult rats. *Exp. Brain Res.*, 15, 430–438 (1972).

Lähdesmäki, P.; and Oja, S. S. On the mechanism of taurine transport at brain cell membranes. *J. Neurochem.*, 20, 1411–1417 (1973).

Lähdesmäki, P.; Pasula, M.; and Oja, S. S. Effect of electrical stimulation and chlorpromazine on the uptake and release of taurine, γ-aminobutyric acid, and glutamic acid in mouse brain synaptosomes. *J. Neurochem.*, 25, 675–680 (1975).

Lineweaver, H.; and Burk, D. The determination of enzyme dissociation constants. *J. Am. Chem. Soc.*, 56, 658–666 (1934).

Murphy, D. L.; and Wyatt, R. J. Reduced monoamine oxidase activity in blood platelets from schizophrenic patients. *Nature*, 238, 225–226 (1972).

Neal, M. J.; Peacock, D. G.; and White, R. D. Kinetic analysis of amino acid uptake by the rat retina in vitro. *Br. J. Pharmac.*, 47, 656P–657P (1973).

Oja, S. S.; Kontro, P.; and Lähdesmäki, P. Transport of taurine in the central nervous system. *Adv. Exp. Med. Biol.*, 69, 237–252 (1976).

Paasonen, M. K. Platelet 5-hydroxytryptamine as a model in pharmacology. *Am. Med. Exp. Biol. Fenn.*, 46, 416–422 (1968).

Page, I. H. The possible singular importance of platelets. In *Serotonin*. YearBook Med. Pub., Inc., Chicago (1968), p. 37.

Pletscher, A. Metabolism transfer and storage of 5-hydroxytryptamine in blood platelets. *Br. J. Pharmacol. Chemotherap.*, 32, 1–16 (1968).

Schmid, R.; Sieghart, W.; and Karobath, M. Taurine uptake in synaptosomal fractions of rat cerebral cortex. *J. Neurochem.*, 25, 5–9 (1975).

Schousboe, A.; Fosmark, H.; and Svenneby, G. Taurine uptake in astrocytes cultured from dissociated mouse brain hemispheres. *Brain Res.*, 116, 158–164 (1976).

Sieghart, W.; and Karobath, M. Evidence for specific synaptosomal localization of exogenous accumulated taurine. *J. Neurochem.*, 23, 911–915 (1974).

Sieghart, W.; and Karobath, M. Uptake of taurine into subcellular fractions of C-6 glioma cells. *J. Neurochem.*, 26, 981–986 (1976).

Snyder, S. H.; Young, A. B.; Bennett, J. P.; and Mulder, A. H. Synaptic biochemistry of amino acids. *Fed. Proc.*, 32, 2039-2047 (1973).

Solomon, H. H.; Ashley, C.; Spirt, N.; and Abrams, W. B. The influence of debrisoquin on the accumulation and metabolism of biogenic amines by the human platelet in vivo and in vitro. *Clin. Pharmaco. Exp. Ther.*, 10, 229-238 (1969).

Starr, M. S. Effects of changes in the ionic composition of the incubation medium on the accumulation and metabolism of ^{3}H-γ amino-butyric acid and ^{14}C-taurine in isolated rat retina. *Biochem. Pharmacol.*, 22, 1693-1700 (1973).

Starr, M. S.; and Voaden, M. J. The uptake, metabolism and release of ^{14}C-taurine by rat retina in vitro. *Vision Res.*, 12, 1261-1269 (1972).

Tuomisto, J. A new modification for studying 5-HT uptake by blood platelets: A re-evaluation of tricyclic antidepressants as uptake inhibitors. *J. Pharm. Pharmac.*, 26, 92-100 (1974).

Zieve, P. D.; and Solomon, H. M. The intracellular pH of the human platelet. *J. Clin. Invest.*, 45, 1251-1254 (1966).

Zieve, P. D.; Solomon, H. M.; and Krevans, J. R. The effect of hematoporphyrin and light on human platelets. *J. Cell. Physiol.*, 67, 271-279 (1966).

Copyright © 1981, Spectrum Publications, Inc.
The Effects of Taurine on Excitable Tissues

CHAPTER 9

Uptake and Stimulated Release of Taurine by Preparations of Cerebral Cortex

G. H. T. Wheler
H. F. Bradford
A. N. Davison
E. J. Thompson

Taurine occurs at high concentrations in neural tissue, for example, at about 8 μmol/g in rat cerebral cortex (Lombardini, 1976). Its role in the cerebral cortex remains obscure, although it has been proposed to act there as a neurotransmitter (Davison and Kaczmarek, 1971; Mandel and Pasantes-Morales, 1978). Its potent neuronal inhibitory properties support this possibility (Curtis and Watkins, 1960, 1965; Okamoto and Quastel, 1973).

Iversen and co-workers have described the biochemical properties associated with amino acid transmitter function (Iversen et al., 1973). These include a high-affinity uptake of the proposed transmitter and a calcium-dependent stimulated release. In view of the slow metabolism of taurine, inactivation by reuptake would be of special importance.

One approach to the study of the possible transmitter role of taurine in the cerebral cortex is to observe the compartment involved in taurine uptake and its stimulated release. Also, the differential actions of L-diaminobutyrate (L-DABA) and β-alanine on neuronal and glial uptake of GABA suggest a means of defining the compartments involved in taurine uptake, as β-alanine is also a competitive uptake inhibitor for taurine (Lähdesmäki and Oja, 1973).

In the work reported here we have further examined the calcium dependence of taurine release for cerebral cortex slices. Also we have used subcellular fractionation and samples of differing neuronal and glial content as an approach to defining the compartments involved in the stimulated release of taurine.

METHODS

Cortex Slices

Topslices of cerebral cortex (0.3 mm thick, 40-80 mg weight) held in Quick Transfer Holders (McIlwain, 1975) were preincubated for 15 min in 5 ml of Krebs-bicarbonate medium and gassed with O_2/CO_2 (95:5, v/v). The slices were transferred to 5 ml of medium containing [^{35}S]taurine at 20 or 540 μM (final specific activity, 48 mCi/mmol). After 5 min incubation, each slice was rapidly rinsed in nonisotopic incubation medium containing taurine at the same final concentration. It was then transferred to 1.5 ml of 10% TCA, homogenized, centrifuged at 1,400 g, and 1 ml taken for radioactivity counting.

Subcellular Fractionation of Slices after [^{35}S]taurine Uptake

Slices of cerebral cortex were incubated as above for 30 min. After rapid rinsing they were homogenized and subfractionated by the method of Bradford et al. (1973). Six pooled slices were homogenized in 9 ml of 0.32 M sucrose at 5°C. The homogenate was then centrifuged at 1,000 g for 10 min in a Beckman L265B ultracentrifuge, to yield supernatant (S1) and pellet (P1). The supernatant was recentrifuged at 20,000 g for 20 min to yield the crude mitochondrial pellet (P2) and combined supernatant and microsomes (S + P3). P2 was resuspended in 2 ml of 0.32 M sucrose, and 1.6 ml was layered on top of a discontinuous gradient of 1.6 ml of 0.8 M sucrose above 1.6 ml of 1.2 M sucrose. The gradient was centrifuged for 1 hr at 76,000 g at 5°C. Three fractions were subsequently collected: P2A, a myelin fraction at the boundary of the 0.32 M and 0.8 M sucrose; P2B, or synaptosome fraction, at the boundary of the 0.8 M and 1.2 M sucrose; and P2C, the mitochondrial pellet resuspended in 0.32 M sucrose. Then, 0.8 ml of each fraction were added to 0.2 ml of 25% TCA, and the radioactivity was counted as before.

Uptake Studies in the Presence of L-DABA or β-Alanine

Cortex slices, C-6 glioma cells, ependymal slices, and synaptosomes were incubated for 5 or 30 min with either L-DABA or β-alanine (final concentration 1 mM), with 20 μM [^{35}S]taurine in the incubation medium. The tissues were tested as described below.

Cortex slices were incubated and the radioactivity counted as described above; synaptosome suspensions and beds were prepared by taking whole cerebral cortex and homogenizing it at 5°C as a 10% (w/v) suspension in 0.32 M sucrose as described elsewhere (Bradford et al., 1973). Synaptosomes were collected for incubation in Krebs-bicarbonate medium, either as suspensions in glass vials or as synaptosome beds. The synaptosome beds consisted of deposits of synaptosomes, sandwiched between nylon gauze, and were prepared as described elsewhere (De Belleroche and Bradford, 1972). Synaptosome beds were incubated as described for 30 min and radioactivity counted as above. Synaptosome suspensions (2–3 mg protein/ml) were incubated as above, and 0.5-ml samples were taken at 5, 10, 20, and 30 min. They were then centrifuged on a table-model centrifuge at 18,000 g for 2 min, and the pellet was extracted with TCA for radioactivity counting. The TCA-insoluble pellet was dissolved in 4.5 N NaOH and assayed for protein (Lowry et al., 1951). C-6 glioma time was suspended in incubation medium using 10 culture plates per 15 ml of medium, giving 0.9 mg protein/ml. This was incubated and treated as with synaptosome suspensions. Ependymal (white matter) slices were obtained by slicing the lining of the lateral ventricle. The slices (30–40 mg fresh weight) were rapidly transferred in rapid transfer holders and treated as for cortex slices.

Release Experiments

Cortex slices were incubated for 30 min in 5 ml of Krebs-bicarbonate medium containing calcium and 20 μM [^{35}S]taurine at 37°C gassed with O$_2$/CO$_2$ (95:5, v/v). They were drained, rinsed in incubation medium containing 20 μM nonisotopic taurine (20 μM), and transferred to 5 ml of incubation medium at 37°C for 15 min. They were then transferred to another beaker and electrically stimulated for 15 min, via the McIlwain Quick Transfer Holder. Stimulation consisted of square wave pulses alternating in polarity: 0.4 sec duration, 50 Hz, and mean current 30–50 mA. Slices were then transferred consecutively to two beakers containing incubation medium as before for 15-min periods. Slices were then rapidly rinsed in Krebs medium, homogenized in 3 ml 0.5 M perchloric acid (PCA), and the radioactivity counted as above. Some slices were preincubated in calcium containing medium with 20 μM [³⁵S]taurine, and after rinsing in medium containing no calcium and 1 mM EGTA, they were incubated and stimulated as before but with a calcium-free medium with 1 mM EGTA.

Release of [³⁵S]taurine from subcellular fractions of cerebral cortex was also studied. Cortex slices were preincubated in medium containing calcium and 20 μM [³⁵S]taurine, as described above. They were stimulated for 20 min in two groups, one at 48 mA mean current and the other at 36 mA mean current using square wave pulses (50 cps and 0.4 msec duration). Following incubation the slices were rapidly rinsed in 0.32 M sucrose at room temperature. They were homogenized and subfractionated as described for cortex slices, and counted for radioactivity as described for cortex slices, and counted for radioactivity as described above.

RESULTS

Uptake of [³⁵S]Taurine into Subcellular Compartments of Cerebral Cortex Slices

Table 9.1 shows the content of [³⁵S]taurine in different subcellular fractions following incubation of cortex slices in isotopic taurine. The [³⁵S]taurine content varied widely. At 20 μM taurine, most of the [³⁵S]taurine was recovered in the microsomes and soluble supernatant (S + P3), and in the crude mitochondrial pellet (P2), which had the next highest taurine content. With-

Table 9.1   Uptake of [³⁵]taurine to subcellular fraction[a]

| | nmol taurine/g wet wt. of slice taken up/30 min | | |
Fraction[b]	20 μM Taurine	540 μM Taurine	Protein content (mg/g)
H	495 ± 9	8796 ± 1061	108.0 ± 1.6
P1	54 ± 4	914 ± 49	31.3 ± 1.2
P2	156 ± 30	2460 ± 796	61.3 ± 1.5
P3 + S	314 ± 50	5783 ± 825	18.7 ± 0.8
P2A	37 ± 14	515 ± 127	3.9 ± 0.3
P2B	83 ± 19	1145 ± 145	24.6 ± 0.3
P2C	23 ± 3	371 ± 67	21.5 ± 1.2

[a]Results are ± SEM of four separate estimations.

[b]Key: H, homogenate; P1, nuclear pellet; P2, crude mitochondrial pellet; P3 + S, microsomes and soluble supernatant; P2A, myelin fraction; P2B, synaptosome fraction; P2C, mitochrondrial pellet.

in P2, the synaptosome fraction (P2B) was the most enriched. A similar pattern was obtained at 540 μM, although slightly more was found in the soluble supernatant fraction.

Table 9.2 shows the recovery of [^{35}S]taurine in the different subcellular fractions. This is given as a percentage of the total isotope present in the homogenate. When protein content was taken into account, a similar distribution of [^{35}S]taurine was observed, described as relative specific activity. The only difference was that they myelin fraction (P2A) showed a higher [^{35}S]taurine content than before, possibly due to leakage of taurine from synaptosomes. When [^{35}S]taurine was described as a percentage of the sedimentable, or particle-bound, [^{35}S]taurine, 41% was found to occur in the synaptosomal fraction at 20 μM; and 33%, at 540 μM. Recoveries of cpm and protein were 103% and 93% respectively.

Taurine and GABA Uptake in the Presence of Diaminobutyrate and β-Alanine

Figure 9.1 shows the effect of 1 mM β-alanine on the uptake of 20 μM taurine in preparations of different neuronal or glial content, and by synaptosome preparations, incubated for 5 or 30 min. After 30 min, β-alanine de-

Table 9.2   Recovery of [^{35}S]taurine in Subcellular Fractions[a]

Percentage recovery fraction		Relative specific activity[b]	Percentage of nonsupernatant taurine
At 20 μM taurine			
P1	10.7 ± 0.9	0.4 ± 0.10	26 ± 2
P2	29.5 ± 1.2	0.5 ± 0.02	74 ± 2
P3 + S	59.9 ± 0.9	3.6 ± 0.10	
P2A	6.6 ± 1.6	1.9 ± 0.40	17 ± 4
P2B	16.5 ± 0.9	0.8 ± 0.03	41 ± 2
P2C	4.5 ± 0.7	0.3 ± 0.03	11 ± 2
At 540 μM taurine			
P1	10.6 ± 1.1	0.4 ± 0.03	20 ± 2
P2	24.7 ± 0.8	0.5 ± 0.01	70 ± 3
P3 + S	64.8 ± 0.6	3.9 ± 0.03	
P2A	5.6 ± 0.8	1.6 ± 0.30	16 ± 3
P2B	12.9 ± 0.9	0.6 ± 0.03	37 ± 2
P2C	4.5 ± 0.7	0.2 ± 0.03	13 ± 2

[a]Results are ± SEM of four separate estimations.

[b]Relative specific activity is percentage of taurine uptake/percentage of protein in the fraction.

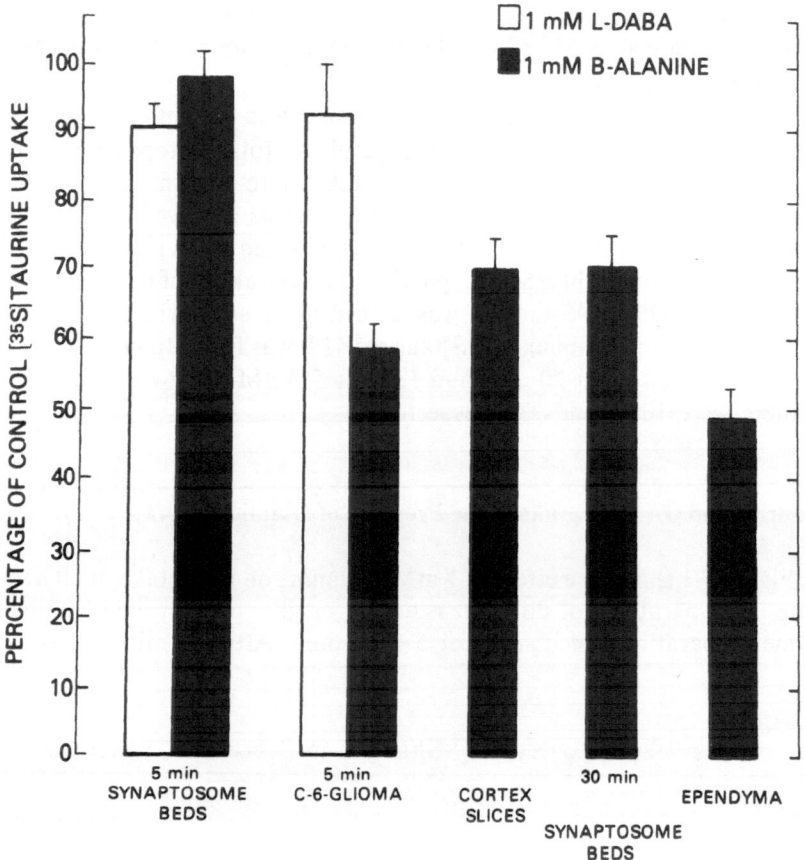

Fig. 9.1. The uptake of [³⁵S]taurine to different tissue samples in the presence of 1 mM L-DABA and 1 mM β-alanine. The time of incubation is described above. Four to ten measurements were made for each value. Vertical bars indicate SEM. Control uptake was in the range 50–80 pmol taurine/g wet wt.

pressed the uptake of [³⁵S]taurine to white matter slices (ependyma) by 50%. In comparison, there was 30% and 28% inhibition of uptake by cortex slices and synaptosome beds, respectively. After 5 min incubation, taurine uptake was not depressed for synaptosome suspensions, but C-6 glioma showed a 41% inhibition of uptake. L-diaminobutyrate (L-DABA) produced little inhibition of taurine uptake by synaptosome suspensions or by C-6 glioma suspensions over a 5-min period.

Figure 9.2 shows the content of [^{14}C]GABA after 5 min incubation with 20 μM [^{14}C]GABA in the presence of 1 mM β-alanine or 1 mM L-DABA. For synaptosome suspensions, L-DABA depressed GABA uptake by 84%. In

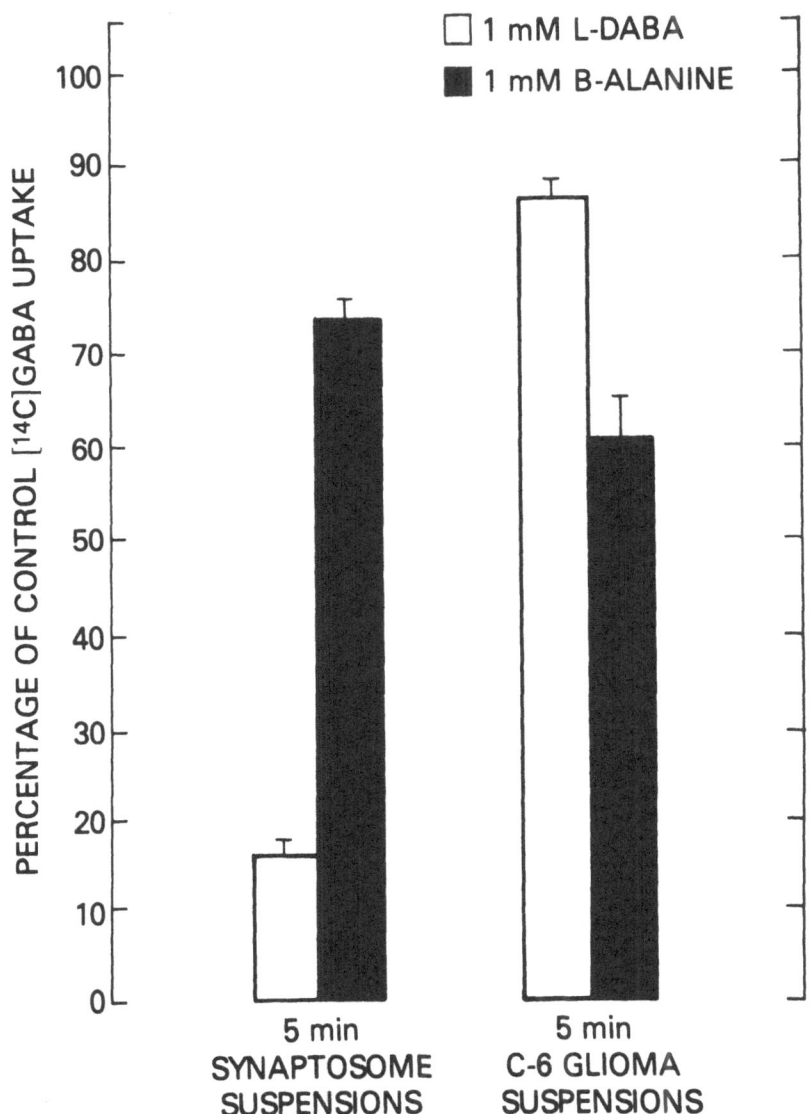

Fig. 9.2.   Comparison of the uptake of [^{14}C]GABA to synaptosomal suspensions and C6-glioma suspensions over a 5-min period of incubation, in the presence of 1 mM L-DABA or 1 mM β-alanine. Each histogram is the mean of four to ten samples. Vertical bars indicate SEM.

comparison, uptake was only depressed by 13% in C-6 glioma suspensions. Thus β-alanine and L-DABA affected these preparations differently, L-DABA having its main effect on [^{14}C]GABA uptake by synaptosome suspensions, and β-alanine mainly affecting [^{14}C]GABA uptake by C-6 glioma suspensions.

Calcium Dependence of [^{35}S]Taurine Efflux in Stimulated Cortex Slices

Figure 9.3 shows the efflux of [^{35}S]taurine from incubated cerebral cortex

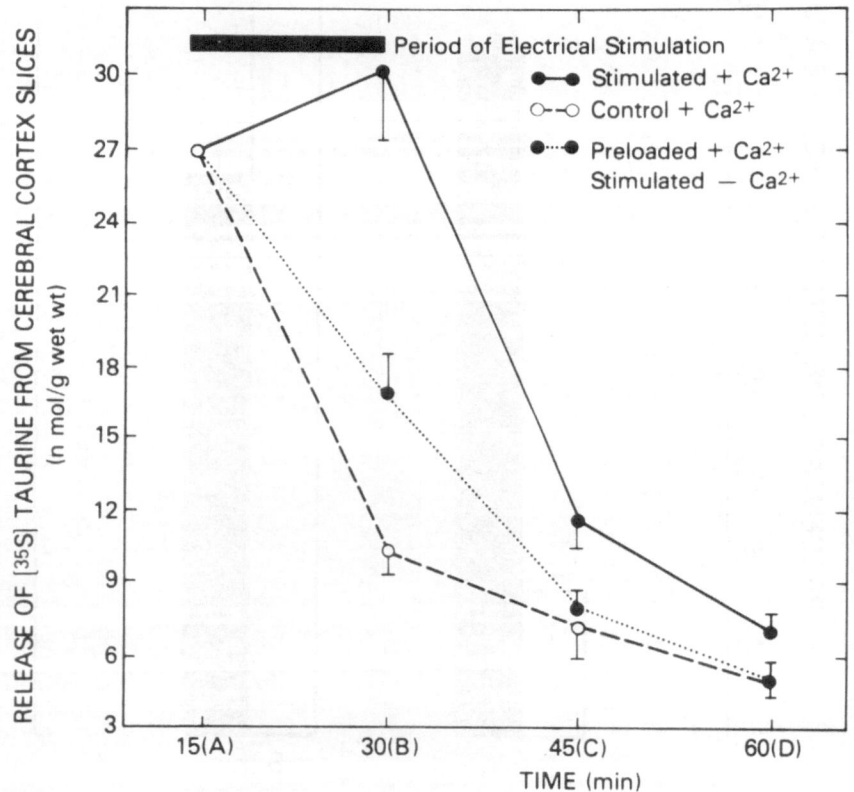

Fig. 9.3. The effect of Ca^{2+} on electrically stimulated [^{35}S]taurine release. Electrical stimulation was for 15 min (period B). Symbols used are: (—●——●—) for electrically stimulated slices in medium containing Ca^{2+}; (— ○ — — ○ —) for control slices in same medium; (··●····●··) for slices preloaded in Ca^{2+}-containing medium and subsequently electrically stimulated in Ca^{2+}-free medium. Vertical lines are SEM from five or more separate measurements. [^{35}S]Taurine efflux was calculated as nanomoles using the specific radioactivity of [^{35}S]taurine present in the incubation medium used for preloading the slices.

slices. Points A, B, C, and D represent 15-min periods of [³⁵S]taurine efflux into consecutive beakers containing incubation medium. Because previous studies had shown that incubation in calcium-free medium decreased uptake of taurine by 30% (Bradford et al., 1976), slices stimulated in calcium-free medium were first preincubated in [³⁵S]taurine in calcium-containing medium. This was done in order to see if calcium dependence of release was still maintained. Figure 9.3 shows that stimulation in calcium-free medium under these conditions was associated with a decrease in the stimulated [³⁵S]taurine release.

Change in [³⁵S]Taurine Content of Subcompartments of Stimulated Cortex Slices

Table 9.3 shows a comparison of stimulated release from subcompartments of slices stimulated at 36 mA mean current.

The change in the synaptosome fraction was 23 nmol/g tissue, and this was commensurate with the increase in content of the incubation medium of 25 nmol/g ($p < 0.05$). The other fraction showing the most change was the soluble cytoplasmic fraction and microsomes (S + P3). This showed a much smaller decrease (6.8 nmol/g), and was not significant ($p < 0.1$). At higher

Table 9.3   [³⁵S]Taurine in subfractions of cortex slices and incubation medium[a]

| Fraction | nmol [³⁵S]taurine/g wet wt. tissue | | |
	Control	Stimulated	Difference[b]
P1[c]	25.7 ± 1.7	25.8 ± 1.9	+0.1
P2[c]	132.0 ± 10.9	99.7 ± 10.6	−32.3 φ
P3 + S[c]	150.0 ± 4.4	143.2 ± 3.3	−6.8
P2A	21.3 ± 3.3	22.1 ± 1.0	+0.8
P2B	78.9 ± 6.4	55.6 ± 7.0	−23.3+
P2C	16.3 ± 1.7	15.8 ± 1.6	−0.5
Release into incubation medium	88.1 ± 11.6	113.2 ± 14.5	+25.33

[a]Slices were stimualted at 36 mA mean current for 20 min, in the presence of Ca²⁺. Release [³⁵S]taurine is corrected for levels in control medium. Results are the mean of four separate estimations ± SEM.

[b]Difference significant with +, $p < 0.05$ or φ, $p < 0.01$.

[c]These represent 93% recovery of homogenate.

currents (48 mA), the soluble fraction showed a 10-fold increase in its loss of taurine and exceeded the synaptosome fractions by a factor of two. Thus at the lower current, the synaptic terminal compartment of the tissue could be contributing the larger proportion of released [^{35}S]taurine, although reuptake during the test period would have decreased the amount of taurine in the medium.

DISCUSSION

Taurine Uptake in Cortex Slices

In considering a possible neurotransmitter role of taurine, because of its slow metabolism (Peck and Awapara, 1967), uptake of taurine would be an important means of terminating such an action of this amino acid. An uptake system for taurine has been described. It is a complex process with saturable and unsaturable components. There are both high- (50–60 μM) and low- (6 mM) affinity saturable systems present (Kaczmarek and Davison, 1972; Lähdesmäki and Oja, 1973; and Lombardini, 1977).

Also important in the consideration of a possible role of taurine is the site of uptake of taurine, as this may suggest the site of taurine release. The nature of the tissue compartments taking up taurine remain uncertain; however, taurine uptake has been reported in glial cells in the retina (Ehinger, 1973), to C-6 glioma in culture (Schrier and Thompson, 1974), and to neuroblastoma cells. Thus it is probable that both neurons and glial cells may actively accumulate taurine.

The fact that the synaptosome subfraction carried the greatest content of [^{35}S]taurine of the crude mitochondrial fraction, indicates that the presynaptic region of the slice does accumulate taurine. This probably represents true uptake, as it has been reported that addition of taurine to homogenate at 5°C did not result in significant changes in the content of taurine in subcellular fractions following a period of incubation with isotopic taurine (Sieghart and Karobath, 1974; Rassin et al., 1977). The results have also shown that there is a large proportion of taurine recovered in the soluble cytoplasmic fraction. This may well represent uptake to glia and neuronal cell bodies.

In order to differentiate between uptake of taurine to neuronal and glial elements, we used β-alanine as a blocker of taurine uptake. β-Alanine has been shown to specifically enter glial cells in the cerebral cortex (Schon and Kelly, 1975; Iversen and Bloom, 1972). β-Alanine also competes with the uptake of taurine and with the uptake of GABA. L-DABA, on the other hand, is taken up specifically by nerve endings (Dick and Kelly, 1975), and it competes with GABA uptake (Iversen and Johnson, 1971). For these reasons we compared the effects of β-alanine and L-DABA on taurine and GABA high-affinity uptake.

β-Alanine was found to depress taurine uptake far more in preparations with a high glial content than in synaptosomes or in cortex tissue (Fig. 9.1). This was especially clear for a 5-min incubation period, but the same pattern was observed when 20-min incubations were used. White matter slices (ependyma) and cortex slices contain both glial and neuronal elements (e.g., nerve fibers) in ependymal slices. Hence only relative differences in the different preparations can be expected. Therefore the greater relative potency of β-alanine in blocking taurine uptake to preparations rich in glial elements may indicate that taurine is largely taken up to synaptosomes rather than to cytoplasmic particles of glial origin, which could contaminate a synaptosome preparation.

The relative effect of β-alanine and L-DABA on [^{14}C]GABA uptake by synaptosome preparations was also studied in order to further assess the purity of the synaptosomes. It was found that L-DABA strongly inhibited [^{14}C]GABA uptake to synaptosome but not to glial preparations. The converse was the case for β-alanine. In view of the reported specific uptake of L-DABA to nerve endings and β-alanine to glial cells, and their competition with GABA uptake, these results suggest that the synaptosome preparations used were not heavily contaminated with particles of glial origin.

It has been reported that β-alanine blocks the uptake of taurine and GABA to synaptosomes by 60–80% (Snodgrass et al., 1973; Schmid et al., 1975). However, this does not agree with the autoradiographic data, which imply a specific uptake of β-alanine to glial cells. A possible explanation of this is that the fraction (P2) studied by Schmid and co-workers was contaminated by glial elements. Also it was not a pure fraction but contained other elements including mitochondria.

Release of Taurine

Although glial cells can release transmitter compounds as well as neurons (Bowery and Brown, 1972), they do not show calcium dependence of release (Sellstrom and Hamberger, 1977) or show atypical calcium dependence. Hence calcium dependence has become an important criteria to distinguish between release for these two cell types. Also nerve trunks do not show calcium dependence (Weinrich and Hammerschlag, 1975). Thus we further examined the calcium dependence of taurine release from cortex slices as a possible index to its release from neurons as opposed to glia.

Because previous studies had shown that omission of calcium from the medium decreased taurine uptake by 30% (Bradford et al., 1976), slices were loaded with [^{35}S]taurine in calcium-containing medium prior to stimulation in calcium-free medium. The stimulated release of [^{35}S]taurine was shown to be calcium dependent (Fig. 9.3). Although the [^{35}S]taurine-stimulated release was less than for medium containing calcium, it was not reduced to the levels of control efflux with or without calcium in the medium. It is therefore possible that some endogenous calcium is still available for the process. Alternatively, the non-calcium-dependent component of stimulated [^{35}S]efflux could be from glial cells.

Slices were subfractionated after electrical stimulation in order to locate the pool of taurine responsible for its release. At the lower stimulating current of 36 mA, the stimulated release was mainly from the synaptosomal pool. At higher currents of 48 mA, however, release of taurine was seen from other fractions including the soluble supernatant fraction. Two conclusions may be drawn from this. First, changes in the soluble supernatant fraction may represent release from other tissue components such as glial cells or neuronal cell bodies. Second, in order to stimulate release from synaptic regions, the stimulus must not be too great.

These results therefore suggest that the stimulated release of taurine in the cerebral cortex slices is from synaptic regions, and that it is released in a calcium-dependent fashion. Calcium dependence of taurine release has also been reported elsewhere. Calcium ionophores have been shown to increase taurine release from the retina (Pasantes-Morales et al., 1974), from the visual cortex (Collins, 1974), and from the olfactory cortex (Berger et al., 1978). In contrast to this, potassium-induced release of taurine from (P2) fractions has been reported to be somewhat reduced in the presence of calcium (Sieghart and Heckl, 1976). The reason for this apparent discrepancy is not yet clear. Factors that may be associated could include the fact that it was a crude fraction, and that it could possibly be contaminated by glial elements.

The above evidence can be related to previous studies of taurine that suggest a possible neurotransmitter role. The enzyme producing taurine, cysteinesulfinate decarboxylase (EC41.1.29), has a synaptic localization (Agrawal et al., 1971), and synaptic vesicles are enriched in taurine (De Belleroche and Bradford, 1973; Rassin et al., 1977). In addition, the existence of a synaptosomal subpopulation that accumulates [^{35}S]taurine has been reported (Sieghart and Karobath, 1974). Taken with these reports, our results further strengthen the case for a role of taurine within synaptosomes, possibly by acting as a neurotransmitter.

ACKNOWLEDGMENT

This work was supported by an MRC project grant.

REFERENCES

Agrawal, H. C.; Davidson, N. A.; and Kazcmarek, L. K. Subcellular distribution of taurine and cysteinsulphinate decarboxylase in developing rat brain. *Biochem. J.*, 122, 759–763 (1977).

Berger, F.; Urban, R. F.; and Mandel, P. Potassium evoked release of [^{14}C]GABA and [^{3}H]taurine from rat ofactory bulb slices. *Neuro. Sci. Lett.*, 8, 1137–1142 (1978).

Bowery, N. G.; and Brown, D. A. γ-Aminobutyric acid uptake by sympathetic ganglia. *Nature, New Biol.*, 238, 89–91 (1972).

Bradford, H. F.; Bennet, G. W.; and Thomas, A. J. Depolarizing stimuli and the release of physiologically active amino acids from suspensions of mammalian synaptosomes. *J. Neurochem.*, 21, 495–505 (1973).

Bradford, H. F.; Davison, A. N.; and Wheler, G. H. T. Taurine and synaptic transmission. In *Taurine*, ed. by R. Huxtable and A. Barbeau. Raven Press, New York, pp. 303–310 (1976).

Collins, G. G. S. The release of endogenous amino acids from rat visual cortex by calcium ions in the presence of calcium iosoptoes X537 and A23187. *J. Neurochem.*, 28, 461–463 (1977).

Curtis, D. R.; and Watkins, J. C. The excitation and depression of spinal neurones by structurally related amino acids. *J. Neurochem.*, 6, 117–141 (1960).

Curtis, D. R.; and Watkins, J. C. The pharmacology of amino acids related to gamma-aminobutyric acid *Pharmac. Rev.*, 17, 347–391 (1965).

De Belleroche, J. S.; and Bradford, H. F. Metabolism of beds of mammalian cortical synaptosomes: Response to depolarizing influences. *J. Neurochem.*, 19, 585–602 (1972).

De Belleroche, J. S.; and Bradford, H. F. Amino acids in synaptic vesicles from mammalian cerebral cortex: A reappraisal. *J. Neurochem.*, 21, 441–451 (1973).

Davison, A. N.; and Kaczmarek, L. K. Taurine-A possible neurotransmitter? *Nature*, 234, 107–108 (1971).

Dick, F.; and Kelly, J. S. L-2,4-Diaminobutyric acid (L-DABA) as a selective marker for inhibitory nerve terminals in rat brain. *Br. J. Pharmacol.*, 53, 439 (1975).

Ehinger, B. Glial uptake of taurine in the rabbit retina. *Brain Res.*, 60, 512–516 (1973).

Iversen, L. L.; and Bloom, F. E. Studies of the uptake ^{3}H-GABA and ^{3}H-glycine in slices and homogenates of rat brain and spinal cord by electron microscopic autoradiography. *Brain Res.*, 41, 131–143 (1972).

Iversen, L. L.; and Johnson, G. A. R. GABA uptake in rat central nervous system: Comparison of uptake in slices and homogenates and the effects of some inhibitors. *J. Neurochem.*, 18, 1939–1950 (1971).

Iversen, L. L.; Kelly, J. S.; Minchin, M.; Schon, F.; and Snodgrass, S. R. Role of amino acids and peptides in synaptic transmission. *Brain Res.*, 62, 567–576 (1973).

Kaczmarek, L. K.; and Davison, A. N. Uptake and release of taurine from rat brain slices. *J. Neurochem.*, 19, 2355–2362 (1972).

Lähdesmäki, P.; and Oja, S. S. On the mechanism of taurine transport at brain cell membranes. *J. Neurochem.*, 20, 1411–1417 (1973).

Lombardini, J. B. Regional and subcellular studies on taurine in the rat CNS. In *Taurine*, ed. by R. Huxtable and A. Barbeau. Raven Press, New York, pp. 311–326 (1976).

Lombardini, J. B. High affinity uptake systems for taurine in tissue slices and synaptosome fractions prepared from various regions of the rat CNS: Correction of transport data by different experimental procedures. *J. Neurochem.*, 29, 305–312 (1977).

Lowry, O. H.; Rosenbrough, N. J.; Farr, A. L.; and Randall, J. R. Protein measurement with the Folin phenol reagent. *J. Biol. Chem.*, 193, 265–275 (1951).

Mandel, P.; and Pasantes-Morales, H. The role of taurine in the central nervous system. In *Reviews of Neuroscience*, ed. by S. Ehrenpreis and I. Kopin. Raven Press, New York, pp. 157–193 (1978).

McIlwain, H. Metabolic experiments with neural tissues. In *Practical Neurochemistry.* Churchill Livingstone, p. 150 (1975).

Okamoto, K.; and Quastel, J. H. Spontaneous action potentials in isolated guinea pig cerebellar slices: Effects of amino acids and conditions affecting sodium and water uptake. *Proc. Royal Soc. Lond. B,* 184, 83–90 (1973).

Pasantes-Morales, H.; Salceda, R.; and Gomez-Puyona. Effect of X537A on the release of amino acids in retina. *Biochim. Biophys. Res. Comm.,* 58, 847–853 (1974).

Peck, E. J.; and Awapara, J. Formation of taurine and isethionic acid in rat brain. *Biochim. Biophys. Acta,* 141, 499–506 (1967).

Rassin, D. K.; Sturman, J. A.; and Gaull, G. E. Taurine in developing rat brain: Subcellular distribution and association with synaptic vesicles of ^{35}S-taurine in maternal, fetal and neonatal rat brain. *J. Neurochem.,* 28, 41–50 (1977).

Schmid R.; Sieghart, W.; and Karobath, M. Taurine uptake in synaptosomal fractions of rat cerebral cortex. *J. Neurochem.,* 25, 5–9 (1975).

Schon, F.; and Kelly, J. S. Selective uptake of ^{3}H-β-alanine by glia: Association with the glial uptake system for GABA. *Brain Res.,* 86, 243–257 (1975).

Schrier, N. K.; and Thompson, E. J. On the role of glial cells in the mammalian nervous system. *J. Biol. Chem.,* 249, 1769–1780 (1974).

Sellstrom, A.; and Hamberger, A. Potassium-stimulated γ-aminobutyric acid release from neurons and glia. *Brain Res.,* 119, 189–198 (1977).

Sieghart, W.; and Heckl, L. Potassium-evoked release of taurine from synaptosomal fractions of rat cerebral cortex. *Brain Res.,* 116, 538–543 (1976).

Sieghart, W.; and Karobath, M. Evidence for specific synaptosomal localization of endogenous accumulated taurine. *J. Neurochem.,* 23, 911–915 (1974).

Snodgrass, S. R.; Hedley-Whyte, T. E.; and Lorenzo, A. V. GABA transport by nerve ending-fractions of rat brain. *J. Neurochem.,* 20, 771–782 (1973).

Weinreich, D.; and Hammerschlag, R. Nerve impulse-enhanced release of amino acids from non-synaptic regions of peripheral and central nerve trunks of bullfrog. *Brain Res.,* 84, 137–142 (1973).

Copyright © 1981, Spectrum Publications, Inc.
The Effects of Taurine on Excitable Tissues

CHAPTER 10

Function and Regulation of Taurine in the Pineal Gland

G. H. T. Wheler
D. C. Klein

INTRODUCTION: TAURINE IN THE PINEAL GLAND

Taurine is highly concentrated in the pineal gland, where it occurs at 20–60 mM (Crabai et al., 1974; Green et al., 1962; Guidotti et al., 1972; La Bella et al., 1968; Vellan et al., 1970). This is higher than the concentration of taurine in any other body tissue, except the neurohypohysis. It is also relatively abundant compared to other amino acids in the pineal gland, comprising 30% of the free amino acids of the adult gland (Nir et al., 1974). In view of these observations, it is surprising that little is known about the role of taurine in this tissue.

Location of Taurine in the Pineal Gland

Some recent reports have suggested that taurine is located in nerve endings in the brain (De Belleroche and Bradford, 1973; Agrawal et al., 1971) and may play a role there as a neurotransmitter or neuromodulator (Davison and Kaczmarek, 1971; Bradford et al., 1976; Mandel and Pasantes-Morales, 1978; Wheler et al., 1979b). In the rat pineal gland, however, it appears not to be located in nerve terminals innervating the pineal gland, but in pinealocytes (Wheler et al., 1979a).

Uptake of Taurine into Pinealocytes

An uptake system for taurine has been found in the pineal gland (Krusz et al., 1977; Grosso et al., 1978). It appears to be comprised of both low (Km = 2 mM) and high (Km = 5.7 μM) affinity components, and is dependent upon sodium. The *in vitro* uptake of taurine from medium containing physiological concentrations (0.2 mM) of the amino acid is blocked by β-alanine, and it appears to have the characteristics of a β-amino acid transport system.

Interestingly, it has recently been found that the pineal taurine can be reduced *in vivo* by 75% of control values within a 14-day period by the inclusion of β-alanine (3%) in the drinking water (Kocsis et al., 1979).

Daily Rhythm in Pineal Taurine

One of the outstanding characteristics of the physiology and biochemistry of the pineal gland is a strong circadian pattern of activity. In all species there is an increase in melatonin production at night, which is the result of adrenergic-cyclic-AMP induction of one of the melatonin-synthesizing enzymes, N-acetyltransferase (Fig. 10.1) (Klein, 1974). In view of this it is interesting that a daily rhythm in pineal taurine has been reported (Leonard et al., 1975; Grosso et al., 1978). Taurine levels increase gradually from a low (approx. 14 mM) at the beginning of the dark period to a high (approx. 24 mM) during the middle of the light period.

REGULATION OF TAURINE RELEASE

In a series of studies, we have recently sought to determine if the release of taurine from the pineal gland is altered when the gland is stimulated by the same mechanism that stimulates melatonin production (Wheler and Klein, 1979a). We used a well-described pineal gland culture system (Wheler et al, 1979a). To study taurine release, pineal glands were first prelabeled in organ culture for 24 hr with 0.25 mM [^{14}C]taurine (specific activity, 5.04 Ci/mol). This dose was chosen as it is similar to the reported plasma concentration of taurine. The glands were then incubated for a second 24-hr period in culture medium without any taurine. Subsequently, glands, which contained about 60 nCi of [^{14}C]taurine, were transferred into culture vessels containing media and compounds of interest for a 0-4-hr test period. [^{14}C]Taurine was measured by conventional techniques. N-Acetyltransferase activity was measured to provide an index of adrenergic activation of the pineal cells.

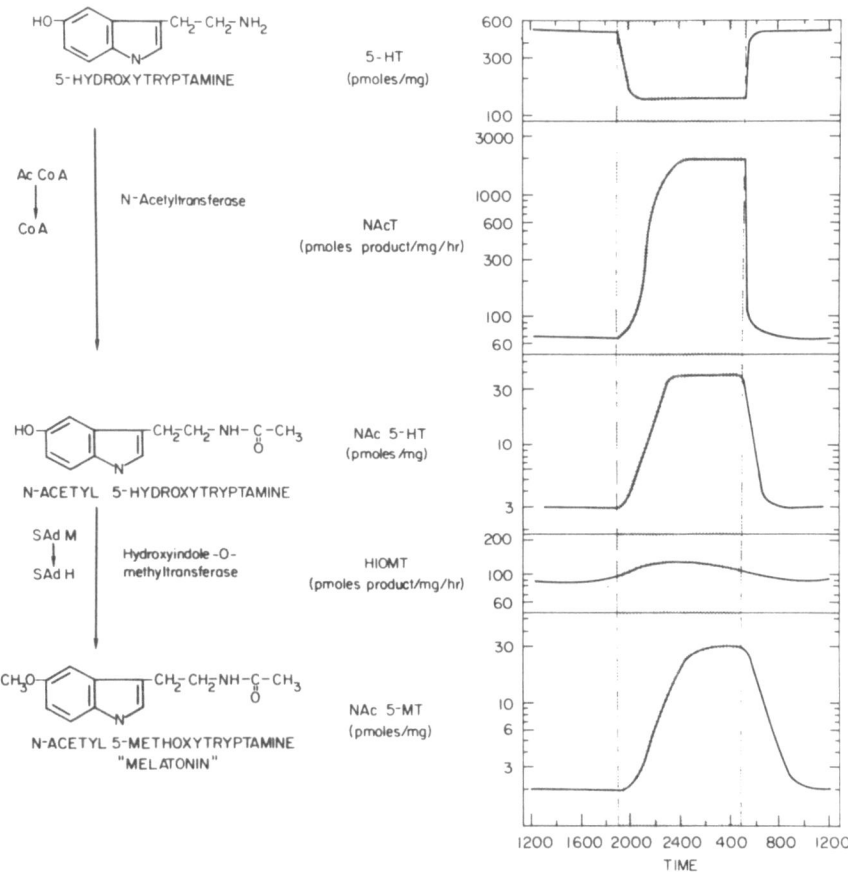

Fig. 10.1. Rhythms in indole metabolism in the rat pineal gland. The metabolic pathway from 5-hydroxytryptamine to melatonin is on the left. The daily variations in the concentration of metabolites and activities of enzymes are on the right. The shaded portion indicates the dark period of the lighting cycle. The data have been abstracted from reports in the literature. Key: AcCoA, acetyl coenzyme A; CoA, coenzyme A; S AdM, S-adenosyl methionine; S AdH, S-adenosyl homocysteine; 5HT, 5-hydroxytryptamine, serotonin; NAcT, N-acetyltransferase; HIOMT, hydroxyindole-O-methyltransferase; NAc 5-MT, N-acetyl 5-methoxytryptamine, melatonin. (From Klein, 1974.)

Norepinephrine-induced Release of [^{14}C]Taurine

L-Norepinephrine treatment was found to produce an increase in [^{14}C]taurine release (Fig. 10.2). This effect was rapid, occurring within 30 min. It occurred several hours prior to the first detectable increase in N-acetyltransferase activity. The dose-response relationship of L-norepinephrine to [^{14}C]taurine release is presented in Fig. 10.3, and that for L-norepinephrine to N-acetyltransferase is presented for comparison. At a concentration of 10^{-7} M norepinephrine there was a significant increase in [^{14}C]taurine release, compared with control release ($p < 0.02$). At this concentration, N-acetyltransferase activity was markedly elevated. The L-norepinephrine-stimulated release of taurine was not due to a nonspecific change in membrane permeability, as comparable release was not seen with [^{14}C]glycine or [^{14}C]α-amino isobutyric acid (Wheler and Klein, 1979a). The identity of the released [^{14}C]taurine was confirmed in three different thin-layer chromatographic systems.

The L-norepinephrine-stimulated [^{14}C]taurine release was blocked better by L-propranolol than by D-propranolol, as was the induction of N-acetyltransferase activity (Table 10.1). L-Norepinephrine was found to be more effective than D-norepinephrine in stimulating both [^{14}C]taurine release and N-acetyltransferase activity. Another β-adrenergic agonist, L-isoproterenol, also stimulated both taurine release and N-acetyltransferase activity. However, other biogenic amines—including dopamine, phenylephrine, tyramine, octopamine, histamine, GABA, serotonin, and carbachol—were relatively ineffective in stimulating [^{14}C]taurine release. These observations suggest that the L-norepinephrine-induced release of [^{14}C]taurine is a specific response mediated by a β-adrenergic receptor. Phenylephrine (1 μM), which did not stimulate taurine release, did stimulate N-acetyltransferase activity (Table 10.1). This was less than 15% of the maximal stimulation by L-norepinephrine (1 μM). It seems probable that a submaximal increase in [^{14}C]taurine release was not detected because of reuptake of the released taurine by the taurine uptake system present in the pineal gland (Grosso et al., 1978; Krusz et al., 1977).

Cyclic-AMP-induced Release of [^{14}C]Taurine from Pinealocytes

An early event in the noradrenergic stimulation of the pineal gland is the increase in the concentration of cyclic-AMP (Axelrod and Zatz, 1977; Klein, 1978). Hence it was of interest to see if cyclic-AMP and its derivatives could increase [^{14}C]taurine release from the pineal gland (Wheler and Klein, 1979b).

Pineal glands were preincubated for 48 hr as before and preloaded with [^{14}C]taurine (specific activity, 5.04 Ci/mol) during the first 24 hr. Glands were then incubated with the test compounds for 4–6 hr. The dose-response relationship of [^{14}C]taurine release and N-acetyltransferase activity with di-

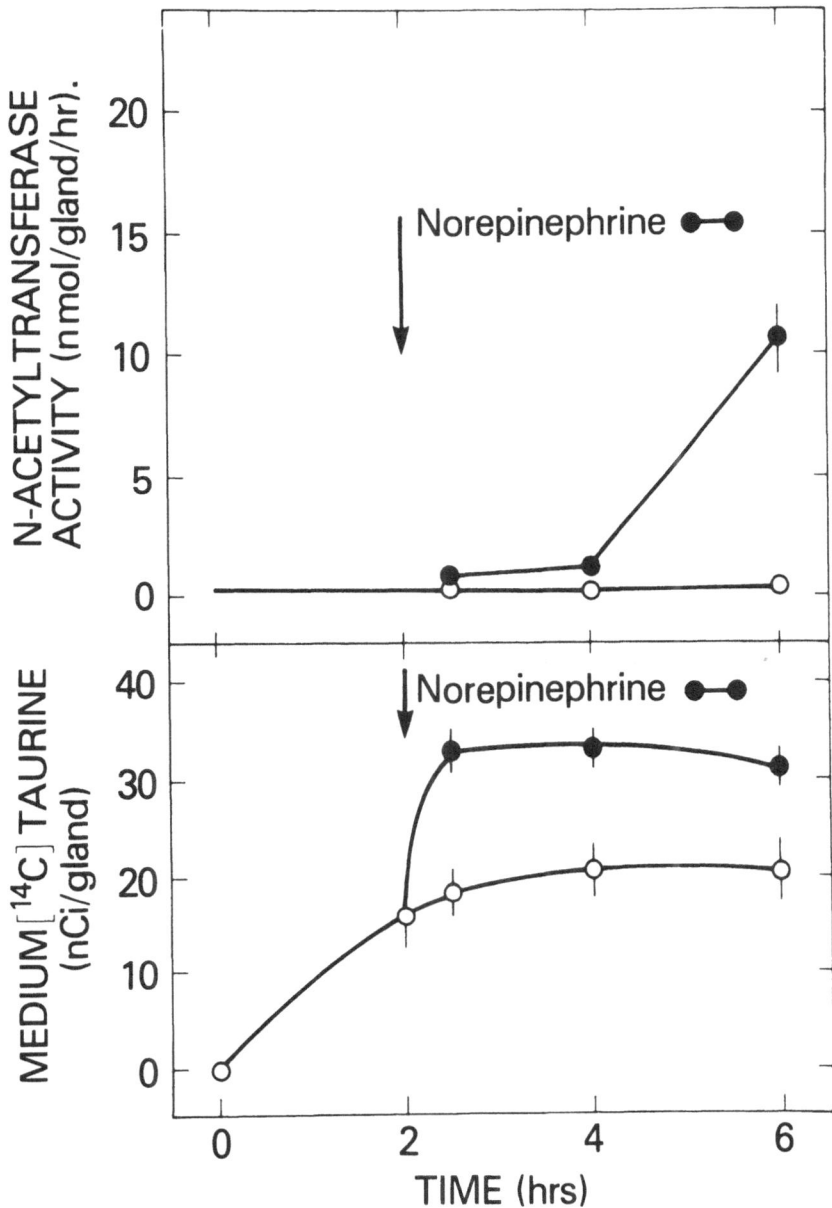

Fig. 10.2. Time course of the stimulation of taurine release by L-norepinephrine. L-Norepinephrine (10^{-6} M) was added 2 hr after the start of the experiment. The bottom box is [^{14}C]taurine release, and the top portion is N-acetyltransferase activity in that experiment. Open circles represent controls; closed circles represent L-norepinephrine treatment. Values are mean ($\pm$ SE) for six glands or three media.

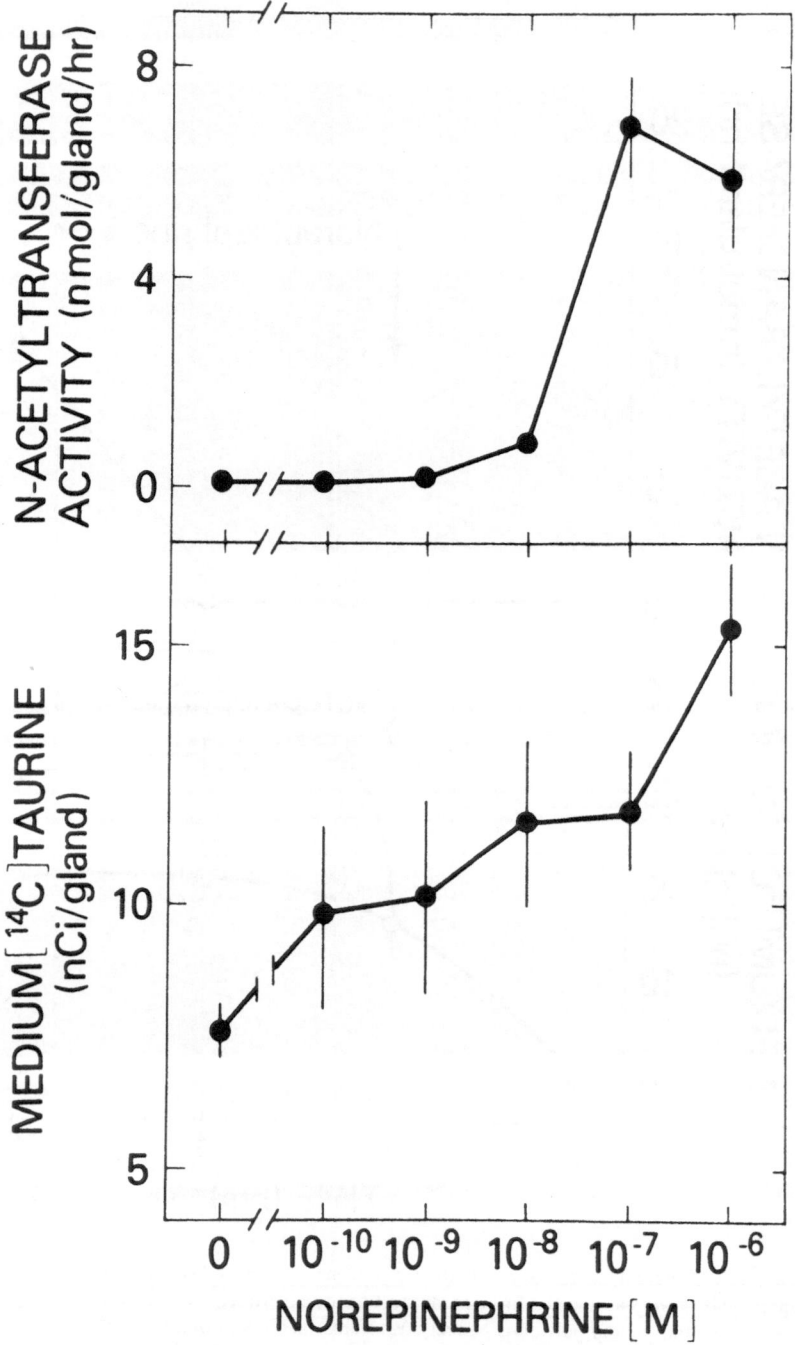

Fig. 10.3. Dose-response relationship of [¹⁴C]taurine release and N-acetyltransferase activity with L-norepinephrine, which was present for the entire 4-hr test period. Values are the means (± SE) for six glands or three media at 4 hr. [¹⁴C]Taurine release from L-norepinephrine (10^{-7} M)-treated glands was significantly greater than from untreated glands ($p < 0.02$).

Table 10.1   Structure activity analysis of biogenic amines for the stimulation of taurine release and N-acetyltransferase activity[a]

Experiment	Test compound	Concentration (μM)	Medium [^{14}C]taurine (nCi/gland)	N-Acetyltransferase activity nmol/gland/hr
I	None		5.1 ± 0.6	0.08 ± 0.03
	L-Norepinephrine	1	25.0 ± 0.2^{b}	6.91 ± 1.02^{b}
	+ D-propranolol	10	24.7 ± 1.9^{b}	6.65 ± 0.37^{b}
	+ L-propranolol	10	$17.5 \pm 1.0^{b,c}$	$1.24 \pm 0.35^{b,c}$
II	None		13.5 ± 0.5	0.17 ± 0.01
	L-Norepinephrine	1	25.8 ± 1.9^{b}	9.32 ± 0.99^{b}
	L-Isoproterenol	1	28.3 ± 1.2^{b}	6.82 ± 1.02^{b}
	Dopamine	1	13.7 ± 1.3	0.49 ± 0.11
	Phenylephrine	1	13.8 ± 0.5	1.24 ± 0.19^{b}
	Tyramine	1	14.7 ± 1.1	0.31 ± 0.03
	Octopamine	1	12.6 ± 2.3	0.80 ± 0.08^{b}
	Histamine	1	16.5 ± 1.6	0.13 ± 0.02
	GABA	1	13.0 ± 1.7	0.16 ± 0.02
	Serotonin	1	12.4 ± 1.2	0.22 ± 0.07
	Carbachol	1	12.8 ± 1.1	0.08 ± 0.01^{b}
III	None		15.7 ± 1.2	0.02 ± 0.01
	L-Norepinephrine	0.1	$28.4 \pm 4.3^{b,d}$	$6.62 \pm 0.61^{b,d}$
	D-Norepinephrine	0.1	15.6 ± 1.7	0.06 ± 0.03

[a]Pineal glands, preincubated with culture medium containing 0.25 mM [^{14}C]taurine as described, were treated in organ culture with the chemical listed for a 4-hr test period. Values are mean $\pm$ SE for six glands or three media.

[b]Significantly different from control values ($p < 0.05$).

[c]Significantly different from values of glands treated with L-norepinephrine and D-propranolol ($p < 0.05$).

[d]Significantly different from values of glands treated with D-norepinephrine ($p < 0.05$).

butyryl cyclic-AMP is presented in Fig. 10.4. The effect of dibutyryl cyclic-AMP (1 mM) was about the same as that seen with 1 μM L-norepinephrine. Another analogue of cyclic-AMP and adenosine were also able to stimulate [^{14}C]taurine release, although cyclic-AMP itself was ineffective (Table 10.2). Other studies have shown that added cyclic-AMP is too rapidly metabolized for its biochemical effects to be readily measured (Klein and Berg, 1970).

The time course of dibutyryl cyclic-AMP (1 mM) stimulation of [^{14}C]taurine release was also studied (Fig. 10.5). The release of [^{14}C]taurine was as rapid as that seen with norepinephrine, with peak values occurring within 20 min. These observations provide evidence that cyclic-AMP could be mediating the rapid norepinephrine-induced release of [^{14}C]taurine.

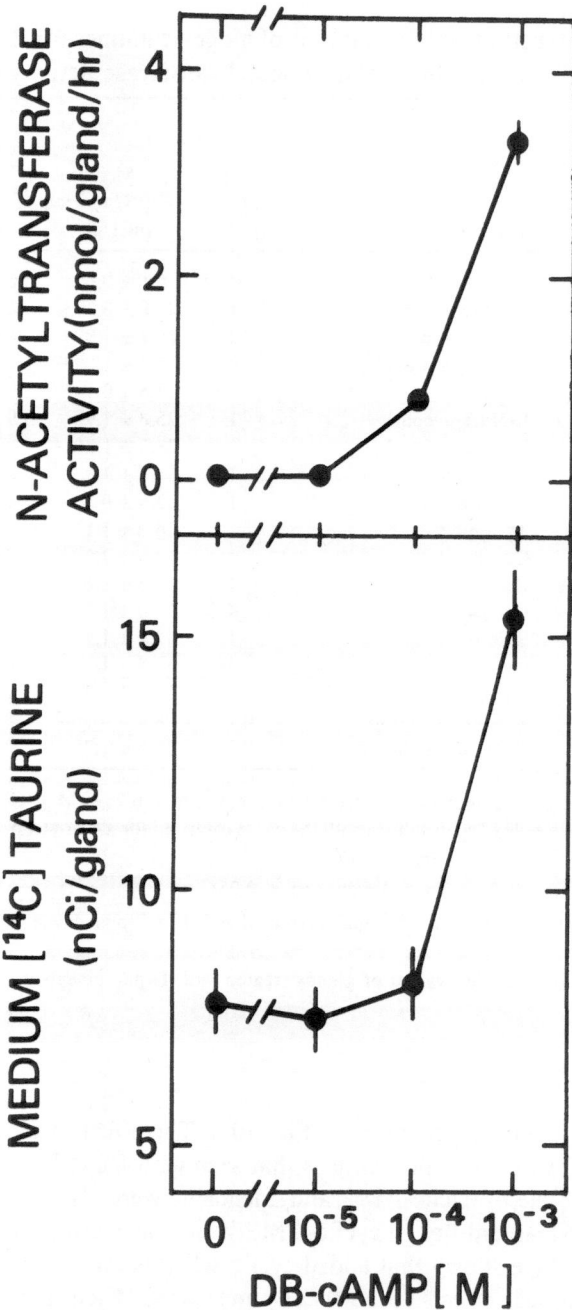

Fig. 10.4.   Dose-response relation ship of [^{14}C]taurine release and N-acetyltransferase activity with dibutyryl cyclic-AMP. Dibutyryl cyclic-AMP was present for the entire 4-hr test period. Values are the means (± SE) for six glands or three media at 4 hr.

Table 10.2   Effects of cyclic-AMP and related compounds on taurine release and of N-acetyltransferase activity[a]

Experiment	Test compound	Concentration (μM)	Medium [^{14}C]taurine (nCi/gland)	N-Acetyltransferase activity nmol/gland/hr
I	None		13.9 ± 2.8	0.19 ± 0.06
	pCl ϕ SH cAMP	10	14.7 ± 0.5	0.46 ± 0.09
		100	21.2 ± 0.2	6.83 ± 1.35[c]
		1,000	31.8 ± 0.3[c]	10.74 ± 1.94[c]
II	None		15.7 ± 1.2	0.02 ± 0.01
	L-Norepinephrine	0.1	28.4 ± 4.3[c]	6.62 ± 0.61[c]
	Cyclic-AMP	10,000	16.4 ± 1.2	1.09 ± 0.16[c]
	Adenosine	100	13.6 ± 0.8	0.08 ± 0.03
		1,000	19.4 ± 1.3	0.02 ± 0.01

[a]Pineal glands, preincubated with culture medium containing 0.25 mM [^{14}C]taurine, were treated in organ culture with the chemical listed for a 4-hr test period. Values are mean ± SE for six glands or three media.

[b]The abbreviations used are: pClϕSH cAMP, para-chloro-phenyl-thio cyclic-AMP; cAMP, cyclic-AMP.

[c]Significantly different from control values ($p < 0.05$).

EFFECT OF TAURINE ON N-ACETYLTRANSFERASE AND MELATONIN PRODUCTION

In view of the release of taurine by norepinephrine and its known effects on excitable membranes in the brain, the effects of extracellular taurine on the pineal gland are of interest.

We incubated pineal glands for 48 hr in organ culture as described (Wheler et al., 1979a). Taurine (1–10 mM) was then added to the tissue culture medium at the start of a 0-12 hr test period. In some cases [^{3}H]tryptophan (0.2 mM, SA 50 mCi/nmol) was added to the medium to observe the effect of taurine on the formation of [^{3}H]melatonin and other potential metabolites of [^{3}H]tryptophan. N-Acetyltransferase and hydroxyindole-O-methyltransferase activities were also measured.

Taurine (10 mM) was found to greatly stimulate both [^{3}H]N-acetylserotonin and [^{3}H]melatonin production, by 40- and 25-fold, respectively, by stimulating N-acetyltransferase activity (Fig. 10.6). Hydroxyindole-O-methyltransferase activity was not altered. This effect was half-maximal at a

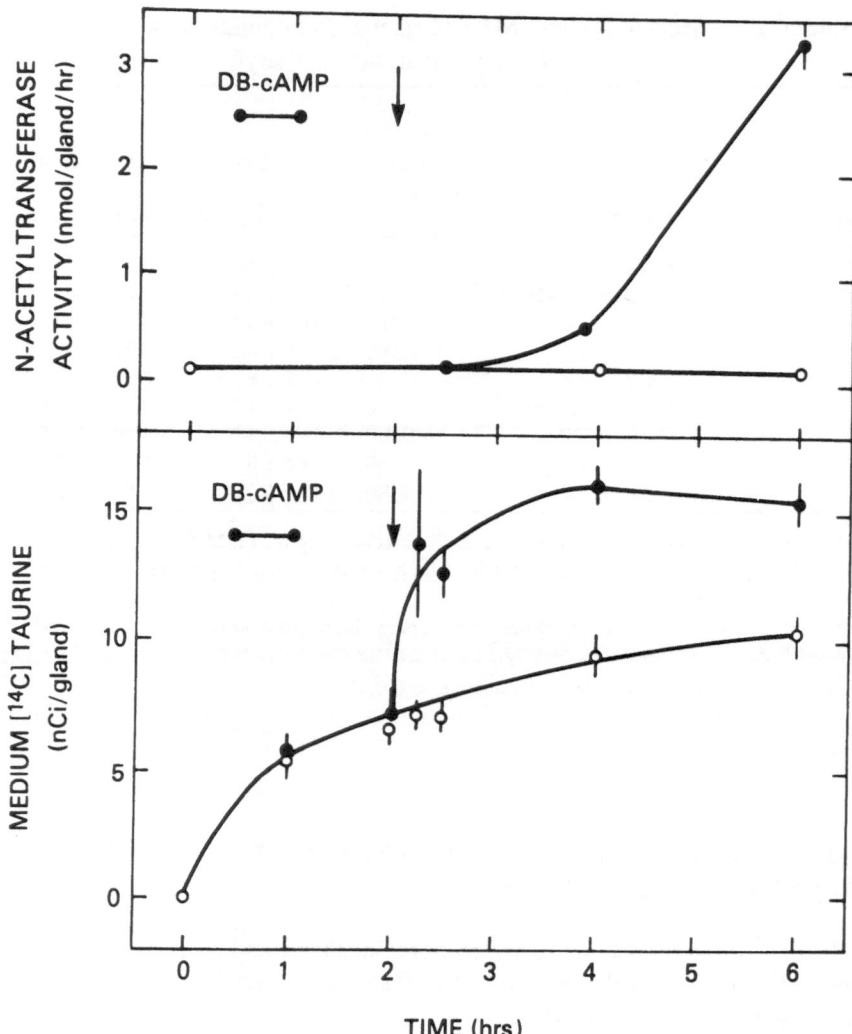

Fig. 10.5. Time course of dibutyryl cyclic-AMP-stimulated [¹⁴C]taurine release and N-acetyltransferase activity. Glands were incubated for 2 hr in control medium. Dibutyryl cyclic-AMP (final concentration 1 mM) was added in 15 μl of 10× concentrated solution, and samples of medium were removed at the times indicated. Diluent was added to control cultures. Several sets of glands were used to prevent removal of more than a total of 30 μl from each dish. Values are the means (± SE) for six media at each time point.

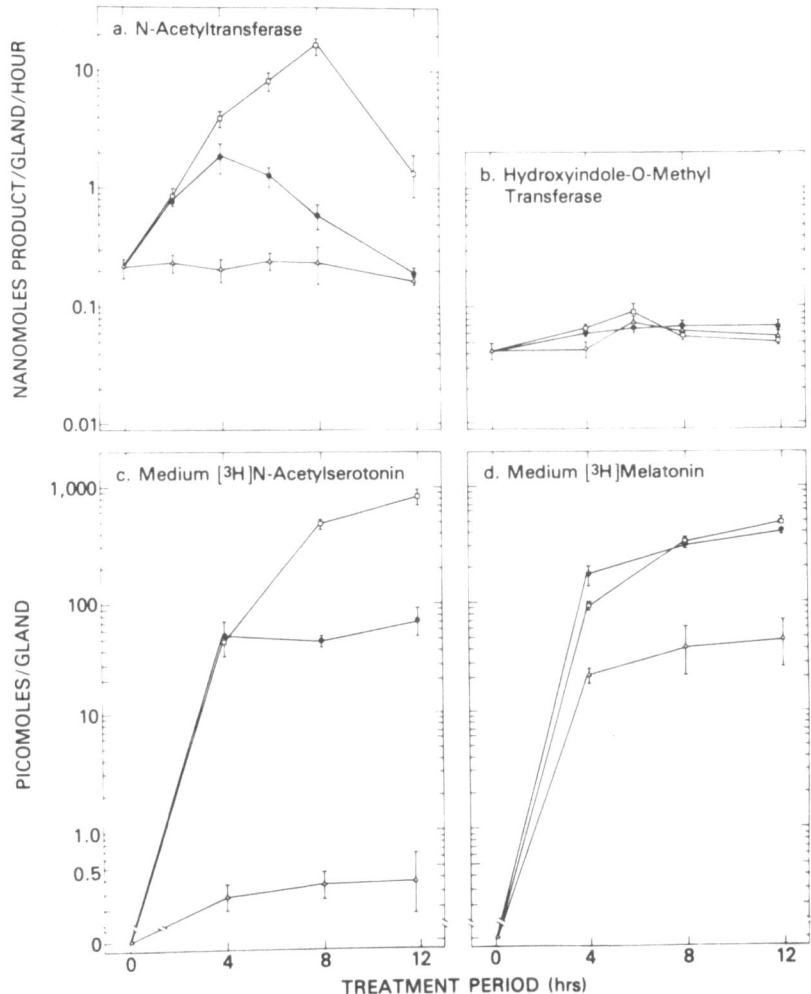

Fig. 10.6.   Time course of the effects of taurine (10 mM) and of L-norepinephrine (1 μM) on: (*a*) N-acetyltransferase activity, (*b*) hydroxyindole-O-methyl-transferase activity, (*c*) medium [^{3}H]N-acetylserotonin, and (*d*) medium [^{3}H]melatonin. The assays used were as described. Each value is the mean (± SE) of the analysis of samples of six glands or three culture media. On the graph ●──● represents 10 mM taurine, □──□ represents 1 μM norepinephrine, and △──△ represents controls.

Table 10.3 Structure activity relationship for stimulation of
N-acetyltransferase activity

Common name	Structure	Activity relative to taurine
Taurine	$NH_2CH_2CH_2SO_3H$	1
3-Aminopropane sulfonic acid	$NH_2CH_2CH_2CH_2SO_3H$	0.06
Isethionic acid	$HOCH_2CH_2SO_3H$	0.03
N-Methyltaurine	$CH_3NHCH_2CH_2SO_3H$	0.03
Cysteic acid	$NH_2(COOH)CHCH_2SO_3H$	0.02
Hypotaurine	$NH_2CH_2CH_2SO_2H$	0.04
β-Alanine	$NH_2CH_2CH_2COOH$	0.29
Cysteinesulfinic acid	$NH_2CH(COOH)CH_2SO_2H$	0.02
Cysteine	$NH_2CH(COOH)CH_2SH$	0.02
Cystathionine	$NH_2CH(COOH)CH_2CH_2S$	0.05
	$NH_2CH(COOH)CH_2$	

concentration of 1 mM taurine. Of several analogues of taurine tested, only
β-alanine stimulated N-acetyltransferase activity to any extent (Table 10.3).
The stimulation by taurine was blocked stereospecifically by L-propranolol
(Table 10.4), indicating that taurine is probably acting via β-adrenergic receptors. Consistent with this conclusion is the finding that taurine was effec-

Table 10.4 Comparison of the effects of L- and D-propranolol on taurine
stimulation of pineal glands in organ culture[a]

Treatment	N-Acetyl-transferase activity (nmol product/ gland/hr)	[³H]Metabolites produced (pmol/gland)	
		Melatonin	N-Acetylserotonin
Control	0.05 ± 0.01	9.5 ± 1.6	2.0 ± 0.1
Taurine, 10 mM	2.41 ± 0.24	170.0 ± 18.9	52.5 ± 0.6
Taurine + D-propranolol	2.23 ± 0.7	144.1 ± 8.4	50.0 ± 15.0
Taurine + L-propranolol	0.08 ± 0.01	11.5 ± 0.9	0.0 ± 0.0

[a]Glands were preincubated for 48 hr as described and then incubated for 4 hr with L- or
D-propranolol (0.3 μM) and 10 mM taurine. Chromatography and N-acetyltransferase assay
were measured as described. Each group consists of the mean ± SE of six separate glands for
enzyme activity values and three separate media for metabolites.

tive in stimulating N-acetyltransferase in denervated pineal glands (Table 10.5).

This effect on melatonin production might be ignored, as 10,000-fold more taurine than norepinephrine is required to produce an equivalent effect. However the high concentration of taurine in the pineal gland (20 mM) and the presence of a release mechanism for taurine suggest that taurine could possibly act in this way if a sufficient concentration of taurine could build up in the extracellular space near β-adrenergic receptors.

It is interesting to note that it has previous123 been suggested that taurine might inhibit melatonin production in the tadpole (Baskin and Dagirman-jian, 1973), an effect that is opposite to what we find in the rat.

POTENTIAL ROLES OF TAURINE IN THE PINEAL GLAND

The available data allow only the formation of a working hypothesis of the role of taurine in the pineal gland. We suspect that taurine could be involved in pineal physiology in two ways. First, it may be involved in the production of norepinephrine-induced membrane hyperpolarization. This event is rapid, blocked by propranolol (Sakai and Marks, 1972), and appears to be mediated by cyclic-AMP (Parfitt et al., 1975). Perhaps taurine release is part of a mechanism of releasing cations as required to establish a new membrane potential. This is consistent with the strong electronegative sulfonate group on taurine, which could complex cations. It is well established that cyclic-AMP stimulates phosphorylation of membrane proteins (Greengard, 1976). Perhaps one of these is a protein that controls the transport of taurine out of

Table 10.5   Effect of denervation on taurine stimulation[a]

Surgical preparation	Drug treatment in organ culture	N-acetyltransferase activity (nmol product/gland/hr)
None	None	0.26 ± 0.01
None	Taurine (10 mM)	4.70 ± 0.91
Sham denervation	None	0.67 ± 0.03
Sham denervation	Taurine (10 mM)	4.70 ± 0.25
Denervation	None	1.15 ± 0.19
Denervation	Taurine (10 mM)	6.13 ± 0.92

[a]Glands were preincubated for 48 hr, then treated for 4 hr in culture. Values are mean ($\pm$ SE) for three glands for controls and four glands for taurine-treated glands.

the cell. Alternatively, the release of taurine may only be a passive reflection of membrane hyperpolarization.

Taurine might also function in the extracellular space. Perhaps an acute increase in the extracellular concentration of taurine will alter β-adrenoreceptor function or the uptake of catecholamines into nerve endings. This idea is supported by our finding that taurine can mimic the effects of norepinephrine by interacting with β-adrenergic receptors.

ACKNOWLEDGMENT

We would like to thank the National Institute of Child Health and Human Development for financial support provided to Dr. G. H. T. Wheler, as a Visiting Fellow at the National Institutes of Health.

REFERENCES

Agrawal, H. C.; Davison, A. N.; and Kaczmarek, L. K. Subcellular distribution of taurine and cysteinsulphinate decarboxylase in developing rat brain. *Biochem. J.,* 122, 759–763 (1971).

Axelrod, J.; Wurtman, R.; and Kelly, D. E. In *The Pineal Gland,* Wurtman, R. J.; Axelrod, J.; and Kelly, D. E., eds. Academic Press, New York (1968), p. 8.

Axelrod, J.; and Zatz, M. The β-adrenergic receptor and the regulation of circadian rhythms in the pineal gland. In *Biochemical Actions of Hormones,* Littwack, G., ed. Academic Press, New York (1977), pp. 249–268.

Baskin, S. I.; and Dagirmajian, R. The effect of taurine on the pigmentation of the bullfrog tadpole. *Comp. Biochem. Physiol.,* 44A, 297–302 (1973).

Bradford, H. F.; Davison, A. N.; and Wheler, G. H. T. Taurine and synaptic transmission. In *Taurine,* Huxtable, R.; and Barbeau, A., eds. Raven Press, New York (1976), pp. 303–310.

Crabai, H.; Sitzer, A.; and Pepeu, G. Taurine concentrations in the neurohypophysis of different species. *J. Neurochem.,* 1091–1092 (1974).

Davison, A. N.; and Kaczmarek, L. K. Taurine—A possible neurotransmitter. *Nature,* 234, 107–108 (1971).

De Belleroche, J. S.; and Bradford, H. F. Amino acids in synaptic vesicles from mammalian cerebral cortex: A reappraisal. *J. Neurochem.,* 21, 441–451 (1973).

Green, J. P.; Day, M.; and Robinson, J. D. Some acidic substances in neoplastic mast cells and in the pineal body. *Eiochem. Pharmac.,* 11, 957–960 (1962).

Greengard, P. Possible role for cyclic nucleotides and phosphorylated membrane proteins in postsynaptic actions of neurotransmitters. *Nature,* 260, 101–108 (1976).

Grosso, D. A.; Bressler, R.; and Benson, B. Circadian rhythm and uptake of taurine by the rat pineal gland. *Life Sci.,* 22, 1789–1798 (1978).

Guidoti, A.; Badiani, G.; and Pepeu, G. Taurine distribution in the cat brain. *J. Neurochem.,* 19, 431–435 (1972).

Klein, D. C. Circadian rhythms in indole metabolism in the rat pineal gland. In *The Neurosciences, Third Study Program,* Schmidt, F. O., ed MIT Press, Cambridge, Mass. (1974), pp. 509–511.

Klein, D. C. The pineal gland: A model of neuroendocrine regulation. In *The Hypothalamus,*

Reichlin, S.; Baldessarini, R. J.; and Martin, J. B., eds Raven Press, New York (1978), pp. 303–327.

Klein, D. C.; and Berg, G. R. Pineal gland: Stimulation of melatonin production by norepinephrine involves cyclic AMP-mediated stimulation of N-acetyltransferase. In *Role of Cyclic AMP in Cell Function*, Greengard, P.; and Costa, E., eds. Advances in Biochemical Psychopharmacology, Vol. 3. Raven Press, New York (1970), pp. 241–263.

Kocsis J. J.; Wheler G. H. T.; and Klein, D. C. Unpublished observations (1979).

Krusz, J. C.; Dix, R. K., and Baskin, S. I. Factors that affect the uptake and endogenous content of taurine in the pineal gland. *Fed. Proc.*, 37, 907 (1977).

La Bella, F.; Vivian, S.; and Queen, G. Abundance of cystathionine in the pineal body: Free amino acids and related compounds of bovine pineal, anterior and posterior pituitary and brain. *Biochem. Biophys. Acta.*, 158, 286–288 (1968).

Leonard, B. E.; Neuhoff, V.; and Tongue, S. R. The effect of chronic administration of D-amphetamine upon circadian changes in amino acids in the pineal gland and pituitary glands of the rat. *J. Neurosci. Res.*, 1, 83–92 (1977).

Mandel, P.; and Pasantes-Morales, H. Taurine in the nervous system. In *Reviews of Neuroscience*, Ehrepreis, S.; and Kopin, I., eds Raven Press, New York (1978), pp. 157–193.

Muramatsu, M.; Kakita, K.; Nakagawa, K.; and Kuriyama, K. A modulating role of taurine on release of acetylcholine and norepinephrine from neural tissues. *Japan J. Pharmacol.*, 28, 259–268 (1978).

Nir, I.; Briel, G.; Dames, W.; and Neuhoff, V. Pineal proteins and free amino acids during ontogenesis in rats. *Neuroendocrinol.*, 14, 34 (1974).

Parfitt, A.; Weller, J. L.; Klein, D. C.; Sakai, K. K.; and Marks, B. H. Blockade by ouabain or elevated potassium concentration of the adrenergic and cAMP-induced stimulation of pineal serotonin N-acetyltransferase activity. *Molec. Pharmacol.*, 11, 241–255 (1975).

Phillis, J. W. An involvement of calcium and Na, K-ATPase in the inhibitory actions of various compounds on central neurons. In *Taurine*, Huxtable, R.; and Barbeau, A., eds. Raven Press, New York (1976), pp. 209–223.

Sakai, K. K.; and Marks, B. H. Adrenergic effects on pineal cell membrane potential. *Life Sci.*, 11, 285–291 (1972).

Vellan, E. J.; Gjessing, L. R.; and Stalsberg, Free amino acids in the pineal and pituitary glands of human brain. *J. Neurochem.*, 17, 699–701 (1970).

Wheler, G. H. T.; and Klein, D. C. Taurine release from the pineal gland is stimulated via a β-adrenergic mechanism. *Brain Res.*, 187, 155–164 (1980).

Wheler, G. H. T.; and Klein, D. C. Cyclic AMP-induced release of [^{14}C]taurine from pinealocytes. *Biochim. Biophys. Res. Comm.*, 90, 22–27 (1979b).

Wheler, G. H. T.; Weller, J. L.; and Klein, D. C. Taurine: Stimulation of N-acetyltransferase activity and melatonin production via a beta-adrenergic mechanism. *Brain Res.*, 166, 65–74 (1979a).

Wheler, G. H. T.; Bradford, H. F.; Davison, A. N.; and Thompson, E. J. Uptake and release of taurine from cerebral cortex slices and their subcellular compartments. *J. Neurochem.*, 33, 331–337 (1979).

Copyright © 1981, Spectrum Publications, Inc.
The Effects of Taurine on Excitable Tissues

CHAPTER 11

Enrichment of Taurine in Synaptosomes and Synaptic Vesicles of Bovine Brain Regions

S. S. Oja
K.-M. Marnela
P. Kontro

The possible role of taurine in synapses—an inhibitory transmitter or modulator—is not yet settled (Oja et al., 1977; Oja and Kontro, 1978). To be able to participate in synaptic transmission, taurine must be present and preferably enriched in the synaptic structures. At least isolated synaptosomes contain appreciable amounts of other putative amino acid neurotransmitters (Sellström et al., 1975; Osborne et al., 1976), but no unanimity obtains with regard to synaptic vesicles. Some workers dispute the existence of any significant amounts of amino acids (Mangan and Whittaker, 1966; Rassin, 1972), whereas some others report the presence of a number of amino acids, including taurine, in synaptic vesicles isolated from the cerebral cortices (De Belleroche and Bradford, 1973; Lähdesmäki et al., 1977). Here we have systematically gauged the amino acid patterns of synaptosomes and synaptic vesicles of functionally different brain areas, using a sensitive amino acid analyzer with good resolution properties. Furthermore, particular care was taken in our analyses to separate taurine reliably from other interfering ninhydrin-positive compounds.

MATERIAL AND METHODS

The bovine brain areas studied were frontal, occipital, and parietal cerebral cortices; cerebellar cortex; caudate and lenticular nuclei; superior colliculi; thalamus; pons; and medulla. The brains were excised from the slaughtered animals and immediately cooled in crushed ice. A minor part of each sample was homogenized in cold 10% trichloroacetic acid solution, centrifuged for 10 min at 1,000 g, and the resulting precipitate was used for protein determination according to Lowry et al. (1951), the supernatant for amino acid analysis. Synaptosomes and synaptic vesicles were isolated with discontinuous sucrose gradients from the major part of the tissue sample according to Whittaker and Barker (1972). In order to increase the purity of synaptic vesicles, an additional 0.3 mol/liter sucrose layer was included in the original gradient. The crude vesicle fraction was further purified by washing once in 5 mmol/liter Tris-EDTA buffer, pH 7.4 (Whittaker, 1969). Samples for protein determinations and amino acid analyses were prepared as above. Some samples were also taken from the other synaptosomal subfractions. Oxygen uptake of the synaptosome fractions was checked as described by Kontro and Oja (1978). It remained constant upon incubation for at least 2 hr.

Synaptosome samples for electron microscopy (Jeol JEM 100C) were prepared as described by Kontro and Oja (1978), and the vesicle fractions, according to Whittaker and Barker (1972). The synaptosome samples were relatively free of membranes or other impurities. Only the fractions obtained from the brain stem structures were morphologically less homogeneous, the pontine samples being of lowest quality. In all samples, however, synaptosomal particles dominated the picture. The synaptic vesicle fractions were all slightly contaminated with membrane structures and occasionally with myelin fragments. The vesicle preparations of the pons were again most heavily contaminated. The membrane fragment impurities can hardly have contained any significant amounts of free amino acids.

Amino acid analyses were made according to Perry et al. (1968) with a Beckman Multichrom M Amino Acid Analyzer equipped with a Honeywell two-channel recorder and an Autolab System AA computing integrator for simultaneous measurements at 440 and 570 nm. To improve the resolution of the strongly acidic ninhydrin-positive compounds at the beginning of the analysis, a second analysis of this region was made with a reduced flow (10 ml/hr) and lowered pH (2.2) of the first eluent buffer. In this way taurine and the phosphates of glyceroethanolamine and ethanolamine were separated. Each sample was analyzed from two to eight times, using different dilutions when necessary. A number of samples were hydrolyzed in sealed ampoules for 24 hr at 383°K with 6 mol/liter HCl, and then subjected again to amino acid analyses.

RESULTS

Altogether 23 ninhydrin-positive compounds were quantitatively analysed, viz., phosphoserine, glyceroethanolamine phosphate, taurine, ethanolamine phosphate, aspartic acid, threonine, serine, glutamic acid, glutamine,
proline, glycine, alanine, valine, cysteine, methionine, isoleucine, leucine,
tyrosine, phenylalanine, γ-aminobutyric acid (GABA), lysine, histidine, and
arginine. The most abundant amino acid in tissue samples in all brain areas
studied was glutamic acid, followed by glutamine, aspartic acid, and GABA
or taurine, depending on the brain region. The concentrations of the amino
acids with postulated synaptic function—viz., taurine, GABA, glycine, glutamic acid, and aspartic acid—are compiled in Table 11.1 with alanine,
serine, and threonine as reference. The taurine content in the frontal cerebral
cortex, pons, and medulla was higher than in the other regions when the
results were calculated per tissue fresh weight.

The amino acid composition of the synaptosomal fractions differed only
slightly from that of the corresponding tissue samples. The most abundant
amino acids were still glutamic acid, taurine, glutamine, and aspartic acid in
almost every brain area. The concentration of taurine was particularly high in
synaptosomes from the caudate and lenticular nuclei (Table 11.1). Also, medullary synaptosomes contained a relatively large amount of taurine. In synaptic vesicles, on the other hand, the free amino acid composition was rather
peculiar. The dominating amino acid in nearly all isolated synaptic vesicle
fractions was taurine, followed by glutamic acid. The concentrations of glutamine, GABA, and aspartic acid were the next, but they were generally
much smaller. Particularly high amounts of taurine were found in vesicles
from the medulla, but vesicles from the superior colliculus and pons contained appreciable amounts of taurine as well. In the cerebellar cortex the
taurine content was lowest.

Table 11.2 sets out the amino acid concentrations of the isolated synaptosome fractions as percentages of the corresponding tissue concentrations. In
order to make such comparisons feasible, the measured amino acid concentrations were first recalculated per protein content in all samples. It appears
that the synaptosomal fractions were nearly always depleted of amino acids.
On only two occasions did the concentration in the isolated synaptosomes
clearly exceed the corresponding tissue concentration, viz., in the case of taurine in the medulla and lenticular nucleus. In two further regions, caudate
nucleus and superior colliculus, the taurine concentrations were of the same
order of magnitude in both samples. In only three random cases was the synaptosomal concentration of any of the seven other amino acids not significantly lower than the tissue concentration.

The enrichment of taurine in synaptic vesicles was striking (Table 11.3).
The vesicular taurine concentration was never less than the corresponding tis-

Table 11.1 Some free amino acids in tissues and synaptosome and synaptic vesicle fractions of bovine brain areas[a]

Amino acid	Concentration (μmol/kg fresh brain)			Concentration (μmol/kg fresh brain)		
	Tissue	Synaptosomes	Vesicles	Tissue	Synaptosomes	Vesicles
	Frontal cerebral cortex			Occipital cerebral cortex		
Taurine	2,400 ± 150	72.3 ± 15.0	6.16 ± 0.84	1,065 ± 15	63.4 ± 10.5	3.36 ± 0.54
GABA	1,300 ± 152	25.9 ± 4.4	0.60 ± 0.02	2,168 ± 83	21.0 ± 3.8	0.14 ± 0.03
Glycine	1,528 ± 225	26.5 ± 4.3	0.47 ± 0.15	1,680 ± 260	29.1 ± 11.0	0.21 ± 0.02
Glutamic acid	6,475 ± 391	132.4 ± 20.0	2.41 ± 0.36	6,044 ± 628	137.2 ± 10.5	0.72 ± 0.21
Aspartic acid	1,799 ± 212	38.0 ± 9.0	0.81 ± 0.77	2,029 ± 272	71.7 ± 2.6	0.30 ± 0.04
Alanine	1,506 ± 96	18.2 ± 3.0	0.62 ± 0.09	1,705 ± 98	40.0 ± 8.2	0.21 ± 0.04
Serine	855 ± 95	20.3 ± 3.4	0.52 ± 0.07	1,238 ± 127	26.4 ± 10.9	0.33 ± 0.08
Threonine	302 ± 54	4.3 ± 1.0	0.10 ± 0.02	425 ± 78	4.9 ± 1.7	0.19 ± 0.03
	[3]	[4]	[3]	[6]	[5]	[3]
	Parietal cerebral cortex			Cerebellar cortex		
Taurine	1,283 ± 250	52.7 ± 6.1	3.84 ± 0.70	1,125 ± 116	63.8 ± 13.0	2.68 ± 0.52
GABA	1,675 ± 237	40.3 ± 8.4	0.56 ± 0.14	1,045 ± 121	23.6 ± 0.5	0.68 ± 0.07
Glycine	1,708 ± 202	38.7 ± 7.8	0.54 ± 0.05	1,031 ± 63	29.5 ± 0.8	0.46 ± 0.02
Glutamic acid	5,799 ± 349	175.7 ± 35.6	2.08 ± 0.21	5,591 ± 239	125.9 ± 3.3	1.80 ± 0.23
Aspartic acid	1,906 ± 204	72.4 ± 12.0	0.76 ± 0.10	1,216 ± 45	37.5 ± 1.4	0.63 ± 0.05
Alanine	1,517 ± 146	31.2 ± 8.1	0.50 ± 0.05	1,300 ± 92	34.5 ± 5.8	0.63 ± 0.05
Serine	957 ± 79	34.3 ± 6.4	0.45 ± 0.04	517 ± 40	24.5 ± 2.2	0.44 ± 0.01
Threonine	375 ± 62	9.1 ± 2.5	0.10 ± 0.01	234 ± 21	10.0 ± 0.9	0.09 ± 0.01
	[5]	[7]	[3]	[10]	[6]	[3]

(Continued)

Table 11.1 Some free amino acids in tissues and synaptosome and synaptic vesicle fractions of bovine brain areas^a (*Continued*)

Concentration (μmol/kg fresh brain)

Amino acid	Caudate nucleus			Thalamus		
	Tissue	Synaptosomes	Vesicles	Tissue	Synaptosomes	Vesicles
Taurine	1,025 ± 192	189.1 ± 14.4	4.21 ± 0.37	893 ± 73	40.3 ± 3.9	5.12 ± 0.32
GABA	3,235 ± 312	228.0 ± 9.8	2.18 ± 0.05	1,567 ± 165	36.2 ± 12.6	0.56 ± 0.18
Glycine	1,619 ± 248	48.7 ± 4.2	0.81 ± 0.06	1,155 ± 182	30.1 ± 8.6	0.62 ± 0.19
Glutamic acid	5,234 ± 140	297.0 ± 30.2	2.20 ± 0.26	5,617 ± 241	89.9 ± 12.7	1.72 ± 0.46
Aspartic acid	1,609 ± 53	96.0 ± 4.0	1.29 ± 0.23	2,047 ± 70	46.1 ± 14.7	0.78 ± 0.18
Alanine	1,175 ± 68	49.3 ± 0.8	0.71 ± 0.07	1,104 ± 85	24.2 ± 6.8	0.50 ± 0.08
Serine	1,180 ± 212	38.7 ± 3.2	0.69 ± 0.08	417 ± 31	18.3 ± 4.2	0.45 ± 0.06
Threonine	226 ± 11	10.7 ± 1.2	0.22 ± 0.07	201 ± 19	6.3 ± 1.3	0.17 ± 0.05
	[7]	[3]	[4]	[8]	[4]	[4]

Concentration (μmol/kg fresh brain)

Amino acid	Lenticular nucleus			Superior colliculus		
	Tissue	Synaptosomes	Vesicles	Tissue	Synaptosomes	Vesicles
Taurine	1,235 ± 91	264.2 ± 48.0	4.34 ± 0.68	1,116 ± 154	44.9 ± 5.7	7.96 ± 1.01
GABA	2,119 ± 351	294.0 ± 58.9	1.21 ± 0.36	3,096 ± 162	15.3 ± 3.1	1.86 ± 0.36
Glycine	1,416 ± 198	51.3 ± 7.2	0.53 ± 0.08	1,272 ± 76	14.1 ± 4.5	0.71 ± 0.07
Glutamic acid	5,411 ± 420	329.7 ± 65.2	1.81 ± 0.47	4,040 ± 136	53.2 ± 9.1	1.80 ± 0.02
Aspartic acid	1,549 ± 79	130.0 ± 26.2	1.11 ± 0.24	2,297 ± 125	19.7 ± 7.1	1.15 ± 0.06
Alanine	1,233 ± 56	49.7 ± 15.8	0.60 ± 0.16	1,154 ± 57	12.3 ± 3.7	0.70 ± 0.06
Serine	645 ± 34	57.0 ± 15.2	0.71 ± 0.14	391 ± 35	7.6 ± 0.4	0.46 ± 0.12
Threonine	241 ± 20	8.0 ± 3.3	0.16 ± 0.01	265 ± 19	2.0 ± 0.4	0.15 ± 0.01
	[9]	[5]	[3]	[11]	[4]	[3]

(*Continued*)

Table 11.1 Some free amino acids in tissues and synaptosome and synaptic vesicle fractions of bovine brain areas[a] (Continued)

Amino acid	Concentration (μmol/kg fresh brain)			Concentration (μmol/kg fresh brain)		
	Tissue	Synapto-somes	Vesicles	Tissue	Synapto-somes	Vesicles
	Pons			Medulla		
Taurine	1,739 ± 222	48.0 ± 8.6	7.68 ± 1.52	1,546 ± 242	110.6 ± 14.4	13.46 ± 1.70
GABA	588 ± 112	7.8 ± 2.8	1.15 ± 0.35	640 ± 29	8.5 ± 2.1	0.77 ± 0.16
Glycine	1,259 ± 81	36.8 ± 9.3	1.68 ± 0.13	1,802 ± 80	9.1 ± 1.1	1.81 ± 0.20
Glutamic acid	4,217 ± 189	53.9 ± 15.9	4.30 ± 0.99	4,281 ± 246	27.0 ± 2.4	3.85 ± 0.54
Aspartic acid	1,991 ± 187	27.2 ± 4.3	2.48 ± 0.45	1,993 ± 86	11.8 ± 0.8	2.50 ± 0.44
Alanine	1,088 ± 56	24.3 ± 5.8	1.25 ± 0.13	1,333 ± 119	5.9 ± 0.4	1.26 ± 0.14
Serine	366 ± 43	31.5 ± 1.5	0.66 ± 0.05	462 ± 53	4.8 ± 0.4	0.95 ± 0.11
Threonine	173 ± 14	9.4 ± 1.8	0.26 ± 0.03	249 ± 28	2.4 ± 0.8	0.19 ± 0.05
	[6]	[5]	[4]	[10]	[4]	[4]

[a]Means ± SEM are given. The number of isolation procedures for each series is given in brackets.

Table 11.2 Enrichment of taurine in synaptosomes of different bovine brain areas in comparison with other amino acids[a]

Brain area	No. of experiments	Percent of the corresponding tissue concentration							
		Taurine	GABA	Glycine	Glutamic acid	Aspartic acid	Alanine	Serine	Threonine
Frontal cerebral cortex	4	43 ± 9	25 ± 3	28 ± 4	26 ± 2	28 ± 3	16 ± 1	40 ± 5	27 ± 3
Occipital cerebral cortex	5	59 ± 8	7 ± 1	10 ± 1	15 ± 1	13 ± 2	15 ± 1	8 ± 1	3 ± 1
Parietal cerebral cortex	7	36 ± 7	16 ± 3	16 ± 4	20 ± 4	14 ± 4	15 ± 3	13 ± 1	15 ± 5
Cerebellar cortex	6	36 ± 6	28 ± 5	40 ± 12	30 ± 7	39 ± 10	33 ± 8	61 ± 19	59 ± 15
Caudate nucleus	3	105 ± 12	44 ± 5	28 ± 5	37 ± 8	36 ± 3	26 ± 5	32 ± 9	27 ± 8
Lenticular nucleus	5	164 ± 38	121 ± 30	41 ± 11	39 ± 13	64 ± 19	38 ± 10	32 ± 5	—
Thalamus	4	40 ± 6	23 ± 5	30 ± 4	26 ± 2	28 ± 5	25 ± 3	67 ± 15	44 ± 8
Superior colliculus	4	103 ± 25	22 ± 5	67 ± 7	58 ± 11	36 ± 4	61 ± 12	145 ± 44	57 ± 18
Pons	5	66 ± 22	42 ± 9	90 ± 13	34 ± 7	44 ± 5	66 ± 22	36 ± 7	66 ± 12
Medulla	4	220 ± 42	37 ± 9	25 ± 6	32 ± 7	28 ± 7	26 ± 3	41 ± 13	40 ± 7

[a]The results (means ± SEM) are calculated from the molar amino acid concentrations per kilogram of protein in samples.

Table 11.3 Enrichment of taurine in synaptic vesicles of different bovine brain areas in comparison with other amino acids[a]

Brain area	No. of experiments	Percent of the corresponding tissue concentration							
		Taurine	GABA	Glycine	Glutamic acid	Aspartic acid	Alanine	Serine	Threonine
Frontal cerebral cortex	3	218 ± 45	38 ± 2	29 ± 4	25 ± 2	34 ± 4	31 ± 2	46 ± 4	30 ± 4
Occipital cerebral cortex	3	167 ± 30	6 ± 1	5 ± 1	12 ± 1	8 ± 1	7 ± 1	11 ± 1	17 ± 2
Parietal cerebral cortex	3	117 ± 39	36 ± 3	28 ± 9	28 ± 5	19 ± 5	30 ± 9	29 ± 10	29 ± 9
Cerebellar cortex	3	147 ± 25	65 ± 16	28 ± 2	33 ± 6	46 ± 13	26 ± 2	77 ± 26	28 ± 7
Caudate nucleus	4	106 ± 16	31 ± 4	26 ± 6	21 ± 3	37 ± 8	23 ± 6	33 ± 12	27 ± 12
Lenticular nucleus	3	381 ± 72	27 ± 7	25 ± 5	12 ± 3	34 ± 3	23 ± 4	87 ± 20	54 ± 11
Thalamus	4	208 ± 42	14 ± 4	18 ± 6	16 ± 4	16 ± 3	20 ± 2	52 ± 13	44 ± 8
Superior colliculus	3	588 ± 97	67 ± 16	48 ± 7	39 ± 2	47 ± 3	56 ± 4	122 ± 23	83 ± 7
Pons	4	110 ± 26	69 ± 13	51 ± 6	43 ± 8	65 ± 13	45 ± 8	73 ± 13	57 ± 9
Medulla	4	276 ± 66	43 ± 13	28 ± 5	29 ± 5	31 ± 6	31 ± 5	52 ± 12	27 ± 5

[a]The results (means ± SEM) are calculated from the molar amino acid concentrations per kilogram of protein in samples.

sue concentration, and it was often even significantly ($p < 0.05$) higher. In the superior colliculus, lenticular nucleus, medulla, frontal cerebral cortex, and thalamus, the synaptic vesicle fractions had more than twice the taurine content of the original tissue samples. On the other hand, the vesicular concentrations of all other amino acids were generally smaller and in no case larger than the tissue concentrations. A comparison of Tables 11.2 and 11.3 suggests that GABA may possess a slight tendency to be more concentrated in the synaptic vesicles than in the synaptoplasm in some brain areas. Taurine is the only amino acid, however, for which such a conclusion can be safely drawn. This inference was also directly confirmed when the amino acids of the synaptoplasmic fractions were analyzed for the control.

DISCUSSION

All tissue, synaptosome, and synaptic vesicle samples contained the same amino acids, albeit in varying amounts. We detected more ninhydrin-positive compounds also in synaptic vesicle fractions than earlier investigators (Rassin, 1972; De Belleroche and Bradford, 1973; Lähdesmäki et al., 1977; Zisapel and Zurgil, 1978). This greater number does not merely reflect a heavier contamination with synaptoplasmic amino acids, since the general amino acid patterns in our synaptosome and synaptoplasm samples were not identical to those of vesicle fractions. Furthermore, the vesicular amino acid pattern remained unaltered when some isolated synaptic vesicles were additionally gel filtrated through Sephadex G-50 columns to eliminate the small-molecular synaptoplasmic contaminants as completely as possible.

The most apparent source of error in the present study is the possible leakage of amino acids out of synaptosomes and synaptic vesicles during their isolation. With more elaborate and lengthy procedures, the purity of the fractions would possibly have further improved, with the expense of loss of material. Our yield of synaptosomal protein was about 3 g/kg fresh tissue in samples from the lower brain stem and 10 g/kg in the others, and that of vesicular protein was generally from 0.1 to 0.2 g/kg. Our figures match earlier theoretical estimates or experimental yields from the mammalian cerebral cortex (Swanson et al., 1973; Lähdesmäki et al., 1977; Nagy et al., 1977). Even if macromolecules may have been better retained, amino acids have partially leaked out, as Tables 11.2 and 11.3 strongly suggest. It has never been established whether or not the neurotransmitter amino acids are stored in synaptic vesicles, even if earlier investigators have also recovered the major part of them in synaptoplasmic rather than vesicular fractions (Mangan and Whittaker, 1966; Rassin, 1972; Rassin et al., 1977; Zisapel and Zurgil, 1978). Our results likewise do not entirely rule out the possibility—though they make it unlikely—that synaptic vesicles would act as storage sites. It should be kept in mind that only a part of the total amount of an amino acid may serve as transmitter, the other part being more involved in intermediary metabolism in general.

In bovine brain tissue the level of taurine was lower than in most other species. Nor were the regional distributions similar: In the bovine pons and medulla the levels were rather high, but in the rat, the brain stem contains, for instance, relatively less taurine (Piha et al., 1962; Collins, 1974). In rat brain synaptosomes the taurine level is also highest in the cerebral cortex or cerebellum (Lombardini, 1976). Taurine differed from all other amino acids, being distinctly enriched in synaptic vesicle fractions. It is obviously more closely associated with or more tightly bound to the vesicles, and it was not lost, or at least was lost to a much lesser degree, during isolation procedures than the other amino acids. A striking enrichment of taurine in synaptic structures in any particular brain area would have suggested that taurine acts as a synaptic transmitter or modulator precisely there. For instance, taurine has recently been proposed as a transmitter in the cerebellum (Frederickson et al., 1978; Okamoto and Namima, 1978). We dare not draw any such conclusions from the present data, however. In spite of some significant differences among the brain areas studied, taurine was enriched in vesicles virtually everywhere. To date, the enrichment in the cerebellum was not strikingly high. In view of the present results, taurine is a ubiquitous constituent of a great number of synaptic vesicles, being possibly bound to vesicular and/or other synaptic membranes. The modulatory interactions of taurine with such membranes may then underlie its inhibitory effects in electrically excitable tissues.

SUMMARY

Taurine, together with other free amino acids, was quantitatively gauged in intact tissues and isolated synaptosome and synaptic vesicle fractions of different bovine brain areas with the aid of a sensitive amino acid analyzer. The brain areas studied were frontal, parietal, and occipital cerebral cortices; cerebellar cortex; caudate and lenticular nuclei; superior colliculi; thalamus; pons; and medulla. The most abundant amino acid in tissue and synaptosome samples was glutamic acid, followed by glutamine, aspartic acid, GABA, and taurine. Taurine was the only amino acid, however, which was enriched in isolated synaptic vesicle fractions. Such an enrichment occurred in samples from all brain areas. It is therefore suggested that taurine is a ubiquitous associate of synaptic membrane structures rather than a specific inhibitory neurotransmitter.

REFERENCES

Collins, G. G. S. The rates of synthesis, uptake and disappearance of [^{14}C]-taurine in eight areas of the rat central nervous system. *Brain Res.*, 76, 447–459 (1974).

De Belleroche, J. S.; and Bradford, H. F. Amino acids in synaptic vesicles from mammalian cerebral cortex: A reappraisal. *J. Neurochem.*, 21, 441–451 (1973).

Frederickson, R. C. A.; Neuss, M.; Morzorati, S. L.; and McBride, W. J. A comparison of the inhibitory effects of taurine and GABA on identified Purkinje cells and other neurons in the cerebellar cortex of the rat. *Brain Res.*, 145, 117–126 (1978).

Kontro, P.; and Oja, S. S. Taurine uptake by rat brain synaptosomes. *J. Neurochem.*, 30, 1297–1304 (1978).

Lähdesmäki, P.; Karppinen, A.; Saarni, H.; and Winter, R. Amino acids in the synaptic vesicle fraction from calf brain: Content, uptake and metabolism. *Brain Res.*, 138, 295–308 (1977).

Lombardini, J. B. Regional and subcellular studies on taurine in the rat central nervous system. In *Taurine*, R. Huxtable and A. Barbeau, eds. Raven Press, New York (1976), pp. 311–326.

Lowry, O. H.; Rosebrough, N. J.; Farr, A. L.; and Randall, R. J. Protein measurement with the Folin phenol reagent. *J. Biol. Chem.*, 193, 265–275 (1951).

Mangan, J. L.; and Whittaker, V. P. The distribution of free amino acids in subcellular fractions of guinea-pig brain. *Biochem. J.*, 98, 128–137 (1966).

Nagy, A.; Várady, Gy.; Joó, F.; Rakonczay, Z.; and Pilc, A. Separation of acetylcholine and catecholamine containing synaptic vesicles from brain cortex. *J. Neurochem.*, 29, 449–459 (1977).

Oja, S. S.; and Kontro, P. Neurotransmitter actions of taurine in the central nervous system. In *Taurine and Neurological Disorders*, A. Barbeau and R. J. Huxtable, eds. Raven Press, New York (1978), pp. 181–200.

Oja, S. S.; Kontro, P.; and Lähdesmäki, P. Amino acids as inhibitory neurotransmitters. *Progr. Pharmacol.*, 1(3), 1–119 (1977).

Okamoto, K.; and Namima, M. Uptake, release and homo- and hetero-exchange diffusions of inhibitory amino acids in guinea-pig cerebellar slices. *J. Neurochem.*, 31, 1393–1402 (1978).

Osborne, R. H.; Duce, I. R.; and Keen, P. Amino acids in "light" and "heavy" synaptosome fractions from rat olfactory lobes and their release by electrical stimulation. *J. Neurochem.* 27, 1483–1488 (1976).

Perry, T. L.; Stedman, D.; and Hansen, S. A versatile lithium buffer elution system for single column automatic amino acid chromatography. *J. Chromatogr.*, 38, 460–466 (1968).

Piha, R. S.; Oja, S. S.; and Uusitalo, A. J. The effect of chlorpromazine on free amino acids in the rat brain. *Ann. Med. Exp. Biol. Fenn.*, 40, Suppl. 5, 1–28 (1962).

Rassin, D. K. Amino acids as putative transmitters: Failure to bind to synaptic vesicles of guinea-pig cerebral cortex. *J. Neurochem.*, 19, 139–148 (1972).

Rassin, D. K.; Sturman, J. A.; and Gaull, G. E. Taurine in developing rat brain: Subcellular distribution and association with synaptic vesicles of [^{35}S]taurine in maternal, fetal and neonatal rat brain. *J. Neurochem.*, 28, 41–50 (1977).

Sellström, Å.; Sjöberg, L.-B.; and Hamberger, A. Neuronal and glial systems for γ-aminobutyric acid metabolism. *J. Neurochem.*, 25, 393–398 (1975).

Swanson, P. D.; Harvey, F. H.; and Stahl, W. L. Subcellular fractionation of postmortem brain. *J. Neurochem.*, 20, 465–475 (1973).

Whittaker, V. P. The synaptosome. In *Handbook of Neurochemistry*, Vol. 2, A. Lajtha, ed. Plenum Press, New York (1969), pp. 327–334.

Whittaker, V. P.; and Barker, L. A. The subcellular fractionation of brain tissue with special reference to the preparation of synaptosomes and their component organelles. In *Methods of Neurochemistry*, Vol. 2, R. Fried, ed. Marcel Dekker, New York (1972), pp. 1–52.

Zisapel, N.; and Zurgil, N. The content of GABA and other amino acids in bovine cerebral cortex synaptic vesicles. *Life Sci.*, 23, 231–236 (1978).

Copyright © 1981, Spectrum Publications, Inc.
The Effects of Taurine on Excitable Tissues

CHAPTER 12

Subcellular Distribution of Intracerebrally Injected Taurine and Certain Other Amino Acids in Mouse Brain

Elsa Kumpulainen
Anneli Karppinen
Kirsi-Marja Marnela
P. Lähdesmäki

The possible role of taurine as a synaptically active substance has recently been investigated; however, its function still remains unclear. Taurine satisfies some of the criteria set for a neurotransmitter (Oja and Lähdesmäki, 1974), but so do a number of other amino acid transmitter candidates. Perhaps only γ-aminobutyric acid (GABA) fulfills all the essential criteria.

Two of the most important criteria for a transmitter are exchange with the intravesicular transmitter pool and release from and uptake into the storage vesicles of nerve terminals. Rassin et al. (1977, 1978) were unable to show any vesicular accumulation of [^{35}S]taurine administered either intravenously or through the diet; however, intracerebral injection may provide a better opportunity for investigating the entry of amino acids into brain cells, since it avoids the complications caused by the blood-brain barrier. We had observed earlier (Karppinen et al., 1979) the distribution of intracerebrally injected [^{35}S]taurine, [^{3}H]lysine, [^{14}C]glutamic acid, and [^{14}C]norleucine in mouse brain subcellular fractions. We have determined the endogenous amino acid concentrations in these fractions and will discuss the accumulation of intracerebrally injected amino acids based on an examination of the ratio of labeled/endogenous amino acids in the fractions.

MATERIAL AND METHODS

A total of 360 adult female NMRI white mice were used. These were anaesthetized by an intramuscular injection of 200 μl of 10% (w/v) chloral hydrate and immobilized on a small animal rack. The skull was opened by drilling a small hole, and one μl of radioactive amino acid was injected into the brains to a depth of 2 mm. These operations are detailed elsewhere (Karppinen et al., 1979). The four labeled amino acids (Radiochemical Centre, Amersham) chosen for injection were: (1) [^{35}S]taurine (specific activity SA, 550 Ci/mol, 20 μCi, and 56 nmol/animal); (2) [^{3}H]lysine (SA, 18 Ci/mmol, 1 μCi, and 60 pmol/animal); (3) [^{14}C]glutamic acid (SA, 275 Ci/mol, 500 nCi, and 1.25 nmol/animal); and (4) [^{14}C]norleucine (SA, 300 Ci/mol, 250 nCi, and 2 nmol/animal).

Twenty hours after injection the mice were killed by decapitation and the brains removed, pooled, weighed, and immediately homogenized in cold 0.32 M sucrose. The homogenates were fractionated according to the method of Whittaker and Barker (1972). The total homogenate was first divided into three fractions (crude nuclear fraction, crude mitochondrial fraction, and soluble cytoplasm, including microsomes). The crude mitochondrial fraction was further subdivided into three fractions (myelin, synaptosomes, and mitochondria). After hypo-osmotic shock and centrifugation, in which the myelin and large mitochondria were sedimented first (Whittaker, 1969), four additional fractions were obtained (soluble synaptoplasm, synaptic vesicles, pooled membrane fractions, and intraterminal mitochondria). Samples for protein and radioactivity determination were taken from each fraction as described elsewhere (Karppinen et al., 1979). Control experiments designed to eliminate the effect of homogenization on the distribution of label in the various subcellular fractions are also described elsewhere (Karppinen et al., 1979). The uniform distribution of radioactivity in the brain during the 20-hr experiments was verified by taking small samples throughout the brain and determining their radioactivity. For the determination of free amino acids in the mouse brain subcellular fractions, the proteins in the fractions were precipitated with 10 % sulfosalicylic acid, and the amino acids in the supernatants were analyzed in an automatic amino acid analyzer (Beckman Multichrom M Amino Acid Analyzer) equipped with a Honeywell two-channel recorder.

RESULTS

The concentrations of free amino acids in the mouse brain subcellular fractions are given in Table 12.1. The initial peaks of the chromatograms, which eluted between the phosphoserine and taurine fractions, were complicated by the appearance of some unidentified, relatively abundant, acidic compounds. The percent distribution of endogenous taurine, glutamic acid, and lysine is given in Table 12.2. The amount of amino acid in the homogenate of the brain was normalized to 100, and the distribution in all other fractions is expressed as a percent of the total homogenate content.

The percent distribution of the intracerebrally injected radioactivity of [^{35}S]taurine, [^{14}C]glutamic acid, [^{3}H]lysine, and [^{14}C]norleucine in the mouse brain subcellular fractions is given in Table 12.3. In order to compensate for the uneven distribution of radioactivity in the brain homogenates, the amount of label of each amino acid in the homogenate was set at 100, and their distribution was calculated as a percent of the total homogenate radioactivity. An excessively high recovery of radioactivity was observed in some experiments. For example, the amount of radioactive lysine and glutamate found in the myelin, mitochondrial, and synaptosomal fractions exceeded the amount found in the crude mitochondrial fraction, although the three subfractions were derived from the crude mitochondrial fraction. Similarly, excessively low recovery was noted in the synaptosomal subfractions.

Table 12.4 gives the ratios of labeled/endogenous taurine, glutamic acid, and lysine in the mouse brain subcellular fractions. Only the synaptosomal fraction seemed to accumulate radioactive amino acids. Glutamic acid was clearly conserved in the synaptic vesicles, but taurine and lysine remained mainly in the soluble synaptoplasm. On the other hand, taurine penetrated the synaptic vesicles very slowly. The ratio of labeled/endogenous glutamate in the soluble synaptoplasm was remarkably low.

DISCUSSION

The four amino acids examined in this study have different metabolic and neurophysiological properties. A neurotransmitter or modulator role has been proposed for taurine (Kaczmarek and Davison, 1972; Oja and Lähdesmäki, 1974), which is metabolically very stable (Lähdesmäki and Korhonen, 1978). Lysine is also metabolized slowly, but it is known to occur in synaptic vesicles (Lähdesmäki and Winter, 1977). Glutamic acid, which is metabolically labile, is apparently an excitatory neurotransmitter with effective trans-

Table 12.1.   Concentrations of endogenous amino acids in mouse brain subcellular fractions[a]

Amino acid	Concentration (μmol/g wet weight) in fractions										
	1	2	3	4	5	6	7	8	9	10	11
Taurine	9.734	1.236	1.226	5.548	0.185	0.273	0.467	0.350	0.0292	0.0155	0.0126
Phosphoserine (+ cysteic and cysteinesulfinic acids)	0.787	0.208	0.108	0.511	0.031	0.067	0.007	0.060	0.0003	0.0003	0.0001
Glycerylphosphoethanolamine (+ unknown)	1.249	0.457	0.286	0.804	0.011	0.230	0.023	0.074	0.0001	0.0005	0.0018
Phosphoethanolamine (+ unknown)	2.045	0.155	0.374	0.943	0.029	0.151	0.047	0.144	0.0003	0.0003	0.0051
Aspartic acid	6.481	1.437	1.059	4.296	0.167	0.485	0.091	0.277	0.0025	0.0017	0.0032
Serine	4.053	0.974	0.523	2.815	0.067	0.250	0.056	0.275	0.0012	0.0009	0.0012
Threonine	0.015	0.001	0.002	0.005	0.001	0.001	0.001	0.002	b	c	b
Glutamic acid	12.330	2.256	0.690	7.003	0.234	0.370	0.321	0.247	0.0061	0.0493	0.0061
Glutamine	7.536	2.125	0.852	3.872	0.129	0.252	0.026	0.109	b	b	0.0006
Glycine	1.404	0.253	0.227	1.072	0.026	0.067	0.026	0.074	0.0008	0.0005	0.0013
Alanine	1.054	0.158	0.155	0.911	0.021	0.048	0.022	0.077	0.0004	0.0004	0.0011
Valine	0.180	0.004	0.042	0.164	c	0.010	0.002	0.005	c	c	c
Leucine	0.005	c	b	c	c	c	b	0.005	c	c	c
Isoleucine	0.005	c	c	c	c	c	b	b	c	c	c
GABA	1.229	0.356	0.247	0.445	0.010	0.090	0.022	0.052	0.0001	0.0003	0.0003
Lysine	0.108	0.022	0.017	0.064	0.010	0.010	0.011	0.0004	0.0006	0.0007	0.0004
Histidine	0.020	0.001	0.005	0.001	b	b	0.001	0.006	b	c	b
Arginine	0.078	0.045	0.028	0.040	b	0.020	0.001	0.008	b	b	b
Cysteine	0.003	0.002	c	c	c	c	b	0.001	c	c	c
Methionine	0.003	c	c	c	c	c	b	c	c	c	c
Phenylalanine	0.005	c	0.001	c	c	b	b	b	c	c	c
Tyrosine	0.010	c	0.001	c	c	b	b	b	c	c	c
α-Aminoisobutyric acid (?)	0.001	c	0.002	c	c	b	b	c	c	c	c

[a]Results are means of five determinations. SD varied from 3 to 7%. Fractions: 1 = total homogenate, 2 = crude nuclear fraction, 3 = crude mitochondrial fraction, 4 = soluble cytoplasm, 5 = myelin, 6 = synaptosomes, 7 = mitochondria, 8 = soluble synaptoplasm, 9 = synaptic vesicles, 10 = synaptic membrane fraction, 11 = intraterminal mitochondria.
[b]Less than 0.001 (0.0001) μmol.
[c]Absent.

Table 12.2. Percent distribution of endogenous taurine, glutamic acid, and lysine in mouse brain subcellular fractions[a]

Subcellular fraction	Percent distribution		
	Taurine	Glutamic acid	Lysine
Total homogenate	100.0	100.0	100.0
Soluble cytoplasm	57.0	56.8	59.3
Crude nuclear fraction	12.7	18.3	20.4
Crude mitochondrial fraction	12.6	5.6	15.7
Myelin	1.9	1.9	9.3
Mitochondria	4.8	2.6	10.2
Synaptosomes	2.8	3.0	9.3
Soluble synaptoplasm	3.6	2.0	0.4
Synaptic vesicles	0.3	0.05	0.6
Synaptic membranes	0.16	0.4	0.6
Intraterminal mitochondria	0.13	0.05	0.4

[a]The values are obtained from Table 12.1. The amount of each amino acid in the total homogenate was taken as 100. The distribution in all other fractions is expressed as a percent of the total homogenate content.

port systems into the synaptosomes (Lähdesmäki et al., 1975) and synaptic vesicles (Lähdesmäki et al., 1977a), and with specific membrane receptors at the synaptic membranes (Lähdesmäki et al., 1977b). Norleucine is inactive in the central nervous system.

Examination of the subcellular recoveries detailed in Table 12.3 indicates that the most consistent data were obtained for norleucine and taurine. One exception is the amount of radioactive taurine in soluble synaptoplasm, which is almost the same as the amount found in the crude mitochondrial fraction. It is not known why taurine repeatedly accumulates in the synaptosomes after hypo-osmotic shock. Similarly, fractionation of the crude mitochondrial fractions yielded excessively high values of radioactive glutamate and lysine in the subfractions, although a similar pattern was obtained for endogenous amino acids.

The work presented here contains several possible sources of error, the two most important being the wide distribution of injected amino acids into various regions of the brain following injection and the metabolism of the amino acids, particularly glutamic acid, during the course of the experiment. A certain fraction of the labeled amino acids enters the capillaries and is rapidly eliminated through the circulation, another portion may reach the

Table 12.3. Percent distribution of intracerebrally injected radioactivity from [^{35}S]taurine, [^{14}C]glutamate, [^{3}H]lysine, and [^{14}C]norleucine in mouse brain subcellular fractions[a]

	Percent distribution			
Subcellular fraction	[^{35}S]-taurine	[^{14}C]-glutamic acid	[^{3}H]-lysine	[^{14}C]-norleucine
Total homogenate	100.0	100.0	100.0	100.0
Soluble cytoplasm	70.3	63.6	46.9	89.0
Crude nuclear fraction	18.3	23.6	26.2	15.0
Crude mitochondrial fraction	14.9	7.3	20.0	2.0
Myelin	2.5	1.9	12.9	0.9
Mitochondria	5.9	3.8	12.9	1.0
Synaptosomes	5.6	5.7	14.6	1.0
Soluble synaptoplasm	10.9	0.8	0.7	0.4
Synaptic vesicles	0.04	1.0	0.5	0.4
Synaptic membranes	0.08	0.5	0.4	0.08
Intraterminal mitochondria	0.04	0.05	0.4	0.2

[a]The subcellular fractions of pooled brains were obtained 20 hr after the injection and their radioactivities measured. The amount of radioactivity in the total homogenate fraction of the pooled brains was taken as 100, and all other fractions were expressed relative to this. Values are means from four series of experiments, each including 20 mice. The SD varied from 12% to 20% of the mean.

brain ventricles and be diluted by distribution into the cerebrospinal fluid, an additional fraction presumably enters the extracellular space, and some radioactivity is found in the neurons and glial cells. This distribution problem results in a loss of radioactivity and can be corrected by use of several animals in an experiment.

Glutamic acid is the most rapidly metabolized amino acid in the central nervous system, being incorporated into proteins and several other compounds. It is also a precursor of both glutamine and GABA, the latter being further metabolized to succinate and carbon dioxide via the tricarboxylic acid cycle. Thus the label that was initially associated with glutamate might become incorporated into a number of other compounds (glutamine, GABA, succinate semialdehyde, and succinate) by the end of the experiment.

In spite of these limitations, the experiments nevertheless clearly demonstrate that glutamate and its metabolites accumulate in the synaptic vesicles,

Table 12.4.   Ratios of labeled/endogenous taurine, glutamic acid, and lysine in mouse brain subcellular fractions[a]

| Subcellular fraction | Ratio of labeled/endogenous amino acids | | |
	Taurine	Glutamic acid	Lysine
Total homogenate	1.0	1.0	1.0
Soluble cytoplasm	1.2	1.1	0.8
Crude nuclear fraction	1.5	1.3	1.3
Crude mitochondrial fraction	1.2	1.3	1.3
Myelin	1.3	1.0	1.4
Mitochondria	1.2	1.5	1.3
Synaptosomes	2.0	1.9	1.5
Soluble synaptoplasm	3.0	0.4	1.8
Synaptic vesicles	0.1	20.0	0.8
Synaptic membranes	0.5	1.3	0.7
Intraterminal mitochondria	0.3	1.0	1.0

[a]The ratios are calculated from the percent distribution of labeled and endogenous amino acids in the fractions presented in Tables 12.2 and 12.3.

while taurine and lysine accumulate somewhat in the soluble synaptoplasm. However, no accumulation of the three amino acids into other subcellular fractions could be demonstrated. Rassin et al. (1977, 1978) and Sturman et al. (1977) were unable to show an accumulation of taurine by subcellular fractions after administration of taurine, either intravenously or through the diet, although labeled taurine appeared in all fractions. Our observations agree with these, except in the case of the soluble synaptoplasm, where a slight accumulation of [35S]taurine could be shown. No injected taurine reached the synaptic vesicles, which are thought to serve as synaptic storage organelles. Therefore, it appears that taurine cannot act synaptically as a transmitter substance. Thus, only glutamate fulfills all of the criteria set for a neurotransmitter.

SUMMARY

Small doses (0.25–20 μCi) of [35S]taurine, [3H]lysine, [14C]glutamic acid, and [14C]norleucine were injected into mice intracerebrally, and the amounts of radioactivity in various subcellular fractions of the brain were analyzed 20hr later. The concentrations of endogenous amino acids in these fractions were measured with an automatic amino acid analyzer, and the ratio of labeled/endogenous amino acids was determined in those fractions that accumulated labeled amino acids. Conservation of taurine, glutamic acid, and lysine was observed only in certain synaptosomal subfractions. [^{35}S]Taurine and [^{3}H]lysine showed a slight accumulation in the soluble synaptoplasm, while [^{14}C]glutamate was clearly conserved in the synaptic vesicles. On the other hand, [^{35}S]taurine penetrated the synaptic vesicles only very slowly. [^{14}C]Norleucine also penetrated the brain cell membranes slowly but was still bound to the synaptic vesicles to a greater extent than was taurine.

ACKNOWLEDGMENTS

The authors are greatly indebted to Mrs. Alli Puirava, B.Sc., for her technical assistance, and to the National Research Council for the Natural Sciences, Finland for financial support.

REFERENCES

Kaczmarek, L. K.; and Davison, A. N. Uptake and release of taurine from rat brain slices. *J. Neurochem.*, 19, 2355–2362 (1972).

Karppinen, A.; Kumpulainen, E.; and Lähdesmäki, P. Synaptosomal accumulation of intracerebrally injected amino acids. *Acta Physiol. Scand.*, 105, 156–162 (1979).

Lähdesmäki, P.; and Korhonen, K. Comparative studies on the degradation of GABA and taurine in the brain. *J. Neurochem.*, 30, 705–711 (1978).

Lähdesmäki, P.; and Winter, R. A peptide containing aspartic acid, glutamic acid and serine in calf brain synaptic vesicles. *Acta Chem. Scand.*, B31, 802–806 (1977).

Lähdesmäki, P.; Pasula, M.; and Oja, S. S. Effect of electrical stimulation and chlorpromazine on the uptake and release of taurine, γ-aminobutyric acid and glutamic acid in mouse brain synaptosomes. *J. Neurochem.*, 25, 675–680 (1975).

Lähdesmäki, P.; Karppinen, A.; Saarni, H.; and Winter, R. Amino acids in the synaptic vesicle fraction from calf brain: Content, uptake and metabolism. *Brain Res.*, 138, 295–308 (1977a).

Lähdesmäki, P.; Kumpulainen, E.; Raasakka, O.; and Kyrki, P. Interaction of taurine, GABA and glutamic acid with synaptic membranes. *J. Neurochem.*, 29, 819–826 (1977b).

Oja, S. S.; and Lähdesmäki, P. Is taurine an inhibitory neurotransmitter? *Med. Biol.*, 52, 138–143 (1974).

Rassin, D. K.; Sturman, J. A.; and Gaull, G. E. Taurine in developing rat brain: Subcellular distribution and association with synaptic vesicles of [^{35}S]taurine in maternal, fetal and neonatal rat brain. *J. Neurochem.*, 28, 41–50 (1977).

Rassin, D. K.; Sturman, J. A.; Hayes, K. C.; and Gaull, G. E. Taurine deficiency in the kitten: Subcellular distribution of taurine and [^{35}S]taurine in brain. *Neurochem. Res.*, 3, 401–410 (1978).

Sturman, J. A.; Rassin, D. K.; and Gaull, G. E. Taurine in developing rat brain: Maternal fetal transfer of [^{35}S]taurine and its fate in the neonate. *J. Neurochem.*, 28, 31–39 (1977).

Whittaker, V. P. The synaptosome. In *Handbook of Neurochemistry*, A. Lajtha, ed., Vol. 2. Plenum Press, New York (1969), pp. 327–364.

Whittaker, V. P.; and Barker, L. A. The subcellular fractionation of brain tissue with special reference to the preparation of synaptosomes and their component organelles. In *Methods of Neurochemistry*, R. Fried, ed., Vol. 2. Marcel Dekker, New York (1972), pp. 1–52.

Copyright © 1981, Spectrum Publications, Inc.
The Effects of Taurine on Excitable Tissues

CHAPTER 13

Taurine in the Retina

M. J. Voaden
A. C. I. Oraedu
J. Marshall
N. Lake

INTRODUCTION

Since the original observations of Kubíček and Dolének (1958), many people have confirmed the predominance of taurine in retinal tissue from a range of species. There is, however, considerable variation in the concentrations reported, with, for example, the values for rat retina ranging between 10 and 90 μmol/g wet wt. Although the differences may relate to techniques, the association of high or low values with a particular method is not obvious, as amino acid analyzers have yielded values (recalculated) of 10–40 μmol/g wet wt. (e.g., Pasantes-Morales et al., 1972; Macaione et al., 1974), whereas results from dansylation range between 14 and 90 μmol/g wet wt. (e.g., Yates and Keen, 1975; Lund Karlsen and Fonnum, 1976). Using double-label dansylation (Snodgrass and Iverson 1973), we routinely find a value of about 50 μmol/g wet wt. for this tissue (Table 13.1) and have been concerned that we might be including a contaminant in our measurements: The derivatives of cysteine, cystine, hypotaurine, cysteamine, cystathionine, and cysteic acid can be distinguished. Separation of dansylated compounds is achieved by chromatography on thin-layer micropolyamide plates, and, at a tissue dilution commensurate with the estimation of other free amino acids in retinal extracts, dansyl taurine is usually seen as a large, highly fluorescent spot. Quantitation is based on the addition of [35]S- or ^{14}C-labeled taurine as internal standard at the start of processing, and on the ratio of this to the level of conjugated ^{3}H-dansyl present (the latter being equivalent to the total taurine

Table 13.1.   Concentration of taurine in the retina[a]

Species	Concentration (μmol/g wet wt.)	Species	Concentration (μmol/g wet wt.)
Rabbit[b]	52 ± 2 [9]	Human periphery[c]	24 ± 2 [3]
Rat[b]	50 ± 2 [23]	Pigeon red spot[b]	21 ± 1 [11]
Cat[b]	43 ± 1 [22]	Pigeon periphery[b]	21 ± 1 [10]
Guinea pig[c]	32 ± 1 [20]	Mouse[c]	16 ± 1 [7]
Baboon	29 ± 1 [3]	Frog[d]	10 ± 0.3 [8]

[a]Results are expressed as the mean ± SEM. The number of estimations for each series is shown in brackets.
[b]Voaden et al. (1977).
[c]Unpublished observations.
[d]Kennedy and Voaden (1974).

content of the sample). Therefore, only portions of the final chromatographic spot need be analyzed. We have found no difference between the fractions of trisected samples. It nevertheless remains possible that another derivative, with virtually identical characteristics to dansyl-taurine, is present.

Because of the variation between "laboratories," species differences are best assessed by comparing values obtained under the same conditions, and Table 13.1 brings together our data. No obvious evolutionary pattern emerges; for example, the rat retina has a high concentration of taurine, whereas that of the mouse is considerably lower (but compare Orr et al., 1976a). The high level of taurine in the rabbit retina is surprising, as with the exception of heart, taurine levels in rabbit tissues are reportedly low compared with those of other mammals (Jacobsen and Smith, 1968).

ENDOGENOUS DISTRIBUTION AND SITES OF UPTAKE

Studies on the intraretinal distribution of taurine in various species have consistently shown high levels in the photoreceptor cell layer, particularly around the nuclei (Orr et al., 1976b; Voaden et al., 1977; Morjaria and Voaden, 1979). This includes the frog retina, which has much of its total tissue taurine present in the photoreceptor cells, concentrations here reaching about 30 mM (Kennedy and Voaden, 1974; Orr et al., 1976b; cf. also the mouse, Orr et al., 1976a). In other species the estimated levels (allowing for 30% extracellular space) range from 26 mM in the periphery of the pigeon retina to 79 mM in the rat (Voaden et al., 1977).

Exogenously applied taurine is taken up into retinas by active, sodium-dependent transport systems that show both high- and low-affinity kinetics (Neal et al., 1973; Starr, 1978). A "nonsaturable" component has also been reported (Schmidt and Berson, 1978). When the sites of uptake are identified by autoradiography, differences are seen according to the substrate concentrations employed. At an exogenous concentration of 5 μM, a level at which about 75% of uptake will be by "high-affinity" processes, taurine enters readily and, in some species, predominantly into all regions of photoreceptor cells, including the synaptic pedicles (Fig. 13.1; Lake et al., 1978): Heavier labeling can sometimes be discerned around the nuclei (Fig. 13.1a). The autoradiographs of Lake et al. (1978) and direct measurements on frogs (Lake et al., 1977) have provided evidence for "high-affinity" carriers in several areas of the retina, including the photoreceptor cells, the pigment epithelium (see also Edwards 1977), and, with species variation, some inner retinal neurones and the glial cells of Müller. At this juncture the presence of taurine in the proximal retina should be emphasized, as, certainly in some species, it is also there at high levels (Orr et al., 1976b; Voaden et al., 1977). A preliminary analysis of "pool" sizes in the rat retina has suggested that 21% of the total taurine may be glial (Müller cells forming about 14% of the total cell volume; Rasmussen, 1972), and that 28% is present in inner retinal neurones and 51% in photoreceptor cells (Voaden, 1978). In the autoradiograph of a rat retina preloaded with ^{3}H-taurine shown in Fig. 13.1a, glial labeling can be discerned but is not prominent. However, it is clearly seen at higher loading levels (Fig. 13.1b; Voaden et al., 1977; Lake et al., 1978).

Wide-ranging species differences exist in the sites of taurine uptake in the inner retina: In the frog, for example, even with more than 500 μM exogenous taurine, glial uptake is barely seen (Kennedy and Voaden, 1976); whereas in the mouse, with 5 μM exogenous taurine, Müller cells are heavily labeled (Fig. 13.1d). In autoradiographs of the frog retina preincubated with 5 μM ^{3}H-taurine (Fig. 13.1c), it is a select population of inner retinal neurones that are labeled. These cells have their perikarya in a position characteristic of bipolar cells, i.e., toward the middle of the inner nuclear layer; the inner synaptic layer is also labeled. At a higher loading level, cells in the position of amacrine interneurones are also delineated (Kennedy and Voaden, 1976; Lake et al., 1977).

Apart from cells that actively accumulate taurine, there are those that stand out because they do not take it up extensively. This is true of the ganglion cells of many of the retinas we have studied and also, with species variation, some amacrine and horizontal cells (see Lake et al., 1978).

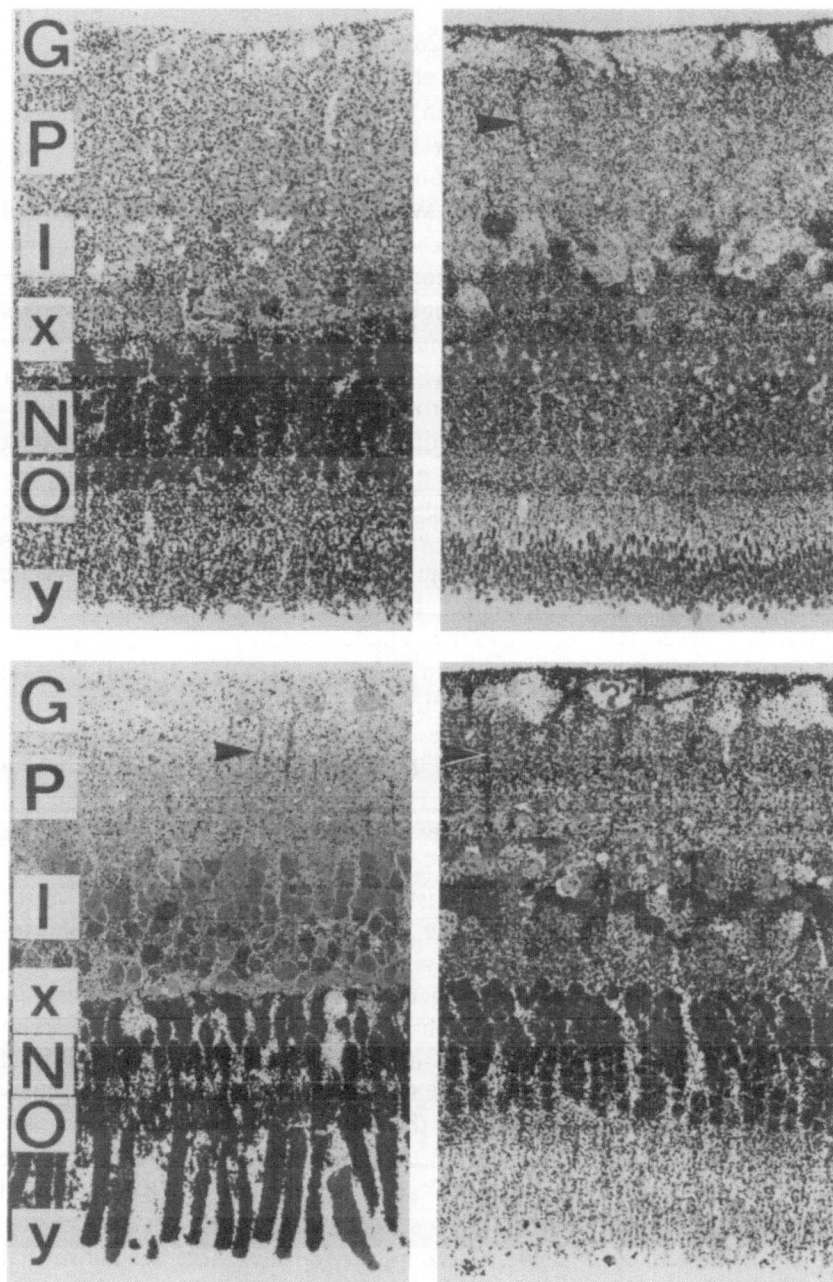

THE PIGMENT EPITHELIUM AND TAURINE TURNOVER IN THE RETINA

The pigment epithelium is a site of the blood-retinal barrier, and, as with other neuroactive amino acids, it effectively restricts the entry of systemic taurine into the neural retina (cf. Figs. 13.2a and b; see also Young, 1969; Miller and Steinberg, 1976). However, in contrast to other amino acids and putative neurotransmitters, radiolabeled taurine is "avidly" accumulated by pigment epithelium both *in vitro* (Fig. 13.2b; Lake et al., 1977, 1978; Edwards, 1977) and *in vivo* (Young, 1969; Lake et al., 1977), and it is then, over days, passed into the neural retina. The entry can occur both by exchange and net uptake, depending on the nutritional status of the animal (Sturman et al., 1978). In a previous study on the entry and homeostasis of radiolabeled taurine in the frog retina, we found no evidence for turnover of the nucleide over a period of 6 weeks (Lake et al., 1977). However, in mice there is a steady loss of preloaded label from the neural retina (Table 13.2), the decrease occurring with a half-life of 11.8 days (calculated from the loss of label between 7 and 56 days after injection). There was no difference between dark-adapted and light-stimulated retinas. Throughout both studies, no labeled metabolites of taurine were detected on silica gel thin-layer chromatograms of the tissue extracts. Turnover of preloaded, labeled taurine has also been observed in the retinas of chickens by Pasantes-Morales et al. (1973), but it was slower in dark-adapted birds. These differences in taurine homeostasis may relate, in part, to species variation but could also depend on diet since food intake by the frogs, in the study of Lake et al. (1977), was minimal, although the mice (Table 13.2) continued to consume their standard laboratory pellets. Taurine turnover in the retinas of kittens maintained on a taurine-free diet also virtually ceases (Sturman et al., 1978), whereas in the normal animal it has a half-life of 9 days.

Fig. 13.1.   Light-microscope autoradiographs showing the sites of ³H-taurine uptake in retinas from (a and b) rat ($\times$ 420), (c) frog ($\times$ 420), and (d) mouse ($\times$ 432). Samples (a), (c), and (d) were incubated for 10 min with 5×10^{-6} M ³H-taurine (18 Ci/mmol; The Radiochemical Centre, Amersham); and sample (b), for 45 min with 1.76×10^{-4} M ³H-taurine (568 mCi/mmol; NEN Chemicals GmbH, Germany). Incubations were in bicarbonate buffered Krebs' or Ringer's solution (frog) at 37°C and 25°C, respectively. Key: O, outer limiting membrane; N, outer nuclear layer; I, inner nuclear layer; P, inner plexiform layer; G, ganglion cell layer. Photoreceptor cells extend from x to y. Müller cells are arrowed. In the retinas incubated with the lower concentration of taurine, uptake has occurred principally in photoreceptor cells and some perikarya in the inner nuclear layer. In the mouse retina (d), Müller cells have also taken up taurine readily, indicating the presence of many "high-affinity" sites. At the higher concentration of taurine, glia are also heavily labeled in the rat retina (b). There is a generally low level of taurine uptake into ganglion cell bodies and some perikarya in the inner nuclear layer. (Data taken from Lake et al., 1978.)

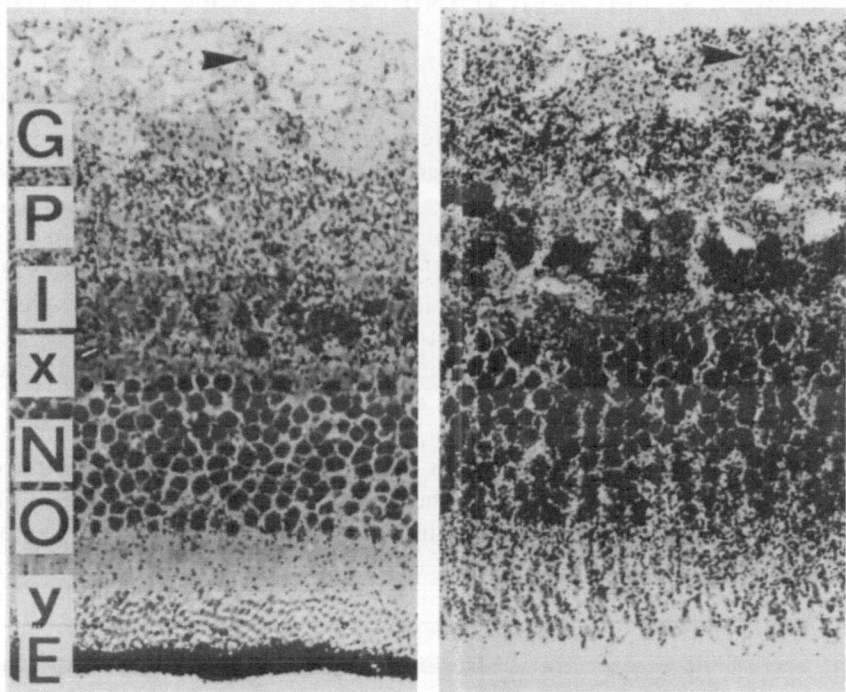

Fig. 13.2. ³H-Taurine uptake by the cat retina incubated (*a*) with and (*b*) without the pigment epithelium. Layers are as designated in Fig. 13.1; E, pigment epithelium (× 575). Taurine has been taken up extensively by the pigment epithelium (*a*), and the presence of this single layer of cells has considerably reduced entry of the amino acid into the neural retina. As in other species, when the neural retina is incubated without the pigment epithelium, the taurine is taken up by photoreceptor cells and specific neurones in the inner nuclear layer. Some uptake has occurred in Müller cells (arrowed). (Data taken from Lake et al., 1978.)

CYSTEINE METABOLISM IN THE RETINA

Although present evidence suggests that systemic taurine may be a major source of retinal taurine, both cysteine oxidase and cysteine sulfinate decarboxylase, enzymes necessary for its synthesis from cysteine, have been found in rat, chicken, and ox retinas (Di Giorgio et al., 1975; Macaione et al., 1976; Pasantes-Morales et al., 1976; Macaione and Di Giorgio, 1977). We therefore considered it of interest to investigate, using autoradiography, the sites of uptake of cysteine in the retinas of various species. For this study, retinas were incubated in Krebs bicarbonate medium for 10 min in the presence of 225 μM exogenous ³⁵S-cysteine. Figure 13.3 shows the results obtained for

Table 13.2.   Turnover of radioactive taurine in the mouse retina *in vivo*[a],[b]

Time after injection (days)	Light stimulated	Dark adapted
0.2	291 ± 18	311 ± 35
1	639 ± 35	548 ± 37
2	1,474 ± 211	1,044 ± 99
7	1,863 ± 148	1,851 ± 263
35	349 ± 27	351 ± 78
56	117 ± 18	98 ± 6

[a] Results are expressed as dpm/mg wet wt. retina ± SEM ($n = 3$). All counts have been corrected for decay of the radioactivity ($t\frac{1}{2}$ of $^{35}S = 87.2$ days).

[b] One μCi/g body wt. of ^{35}S taurine (79 mCi/mmol) (The Radiochemical Centre, Amersham) was injected intraperitoneally into 8-week-old mice. Light-stimulated animals were maintained under a regime of 12 hr dark and 12 hr light (a 40 W tungsten lamp was located 50 cm above the cage). Dark-adapted animals were kept in the dark before use and were injected under a dim, red light. They were then kept in total darkness until sacrifice. Tissue extracts were chromatographed on 20 cm × 20 cm TLC plastic sheets with layered silica gel (Merck Darmstadt) using propanolol/ammonia/acetone (7:3:1 v/v/v) as solvent. Carrier amino acids were also plated and the chromatograms developed with ninhydrin in butanol. No labeled metabolites of taurine were detected.

baboon and guinea pig retinas. In the autoradiograph of the baboon tissue, the predominant label is present over specific cells in the inner retina: Although label is present in photoreceptor cells, there is no discernible difference between rods and cones. In contrast, in the guinea pig (Fig. 13.3b), cone photoreceptor cells are more heavily labeled. Of the other species with well-defined rods and cones that have been studied, cats present a similar picture to the guinea pig, whereas autoradiographs obtained with the frog retina resemble more those of the baboon (Lake and Marshall, 1978).

The metabolic products of cysteine in these retinas have not been identified. However, in a separate series of experiments in which portions of baboon and guinea pig tissue were incubated for 30 min with ^{35}S-cysteine (initial concentration 70 μM), the formation of a labeled product, which ran on TLC in the position characteristic of taurine, was observed. Chromatography was done on glass fiber sheets, impregnated with silica gel (Gelman, Michigan), and the solvent used was propanolol/ammonia/acetone (7:3:1, v/v/v). In the extracts of baboon retina, $5.3 \pm 0.7\%$ ($n = 4$) of the total label (1.24×10^5 dpm/mg wet wt. retina) ran with taurine, whereas in the guinea pig (total label 1.60×10^5 dpm/mg wet wt. retina) $12.1 \pm 1.1\%$ ($n = 5$) ran in this position.

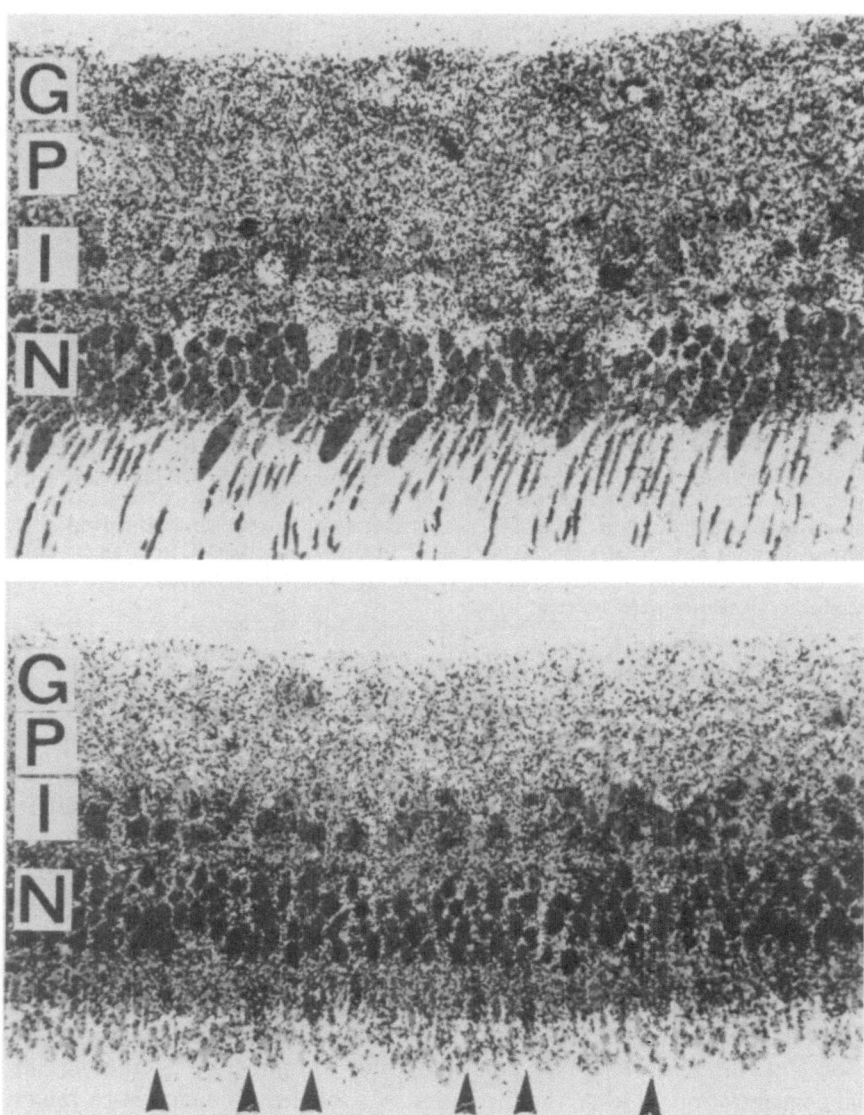

Fig. 13.3. Light-microscope autoradiographs showing the sites of ^{35}S-cysteine uptake in retinas from (a) baboon ($\times$ 360) and (b) guinea pig ($\times$ 540). Samples were incubated for 10 min with 225 μM ^{35}S-cysteine (89 mCi/mmol, The Radiochemical Centre, Amersham). Incubations were in Krebs' bicarbonate medium at 37°C. Layers are as designated in Fig. 13.1. Grain scatter is inevitably high because of the energy of the emitted β-particles from the ^{35}S nucleide. However, some localization can be discerned. In the baboon retina, a heavier grain density is present over occasional perikarya in the middle of the inner nuclear layer; there is no discernible difference between rods and cones. In contrast, in the guinea pig, cone photoreceptor cells (arrowed) are heavily labeled.

The localization of this metabolism is not known. However, the observations merit further study, as Schmidt et al. (1976) have observed that photoreceptor cells degenerate in taurine-deficient cats, and that administration of cysteine will not prevent the degeneration (Berson et al., 1976). Moreover, it is the cone-rich regions that are first affected.

FUNCTIONS OF TAURINE IN THE RETINA

Neurotransmission

Taurine is neuroactive and, in the *Necturus* retina, hyperpolarizes "on-center" bipolar cells but has no effect on the "off-center" variety (see Cohen, 1978). It might, therefore, function as a specific photoreceptor neurotransmitter. However, its widespread distribution in photoreceptors and its uptake into both rods and cones suggest a more general role, present evidence favoring glutamate and/or aspartate as the more likely neurotransmitter candidates for at least some photoreceptor cells (Wu and Dowling, 1978; Voaden, 1979).

The uptake of taurine by specific higher order retinal neurones—e.g., in the frog (Fig. 13.1c; see also Lake et al., 1978)—suggests that these cells may be taurinergic, particularly as the inner synaptic layer of the retina is also labeled. However, in the studies of Kennedy and Voaden (1976), light stimulation and the application of 40 mM K^+ had no effect on the spontaneous efflux of radioactive taurine (preloaded at an exogenous level of 1.0 μM) from the frog retina. Electrical stimulation increased efflux of the radioactivity, but the effect was not calcium sensitive. In these studies most of the accumulated taurine would have been in photoreceptor cells (cf. Fig. 13.1c). It is possible, therefore, that release of this pool masked a response, perhaps with different characteristics, from more proximal neurones.

Physiological evidence in favor of a transmitter role for taurine is its selective effects, both on the b-wave of the chicken retina (Mandel et al., 1976) and on ganglion cell activity in the rabbit (Cunningham and Miller, 1976). In addition, prolonged release of preloaded and endogenous taurine from the chicken retina has been observed on light stimulation (Mandel et al., 1976), and transient releases of labeled taurine at both the start and cessation of a light stimulus have been found for cat and rat retinas (Schmidt, 1978).

Membrane Stabilization and Light Damage

The prolonged exposure of animals to even moderate levels of illumination causes irreversible damage to their photoreceptor cells (Noell et al., 1966; Kuwabara and Gorn, 1968; O'Steen and Anderson, 1972), a main characteristic of the early stages of the lesion being disorganization and vesiculation of the outer limb membranes. As there is evidence suggesting that taurine may help to stabilize membranes (e.g., Huxtable and Bressler, 1973), we thought it of interest to investigate taurine homeostasis in rod photoreceptor cells during the progression of a light-induced lesion. Albino wistar rats were studied and were exposed to a photographic light box (Kodak cold light illuminator, series 2), located at the ceiling of their cage. The box had a sheet of cinemoid frosted glass on its surface and contained two 41.5 cm, 15 W Atlas tropical daylight fluorescent tubes (6,500 k). The cage dimensions were $45 \times 30 \times 16$ cm, and at most, three rats were exposed at one time. The luminance of various areas of the cage, measured with an SE1 photometer was: ceiling, 645 nits; walls, 140 nits; and floor, 64 nits.

Disorganization of photoreceptor outer limbs was detected after 30 hr continuous illumination of the animals. Disk membranes were grossly disorientated and extensively vesiculated, but, in most cases, the degenerative changes did not involve the boundary membrane of the cell. There was a massive increase in phagocytosis of outer limbs by the pigment epithelium. After 18 hr exposure, retinas appeared normal apart from a slight disorganization of the newly forming disks at the base of the outer limb. However, if, at this stage, they were returned to a normal environment, membrane disorganization in the photoreceptor cells progressed, and, after a 3-day "recovery" period the outer limbs exhibited a typical light-damage response. Disk membranes were ill-defined in cross-section, and groups of disks were twisted and at random angles within the boundary membrane of the cells. Such internal conformations resulted in outer segments becoming shorter and, in some cases, almost spherical.

The endogenous levels of retinal taurine after 18 and 48 hr of light exposure are shown in Table 13.3. A significant reduction was observed after 48 hr, but at 18 hr the results were erratic and, as a group, nonsignificant: The concentration of taurine was greatly reduced in retinas left for a 3-day "recovery" period, but the reduction was never more than about 50%. A specific pool of taurine may have been affected, therefore. The results from a preliminary study of bisected retinas (Table 13.4) suggest that this is in the photoreceptor cells.

The reduction in taurine appears specific in that the concentrations of other amino acids remain normal or are increased (Table 13.5). Protein breakdown could explain the raised levels of glutamate, aspartate, glycine, and glutamine observed in tissue analyzed immediately after 48 hr light exposure, particularly as GABA, a "nonprotein" amino acid like taurine, was present at normal levels. However, an increased concentration of GABA has been observed at earlier times of exposure (Oraedu et al., 1980).

So far we have been unable to distinguish cause and effect as regards the

Table 13.3. Concentration of taurine in the retinas of rats exposed to fluorescent light[a]

	Period of exposure to the experimental light source (hr)		
Time to sacrifice	0[b]	18	48
Immediate	67.2 ± 1.3 [60]	59.5 ± 4.5 [18]	55.5 ± 2.0[c] [8]
3 days "recovery"	—	31.2 ± 2.6[d] [24]	44.4 ± 3.3[d] [10]

[a]Results are expressed as the means ± SEM and represent nmol/3.0 mm diameter disk of retina, approximate wet wt. 1.4 mg. The number of estimations for each series is shown in brackets.
[b]Light-adapted control.
[c]$p < 0.01$.
[d]$p < 0.001$.

reduction in taurine and the light-damage response. However, we have not been able to prevent the loss of taurine by parenteral administration of the amino acid before, during, and after light exposure (Table 13.6). Again, this result cannot be considered definitive, as the levels reaching the retina were only about 6% of the normal endogenous pool.

Table 13.4. Effect of continuous exposure to fluorescent light on the concentration of taurine in bisected rat retinas[a]

Section	Light-adapted control	48-hr exposure[b] plus 3-day "recovery"
Photoreceptor section	48.2 ± 1.4	29.7 ± 2.5[c]
Inner retina	16.2 ± 2.8	14.6 ± 1.0

[a]Results are expressed as the means ± SEM ($n = 4$) and represent nmol/3.0 mm diameter disk of retina. Isolated retinas were floated photoreceptors downward onto filter paper and then bisected (Arden and Ernst, 1971). Three-millimeter disks were trephined from the drained retina/filter paper preparations. These were placed, ganglion cells downward into 80 μ-depth, freshly milled, 3.5-mm diameter wells drilled in blocks of aluminium. The base of each block was then stood in melting isopentane ($-80°C$), and then exposed, frozen tissue and filter paper sliced away. Both the exposed tissue and that remaining in the well were recovered and separately analyzed.
[b]See text for explanation.
[c]$p < 0.001$.

Table 13.5. Endogenous amino acids in the retinas of rats exposed to fluorescent light[a]

Amino acid	Light-adopted control ($n = 60$)	48-hr exposure ($n = 8$)	48-hr exposure plus 3-day "recovery" ($n = 6$)
Taurine	67.2 ± 1.3	55.5 ± 2.0[c]	44.4 ± 3.3[d]
GABA	4.3 ± 0.1	4.8 ± 0.2	4.0 ± 0.1
Glycine	3.6 ± 0.1	5.5 ± 0.3[d]	4.1 ± 0.2[b]
Glutamate	8.9 ± 0.2	11.9 ± 0.6[d]	8.2 ± 0.4
Aspartate	4.3 ± 0.1	6.7 ± 0.4[d]	4.2 ± 0.1
Glutamine	6.5 ± 0.2	7.6 ± 0.5[b]	6.0 ± 0.3

[a]Results are expressed as the means ± SEM and represent nmol/3 mm diameter disk of retina.
[b]$p < 0.05$.
[c]$p < 0.01$.
[d]$p < 0.001$.

Membrane Stabilization and ATPase Activities

Taurine influences tissue cation fluxes, perhaps via membrane "stabilization" (Grosso and Bressler, 1976; McBroom and Welty, 1977), and stimulatory and/or protective effects on ATPase activities have been observed (e.g., Huxtable and Bressler, 1973; Igisu et al., 1976).

As in other CNS tissues, ATPase activities are prominent in the retina and subserve many important functions. In particular, Na^+/K^+ ATPase activity is present in photoreceptor inner limbs (e.g., Berman et al., 1977) and may be

Table 13.6. Effect of parenteral taurine on the concentration of endogenous taurine in the retinas of rats exposed to fluorescent light[a]

State of animal	Plus saline	Plus taurine[b]
Light-adopted control	65.7 ± 2.3 [11]	69.1 ± 2.3 [12]
30-hr exposure plus 3-day "recovery"	45.9 ± 3.3 [6]	42.8 ± 5.3 [6]

[a]Results are expressed as the mean ± SEM and represent nmol/3 mm diameter disk of retina. The number of estimations for each series is shown in brackets.
[b]0.5 ml, 1.0 M, taurine was injected intraperitoneally at the start of the light-exposure period. Injections were repeated after 9, 21, 54 and 78 hr. Separate tracer studies with ^{35}S-taurine showed that 0.008% of the injected label entered the retina within 24–48 hr of injection.

Table 13.7. Effect of taurine on ATPase activities in the rat retina[a]

ATP-ase activity	PO₄ formed (nmol/min/mg protein)		
	Normal	Added taurine (25 mM)	Rats exposed to fluorescent light light for 48 hr
Na^+/K^+	12.7 ± 0.5	17.5 ± 1.7[b]	13.9 ± 1.0
Mg^{2+}	5.8 ± 0.4	7.8 ± 0.4[b]	6.8 ± 0.6
Ca^{2+}	3.4 ± 0.6	2.6 ± 0.5	4.3 ± 0.5
Mg^{2+}/Ca^{2+}	5.1 ± 0.5	7.7 ± 0.2[c]	5.3 ± 0.8

[a]Results are expressed as mean ± SEM (4–6 estimations). Assay methods were based on those of Sack and Harris (1977). Two retinas were homogenized in 1.5 ml H_2O ± 25 mM taurine, and the resultant mixture stood in an ice bath for 1 hr. Aliquots (0.1 ml) of homogenate were then added to 0.9 ml of one of the following solutions (final concentrations): (1) 0.1 mM EDTA, 3 mM $MgCl_2$, 0.05 M Tris-HCl buffer (pH 7.4), 2 mM Tris-ATP, 1.35 mM ouabain; (2) 0.1 mM EDTA, 0.15 M NaCl, 0.01 M KCl, 3 mM $MgCl_2$, 0.05 M Tris-HCl buffer (pH 7.4), 2 mM Tris-ATP; (3) 0.1 mM EDTA, 1 mM $CaCl_2$, 0.05 M Tris-HCl (pH 7.4), 2 mM Tris-ATP, 25 λ ethanol; (4) As (1) but with 1 mM $CaCl_2$. After a 25-min incubation at 37°C, the reaction was terminated by the addition of 4 ml of a solution containing 9% trichloracetic acid, 1.3% ascorbic acid, and 0.015% EDTA. The samples were centrifuged and aliquots of the supernatant assayed for phosphate according to Bonting et al. (1961). Na^+/K^+ ATPase activity was obtained by subtracting (1) from (2); Mg^{2+} ATPase, directly from assay (1); Ca^{2+} ATPase, from (3); and Ca^{2+}/Mg^{2+} ATPase activity, from (4).

[b]$p < 0.05$.

[c]$p < 0.001$.

involved in the generation of the sodium current that flows between the inner and outer limbs of dark-adapted photoreceptor cells (Hagins and Yoshi-kami, 1975), forming an essential part of the transduction mechanism. In addition, Ca^{2+}- and/or Mg^{2+}-dependent ATPases may also be involved in transduction processes (Sack and Harris, 1977; Berman et al., 1977; Thacher, 1978). We have therefore investigated the effect of the presence of taurine on ATPase activities in retinal homogenates. The results are shown in Table 13.7. There were significant increases in Na^+/K^+, Mg^{2+}, and Ca^{2+}/Mg^{2+} ATPase activities in homogenates prepared in the presence of 25 mM taurine. Whether these result from a stimulation (measurements being done in the presence of 2.5 mM taurine) or a preservation of ATPase activity remains to be elucidated. The lack of an effect on Ca^{2+} ATPase is intriguing, as several observations suggest that taurine may be involved in calcium homeostasis (Huxtable and Bressler, 1973; Dolara et al., 1976). There was no change in ATPase activities in the retinas of rats exposed to light for 48 hr

(Table 13.7), a time when there is a significant depletion of taurine, perhaps from photoreceptor cells (Table 13.3).

CONCLUSIONS

The existence of such high levels of taurine in the retina, particularly in photoreceptor cells, suggests that it has an important function. However, the precise nature of this remains to be elucidated. Evidence from autoradiography suggests the possibility of a transmitter role in some higher order retinal neurones (e.g., in the frog), but the results from release studies are equivocal.

REFERENCES

Arden, G. B.; and Ernst, W. A method of determining photoreceptor ion contents. *J. Physiol.,* 216, 5p-7p (1971).

Berman, A.L.; Azimova, A. M.; and Gribakin, F. G. Localization of Na^+, K^+-ATP ase and Ca^{2+} activated, Mg^{2+}-dependent ATP ase in retinal rods. *Vision Res.,* 17, 527-536 (1977).

Berson, E. L.; Hayes, K. C.; Rabin, A. R.; Schmidt, S. Y.; and Watson, G. Retinal degeneration in cats fed casein. II: Supplementation with methionine cysteine or taurine. *Invest. Ophthal.,* 15, 52-58 (1976).

Bonting, S. L.; Simon K. A.; and Hawkins, N. M. Studies on sodium potassium-activated adenosine triphosphatase. *Arch. Biochem. Biophys.,* 95, 416-423 (1961).

Cohen, A. I. Retinal organization and function: Possible roles for taurine. In *Taurine and Neurological Disorders,* A. Barbeau and R. T. Huxtable, eds. Raven Press, New York (1978), pp. 249-264.

Cunningham, R.; and Miller, R. F. Taurine: Its selective action on neuronal pathways in the rabbit retina. *Brain Res.,* 117, 341-345 (1976).

Di Giorgio, R. M.; Tucci, G.; and Macaione, S. Cysteine oxidase activity in rat retina during development. *Life Sciences,* 16, 429-436 (1975).

Dolara, P.; Agresti, A.; Giotti, A.; and Sorace, E. The effect of taurine on calcium exchange of sarcoplasmic reticulum of guinea pig heart studied by means of dialysis kinetics. *Can. J. Physiol. Pharmacol.,* 54, 529-533 (1976).

Edwards, R. B. Accumulation of taurine by cultured retinal pigment epithelium of the rat. *Invest. Ophthal.,* 16, 201-208 (1977).

Grosso, D. S.; and Bressler, R. Taurine and cardiac physiology. *Biochem. Pharmacol.,* 25, 2227-2232 (1976).

Hagins, W. A.; and Yoshikami, S. Ionic mechanisms in excitation of photoreceptors. *Ann. N.Y. Acad. Sci.,* 264, 314-325 (1975).

Huxtable, R.; and Bressler, R. Effect of taurine on a muscle intracellular membrane. *Biochim. Biophys. Acta,* 323, 573-583 (1973).

Igisu, H.; Izumi, K.; Goto, I.; and Kina, K. Effects of taurine on the ATPase activity in the human eythrocyte membrane. *Pharmacology,* 14, 362-366 (1976).

Jacobsen, J. G.; and Smith, L. H., Jr. Biochemistry and physiology of taurine and taurine derivatives. *Physiol. Rev.,* 48, 424-511 (1968).

Kennedy, A. J.; and Voaden, M. J. Distribution of free amino acids in the frog retina. *Biochem. Soc. Trans.*, 2, 1256–1258 (1974).

Kennedy, A. J.; and Voaden, M. J. Studies on the uptake and release of radioactive taurine by the frog retina. *J. Neurochem.*, 27, 131–137 (1976).

Kubíček, R.; and Dolének, A. Taurine et acides aminés dans la rétine des animaux. *J. Chromatography*, 1, 266–268 (1958).

Kuwabara, T.; and Gorn, R. A. Retinal damage by visible light. *Arch. Ophthalmol.*, 79, 69–78 (1968).

Lake, N.; and Marshall, J. The uptake of radioactivity by cone photoreceptor cells in retinas incubated with ^{35}S-cysteine. *Proc. Can. Fed. Biol. Soc.*, 21, 53 (1978).

Lake, N.; Marshall, J.; and Voaden, M. J. The entry of taurine into the neural retina and pigment epithelium of the frog. *Brain Res.*, 128, 497–503 (1977).

Lake, N.; Marshall, J.; and Voaden, M. J. High affinity uptake sites for taurine in the retina. *Exp. Eye Res.*, 27, 713–718 (1978).

Lund Karlsen, R.; and Fonnum, F. The toxic effect of sodium glutamate on rat retina: Changes in putative transmitters and their corresponding enzymes. *J. Neurochem.*, 27, 1437–1441 (1976).

Macaione, S.; and Di Giorgio, R. M. Subcellular distribution of cysteine oxidase activity in ox retina. *Life Sciences*, 20, 617–622 (1977).

Macaione, S.; Ruggeri, P.; DeLuca, G., and Tucci, G. Free amino acids in developing rat retina. *J. Neurochem.*, 22, 887–891 (1974).

Macaione, S.; Tucci, G.; DeLuca, G.; and Di Giorgio, R. M. Subcellular distribution of taurine and cysteine sulphinate decarboxylase activity in ox retina. *J. Neurochem.*, 27, 1411–1415 (1976).

Mandel, P.; Pasantes-Morales, H.; and Urban, P. F. Taurine, a putative transmitter in retina. In *Transmitters in the Visual Process*, S. L. Bonting, ed. Pergamon Press, Oxford (1976), pp. 89–105.

McBroom, M. J.; and Welty, J. D. Effects of taurine on heart calcium in the cardiomyopathic hamster. *J. Mol. and Cell. Cardiol.*, 9, 853–858 (1977).

Miller, S.; and Steinberg, R. H. Transport of taurine, L-methionine and 3-O-methyl-D-glucose across frog retinal pigment epithelium. *Exp. Eye Res.*, 23, 177–190 (1976).

Morjaria, B.; and Voaden, M. J. The formation of glutamate, aspartate and GABA in the rat retina: Glucose and glutamine as precursors. *J. Neurochem.*, 33, 541–551 (1979).

Neal, M. J.; Peacock, D. G.; and White, R. D. Kinetic analysis of amino acid uptake by the rat retina *in vitro. Br. J. Pharmac.*, 47, 656–657P (1973).

Noell, W. K.; Walker, V. S.; Kang, B. S.; and Berman, S. Retinal damage by light in rats. *Invest. Ophthal.*, 5, 450–473 (1966).

Oraedu, A. C. I.; Voaden, M. J.; and Marshall, J. Photochemical damage in the albino rat retina: morphological changes and endogenous amino acids. *J. Neurochem.*, 35, 1361–1369 (1980).

Orr, H. T.; Cohen, A. I.; and Carter, J. A. The levels of free taurine, glutamate, glycine and γ-aminobutyric acid during the post-natal development of the normal and dystrophic retina of the mouse. *Exp. Eye Res.*, 23, 377–384 (1976a).

Orr, H. T.; Cohen, A. I.; and Lowry, O. H. The distribution of taurine in the vertebrate retina. *J. Neurochem.*, 26, 609–611 (1976b).

O'Steen, W. K.; and Anderson, K. V. Photoreceptor degenerations after exposure of rats to incandescent illumination. *Z. Zellforsch.*, 127, 306–313 (1972).

Pasantes-Morales, H.; Klethi, J.; Ledig, M.; and Mandel, P. Free amino acids of chicken and rat retina. *Brain Res.*, 41, 494–497 (1972).

Pasantes-Morales, H.; Klethi, J.; Ledig M.; and Mandel, P. Influence of light and dark on the free amino acid pattern of the developing chick retina. *Brain Res.*, 57, 59–65 (1973).

Pasantes-Morales, H.; López-Colomé, A. M.; Salceda, R.; and Mandel, P. Cysteine sulphinate decarboxylase in chick and rat retina during development. *J. Neurochem.*, 27, 1103–1106 (1976).

Rasmussen, K.-E. A morphometric study of the Müller cell cytoplasm in the rat retina. *J. Ultrastructure Res.*, 39, 413–429 (1972).

Sack, R. A.; and Harris, C. M. Ca^{2+}dependent ATPase activity of bovine receptor cell outer segment. *Nature*, 265, 465–466 (1977).

Schmidt, S. Y. Taurine fluxes in isolated cat and rat retinas: Effects of illumination. *Exp. Eye Res.*, 26, 529–535 (1978).

Schmidt S. Y.; and Berson, E. L. Taurine uptake in isolated retinas of normal rats and rats with hereditary retinal degeneration. *Exp. Eye Res.*, 27, 191–198 (1978).

Schmidt, S. Y.; Berson, E. L.; and Hayes, K. C. Retinal degeneration in cats fed casein. I: Taurine deficiency. *Invest. Ophthal.*, 15, 47–52 (1976).

Snodgrass, S. R.; and Iversen, L. L. A sensitive double isotope derivative assay used to measure release of amino acids from brain *in vitro*. *Nature*, 241, 154–156 (1973).

Starr, M. S. Uptake of taurine by retina in different species. *Brain Res.*, 151, 604–608 (1978).

Sturman, J. A.; Rassin, D. K.; Hayes, K. C.; and Gaull, G. E. Taurine deficiency in the kitten: Exchange and turnover of ^{35}S taurine in brain, retina and other tissues. *J. Nutrition*, 108, 1462–1476 (1978).

Thacher, S. M. Light-stimulated, magnesium dependent ATPase in toad retinal rod outer segments. *Biochemistry*, 17, 3005–3011 (1978).

Voaden, M. J. The localization and metabolism of neuroactive amino acids in the retina. In *Amino Acids as Neurotransmitters*, F. Fonnum, ed. Plenum Press, New York and London (1978), pp. 257–274.

Voaden, M. J. The chemical specificity of neurones in the retina. *Prog. Brain Res.*, 51, 389–402 (1979).

Voaden, M. J.; Lake, N.; Marshall, J.; and Morjaria, B. Studies on the distribution of taurine and other neuroactive amino acids in the retina. *Exp. Eye Res.*, 25, 249–257 (1977).

Wu, S. M.; and Dowling, J. E. L-Aspartate: Evidence for a role in cone photoreceptor synaptic transmission in the carp retina. *Proc. Nat. Acad. Sci.*, 75, 5205–5209 (1978).

Yates, R. A.; and Keen, P. The effect of optic stalk section on the amino acid content of rat retina. *Brain Res.*, 99, 166–169 (1975).

Young, R. W. Morphology, function and clinical characteristics. In *The Retina*, B. R. Straatsma, M. O. Hall, R. A. Allen, and F. Crescitelli, eds. University of California Press, Los Angeles (1969), pp. 177–210.

Copyright © 1981, Spectrum Publications, Inc.
The Effects of Taurine on Excitable Tissues

The Effect of Taurine on ^{45}Calcium Transport by Retinal Subcellular Fractions

H. Pasantes-Morales
R. M. Ademe
A. M. López-Colomé

The presence of particularly high amounts of taurine in the retina, as well as its distribution in the retina of several species, has been well established (Kennedy and Voaden, 1974; Yates and Kenn, 1976; Orr et al., 1976). Yet investigations up to now have not clarified the role of taurine in retinal function.

The depressant action of taurine, widely detected in the central nervous system (Krnjević and Puil, 1976), is also observed in the retina. Taurine exerts a profound inhibitory effect on the b-wave of the electroretinogram (Pasantes-Morales et al., 1973a). This result, together with the reported release of taurine from the retina upon the action of nerve excitatory agents (Pasantes-Morales et al., 1976), has suggested its involvement as an inhibitory neurotransmitter in the retina (Mandel et al., 1976). However, the identification of specific postsynaptic receptors in the retina or in other regions of the central nervous system, either by means of physiological or biochemical approaches, has not been achieved. Attempts in this laboratory to demonstrate the interaction of taurine with retinal membranes, corresponding to the binding of neurotransmitters to postsynaptic receptors, have been unsuccessful (López-Colomé and Pasantes-Morales, 1979). To date, we have only detected taurine binding to sites on the retinal membranes which resemble transport sites. These transport sites are widely distributed in the membranes derived from several retinal layers.

The highest concentration of retinal taurine is in the photoreceptors and not in the synaptic zones (Orr et al., 1976), thus suggesting that if taurine is serving a neurotransmitter role in the synaptic layers, an alternate function should be envisaged for the taurine present in the photoreceptors. The observed release of taurine evoked by light stimulation from the whole retina (Pasantes-Morales et al., 1973b), as well as from isolated rod outer segments (Salceda et al., 1977), suggests a role for taurine related to photoexcitation.

Although the possibility that taurine is involved in more than one action in the retina cannot be ruled out, we favor the view of a single action for taurine throughout all the retinal cells. This action might be related to a basic common mechanism which underlies photoexcitation in photoreceptors, namely stimulus-secretion coupling in nerve terminals and probably also contractile phenomena which eventually occur in the retina. This suggestion is supported by the existence of remarkable similarities between the mechanisms of stimulation-secretion coupling and excitation-contraction coupling, and coincidently, the highest taurine concentration is found in nervous, contractile, and secretory tissues (Jacobsen and Smith, 1968). Calcium ion translocations appear to be a common feature in all these functions (Rubin, 1974). Since taurine has been implicated in modifications of calcium kinetics in contractile tissues (Dolara et al., 1973), the idea that it may be involved in the modulation of calcium fluxes in retina is an attractive possibility. In the present work, we describe some of our recent data concerning the effect of taurine on calcium transport by retinal subcellular fractions, which may provide some experimental support for this hypothesis.

CALCIUM TRANSPORT BY RETINAL SUBCELLULAR FRACTIONS

Rod outer segments (ROS), isolated from chick or frog retinas, and a crude synaptosomal fraction, obtained from the chick retina, were used throughout this work. The experimental procedure for subcellular fractionation is described in Table 14.1.

ROS and synaptosomes, when incubated in a Krebs-bicarbonate medium containing $^{45}CaCl_2$, rapidly accumulate ^{45}Ca. Saturation of ^{45}Ca transport is very rapid; values close to maximum are attained after 2–3 min of incubation; asymptotic values are reached after 5 min of incubation. Maximal calcium uptake was calculated to be 1.0–1.5 nmol/mg protein in synaptosomes and 3.5–4.0 nmol/mg protein in frog ROS (Table 14.1).

The main features of calcium transport by retinal subcellular fractions are summarized in Table 14.1. ^{45}Ca accumulation by ROS or synaptosomes is not stimulated by ATP; it is only slightly sensitive to ruthenium red, but it is markedly decreased by incubation at 4°C. Calcium transport is not in-

Table 14.1.   Factors affecting ^{45}Ca transport by synaptosomes and rod outer segments isolated from frog and chick retina[a]

| Treatment | ^{45}Ca uptake (nmol/mg protein) | |
	Synaptosomes	ROS
Control: 25°C, 1 mM ATP	—	3.45 ± 0.16 (12)
37°C, 1 mM ATP	1.12 ± 0.07 (16)	—
4°C, 1 mM ATP	0.27 ± 0.02 (4)	1.18 ± 0.13 (4)
Without ATP	1.10 ± 0.12 (4)	3.24 ± 0.32 (4)
Ouabain (100 μM)	0.98 ± 0.08 (4)	3.53 ± 0.29 (3)
Ruthenium red (10 μM)	—	3.17 ± 0.33 (3)
(100 μM)	0.97 ± 0.07 (4)	1.69 ± 0.33 (3)
KCl (68.5 mM)	1.06 ± 0.09 (4)	—
Low-sodium medium	1.98 ± 0.12 (4)	—

[a]Chick retinas were homogenized in 0.32 M sucrose, and primary fractions were obtained by centrifugation of the homogenate at 900 g to obtain the crude nuclear fraction (P_1); and then the supernatant at 9 000 g, 20 min, to obtain the crude synaptosomal fraction (P_2). P_2 contained mainly synaptosomes with only few mitochrondria and was the fraction referred to as "synaptosomes" throughout this work. Rod outer segments (ROS) were obtained from frog or chick retinas by gentle brushing of retinas; the detached ROS were sedimented by centrifugation at 900 g, 10 min, at 4°C. ROS were more easily detached from frog than from chick retinas, and unless otherwise indicated, synaptosomes were obtained from chick retinas; and ROS, from frog retinas. ROS or synaptosomes were resuspended in a Krebs-bicarbonate medium (118 mM NaCl, 4.7 mM KCl, 1.17 mM MgSO$_4$, 1.2 mM KH$_2$PO$_4$, 25 mM NaHCO$_3$, 5.6 mM glucose), pH 7.4, and aliquots of the suspension containing 0.5–1 mg protein were incubated for 5 min with a mixture containing 1 mM ATP, 2 μCi of ^{45}CaCl$_2$, and 2.5 mM unlabeled CaCl$_2$ in a final volume of 1 ml. At the end of the incubation, aliquots of 0.3 ml were withdrawn and centrifuged in a Beckman microfuge; the sediment was solubilized with NCS (tissue solubilizer, Amersham) and radioactivity measured after addition of Tritosol (Fricke, 1975). Fractions were exposed to ouabain, 100 μM ruthenium red, and a high KCl, low-sodium medium from the beginning of the incubation. In experiments with 100 μM ruthenium red, fractions were preincubated for 10 min with the dye. In the low-sodium medium, isoosmolar choline chloride replaced NaCl, and synaptosomes were preincubated for 10 min in a Na-containing medium and then incubated in the low-Na medium (Blaustein and Oborn, 1975). Values are means ± SEM of the number of experiments indicated in parentheses.

creased by exposure of synaptosomes to depolarizing potassium concentrations. Observed ^{45}Ca transport is unaffected by ouabain or by potassium deprivation; however, it is activated by lowering external sodium concentration. The properties of ^{45}Ca transport observed in retinal preparations are in agreement with the observations of Schnetkamp et al. (1977) on calcium uptake by isolated cattle ROS and with those of Blaustein and Oborn (1975) in mammalian synaptosomes.

The features exhibited by this calcium accumulation mechanism seem to correspond to those of the Na–Ca exchange process described in excitable tissues. This mechanism appears to be different from the calcium transport process linked to oxidative phosphorylation in mitochondria (Carafoli and Crompton, 1978) and also from the depolarizing-induced calcium entry into nerve terminals (Blaustein and Oborn, 1978). The exact mechanism for this calcium transport, and its relationship to sodium fluxes, has not been completely elucidated. Some of its properties, such as its insensitivity to ouabain or to potassium deprivation, suggest that it is not directly linked to a coupled Na–K pump but probably involves a Na–Ca exchange. Observations of Blaustein and Russell (1975) in the squid giant axon, and of Carafoli and Crompton (1978) in heart mitochondria, are consistent with the existence of a mobile carrier mechanism which can exchange intracellular calcium for extracellular sodium. It has also been proposed that Ca–Ca exchange also may be carried out by the same mechanism. The stoichiometry between sodium and calcium exchange is believed to be 3 Na for 1 Ca (Blaustein and Russell, 1975; Carafoli and Crompton, 1978).

The source of the energy necessary for this coupled exchange has not been clarified. The Na-dependent calcium transport appears to utilize ATP under certain conditions (DiPolo, 1974); however, experiments with dialyzed axons in an ATP-free media, or with axons exposed to metabolic poisons (DiPolo, 1973; Blaustein and Russell, 1978), indicate that the Na–Ca exchange can operate in the virtual absence of ATP. The energy for this coupled exchange may be provided by the Na electrochemical gradient rather than by ATP directly (Blaustein and Russell, 1975).

EFFECT OF TAURINE ON ^{45}CA ACCUMULATION BY RETINAL SUBCELLULAR FRACTIONS

Synaptosomes

The presence of taurine (25 mM) in a Krebs-bicarbonate incubation medium produces a rapid inhibition of ^{45}Ca accumulation by retinal synaptosomes. The effect of taurine is observed at the shortest time examined (1 min) and persists throughout the incubation period (Fig. 14.1). By 5 min, taurine reduced ^{45}Ca uptake by nearly 70%. This action of taurine is also observed at lower concentrations, from 2.5 to 10 mM; taurine seems to reduce ^{45}Ca accumulation in a dose-dependent manner (Fig. 14.2). In order to test the specificity of this inhibition of calcium transport, the effect of various amino acids known to exert an action on neuronal activity, and having structural analogies with taurine, were examined; GABA, glycine, β-alanine, and

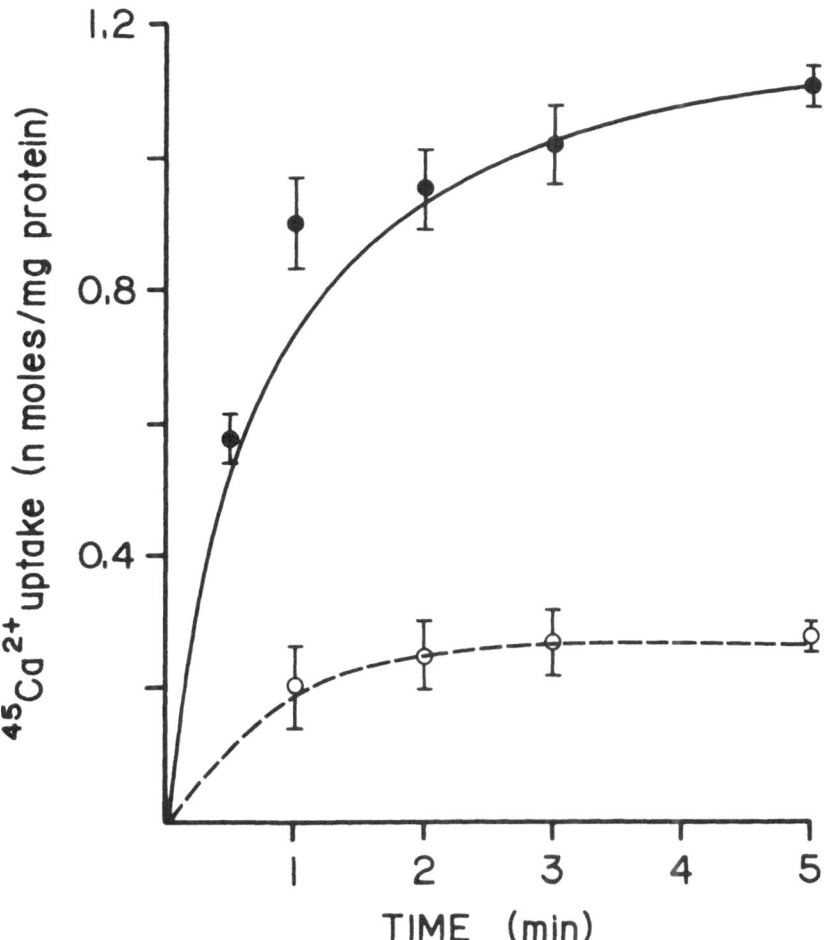

Fig. 14.1. The effect of taurine (25 mM) on ^{45}Ca accumulation by chick retinal synaptosomes at different incubation times. Taurine at a concentration of 25 mM was present in an incubation mixture containing Krebs-bicarbonate medium (see Table 14.1), 1 mM neutralized ATP, 2.5 mM $CaCl_2$, 2 μCi of $^{45}CaCl_2$, and 0.1–1.0 mg of protein/ml. Aliquots of 0.3 ml of the suspension were withdrawn at the indicated incubation times. Values are means ± SEM of 4–6 separate experiments. Key: Control (●—●), taurine (O--O).

glutamate were tested at a concentration of 10 mM. GABA and glutamate were found not to alter ^{45}Ca transport into synaptosomes, whereas glycine and β-alanine reduced it 15% and 25%, respectively. By comparison, 10 mM taurine reduced ^{45}Ca uptake by 50% (Table 14.2). Modification in the ionic

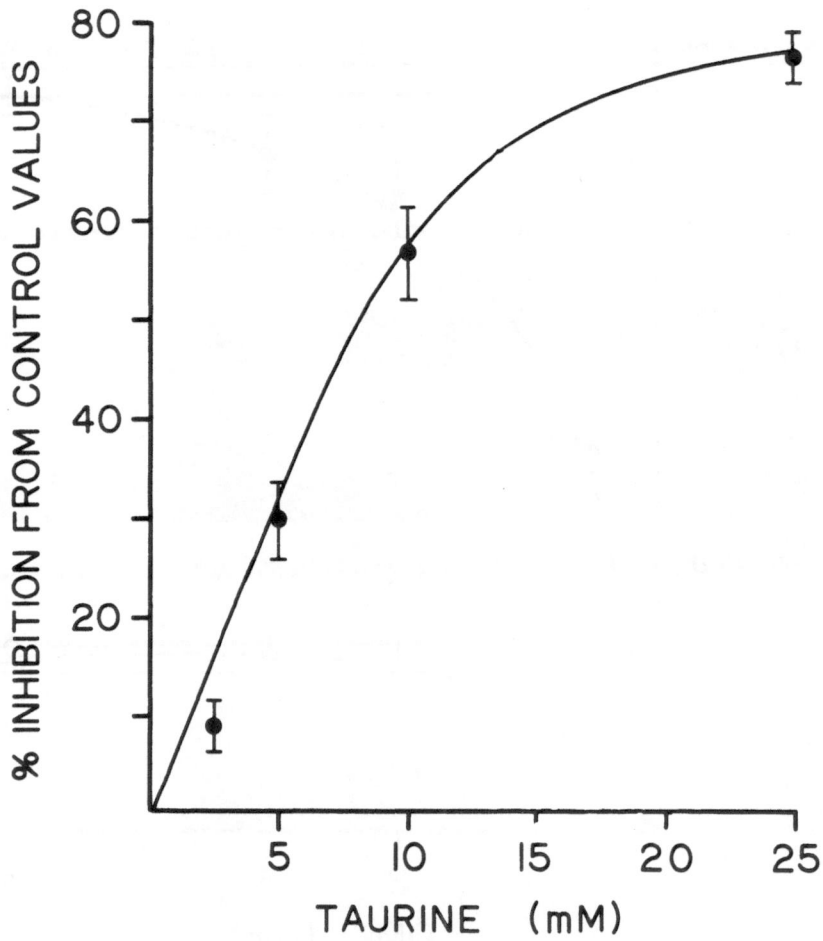

Fig. 14.2. The effect of different concentrations of taurine on ^{45}Ca accumulation by chick retinal synaptosomes. ^{45}Ca uptake in incubation flasks containing 2.5, 5, 10, and 25 mM taurine was measured as described in Fig. 14.1. Values are means ± SEM of 4–6 separate experiments.

composition of the incubation medium produced changes in the effect of taurine on ^{45}Ca transport; replacement of bicarbonate buffer by TRIS, HEPES, or phosphate buffer completely abolished the inhibitory action of taurine. In a sodium-free medium, prepared by replacement of NaCl by choline chloride and $NaHCO_3$ by $KHCO_3$, the taurine effect was still observed; however, in a medium containing only isotonic sucrose and sodium bicarbonate, the taurine effect was no longer apparent (Table 14.3). These results indicate that

Table 14.2. Effect of various amino acids on ^{45}Ca transport by retinal synaptosomes and rod outer segments[a]

Amino acid	^{45}Ca uptake (nmol/mg protein)	
	Synaptosomes	ROS
Control	1.12 ± 0.07 (16)	3.45 ± 0.16 (12)
GABA	1.11 ± 0.12 (4)	3.29 ± 0.19 (4)
Glycine	0.95 ± 0.09 (4)	3.28 ± 0.23 (4)
β-Alanine	0.84 ± 0.09 (4)	3.59 ± 0.28 (4)
Glutamate	1.17 ± 0.13 (4)	—
Taurine	0.56 ± 0.03 (6)	2.41 ± 0.20 (6)

[a]Chick retinal synaptosomes and frog ROS were obtained as described in Table 14.1. Incubation in a Krebs-bicarbonate medium containing 2 μCi of $^{45}CaCl_2$, 1 mM ATP, and 0.5–1.0 mg protein was carried out at 37°C for the synaptosomal fraction and at 25°C for ROS. The tested amino acids, at 10 mM concentration, in a final volume of 1 ml, were present from the beginning of the incubation. Values are means ± SEM of the number of experiments indicated in parentheses.

Table 14.3. Effect of different incubation media on the inhibitory effect of taurine on ^{45}Ca transport by chick retinal synaptosomes[a]

Incubation Medium	^{45}Ca uptake (nmol/mg protein)	
	Control	25 mM taurine
Krebs-bicarbonate	1.12 ± 0.07 (16)	0.37 ±00.01 (6)
Krebs-phosphate	1.16 ± 0.12 (4)	1.09 ± 0.10 (4)
Krebs-TRIS	1.24 ± 0.14 (9)	1.29 ± 0.12 (9)
Sucrose-HEPES	1.09 ± 0.09 (7)	1.12 ± 0.16 (7)
Sucrose-bicarbonate	0.78 ± 0.07 (5)	0.76 ± 0.09 (5)

[a]The synaptosomal fraction was prepared as described in Table 14.1. Krebs-bicarbonate medium contained in 118 mM NaCl, 4.7 mM KCl, 1.2 mM KH_2PO_4, 1.17 mM $MgSO_4$, 25 mM $NaHCO_3$, and 5.6 mM glucose; Krebs-phosphate and Krebs-Tris media were similarly prepared except that 25 mM Na_2HPO_4 and 25 mM Tris, respectively, were used in place of $NaHCO_3$. Sucrose-HEPES and sucrose-bicarbonate media contained only isotonic sucrose and 25 mM HEPES or $NaHCO_3$, respectively. In all cases, the pH of the media was adjusted to 7.4. Synaptosomes were incubated for 5 min at 37°C in the different media, in the presence of 2.5 mM $CaCl_2$ and 2 μCi of $^{45}CaC_2$. Values are means ± SEM of the number of experiments indicated in parentheses.

the presence of bicarbonate is necessary for the action of taurine on ^{45}Ca transport. Therefore, the effect of varying sodium bicarbonate concentrations was further examined. Results in Fig. 14.3 show that increasing bicarbonate concentration produces an increased ^{45}Ca accumulation by retinal synaptosomes. The percent of taurine inhibition also progressively increases, suggesting that the effect of the amino acid is exerted on a calcium accumulation process stimulated by bicarbonate.

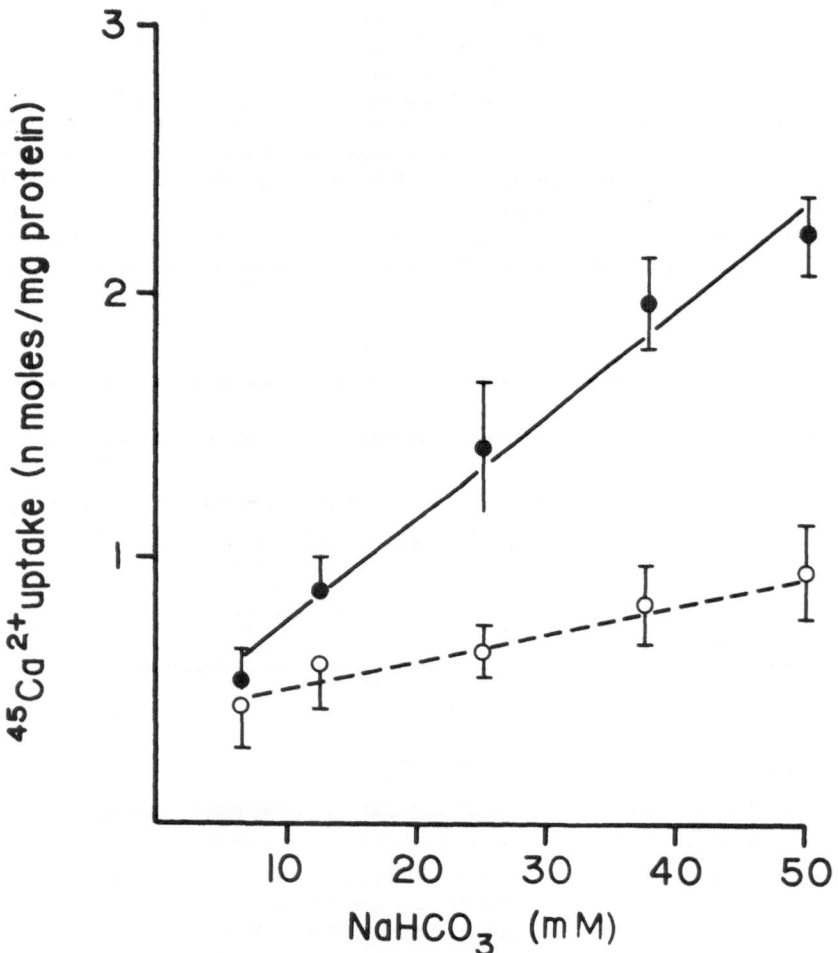

Fig. 14.3. The effect of taurine on ^{45}Ca accumulation measured at different concentrations of bicarbonate. ^{45}Ca accumulation was measured in a Krebs medium containing between 7.5 and 50 mM NaHCO$_3$, in the presence or absence of 25 mM taurine. The medium was gassed with a mixture of 95% O$_2$–5% Co$_2$ yielding a pH of 7.4. Values are means ± SEM of 6 separate experiments. Key: Control (●—●), taurine (O--O).

The taurine effect seems to be exerted on ^{45}Ca transport rather than on ^{45}Ca binding to synaptosomal membranes; in synaptosomes submitted to osmotic shock, taurine does not affect the retention of radioactivity by the remaining membranes. Exogenous taurine does not seem to affect ^{45}Ca efflux from synaptosomes previously loaded with labeled calcium. The efflux pattern of radioactivity released in the presence or absence of taurine was identical (data not shown).

Rod Outer Segments

An effect of taurine on calcium transport in isolated ROS similar to that observed in synaptosomes would be of particular interest. An involvement of taurine in the process of photoexcitation has been suggested by the observation that taurine is released from isolated ROS upon stimulation by light flashes (Salceda et al., 1977). On the other hand, calcium redistribution appears to be involved in the mechanism leading to photoreceptor hyperpolarization subsequent to the activation of rhodopsin (Bonting and Daemen, 1976). Calcium has been considered as the direct intracellular transmitter between the light-activated rhodopsin in the disks and the sodium channels in the plasmalemma. It seems to be established that photoreceptor hyperpolarization caused by light is produced by the decrease of rod outer segment membrane conductance to sodium ions after light activation (Tomita, 1970; Hagins et al., 1970; Penn and Hagins, 1972) Korenbrot and Cone, 1977). However, the light-sensitive molecule, rhodopsin, is located in the disks, 100–200 nm away from the rod outer membrane, where changes in ionic membrane permeability occur. The calcium hypothesis states that light induces the release of calcium bound to disks into the cytoplasm and that calcium then diffuses to the plasmalemma, thereby decreasing sodium conductance (Hagins, 1972; Hagins and Yoshikami, 1974). The initial electrophysiological evidence supporting the calcium transduction hypothesis of photoexcitation (Hagins and Yoshikami, 1974; Brown and Pinto, 1974; Brown et al., 1997) has not been definitely confirmed by biochemical observations. Some discrepancies have been reported regarding calcium content in light- and dark-adapted ROS (Hemminki, 1975; Szuts and Cone, 1977; Hendriks et al., 1977). The demonstration of light-stimulated calcium release from ROS, as required by the calcium hypothesis, is also controversial (Kaupp and Junge, 1977; Smith et al., 1977; Liebman, 1978). An alternate hypothesis (Farber et al., 1978) involves cyclic-GMP as a messenger between the bleaching of rhodopsin and the control of sodium permeability in ROS membrane. According to this hypothesis, light activation of rhodopsin stimulates cyclic-GMP phosphodiesterase, causing a rapid fall in the cyclic-GMP content of

ROS (Goridis et al., 1974, 1977). This reduction in cyclic-GMP levels would favor the dephosphorylation of a soluble protein (Chader et al., 1975; Lolley et al., 1977), which would be responsible for the closure of sodium channels, probably by modifying calcium binding (Farber et al., 1978; Weller et al., 1975). In any event, as redistribution of calcium seems to occur during light excitation, taurine release observed under this condition may be related to an action of this amino acid on calcium fluxes.

Our results show that the taurine action on ^{45}Ca transport by isolated ROS is similar to that observed in synaptosomes. At concentrations varying from 2.5 to 25 mM, taurine significantly reduces ^{45}Ca uptake by ROS (Fig. 14.4). This effect seems to be specific for taurine; it is not produced by GABA, glycine, or β-alanine (Table 14.2). Taurine inhibitory action on ^{45}Ca uptake is observed throughout a pH range of 6.4 to 8.3; maximal effect is observed at pH 7.4. As found for synaptosomes, taurine inhibition of ^{45}Ca transport is not observed when sodium bicarbonate is replaced by TRIS in the incubation medium (Table 14.3). A net uptake of calcium has been described when ROS are previously calcium depleted by addition of EGTA to the isolation medium (Schnetkamp et al., 1977). Under these conditions, the effect of taurine on ^{45}Ca entry is magnified, inhibiting calcium transport to less than 20% of the control values. ATP does not increase calcium uptake by calcium-depleted ROS, but its presence in the incubation medium at a concentration of 1 mM partially reversed the inhibitory action of taurine on ^{45}Ca transport.

Taurine seems to modify ^{45}Ca uptake through the outer ROS membrane, as well as across the membrane of the disks. In ROS, which have been submitted to hypo-osmotic shock to disrupt the outer membrane and expose the inner disks, ^{45}Ca transport to the remaining disks is also reduced by taurine (Table 14.4). When ROS are disrupted by hypotonic shock and ^{45}Ca binding is measured at 4°C, taurine does not modify the incorporation of labeled calcium, suggesting that its action is exerted on an energy-dependent process.

CONCLUDING REMARKS

Taurine effects on ^{45}Ca transport appear to be similar in synaptosomes and in ROS isolated from the retina. The characteristics of the taurine-induced decrease on calcium accumulation suggest that taurine is acting by the same mechanism in both preparations. Taurine does not seem to bind calcium in the form of a chelate, because at physiological pH no chelation is possible between protonated taurine and calcium (Dolara et al., 1978). A calcium-taurine interaction is not observed in the Krebs-bicarbonate medium in which all the effects of taurine were studied; the amount of calcium free in the medium, followed spectrophotometrically in a double-beam spectrometer us-

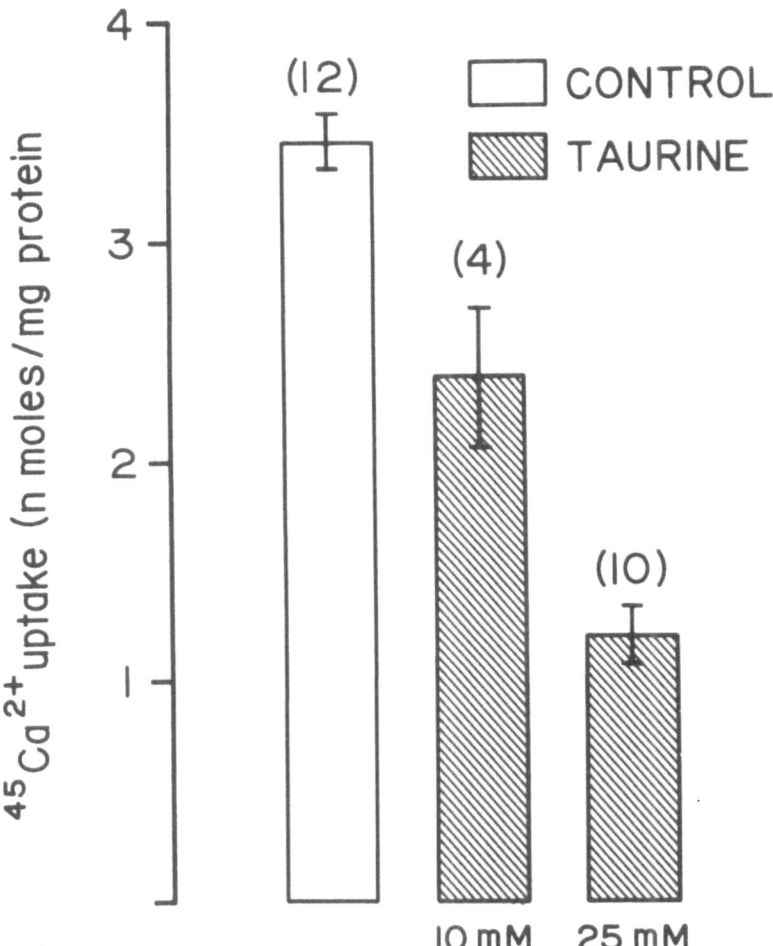

Fig. 14.4. The effect of different concentrations of taurine on ^{45}Ca accumulation by isolated frog rod outer segments. Rod outer segments (0.7–1.0 mg protein) were incubated in 1 ml of Krebs-bicarbonate medium, at 25°C, for 5 min. The incubation mixture contained 1 mM ATP, 2.5 mM $CaCl_2$, and 2 μCi of $^{45}CaCl_2$. Values are means ± SEM of the number of experiments indicated in parentheses.

ing murexide as an indicator, is not modified by 25 mM taurine. Therefore, the taurine action seems to be related to a calcium transport mechanism involving the membrane of ROS and synaptosomes.

The observed effect of taurine seems to require the presence of bicarbonate

Table 14.4.   Effect of taurine on ^{45}Ca transport by osmotically shocked frog retinal rod outer segments[a]

Temperature	^{45}Ca uptake (nmol/mg protein)	
	Control	25 mM taurine
25°C	2.03 ± 0.21 (4)	1.03 ± 0.12 (6)
4°C	1.06 ± 0.12 (4)	0.97 ± 0.08 (6)

[a]Isolated rod outer segments were submitted to hypotonic shock by addition of 4 volumes of water, followed by vigorous shaking in a vortex for 10 sec. ^{45}Ca accumulation was measured in Krebs-bicarbonate medium at the indicated temperatures. Values are means ± SEM of the number of experiments indicated in parentheses.

in the incubation medium. It has been reported that retinal respiration, as well as other parameters of the functional state of the retina, is markedly inhibited in the absence of bicarbonate (Cohen and Noell, 1960; Riley and Voaden, 1970). Moreover, it has been observed that omission of bicarbonate and replacement with either phosphate or TRIS to maintain constant pH, produces a selective loss of the b-wave and a decline of the a-wave of the electroretinogram (Winkler et al., 1972). Therefore, the presence of bicarbonate seems to be necessary for the maintenance of adequate conditions of retinal function. The finding that the taurine effect on ^{45}Ca transport is only observed in the presence of bicarbonate suggests that it may be related to mechanisms acting under physiological conditions.

Calcium transport affected by taurine appears to correspond to the Na-Ca, Ca-Ca exchange mechanism described by Blaustein and Hodgkin (1969) and by Baker et al. (1969). This calcium exchange has been identified in a variety of excitable tissues. It has been demonstrated in squid and crab axons, in rabbit nerve, in smooth and cardiac muscle preparations, in secretory tissues, and in isolated nerve endings (Baker, 1978; Blaustein and Russell, 1975; Thorn et al., 1978; Rink, 1977; Blaustein, 1974; Blaustein and Oborn, 1975). In excitable tissues, the Na-Ca exchange has been partially implicated in the long-term regulation of intracellular calcium levels (Blaustein and Oborn, 1975). In nervous and secretory tissues, calcium that enters presynaptic terminals during depolarization, or secretory cells during activity appears to be sequestered within a few milliseconds by intracellular organelles or binding sites. Thus, the nerve terminal or secretory cell appears to rapidly attain resting calcium levels (Blaustein et al., 1978). Subsequently, calcium is slowly released from the storage sites and is extruded until a normal calcium distribution is reachieved. The Na-Ca exchange process is thought to be involved with calcium extrusion, which is necessary to maintain extremely low levels of intracellular calcium. In muscle, the importance of this Na-Ca exchange is emphasized by the observation that a high proportion of calcium efflux from mammalian cardiac preparations is due to Na-Ca exchange (Reuter, 1970).

Taurine is maximally concentrated in contractile, nervous, and secretory tissues, all of which share the above-described mechanism for intracellular calcium regulation; it may therefore be speculated that taurine is involved in these tissues through its effect on Na-Ca exchange. We have observed that taurine decreases ^{45}Ca transport in brain synaptosomes as it does in retinal fractions (Pasantes-Morales and Gamboa, 1980. It seems possible that the main role of taurine in excitable tissues is related to calcium redistribution.

ACKNOWLEDGMENTS

We thank Dr. R. Tapia for calcium measurements in the double-beam spectrometer. This work was supported in part by Grants No. 1 R01 EY025 40-04 from the National Eye Institute and No. 1621 from Consejo Nacional de Ciencia y Tecnología (CONACyT).

REFERENCES

Baker, P. F. The regulation of intracellular calcium in giant axons of *Loligo* and *Myxicola. Ann. N.Y. Acad. Sci.*, 307, 250–268 (1978).

Baker, P. F.; Blaustein, M. P., Hodgkin, A. L.; and Steinhardt, R. A. The influence of calcium on sodium efflux in squid axons. *J. Physiol.*, 200, 431–450 (1969).

Blaustein, M. P. The interrelationship between sodium and calcium fluxes across cell membranes. *Rev. Physiol. Biochem. Pharmacol.*, 70, 33–82 (1974).

Blaustein, M. P.; and Hodgkin, A. L. The effect of cyanide on the efflux of calcium from squid axons. *J. Physiol.*, 200, 497–513 (1969).

Blaustein, M. P.; and Oborn, C. J. The influence of sodium on calcium fluxes in pinched off nerve terminals in vitro. *J. Physiol.*, 247, 657–686 (1975).

Blaustein, M. D.; and Russell, J. M. Sodium-calcium exchange and calcium-calcium exchange in internally dialyzed squid giant axons. *J. Memb. Biol.*, 22, 313–328 (1975).

Blaustein, M. P.; Ratziaff, R. W.; Kendrick, N. C.; and Schweitzer, E. J. Calcium buffering in presynaptic terminals. I: Evidence for involvement of a nonmitochondrial ATP-dependent sequestration mechanism. *J. Gen. Physiol.*, 72, 16–41 (1978).

Bonting, S. L.; and Daemen, F. J. F. Calcium as a transmitter in photoreceptor cells. In *Transmitter in the Visual Process*, S. L. Bonting, ed. Pergamon Press, Oxford (1976), pp. 59–89.

Brown, J. E.; Coles, J. A.; and Pinto, L. H. Effects of injections of calcium and EGTA into the outer segments of retinal rods of *Bufo marinus. J. Physiol.*, 269, 707–722 (1977).

Brown, J. E.; and Pinto, L. H. Ionic mechanism for the photoreceptor potential of the retina of *Bufo marinus. J. Physiol.*, 236, 575–591 (1974).

Carafoli, E.; and Crompton, M. The regulation of intracellular calcium by mitochondria. *Ann. N. Y. Acad. Sci.,* 307, 269–284 (1978).

Chader, G. J.; Fletcher, R. T.; and Krishna, G. Light-induced phosphorylation of rod outer segments by guanosine triphosphate. *Biochem. Biophys. Res. Commun.,* 64, 535–538 (1975).

Cohen, L. H.; and Noell, W. K. Glucose catabolism of the rabbit retina before and after development of visual function. *J. Neurochem.,* 5, 253–262 (1960).

DiPolo, R. Calcium efflux from externally dialyzed squid axons. *J. Gen. Physiol.,* 62, 575–589 (1973).

DiPolo, R. Effect of ATP on the calcium efflux in dialyzed squid giant axons. *J. Gen. Physiol.,* 54, 503–519 (1974).

Dolara, P.; Agresti, A.; Giotti, A.; and Pasquini, G. Effect of taurine on calcium kinetics of guinea pig heart. *Eur. J. Pharmacol.,* 24, 352–358 (1973).

Dolara, P.; Franconi, F.; Giotti, A.; Basosi, R.; and Valensin, G. Taurine-calcium interaction measured by means of ^{13}C nuclear magnetic resonance. *Biochem. Pharmacol.,* 27, 803–804 (1978).

Farber, D. B.; Brown, B. M.; and Lolley, R. N. Cyclic GMP: Proposed role in visual cell function. *Vision Res.,* 18, 497–499 (1978).

Fricke, U. Tritosol: A new scintillation cocktail based on Triton X 100. *Anal. Biochem.,* 63, 555–558 (1975).

Goridis, C.; Virmaux, N.; Cailla, H. L.; and Delaage, M. A. Rapid light-induced changes of retinal cyclic GMP levels. *FEBS Lett.,* 49, 167–169 (1974).

Goridis, C.; Urban, P. F.; and Mandel, P. The effect of flash illumination on the endogenous cyclic GMP content of isolated frog retinae. *Exp. Eye Res.,* 24, 171–177 (1977).

Hagins, W. A. The visual process: Excitatory mechanisms in the primary receptor cells. *Ann. Rev. Biophy. Bioeng.,* 1, 131–158 (1972).

Hagins, W. A.; and Yoshikami, S. A role for Ca^{++} in excitation of retinal rods and cones. *Exp. Eye Res.,* 18, 299–305 (1974).

Hagins, W. A.; Penn, R. D.; and Yoshikami, S. Dark current and photocurrent in retinal rods. *Biophys. J.,* 10, 380–412 (1970).

Hemmiki, D. Light-induced decrease in calcium binding to isolated bovine photoreceptors. *Vision Res.,* 15, 69–72 (1975).

Hendriks, T. H.; VanHaard, P. M. M.; Daemen, F. J. M.; and Bonting, S. L. Biochemical aspects of the visual process. XXXV: Calcium binding by cattle rod outer segment membranes studied by means of equilibrium dialysis. *Biochim. Biophys. Acta,* 467, 175–184 (1977).

Jacobsen, J. G.; and Smith, L. H. Biochemistry and physiology of taurine and taurine derivatives. *Physiol. Rev.,* 48, 424–511 (1968).

Kaupp, U. B.; and Junge, W. Rapid calcium release by passively loaded retinal discs on photoexcitation. *FEBS LETT.,* 81, 229–232 (1977).

Kennedy, A. J.; and Voaden, M. J. Free amino acids in the photoreceptor cells of the frog retina. *J. Neurochem.,* 23, 1093–1095 (1974).

Korenbrot, J. I.; and Cone, R. A. Dark ionic flux and the effect of light in isolated rod outer segments. *J. Gen. Physiol.,* 60, 20–45 (1972).

Krnjević, K.; and Puil, E. Electrophysiological studies on actions of taurine. In *Taurine,* R. Huxtable and A. Barbeau, eds. Raven Press, New York (1976), pp. 179–190.

Liebman, P. A. Rod disk calcium movement and transduction: A poorly illuminated story. *Ann. N.Y. Acad. Sci.,* 307, 643–644 (1978).

Lolley, R. N.; Brown, B. M.; and Farber, D. B. Protein phosphorylation in rod outer segments from bovine retina: Cyclic nucleotide-activated protein kinase and its endogenous substrate. *Biochem. Biophys. Res. Commun.,* 78, 572–578 (1977).

López-Colomé, A. M.; Pasantes-Moreales, H. Taurine interactions with retinal membranes. *J. Neurochem.,* 34, 1047–1052 (1980).

Mandel, P.; Pasantes-Morales, H.; and Urban, P. F. Taurine, a putative transmitter in retina. In *Transmitters in the Visual Process*, S. L. Bonting, ed. Pergamon Press, Oxford (1976), pp. 89–105.

Orr, H. T.; Cohen, A. I.; and Lowry, O. H. The distribution of taurine in the vertebrate retina. *J. Neurochem.*, 26, 609–611 (1976).

Pasantes-Morales, H.; Bonaventure, N.; Wioland, N.; and Mandel, P. Effect of intravitreal injections of taurine and GABA on chicken electroretinogram. *Int. J. Neurosci.*, 5, 235–241, (1973a).

Pasantes-Morales, H.; Urban, P. F.; Klethi, J.; and Mandel, P. Light stimulated release of ^{35}S-taurine from chicken retina. *Brain Res.*, 51, 375–378 (1973b).

Pasantes-Morales, H.; Salceda, R.; and López-Colomé, A. M. The role of taurine in retina: Factors affecting its release. In *Taurine*, R. Huxtable and A. Barbeau, eds. Raven Press, New York (1976), pp. 191–200.

Pasantes-Morales, H.; and Gamboa, A. Effect of taurine on ^{45}Ca^{2+} accumulation in rat brain synaptosomes. *J. Neurochem.*, 34, 244–246 (1980).

Penn, R. D.; and Hagins, W. A. Kinetics of the photocurrent of retinal rods. *Biophys. J.*, 12, 1073–1094 (1972).

Reuter, H. Kinetic aspects of calcium current in ventricular myocardial fibres. In *Calcium and Cellular Function*, A. W. Cuthbert, ed. Macmillan Press, London (1970), p. 261.

Riley, M. V.; and Voaden, M. J. The metabolism of the isolated retina: Some effects of calcium and potassium. *Ophthalmol. Res.*, 1, 58–64 (1970).

Rink, T. J. The influence of sodium and calcium movements and catecholamine release in thin slices of bovine adrenal medulla. *J. Physiol.*, 266, 297–325 (1977).

Rubin, R. P. *Calcium and the Secretory Process.* Plenum Press, New York (1974), pp. 1–124.

Salceda, R.; López-Colomé, A. M.; and Pasantes-Morales, H. Light stimulated release of ^{35}S taurine from frog retinal rod outer segments. *Brain Res.*, 135, 186–191 (1977).

Schnetkamp, P. P. M., Daemen, F. J. M.; and Bonting, B. L. Biochemical transport of the visual process. XXXVI: Calcium accumulation in cattle rod outer segments: Evidence for a calcium-sodium exchange carrier in the rod sac membrane. *Biochim. Biophys. Acta*, 468, 259–270 (1977).

Smith, G. H.; Fager, R. S.; and Litman, B. J. Light-activated calcium release from sonicated bovine retinal rod outer segment disks. *Biochemistry*, 7, 1399–1405 (1977).

Szuts, E. Z.; and Cone, R. A. Calcium content of frog rod outer segments and discs. *Biochim. Biophys. Acta*, 468, 194–208 (1977).

Thorn, N. A.; Russell, J. T.; Torp-Pedersen, C.; and Treimen, M. Calcium and neurosecretion. *Ann. N.Y. Acad. Sci.*, 307, 618–639 (1978).

Tomita, T. Electrical activity of vertebrate photoreceptors. *Q. Rev. Biophys.*, 3, 179–222 (1970).

Weller, M.; Virmaux, N.; and Mandel, P. Role of light and rhodopsin phosphorylation in control of permeability of retinal rod outer segment disks to Ca^{2+}. *Nature*, 256, 68–70 (1975).

Winkler, B. S.; Simson, V.; and Benner, J. Importance of bicarbonate in retinal function. *Invest. Ophthal. Vis. Sci.*, 161, 766–779 (1977).

Yates, R. A.; and Keen, P. The distribution of free amino acids in subdivisions of frog and rat retinae obtained by a new technique. *Brain Res.*, 107, 117–126 (1976).

Copyright ©1981, Spectrum Publications, Inc.
The Effects of Taurine on Excitable Tissues

CHAPTER 15

Taurine in Retinas of Taurine-deficient Cats and RCS Rats

Susan Y. Schmidt

Cats fed a taurine-free, casein diet develop retinal taurine deficiency and, subsequently, photoreceptor cell death (Berson et al., 1976; Schmidt et al., 1976; Schmidt et al., 1977). Supplementation of this diet with methionine, cysteine, inorganic sulfate, vitamin B_6, or vitamin B_6 plus cysteine did not prevent development of retinal taurine deficiency and retinal malfunction. A synthetic amino acid diet devoid of casein and taurine also resulted in retinal taurine deficiency and retinal malfunction. Only taurine-containing diets (i.e., chow or casein plus taurine) preserved normal retinal taurine concentrations and electroretinogram (ERG) amplitudes. These findings have firmly established a role for exogenous taurine in maintaining normal retinal function in the cat.

In taurine-deficient cats the decreases in retinal taurine concentrations and reductions in ERG amplitudes have been closely correlated (Fig. 15.1). This correlation could be demonstrated prior to detectable cell death as measured by changes in retinal deoxyribonucleic acid (DNA) concentrations. Reductions in retinal taurine concentrations below 50% of normal were associated with the appearance of abnormal granularity in the area centralis and reduction in retinal DNA concentrations. At a time when retinal taurine concentrations were reduced 70–80% below normal, the ERG responses were small or nondetectable, and ultrastructural studies showed photoreceptor cell death (Hayes et al., 1975).

In vitro studies (Schmidt, 1979) have shown that homogenates of cat livers fail to convert ^{35}S-cysteine and ^{35}S-cysteic acid to taurine. Under identical conditions homogenates of rat livers synthesized appreciable amounts of ^{35}S-

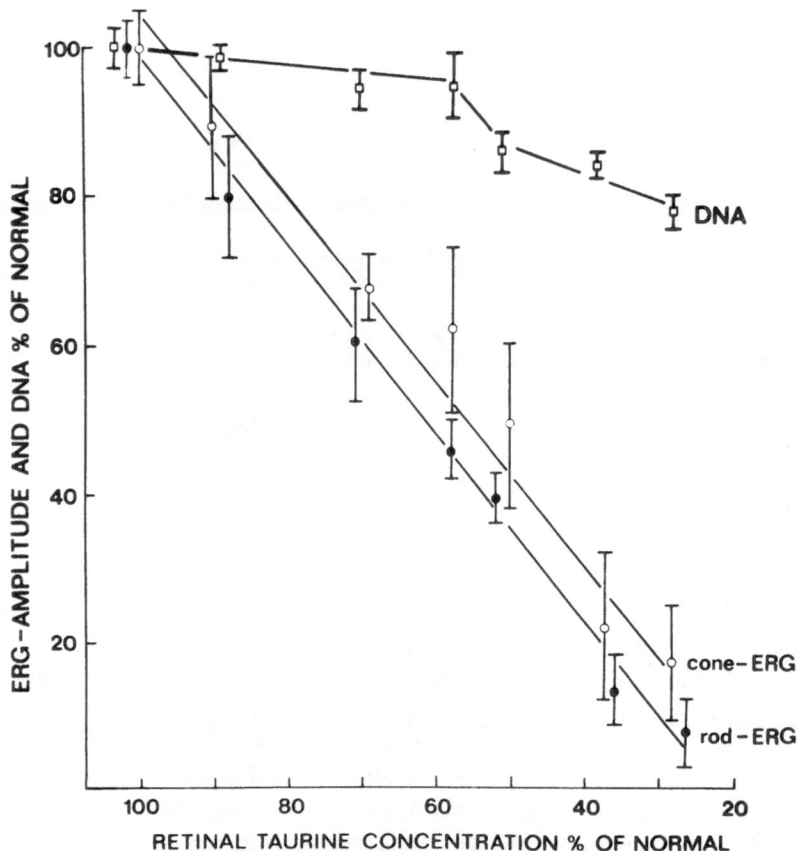

Fig. 15.1. Peak-to-peak ERG amplitudes and retinal DNA concentrations related to retinal taurine concentration in 10 control and 38 taurine-deficient cats. Controls were considered to have 100% retinal taurine concentration, ERG amplitude, and retinal DNA concentration. Taurine-deficient cats were divided into six groups of four to eight, according to the amount of taurine in their retinas. For each group, average amplitudes (mean ± SEM) for rod (black dots) and cone (circles) responses, and average DNA values (squares) are presented. The coefficients of correlation for rod ERG amplitude and cone ERG amplitude to retinal taurine concentration were 0.90 and 0.84, respectively. (Reproduced with permission from Schmidt, Berson, Watson, and Huang, *Invest. Ophthalmol.*, 16, 673–678, 1977.)

taurine (0.01–0.03 nmol/mg protein/min).[1] The failure to detect measurable taurine synthesis *in vitro* in cat livers is consistent with previous findings of

[1]For studies of taurine synthesis from cysteic acid, incubations were conducted according to MacDonnell and Greengard (1975) in the presence of 20 mM ^{35}S-cysteic acid. For studies of taurine synthesis from ^{35}S-cysteine, the incubations were done according to Misra and Olney (1975). Pyridoxal-5′-phosphate (0.5 mM) was present in all incubations. Taurine synthesis was evaluated by thin-layer chromatography as previously described (Schmidt et al., 1977).

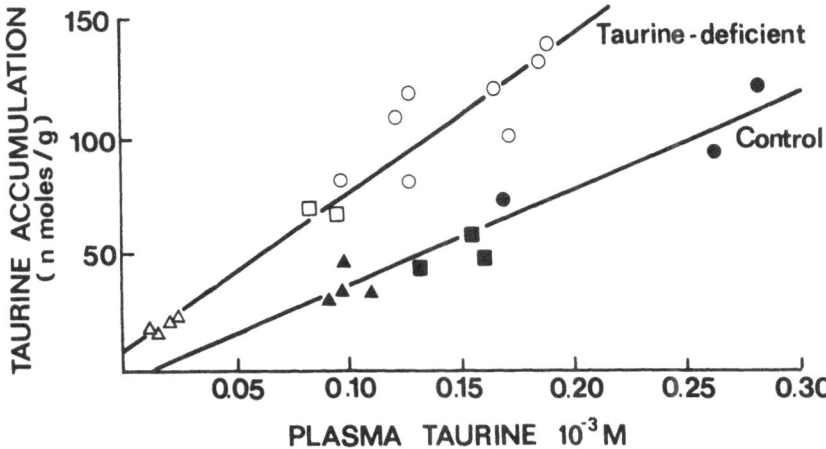

Fig. 15.2. Taurine accumulation (nmol/g wet weight) by the retina of control (▲, ■, ●) and slightly taurine-deficient cats (△, □, O), related to plasma taurine concentration. The cats were injected with either ^{3}H-taurine (▲, △), ^{35}S-taurine (■, □) or ^{14}C-taurine (●, O). The coefficients of correlation are 0.81 for taurine-deficient cats and 0.90 for the control group. (Reproduced with permission from Schmidt, Berson, Watson, and Huang, *Invest. Ophthalmol.*, 16, 673–678, 1977.)

low levels of cysteine decarboxylase activity in cat livers (Jacobsen et al., 1964; Hardison et al., 1977). No taurine synthesis could be detected in retinal homogenates of either cats or rats (Schmidt, 1979). This suggests that in both species, retinal taurine concentrations are maintained by uptake of taurine from the plasma and not by synthesis *in situ* in the retina.

The uptake of taurine into the outer nuclear layer (where photoreceptor cell bodies and Müller cell processes are located) undoubtedly occurs at least in part via transport of taurine from the plasma across the pigment epithelium. The pigment epithelium has been shown to be an active site of taurine accumulation *in vivo* in rats and frogs (Young, 1969), and in tissue culture (Edwards, 1977). Schmidt et al. (1977) have shown that 48–72 hr after injection of radioactive taurine of differing specific activities, the accumulation of labeled taurine by the retina could be correlated with plasma taurine concentration; accumulation was higher in slightly taurine-deficient retinas than in normal retinas for a given plasma taurine concentration (Fig. 15.2). Studies with tritiated taurine of high specific activity showed that uptake of taurine into the retina could be detected even when plasma levels of taurine were reduced to 10–20% of normal in slightly taurine-deficient cats; this could explain why retinal levels of taurine remained nearly normal as plasma concentrations of taurine were decreasing in these animals. In control cats, retinal taurine concentrations are maintained *in vivo* against a 400-fold gradient (taurine concentrations in retina and plasma were about 40 mM and 0.1 mM, respectively). Moreover, in the slightly taurine-deficient cat, the retina has a capacity to concentrate taurine against a 1,600-fold gradient (taurine concentrations in retina and plasma were about 32 mM and 0.02 mM, respectively) (Schmidt, 1979).

Kinetic analysis of taurine uptake by isolated cat retinas (Fig. 15.3) revealed that two mechanisms exist for uptake of taurine. One process, designated as a high-affinity uptake mechanism, has an apparent Michaelis constant (Km) of 50 μM and a maximal velocity (Vmax) of 0.55 nmol/mg dry wt./min (Schmidt, 1979). The second process, designated as a nonsaturable

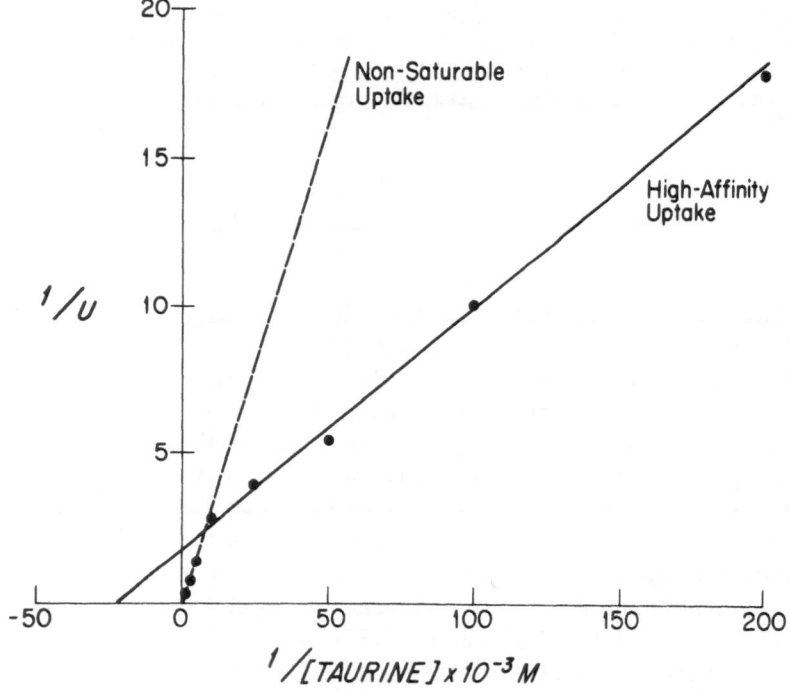

Fig. 15.3. Representative double reciprocal plot of ^{3}H-taurine uptake (U) vs. taurine concentration in the medium in isolated normal control cat retinas. The high-affinity uptake mechanism (solid line) was derived with least-square analysis to best approximate the data point for the linear portion of the double reciprocal plot (between 0.005 and 0.10 mM taurine in the medium). At concentrations above 0.1 mM taurine in the medium, the data points for the nonsaturable mechanism approach zero (hatched line).

mechanism, has a rate constant of 3.20 nmol/mg/dry wt./min/mM taurine in the medium. In isolated retinas from slightly taurine-deficient cats, the Km value was 29 μM, suggesting an increased affinity for taurine in these retinas. The high-affinity uptake mechanism, in contrast to the nonsaturable mechanism, could not be detected in isolated cat retinas after the photoreceptor cells had degenerated at late stages of taurine deficiency. Comparison of taurine uptake in normal and photoreceptorless retinas shows that the high-affinity uptake mechanism is inhibited to a greater extent than the nonsaturable mechanism by reduced temperature, ouabain (100 μM), and omission of glucose from the incubation medium (Table 15.1). At low concentrations (0.005–0.010 mM) of taurine in the media, 60–70% of taurine uptake was due to the high-affinity uptake mechanism and 30–40% to the nonsaturable mechanism. Autoradiograms of isolated retinas of control and slightly taurine-deficient cats incubated in dim light in media containing low concentrations of ^{3}H-taurine showed that accumulation of ^{3}H-taurine was greatest over the photoreceptor cell layer (Schmidt and Szamier, 1978; Lake et al., 1978). Microdissection of the same incubated cat retinas (Schmidt, 1979) showed that 60–70% of the retinal radioactivity was in the outer retina (Fig. 15.4), where endogenous taurine is also known to be concentrated (Cohen et al., 1973; Kean and Yates, 1974; Kennedy et al., 1977; Orr et al., 1976; Schmidt et al., 1976). Within the inner retina, uptake has been observed in Müller cells, and in some horizontal cells and amacrine cells (Lake et al., 1978).

In contrast to the taurine-deficient cat, in which retinal taurine deficiency precedes photoreceptor cell death, decreases in retinal taurine concentrations in the RCS/p rat occur simultaneously with photoreceptor cell death

Table 15.1 Effects of metabolic inhibitors on the high-affinity and nonsaturable mechanisms for taurine uptake by isolated cat retinas[a]

Conditions of incubation	Normal retinas	Photoreceptorless retinas
37°C (control)	100 ± 8.2	100 ± 8.6
34°C	66 ± 3.2	77 ± 3.7
30°C	23 ± 6.7	54 ± 5.0
Ouabain (100 μM)	50 ± 3.5	69 ± 5.0
No glucose	54 ± 4.1	76 ± 6.2

[a]Sections of retinas were incubated in media containing ^{3}H taurine (0.005 mM). The values are expressed as percentages of control and represent the means ± SEM for three to six separate incubations. The uptake rates were, respectively, 0.059 ± 0.005 and 0.025 ± 0.0024 nmol/mg dry wt./min for normal and photoreceptorless retinas.

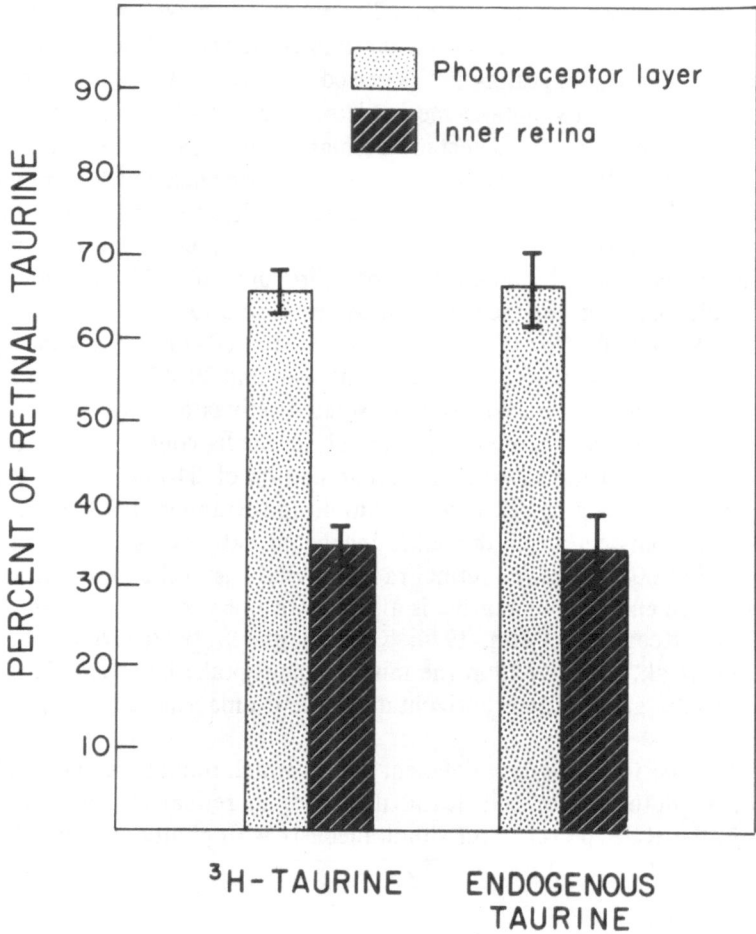

Fig. 15.4. Distribution of labeled taurine taken up from the medium and endogenous taurine in the microdissected photoreceptor cell layer and inner retina. Isolated cat retinas were incubated for 10 min in dim light in media containing 0.05 mM ^{3}H-taurine. The vertical lines within the bars indicate ± SEM for six different experiments.

after the third week of postnatal life. During early postnatal life, the RCS rat retina develops comparably to that of normal rats; the retina of 21–23-day-old RCS rats is similar to that of the normal rat with respect to the thickness of the outer nuclear layer, retinal DNA and taurine concentrations, and ERG amplitude. After the 23rd postnatal day, the photoreceptor cells begin to degenerate, and thereafter the reductions in retinal taurine content (Fig. 15.5)

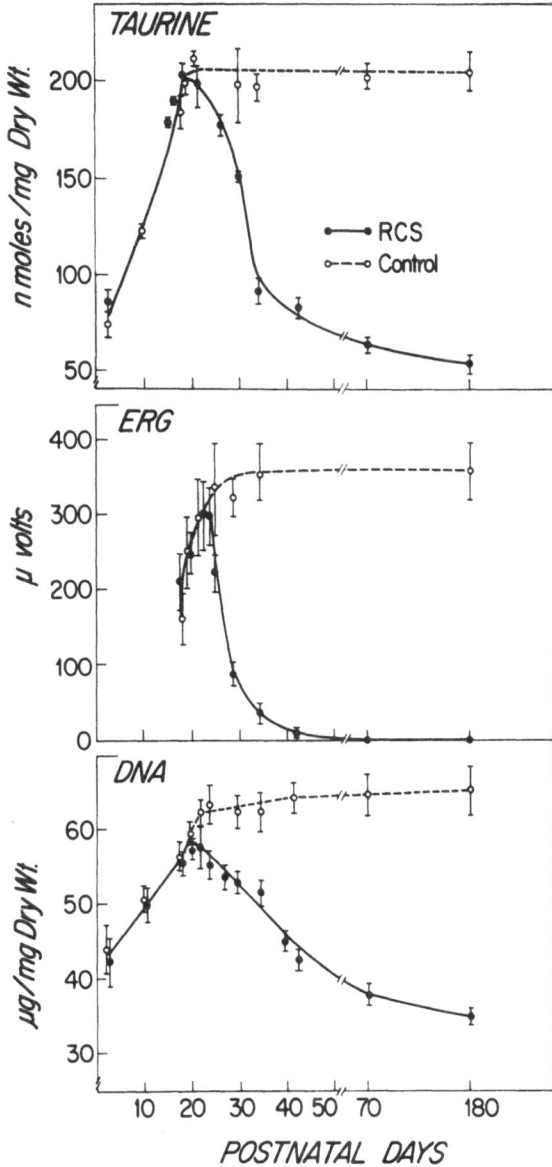

Fig. 15.5. Retinal taurine concentrations, ERG amplitudes, and retinal DNA concentrations at various postnatal ages in normal Long-Evans rats and pigmented RCS rats with hereditary retinal degeneration. Each data point and vertical bar represents the mean ± SEM for measurements from three to five rats. (Reproduced with permission from Schmidt and Berson, *Exp. Eye Res.,* 27, 191–198, 1978.)

can be correlated with reduction in the thickness of the outer nuclear layer, decrease in retinal DNA concentration, and a decline in ERG amplitude.

As discussed above for the cat retina, two processes exist for the uptake of taurine in isolated rat retinas. The high-affinity uptake mechanism has an apparent Km value of about 50 μM (Schmidt and Berson, 1978; Neal et al., 1973) and a Vmax of 0.31 nmol/mg/min. The nonsaturable uptake mechanism has a rate constant of 0.63 nmol/mg dry wt./min/mM taurine in the medium (Schmidt and Berson, 1978). The high-affinity uptake mechanism in the rat retina is inhibited to a greater extent than the nonsaturable mechanism by ouabain and reduced temperature. Retinas from normal rats studied at ages 22–180 days and from RCS rats studied at ages 22–45 days show both mechanisms at a time when photoreceptor cell function is detectable *in vivo* with the ERG. Retinas from 180-day-old photoreceptorless RCS rats retain the nonsaturable uptake mechanism at a time when no photoreceptor function can be detected in the ERG; this again demonstrates that the high-affinity mechanism for taurine uptake depends on the presence of viable photoreceptor cells (Schmidt and Berson, 1978).

The high-affinity uptake mechanism, but not the nonsaturable mechanism, is lost in the isolated normal rat retina when the photoreceptor cells are mechanically disrupted and is also lost when the isolated normal rat retina is maintained in room light for 20 min or in darkness for 60 min, conditions associated with a large efflux of taurine from the isolated retina and presumed photoreceptor cell death (Schmidt, 1978).

In addition to the high-affinity uptake mechanism for taurine, light-evoked fluxes of taurine have also been associated with the photoreceptor cells (Schmidt, 1978). Onset and cessation of illumination have been shown to be associated with a prompt transient release followed by reuptake of taurine by isolated retinas incubated in media containing labeled taurine. Light-evoked taurine fluxes are also observed in the isolated retinas from normal rats and 30-day-old RCS rats but cannot be demonstrated in the photoreceptorless retinas from 180-day-old RCS rats.

Studies with taurine-deficient cats have provided a new approach to the study of the cell biology of photoreceptor cells. The mechanism by which retinal taurine deficiency leads to photoreceptor cell death remains to be defined. The close correlation of retinal taurine deficiency with peak-to-peak ERG amplitudes in taurine-deficient cats suggests that taurine deficiency may have some effect on the ionic fluxes of Na and K involved in the generation of the ERG. Since the generation of the ERG depends on hyperpolarization of photoreceptor cells and depolarization of Müller cells, it is possible that taurine deficiency has led to abnormal ionic concentrations (Na and K) in photoreceptor and Müller cells. This possibility is currently under investigation.

ACKNOWLEDGMENTS

This work was supported by Research Grant EY01687 from the National Eye Institute and by grants from the National Retinitis Pigmentosa Foundation, Baltimore, Md., and the George Gund Foundation, Cleveland, Ohio.

REFERENCES

Berson, E. L.; Hayes, K. C.; Rabin, A. R.; Schmidt, S. Y.; and Watson, G. Retinal degeneration in cats fed casein: II. Supplementation with methionine, cysteine or taurine. *Invest. Ophthalmol.*, 15, 52–58 (1976).

Cohen, A. I.; McDaniel, M.; and Orr, H. Absolute levels of some free amino acids in normal and biologically fractionated retinas. *Invest. Ophthalmol.*, 12, 686–93 (1973).

Edwards, R. B. Accumulation of taurine by cultured retinal pigment epithelium of the rat. *Invest. Ophthalmol. & Vis. Sci.*, 16, 201–208 (1977).

Hardison, W. G. M.; Wood, C. A.; and Proffitt, J. H. Quantification of taurine synthesis in the intact rat and cat liver. *Proc. Soc. Exp. Biol. Med.*, 155, 55–58 (1977).

Hayes, K. C.; Rabin, A. R.; and Berson, E. L. An ultrastructural study of nutritionally induced and reversed retinal degeneration in cats. *Am. J. Pathol.*, 78, 505–524 (1975).

Jacobsen, J. G., Thomas, L. L.; and Smith, L. H., Jr. Properties and distribution of mammalian L-cysteine sulfinate carboxylases. *Biochem. Biophys. Acta*, 85, 103–116 (1964).

Keen, P.; and Yates, R. A. Distribution of amino acids in subdivided rat retinae. *Br. J. Pharmacol.*, 52, 118P (1974).

Kennedy, A. J.; Neal, M. J.; and Lolley, R. N. The distribution of amino acids within the rat retina. *J. Neurochem.*, 29, 157–159 (1977).

Lake, N.; Marshall, J.; and Voaden, M. J. High affinity uptake sites for taurine in the retina. *Exp. Eye Res.*, 27, 713–718 (1978).

MacDonnell, P.; and Greengard, O. The distribution of glutamate decarboxylase in rat tissues: Isotopic vs fluorometric assays. *J. Neurochem.*, 24, 615–618 (1975).

Misra, C. H.; and Olney, J. W. Cysteine oxidase in brain. *Brain Res.*, 97, 117–126 (1975).

Neal, M. J.; Peacock, D. G.; and White, R. D. Kinetic analysis of amino acid uptake by the rat retina in vitro. *Br. J. Pharmacol.*, 47, 656–670 (1973).

Orr, H. T.; Cohen, A. I.; and Lowry, O. H. The distribution of taurine in the vertebrate retina. *J. Neurochem.*, 26, 609–611 (1976).

Schmidt, S. Y. Taurine fluxes in isolated cat and rat retinas: Effects of illumination. *Exp. Eye Res.*, 26, 529–535 (1978).

Schmidt, S. Y. Unpublished observations (1979).

Schmidt, S. Y.; and Berson, E. L. Taurine uptake in isolated retinas of normal rats and rats with hereditary retinal degeneration. *Exp. Eye Res.*, 27, 191–198 (1978).

Schmidt, S. Y.; and Szamier, R. B. Taurine uptake by the photoreceptor cells in the isolated cat retina. Abstract, Association for Research in Vision and Ophthalmology, Sarastota, Florida, (1978).

Schmidt, S. Y.; Berson, E. L.; and Hayes, K. C. Retinal degeneration in cats fed casein: I. Taurine deficiency. *Invest. Ophthalmol.*, 15, 47–52 (1976).

Schmidt, S. Y.; Berson, E. L.; Watson, G., and Huang, C. Retinal degeneration in cats fed casein: III. Taurine deficiency and ERG amplitudes. *Invest. Ophthalmol.*, 16, 673–678 (1977).

Young, R. W. The organization of vertebrate photoreceptor cells. In *The Retina: Structure, Function and Clinical Characteristics*, Straatsma, B.; Allen, R.; Hall, M.; and Crescitelli, F., eds. University of California Press, Los Angeles (1969), pp. 177-209.

Copyright © 1981, Spectrum Publications, Inc.
The Effects of Taurine on Excitable Tissues

CHAPTER 16

Axonal Transport of Taurine

John A. Sturman

Taurine is present in high concentrations in mammalian brain and in especially high concentrations in developing brain, in which it is the free amino acid present in the greatest concentration (Sturman et al., 1978). Little is known about the functions of taurine, other than its role in bile acid conjugation, an observation made over a century ago (Strecker, 1849) and repeatedly confirmed (Danielsson, 1963). Taurine has been proposed as an inhibitory neurotransmitter in brain and retina (Mandel and Pasantes-Morales, 1978; Oja and Kontro, 1978; Mandel et al., 1976), although it is present in far greater amounts than would be required for such a neurophysiological function. An alternative function of taurine may be that of membrane stabilization. It has been shown that taurine increases the permeability of lobster giant axon membranes to potassium and chloride but not to sodium (Gruener and Bryant, 1975). Although such ion fluxes are consistent with the actions of an inhibitory neurotransmitter, the observations were made on a region of lobster axon void of synaptic receptors. Taurine has been implicated in some forms of epilepsy, both in animal models and in man (Barbeau et al., 1975; Van Gelder, 1976, 1978). Frequently the concentration of taurine at the epileptic focus is lower than that in the surrounding tissue, and this change is often accompanied by disturbances in glutamic acid metabolism. Taurine therapy of epilepsy has been tested clinically with some success. A role for taurine in developing brain has been suggested in addition to any functional role it may have in the mature brain (Sturman and Gaull, 1975; Sturman et al., 1978). More recently, taurine has been demonstrated to be involved in maintaining the structural integrity of the photoreceptor cells of the retina of the cat (Hayes et al., 1975; Schmidt et al., 1976), at least; and to be necessary for maintaining the normal lattice array of tapetal rods within cells of the tapetum lucidum cellulosum of the cat (Wen et al., 1979).

Taurine has recently been demonstrated to be transported axonally in goldfish optic nerves (Ingoglia et al., 1976, 1978), and this function has been examined more extensively in optic nerves of developing rabbits and rats (Sturman, 1979; Politis and Ingoglia, 1979). The process of axonal transport was discovered by Weiss and Hiscoe (1948) when they observed that material accumulated proximal to the site of a constriction in a nerve. Based on this observation they postulated that materials from the cell body flowed along the axon to the nerve terminals. This hypothesis was subsequently confirmed and extended following the advent of the use of radioactive chemicals in biology. Such experiments have demonstrated that various components of the cytoskeleton of the axon are continually being replaced with material synthesized in the cell body. The same process is used also for the delivery of compounds involved in neurotransmission to the nerve terminals. Thus, to date, it has been demonstrated that a multitude of proteins, nucleic acids, phospholipids, and small molecules are transported from the cell body along the axon to the nerve terminals. The usual design is to introduce the radioactive compound, commonly an amino acid, into the vicinity of the cell body, and to examine the time course and nature of the radioactivity in the axon and/or nerve terminals. The visual system is especially suited to studies of axonal transport for a number of reasons. The eye is readily accessible for injection of material into the vitreous fluid, which then has easy access to the perikarya of the retinal ganglion cells, the axons of which emerge from the eye through the optic disk into the optic nerve. The optic nerve is a clearly defined bundle of fibers that terminate in easily recognized, well-defined brain nuclei, in the optic tectum in lower vertebrates, such as the goldfish, and in the lateral geniculate body or superior colliculus in mammals, such as the rabbit. Furthermore, a relatively large volume can be injected into the vitreous fluid, and its dispersal is limited to a relatively constant volume.

In the experiments described in this chapter, radioactive taurine (or other radioactive compounds) was injected into the right eye of goldfish, rabbits, or rats. Since the optic axons of the goldfish cross completely at the optic chiasm, and all but a few of the optic axons of the rabbit and rat cross at the optic chiasm, radioactivity in the tectum, lateral geniculate body, or superior colliculus ipsilateral to the injected eye represents material that has arrived by the general circulation, whereas radioactivity in the contralateral tectum, lateral geniculate body, or superior colliculus represents material that has arrived by migration along the optic axons as well as by the general circulation (Fig. 16.1). The amount of radioactivity migrating via the optic axons is therefore calculated by subtracting the radioactivity found in the ipsilateral visual structure from that found in the corresponding contralateral visual structure.

The first demonstration of axonal migration of taurine was in the goldfish visual system (Ingoglia et al., 1976). Within 24 hr after intravitreal injection

OPTIC TECTUM

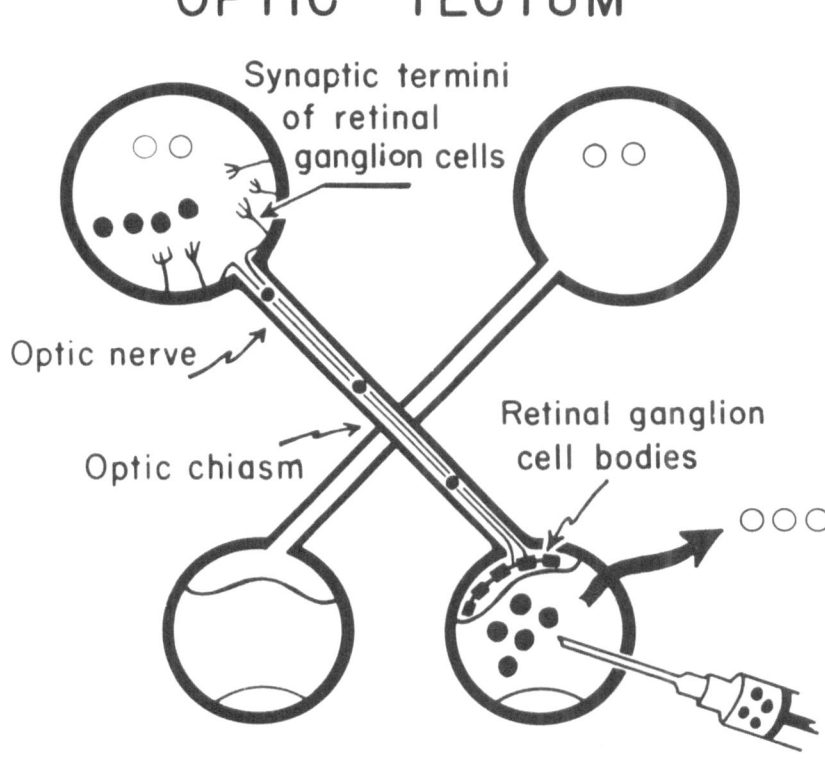

Fig. 16.1. Schematic diagram of the goldfish visual system illustrating the use of intravitreal injection of labeled material for investigating axonal transport phenomenon. Labeled material can enter the general circulation (open circles) and accumulate equally at both optic tecta. Because the optic axons of the goldfish cross completely at the optic chiasm, labeled material that migrates via the optic axons (closed circles) reaches only the contralateral optic tectum. Thus the difference in radioactivity between the contralateral and ipsilateral optic tecta represents that labeled material which has migrated via the optic axons, presumably by axonal transport. The situation is similar for the rabbit and the rat since most of the optic axons in these species cross completely at the optic chiasm. In mammals, the optic axons terminate in the lateral geniculate body or superior colliculus, rather than in the optic tectum, but the principle remains the same.

of [^{35}S]taurine, large differences were obtained in the amounts of radioactivity present in the contralateral and ipsilateral optic tecta (Fig. 16.2). Furthermore, this radioactivity was all present in the soluble fraction of a

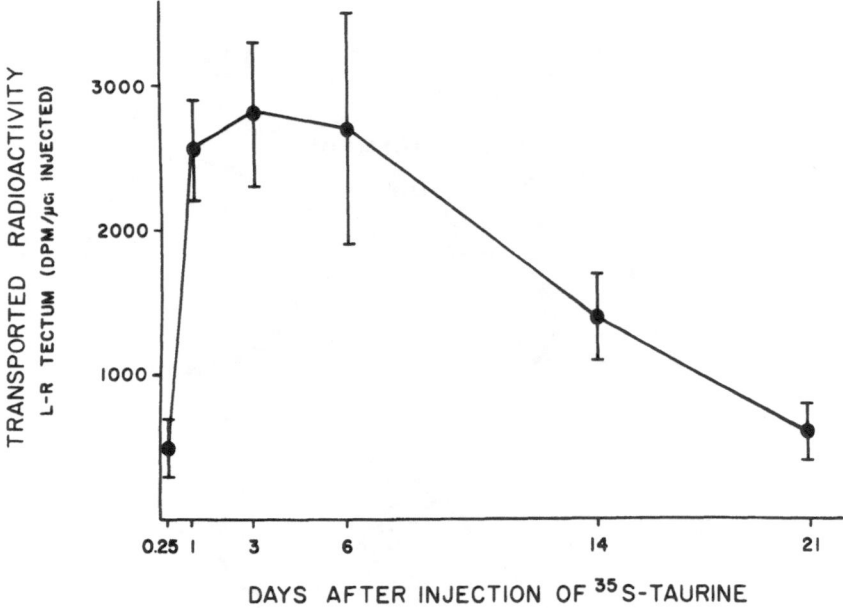

Fig. 16.2. Axonal transport of [³⁵S]taurine in the goldfish visual system. In this case, 1 μCi [³⁵S]taurine in 4 μl saline was injected into the right eye of goldfish. TCA-soluble radioactivity was extracted from each tectum and measured. Each point represents the mean ± SEM of left minus right tectum radioactivity from six goldfish. Taken from Ingoglia et al. (1976), with permission.

100% trichloroacetic acid (TCA) extract, and analysis using an automatic amino acid analyzer in conjunction with a flow-cell scintillation spectrometer (Sturman, 1973) demonstrated that the radioactivity remained in the form of taurine. Similar experiments in which [³⁵S]cysteine or [³⁵S]cystathionine, metabolic precursors of taurine, was injected intravitreally also resulted in greater amounts of radioactivity in the contralateral optic tectum (Ingoglia et al., 1976, 1978) (Fig. 16.3). In these experiments, radioactivity was present both in the TCA-soluble fraction and in the TCA-insoluble fraction. The radioactivity in the TCA-insoluble fraction represents [³⁵S]cysteine incorporated into proteins in the retinal ganglion cells and transported axonally. The TCA-soluble radioactivity was present only in the form of [³⁵S]taurine and [³⁵S]inorganic sulfate, which were present also in the retina. Other radioactive compounds were also present in retina, presumptively identified as oxidized and reduced glutathione. The radioactive inorganic sulfate presumably reached the tectum through the general circulation since it has been shown that free inorganic sulfate is not transported axonally (Elam et

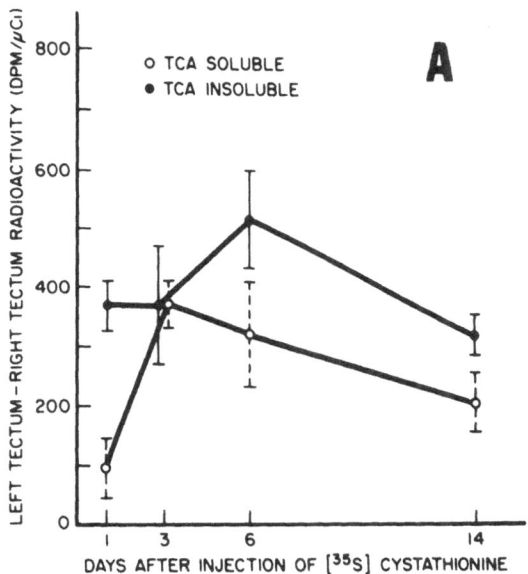

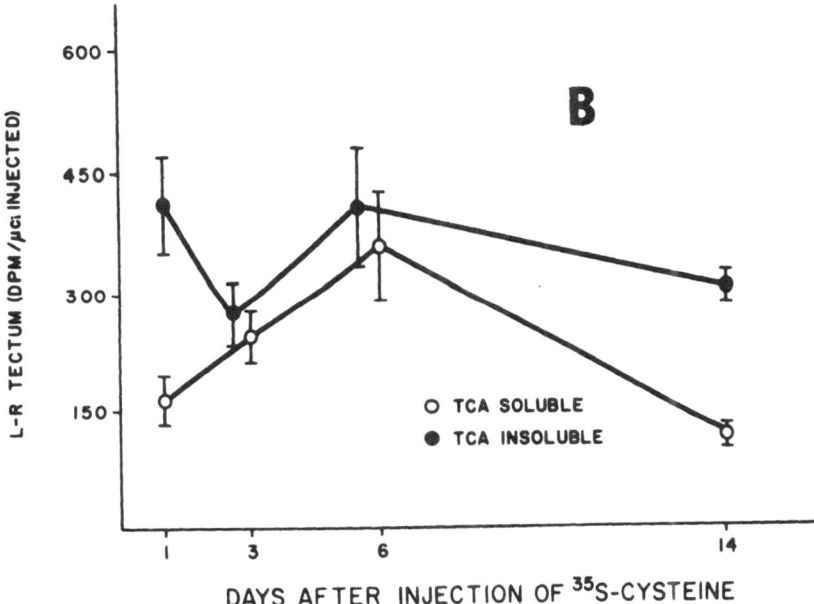

Fig. 16.3. Radioactivity transported in the goldfish visual system after the intravitreal injection of (a) [³⁵S]cystathionine or (b) [³⁵S]cysteine. In each case 1 μCi of labeled material in 4 μl saline was injected into the right eye of goldfish. TCA-soluble (open circles) and TCA-insoluble (closed circles) radioactivity was extracted from each tectum and measured. Each point represents the mean ± SEM of left minus right tectum radioactivity from six goldfish. The radioactivity in the TCA-soluble fractions is composed primarily of taurine. Adapted from Ingoglia et al. (1976, 1978), with permission.

al., 1970, 1971; Karlsson and Linde, 1977). These experiments demonstrated that only taurine, and not its metabolic precursors, migrates axonally. Since other free amino acids are not generally transported axonally (Di Giamberadino, 1971; Elam and Agranoff, 1971; Neale et al., 1974; Karlsson, 1977), taurine is unique in this respect.

The movement of any material from the retina to the optic tectum may be the result of intra- or extra-axonal diffusion or axonal transport. Intra-axonal diffusion seems unlikely since molecules of similar size—such as inorganic sulfate, amino acids, and the diamine putrescine—have been shown not to diffuse to the tectum in the time period studied (Elam et al., 1970; Elam and Agranoff, 1971; Ingoglia et al., 1977). Extra-axonal diffusion can be ruled out since when [^{14}C]mannitol was injected intravitreally, there was no difference in radioactivity in the contralateral and ipsilateral tecta (Ingoglia et al., 1978) (Fig. 16.4).

Examination of radioactivity reaching the contralateral tectum shortly after an intravitreal injection of [^{35}S]taurine or [^{3}H]leucine in goldfish indicated a similarity between taurine transport and the fast component of protein transport (Fig. 16.4). The possible association of these phenomena was examined further by testing the effects of cycloheximide, an inhibitor of protein synthesis, and vinblastine, an inhibitor of protein transport (which presumably acts by disassembling microtubules), on the axonal transport of taurine and of proteins. In these experiments, cycloheximide was injected into the right eye of goldfish 1 hr prior to the injection of [^{35}S]taurine and [^{3}H]proline, whereas vinblastine was injected 2 hr after the injection of [^{35}S]taurine and [^{3}H]proline. Cycloheximide inhibited protein synthesis in the retina, as expected, as well as the subsequent rapid axonal transport of protein (Fig. 16.5). It decreased the uptake of taurine by retina as well as the amount of taurine transported axonally. Vinblastine did not decrease the amount of protein synthesized in the retina but did decrease the amount transported axonally (Fig. 16.5). It decreased the amount of taurine taken up by the retina and the amount of taurine transported axonally. Although these results are somewhat equivocal, there is the suggestion of some association between axonal transport of proteins and of taurine.

Further characterization of the axonally transported taurine was attempted by examining its subcellular distribution in homogenates of tecta. For these studies, the subcellular distribution of [^{35}S]taurine 24 hr after the intravitreal injection of [^{35}S]taurine was compared, in separate experiments, with the subcellular distribution of radioactive protein after the concurrent intravitreal injection of [^{3}H]proline, and with the subcellular distribution of [^{3}H]taurine after the concurrent intracranial injection of [^{3}H]taurine. The greatest proportions of the axonally transported proteins labeled with [^{3}H]proline were present in the crude mitochondrial and synaptosomal frac-

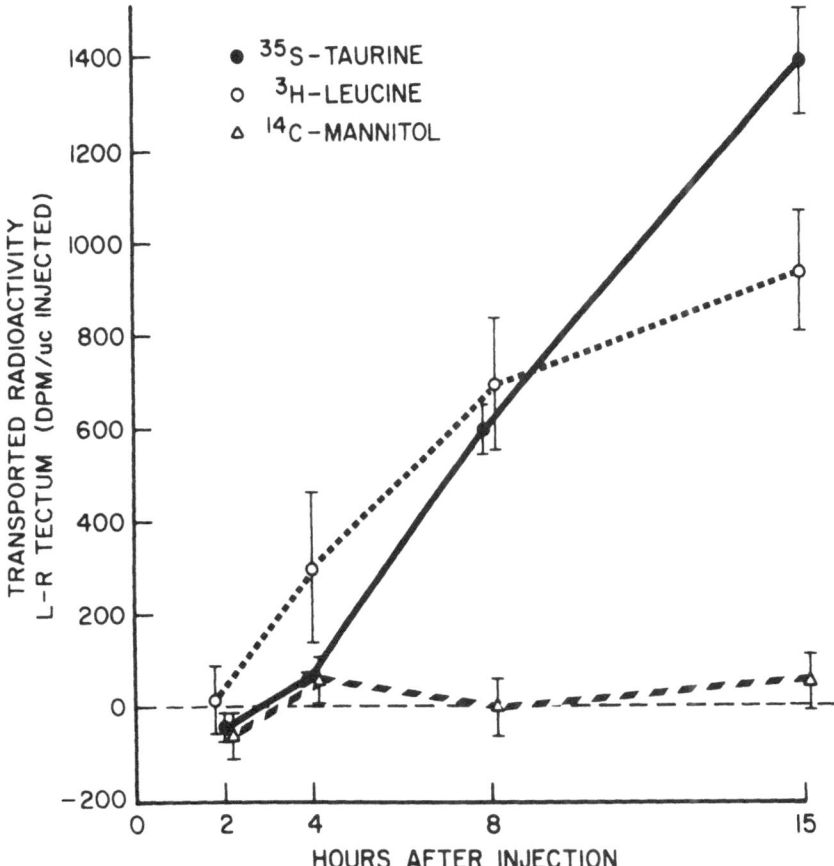

Fig. 16.4. Radioactivity transported in the goldfish visual system after the intravitreal injection of [^{35}S]taurine, [^{3}H]leucine, or [^{14}C]mannitol. In each case, 1 μCi of labeled material in 4 μl saline was injected into the right eye of goldfish. Each tectum was solubilized and radioactivity measured. Each point represents the mean $\pm$ SEM of left minus right tectum radioactivity from four to six goldfish. Adapted from Ingoglia et al. (1978), with permission.

tion, P_2, and in the crude nuclear fraction P_1 (Fig. 16.6). The subfractions prepared from P_2 that contained most of the radioactivity were the disrupted synaptosomal fraction, H, and the intact synaptosomal fraction, B (Fig. 16.6b). This result indicates that most of the axonally transported protein was incorporated into the membranes at the nerve ending, as has been suggested by other similar experiments (Elam and Agranoff, 1971; Elam et al., 1971). The bulk of the [^{35}S]taurine remained in the supernatant fraction, although some was present in all of the other crude fractions, the most enriched being

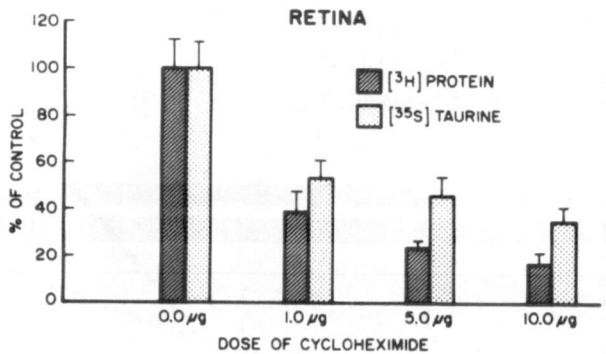

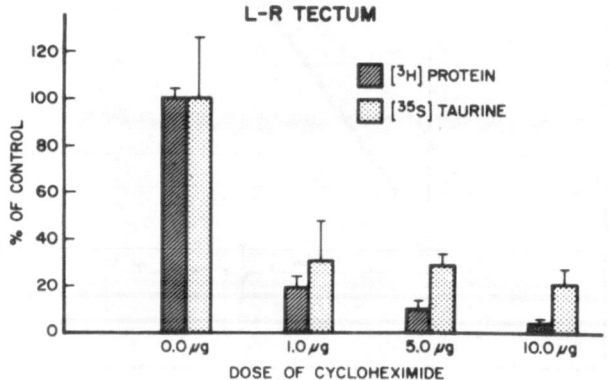

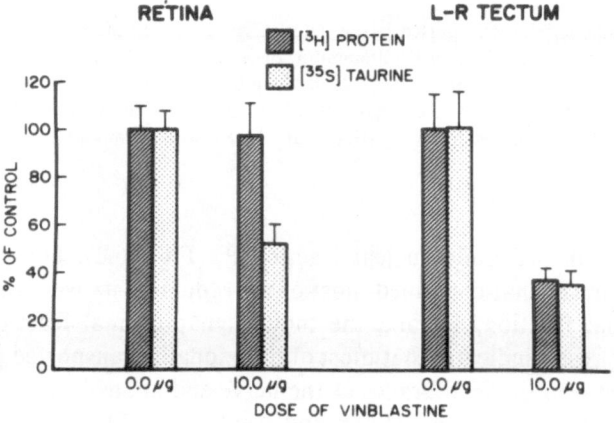

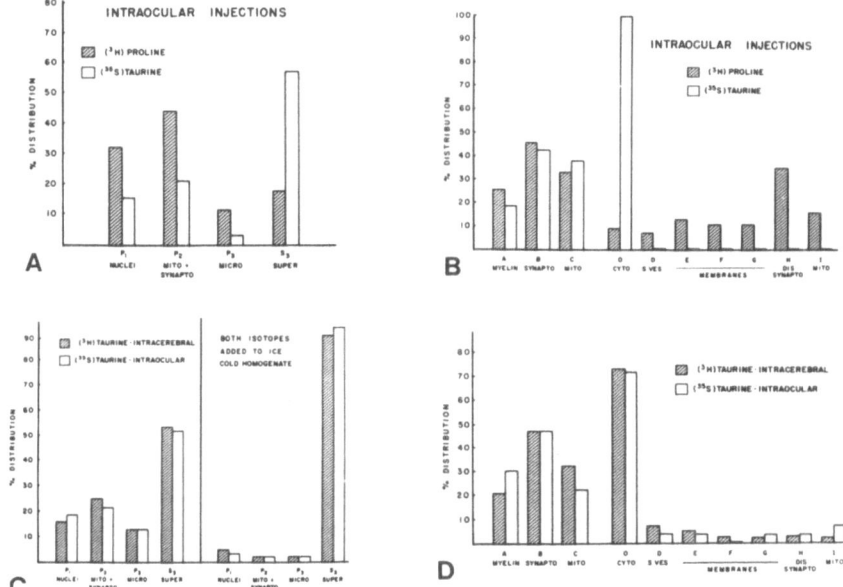

Fig. 16.6. (*a* and *b*) Subcellular distribution of [³⁵S]taurine and [³H]proteins in goldfish tecta after intravitreal injection of [³⁵S]taurine and [³H]proline. (*c* and *d*) Subcellular distribution of [³⁵S]taurine and [³H]taurine in goldfish tecta after intravitreal injection of [³⁵S]taurine and intracranial injection of [³H]taurine. Fractionations were performed on tecta 24 hr after injection of labeled material. Results are presented in each case as percent distribution. Adapted from Ingoglia et al. (1978), with permission.

the crude mitochondrial and synaptosomal fraction, P_2. The subfractions prepared from P_2 indicated that the [³⁵S]taurine was associated with the enriched synaptosomal fraction, B, and with the soluble fraction, S_3. The subcellular distribution of [³H] taurine after intracranial injection was identical to that of [³⁵S]taurine after intravitreal injection (Figs. 16.6c and d). Thus taurine administered by either route has apparently equal access to the various subcellular fractions of tecta. The results of a further experiment in

Fig. 16.5. Effect of cycloheximide or vinblastine on the axonal transport of taurine and proteins in the goldfish visual system. Cycloheximide was injected into the right eye of goldfish 1 hr prior to the injection of [³⁵S]taurine and [³T]proline; vinblastine was injected into the right eye of goldfish 2 hr after injection of ¹³⁵S]taurine and [³H]proline. The fish were killed 1 day later and the radioactivity in the right retina and each tectum measured after solubilization. Values represent the means ± SEM from six goldfish expressed as percent of control. Taken from Ingoglia et al. (1978), with permission.

which [^{35}S]taurine and [^{3}H]taurine were added to a cold homogenate prior to fractionation ensured that the subcellular distribution of taurine was a real phenomenon and not simply an artifact of the preparation procedure (Figs. 16.6c and d). A possible explanation of these results is that intracranially injected taurine remains predominantly in neuronal and glial cytoplasm, and that taurine transported axonally remains loosely bound in the axoplasm of optic axons within the tecta. Such an interpretation is consistent with the results obtained, although by no means proved by them.

This explanation is supported by the results of an additional experiment. [^{35}S]Taurine was injected intravitreally into three groups of goldfish, one group killed 24 hr after injection, a second group having both optic nerves transected 24 hr after injection and killed 6 days later, and a third group killed 7 days after injection. Most of the labeled taurine that reached the tecta within 24 hr disappeared when the optic nerves degenerated (Fig. 16.7). This result suggests that the axonally transported taurine remains largely within the optic axons and does not diffuse appreciably into surrounding glial cells. Even when the optic axons degenerate, the axonal taurine released is not incorporated into other cellular pools within the tecta but is rapidly removed.

More recently, the axonal transport of taurine has been examined in the visual system of the developing rabbit and rat (Sturman, 1979; Politis and Ingoglia, 1979). [^{35}S]Taurine was injected intravitreally in the right eye of 10-day-old, 20-day-old, and adult rabbits, and the radioactive taurine present in the contralateral optic tract, lateral geniculate body, and superior colliculus in excess of the radioactivity present in the corresponding ipsilateral components, was determined at various times after injection. The same rabbits were used to determine the axonal transport of labeled proteins after the concurrent intravitreal injection of [^{3}H]proline. Considerably more taurine was transported axonally in the visual system of neonatal rabbits than in the visual system of adult rabbits (Fig. 16.8). The [^{35}S]taurine arrived in the various components of the visual system as a peak, which has only one discernible component, in contrast to the two components of radioactivity observed for axonally transported proteins (Fig. 16.9). The latter correspond to the well-described fast and slow axonal transport processes. The maximum of the peak of [^{35}S]taurine is similar in the different regions of rabbits of the same age, although the time of arrival of the peak varies approximately with the distance of the region from the retina. When the maxima are expressed as a ratio of neonatal to adult values, the difference in axonally transported taurine with age is clearly evident (Table 16.1). It is of note that, with the exception of optic tract of 10-day-old rabbits, no significant migration of [^{35}S]taurine could be measured within 6 hr after intravitreal injection, a time at which labeled protein could be measured readily in all structures, indicating that taurine is not transported as rapidly as the fastest component of proteins in

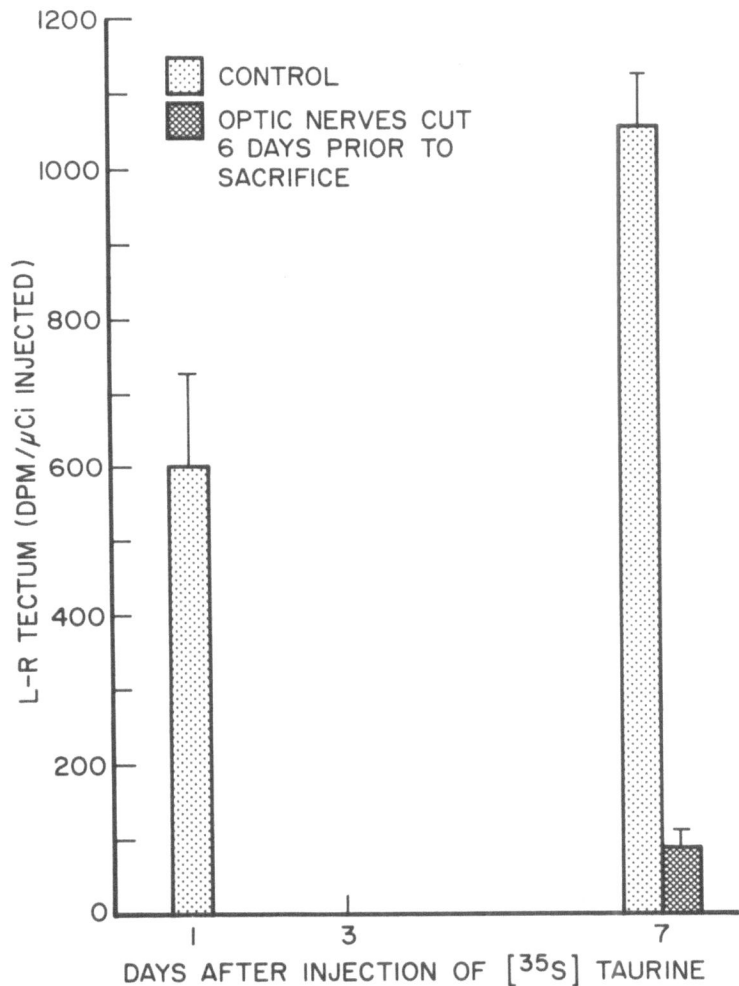

Fig. 16.7. Axonally transported [³⁵S]taurine in goldfish optic tecta (left minus right tectum radioactivity) 1 day and 7 days after intravitreal injection of [³⁵S]taurine, and in optic tecta from goldfish in which the optic nerves were transected 1 day after injection and the fish killed 6 days later. Each value represents the mean ± SEM from six goldfish. Taken from Ingoglia et al. (1978), with permission.

this system. The estimated rates of axonal transport, using the arrival of [³⁵S]taurine in the optic tract as the basis, suggest that taurine is transported at a rate intermediate between that of the fast and slow components of protein transport (Table 16.2).

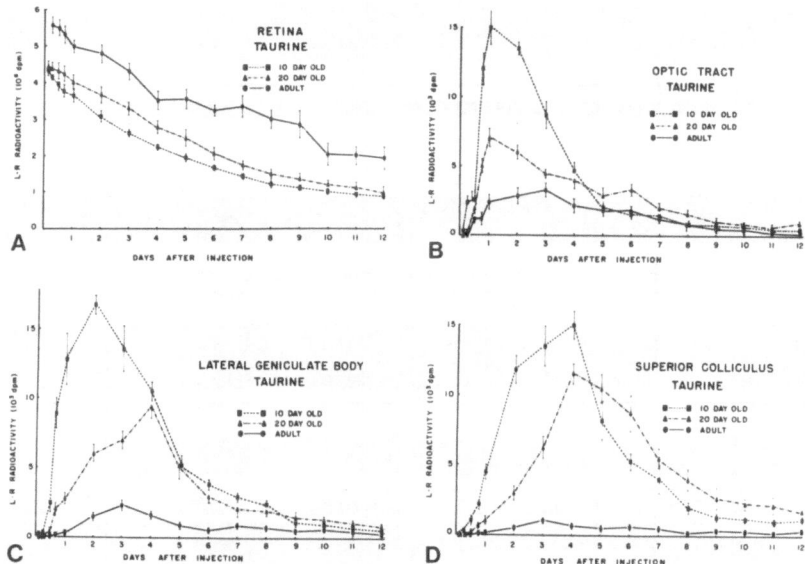

Fig. 16.8. Axonal transport of taurine in the developing rabbit visual system. (a) [³⁵S]Taurine in the right retina, (b) left minus right optic tract (c) left minus right lateral geniculate body, and (d) left minus right superior colliculus of rabbits at various times after the intravitreal injection of 5 μCi [³⁵S]taurine into the right eye. Each point represents the mean ± SEM from six neonatal rabbits or four adult rabbits. Taken from Sturman (1979), with permission.

These results in the rabbit visual system suggest that axonal transport is not linked directly to the axonal transport of proteins, as was suggested by the results in the goldfish visual system. This is an indication of yet another neurobiological difference between mammalian central nerves and lower vertebrate central nerves. The absence of a direct link between axonal transport of taurine and that of proteins in the rabbit visual system is supported by the results obtained after an intravitreal injection of cycloheximide 2 hr prior to the intravitreal injection of [³⁵S]taurine and [³H]proline. As expected, protein synthesis in the retina was severely reduced, and the amount of radioactive proteins reaching the optic tract, lateral geniculate body, and superior colliculus was generally reduced to an even greater extent (Table 16.3). Neither the amount of [³⁵S]taurine accumulated by the retina nor the amount of [³⁵S]taurine reaching the optic tract, lateral geniculate body, and superior collicus was reduced to the extent noted for labeled proteins. This result also differs from the results of the similar experiment in the goldfish visual system described earlier.

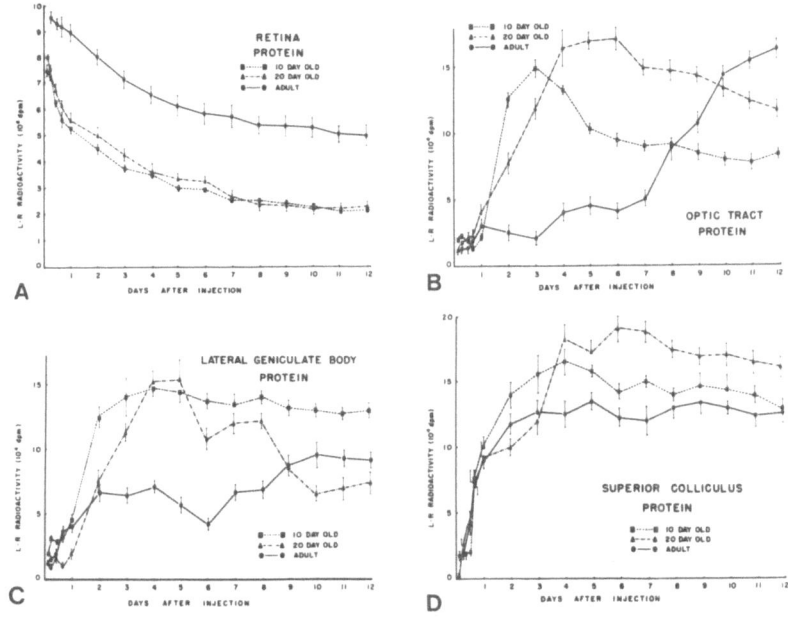

Fig. 16.9. Axonal transport of proteins in the developing rabbit visual system. (*a*) [³H]Proteins in the right retina, (*b*) left minus right optic tract, (*c*) left minus right lateral geniculate body, and (*d*) left minus right superior colliculus of rabbits at various times after the intravitreal injection of 10 μCi [³H]proline into the right eye. Each point represents the mean ± SEM from six neonatal rabbits or four adult rabbits. Taken from Sturman (1979), with permission.

Table 16.1 Axonal transport of [³⁵S]taurine in the visual system of developing rabbits compared to adult rabbits[a]

	Age	
	10-day-old	20-day-old
Optic tract	4.26	1.97
Lacteral geniculate body	7.26	4.13
Superior colliculus	13.73	10.45

[a]Results represent the maximum amount transported in each case as a fraction of the maximum amount transported in the same structure in adult rabbits. Each value is calculated from the mean of six developing rabbits and four adult rabbits. Taken from Sturman (1979), with permission.

Table 16.2   Rates of axonal transport of taurine and proteins in the visual system of developing and adult rabbits[a]

	Taurine[b]	Protein[b]	
		Fast	Slow
10-day-old	32	128	4
20-day-old	20	152	3
Adult	30	208	2

[a]Adapted from Sturman (1979), with permission.
[b]Millimetres per day.

A similar investigation of axonal transport of taurine has been reported in the developing rat visual system (Politis and Ingoglia, 1979). The first experiment demonstrated that the greatest amount of [^{35}S]taurine arrived at the lateral geniculate body of 1-day-old, 4-day-old, and 40-day-old rats by 24 hr after the intravitreal injection of [^{35}S]taurine (Fig. 16.10). A greater number of different age groups were then investigated, measuring the amount of [^{35}S]taurine arriving at the lateral geniculate body 24 hr after the intravitreal injection of [^{35}S]taurine. These experiments demonstrated that the amount of [^{35}S]taurine migrating axonally varied considerably with the age of the rat, being greatest in 12-day-old rats (Fig. 16.11a). The amount of [^{35}S]taurine accumulated by the retina also varied considerably with age, suggesting that the age-related differences in axonally transported taurine might be a result of developmental variations in retinal uptake of [^{35}S]taurine (Fig. 16.11b). These authors considered using the [^{35}S]taurine accumulated by the retina as an index of retinal ganglion cell uptake of [^{35}S]taurine, and validated it with the following experiment. [^{3}H]Taurine was injected intravitreally in 4-day-old, 11-day-old, and 40-day-old rats, and 24 hr later the retinae were prepared for autoradiographic analysis. Quantification of the grain distribution in all of the intraretinal layers of rats injected at 11 days of age and at 40 days of age showed closely similar results (Fig. 16.12). In the 4-day-old rats, only the grains in the ganglion cell layers were reported, and the 12% of total grains found at this age was similar to the proportion found at the other ages. These results indicate that a correction based on total [^{35}S]taurine accumulated by the retina can be used with justification. Results corrected in this fashion indicate that a much greater amount of [^{35}S]taurine is transported axonally in neonatal rats than in adult rats and young rats beyond 12 days of age (Fig. 16.11c).

Table 16.3 Effect of cycloheximide on axonal transport of taurine and proteins in the developing rabbit visual system[a]

	18 hr		1 day		3 days		5 days	
	Taurine	Proteins	Taurine	Proteins	Taurine	Proteins	Taurine	Proteins
10-day-old rabbits								
Retina	91	28	88	32	76	29	90	18
Optic tract	62	7	54	14	64	9	86	13
Lateral geniculate body	48	6	38	11	61	9	65	11
Superior colliculus	32	3	23	7	69	12	63	10
20-day-old rabbits								
Retina	89	29	86	28	80	27	71	22
Optic tract	76	13	74	20	59	25	63	28
Lateral geniculate body	65	14	63	15	66	11	75	15
Superior colliculus	67	11	64	16	77	18	75	19

[a]Results are expressed as percent of control, and each value represents the mean of six rabbits killed 18 hr, 1 day, 3 days, or 5 days after the intravitreal injection of [^{35}S]taurine and [^{3}H]proline. Cycloheximide (50 μg in 10 μl 0.9% saline) was injected 2 hr prior to injection [^{35}S]taurine and [^{3}H]proline. Taken from Sturman (1979), with permission.

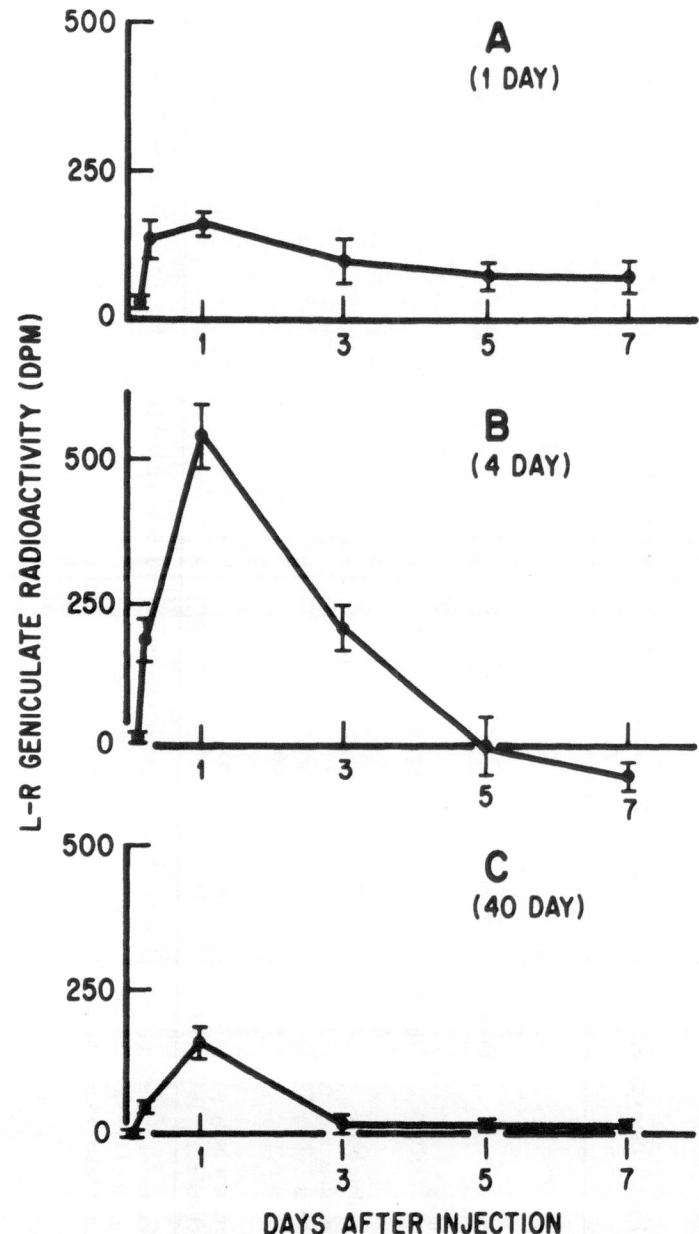

Fig. 16.10. Axonal transport of taurine in the developing rat visual system. Left minus right [³⁵S]taurine in lateral geniculate body at various times after the intravitreal injection of 0.3 μCi [³⁵S]taurine in 1.5 μl saline into the right eye. Each point represents the mean ± SEM from six rats of each age. Adapted from Politis and Ingoglia (1979), with permission.

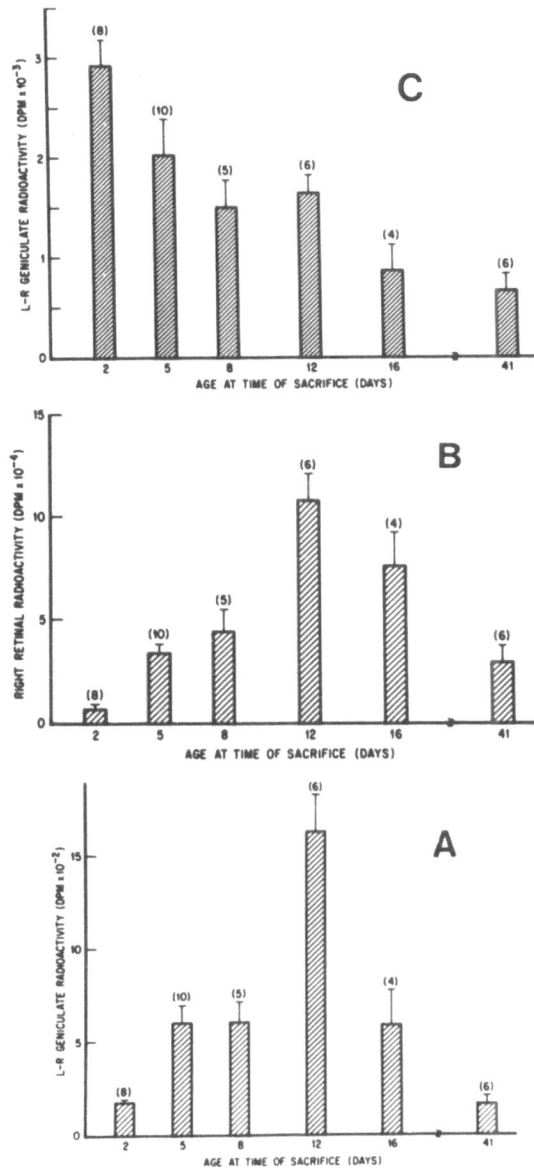

Fig. 16.11. Axonal transport of taurine in the developing rat visual system. (*a*) Left minus right [35S]taurine in lateral geniculate body 24 hr after the intravitreal injection of [35S]taurine into the right eye of rats of various ages. (*b*) [35S]Taurine in the right retina of the same rats. (*c*) Results from (*a*) corrected for radioactivity in right retina. Each point represents the mean ± SEM from the number of rats in parentheses. Taken from Politis and Ingoglia (1979), with permission.

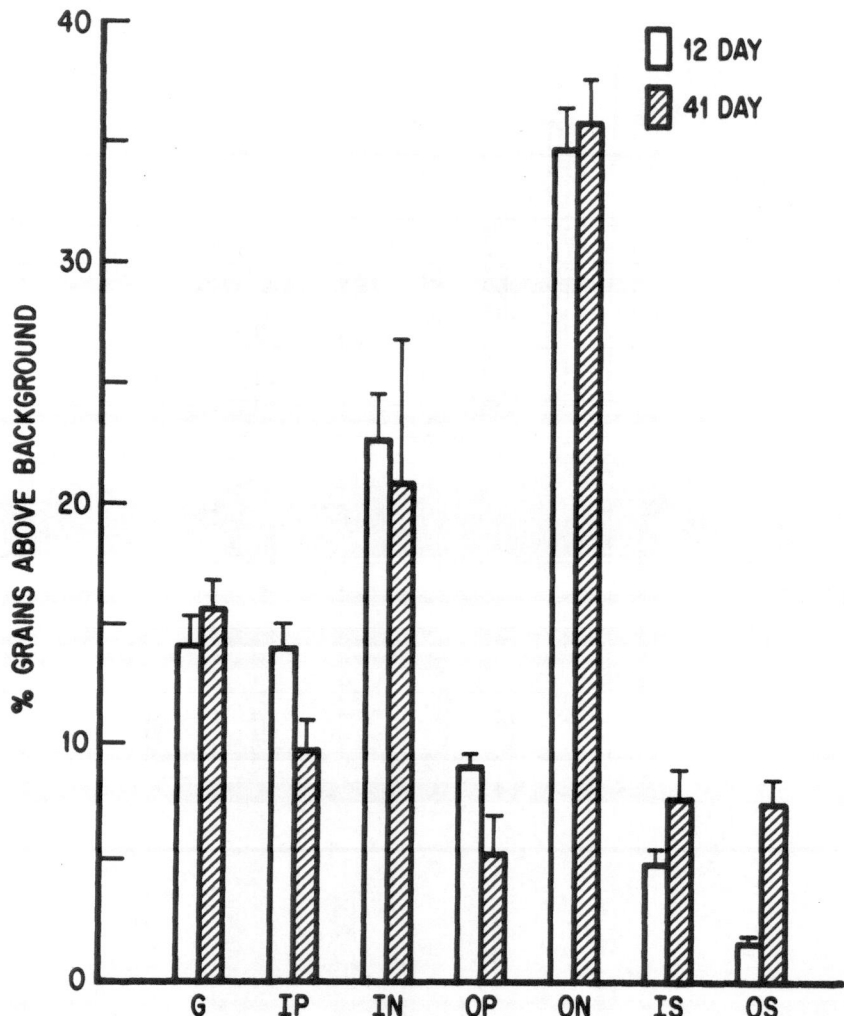

Fig. 16.12. Intraretinal distribution of grains 24 hr after the intravitreal injection of 1 μCi [³H]taurine. The average proportion of grains in each layer was determined from 10 sections of each retina. Each value represents the mean ± SEM from three retinas of each age. Key: G, ganglion cell layer; IP, inner plexiform layer; IN, inner nuclear layer; OP, outer plexiform layer; ON, outer nuclear layer; IS, inner segments of photoreceptors; OS, outer segment of photoreceptors. Taken from Politis and Ingoglia (1979), with permission.

The results of these studies of axonal transport of taurine, both in the developing rabbit visual system and in the developing rat visual system, indicate that more taurine is transported from the cell body along the optic axons prior to and during the period of formation of synapses in the lateral geniculate body and superior colliculus than is transported after the period of synaptogenesis. The function of axonally transported taurine in developing mammalian nerves is unknown, although it is tempting to speculate that taurine may regulate electrical activity and thus facilitate development of axons and formation of synaptic connections. In support of this suggestion, a number of studies have indicated that developing and regenerating axons contain neurotransmitters prior to their making synaptic connections (Champlain et al., 1970; Olson and Malmfors, 1970; Sachs et al., 1970; Coyle and Axelrod, 1971; Cheah and Geffen, 1973; Karlstrom and Dahlstrom, 1973), and that growing axons have somewhat different electrical properties compared to mature axons (Naka, 1964). Thus the large amounts of taurine transported axonally prior to and during synaptogenesis may serve to stabilize the electrical properties of the axons until formation of synapses has been completed. At this time, axonal transport of taurine from the cell body is reduced to allow normal functioning of the synapses, presumably in response to some signal from the nerve terminals, such as might be carried by retrograde axonal transport. The results of these studies are obtained exclusively from the regions of the brain in which the optic axons terminate, but they could be representative of the brain as a whole, and perhaps provide the explanation for the high concentrations of taurine in developing mammalian brain.

ACKNOWLEDGMENT

Studies performed in the author's laboratory were supported by the Office of Mental Retardation and Developmental Disabilities of the State of New York, and by Public Health Service Grant HD-11129 from the National Institute of Health.

REFERENCES

Barbeau, A.; Inoue, N.; Tsukada, Y.; and Butterworth, R. The neuropharmacology of taurine. *Life Sci.*, 17, 669–678 (1975).

Champlain, J.; Malmfors, T.; Olson, L.; and Sachs, C. Ontogenesis of peripheral adrenergic neurons in the rat: Pre- and postnatal observations. *Acta Physiol. Scand.*, 80, 276–288 (1970).

Cheah, T. B.; and Geffen, L. B. Effects of axonal injury on norepinephrine, tyrosine hydroxylase, and monoamine oxidase levels in sympathetic ganglia. *J. Neurobiol.*, 4, 433–452 (1973).

Coyle, J. T., and Axelrod, J. Development of the uptake and storage of L-^{3}H norepinephrine in the rat brain. *J. Neurochem.*, 18, 2061–2078 (1971).

Danielsson, H. Present states of research on catabolism and excretion of cholesterol. *Advan. Lipid Res.*, 1, 335–385 (1963).

Di Giamberadino, L. Independence of the rapid axonal transport of protein from the flow of free amino acids. *Acta Neuropath.*, Suppl. V, 132–135 (1971).

Elam, J. S.; and Agranoff, B. W. Rapid transport of proteins in the optic system of the goldfish. *J. Neurochem.*, 18, 375–387 (1971).

Elam, J. S.; Neale, E. A.; and Agranoff, B. W. Rapid axonal transport of sulfated mucopolysaccharide proteins. *Science*, 170, 458–460 (1970).

Elam, J. S.; Neale, E. A.; and Agranoff, B. W. Axonal transport in the goldfish visual system. *Acta Neuropath.*, Suppl. V, 257–266 (1971).

Gruener, R.; and Bryant, H. J. Excitability modulation by taurine: Action on axon membrane permeabilities. *J. Pharmacol. Exp. Ther.*, 194, 514–521 (1975).

Hayes, K. C.; Carey, R. E.; and Schmidt, S. Y. Retinal degeneration associated with taurine deficiency in the cat. *Science*, 188, 949–951 (1975).

Ingoglia, N. A.; Sturman, J. A.; Lindquist, T. D.; and Gaull, G. E. Axonal migration of taurine in the goldfish visual system. *Brain Res.*, 115, 535–539 (1976).

Ingoglia, N. A.; Sturman, J. A.; and Eisner, R. A. Axonal transport of putrescine, spermidine and spermine in normal and regenerating goldfish optic nerves. *Brain Res.*, 130, 433–445 (1977).

Ingoglia, N. A.; Sturman, J. A.; Rassin, D. K.; and Lindquist, T. D. A comparison of the axonal transport of taurine and proteins in the goldfish visual system. *J. Neurochem.*, 31, 161–170 (1978).

Karlsson, J. O. Is there an axonal transport of amino acids? *J. Neurochem.*, 29, 615–617 (1977).

Karlsson, J. O.; and Linde, A. Axonal transport of [^{35}S]sulfate in retinal ganglion cells of the rabbit. *J. Neurochem.*, 28, 293–297 (1977).

Karlstrom, L.; and Dahlstrom, A. The effect of different types of axonal trauma on the synthesis and storage of amine storage granules in the rat sciatic nerve. *J. Neurobiol.*, 4, 191–200 (1973).

Mandel, P.; and Pasantes-Morales, H. Taurine in the nervous system. *Rev. Neuroscience*, 3, 157–193 (1978).

Mandel, P.; Pasantes-Morales, H.; and Urban, P. F. Taurine, a putative neurotransmitter in retina. In *Transmitters in the Visual Process*, S. L. Bonting, ed. Pergamon Press, New York (1976), pp. 89–105.

Naka, K. Electrophysiology of fetal spinal cord: 1. Action potentials of motor neurons. *J. Gen. Physiol.*, 47, 1003–1022 (1964).

Neale, J. H.; Elam, J. S.; Neale, E. A.; and Agranoff, B. W. Axonal transport and turnover of proline- and leucine-labeled protein in the goldfish visual system. *J. Neurochem.*, 23, 1045–1055 (1974).

Oja, S. S.; and Kontro, P. Neurotransmitter actions of taurine in the central nervous system. In *Taurine and Neurological Disorders*, A. Barbeau and R. J. Huxtable, eds. Raven Press, New York (1978), pp. 181–200.

Olson, L.; and Malmfors, T. Growth characteristics of adrenegic nerves in the adult rat. *Acta Physiol. Scand.*, Suppl. 348, 1–112 (1970).

Politis, M. J.; and Ingoglia, N. A. Axonal transport of taurine along neonatal and young adult rat optic axons. *Brain Res.*, 166, 221–231 (1979).

Sachs, C.; Champlain, J.; Malmfors, T.; and Olson, T. Postnatal development of noradrenalin uptake in adrenegic neurons: *In vitro* isotope studies of different rat tissues with or without pretreatment with drugs. *Europ. J. Pharmacol.*, 9, 67–79 (1970).

Schmidt, S. Y.; Berson, E. L.; and Hayes, K. C. Retinal degeneration in cats fed casein: 1. Taurine deficiency. *Invest. Ophthalmol.*, 15, 47–52 (1976).

Strecker, A. Beobachtungen über die Galle verschiedener Thiere. *Ann. Chim.*, 70, 149–197 (1849).

Sturman, J. A. Taurine pool sizes in the rat: Effects of vitamin B-6 deficiency and high taurine diet. *J. Nutr.*, 102, 1566–1580 (1973).

Sturman, J. A. Taurine in the developing rabbit visual system: Changes in concentration and axonal transport including a comparison with axonally transported proteins. *J. Neurobiol.*, 10, 221–237 (1979).

Sturman, J. A.; and Gaull, G. E. Taurine in the brain and liver of the developing human and monkey. *J. Neurochem.*, 25, 831–835 (1975).

Sturman, J. A.; Rassin, D. K.; and Gaull, G. E. Taurine in the development of the central nervous system. In *Taurine and Neurological Disorders*, A. Barbeau and R. J. Huxtable, eds. Raven Press, New York (1978), pp. 49–71.

Van Gelder, N. M. Rectification of abnormal glutamic acid levels by taurine. In *Taurine*, R. Huxtable and A. Barbeau, eds. Raven Press, New York (1976), pp. 293–302.

Van Gelder, N. M. Glutamic acid and epilepsy: The action of taurine. In *Taurine and Neurological Disorders*, A. Barbeau and R. J. Huxtable, eds. Raven Press, New York (1978), pp. 287–402.

Weiss, P.; and Hiscoe, H. B. Experiments on the mechanisms of nerve growth. *J. Exp. Zool.*, 107, 315–395 (1948).

Wen, G. Y.; Sturman, J. A.; Wisniewski, H. M.; Lidsky, A. A.; Cornwell, A. C.; and Hayes, K. C. Tapetum disorganization in taurine-depleted cats. *Invest. Ophthalmol. Vis. Sci.*, 18, 1201–1206 (1979).

Discussion

DR. NARANJAN DHALLA (University of Manitoba): Dr. Pasantes-Morales, I have two questions. First, what is the role of ATP in calcium uptake by synaptosomes? Second, the depression in calcium uptake by taurine would lead to depletion in the intracellular concentration of free calcium. It is known that several interventions which increase the intracellular concentration of calcium are known to depress tissue excitability. Would you explain how taurine can depress excitability under conditions where it inhibits calcium uptake?

DR. HERMINIA PASANTES-MORALES (National University of Mexico): Concerning the first question, calcium transport into synaptosomes does not require ATP because the calcium gradient favors its entry. ATP-dependent calcium transport is a mitochondrial process. In answer to your second question, I believe that the depressant effect of taurine and the action of taurine on calcium transport occur in different compartments. My feeling is that the taurine-depressant effect on normal activity is due to the combination of taurine with receptors which probably belong to either the glycine-like or the GABA-like receptors in the postsynaptic terminal, whereas its effect on calcium probably occurs at the presynaptic level.

DR. MARY VOADEN (University of London): Do you feel that light-stimulated release of taurine is associated with the entry of calcium into the intracellular environment of the receptor's outer rim?

DR. HERMINIA PASANTES-MORALES (National University of Mexico): I believe that in the outer segment of photoreceptors, taurine might regulate calcium redistribution following photoexcitation. Since calcium has been considered an intracellular transmitter between light-activated rhodopsin and plasmalemmal sodium channels, calcium redistribution may be very important in maintaining the excitation and resting states of the photoreceptors.

DR. MARY VOADEN (University of London): Your data do not seem to fit well into the current schemes of excitation and recovery. If taurine is inhibiting calcium redistribution, which it seems to be, then it presumably would inhibit the recovery processes following light stimulation.

DR. HERMINIA PASANTES-MORALES (National University of Mexico): This problem probably arises because we were not using physiological conditions; exogenous taurine levels were increased, whereas taurine is found intracellularly. We are planning to investigate the effect of taurine on the inner disks of the photoreceptors. In these studies we will be using more physiological conditions because taurine is localized outside the disks. Nevertheless, it is interesting that the effects of taurine may be bilateral, occurring on both sides of the membrane; and to consider how different ionic environments on each side of the membrane may affect responses to taurine. I mentioned that when the experiments were carried out in a high-potassium, low-sodium medium, calcium uptake increased in the presence of taurine. This suggests that the taurine effects could vary, depending on conditions in and around the cells.

DR. SUSAN SCHMIDT (Harvard University): Dr. Klein, is adrenergic-stimulated efflux of taurine from the pineal gland related to changes in ionic currents of depolarization?

DR. DAVID KLEIN (National Institutes of Health): We know that as soon as catecholamines are added, the cell rapidly becomes hyperpolarized, and that this occurs within the time frame in which taurine is released. The efflux of taurine could possibly occur by one of two mechanisms; it could be transported through the beta-amino acid pump or cotransported with an ion. Since the sulfonic acid moiety of taurine has a great capacity to bind ions, we want to compare sodium and potassium fluxes with taurine fluxes.

DR. MICHAEL WILLIAMS (Merck Research Institute): Dr. Klein, have you examined the effect of taurine on the binding of beta-adrenergic antagonists such as dihydroalprenolol or iodohydroxybenzylpindolol?

DR. DAVID KLEIN (National Institutes of Health): Yes, we have found that each cell contains about 10,000 beta receptors. Since high concentrations of taurine can displace iodohydroxybenzylpindolol from these receptors, it suggests that taurine does interact with the beta receptors.

DR. MICHAEL WILLIAMS (Merck Research Institute): Are there presynaptic beta receptors?

DR. DAVID KLEIN (National Institutes of Health): I don't know about presynaptic beta receptors, but there are presynaptic receptors that control the alpha receptors. There is also an adrenergic binding protein that is specific for norepinephrine uptake. Thus, the cell contains two presynaptic sites where taurine potentially could act.

DR. SUSAN SCHMIDT (Harvard University): I have two questions. First, has the taurine content of the platelets ever been estimated; and, second, is there a sizable amount of taurine released from the platelets under your experimental conditions?

DR. CHERYL NAUSS-KAROL (Rutgers University): To answer your first question, the concentration of taurine in the platelets is about 20 mM. This intracellular taurine is avidly retained by the cell and appears to be released only when the cells are lysed.

DR. WALTER LOVENBERG (National Institutes of Health): Does incubating platelets with taurine have any effect on their aggregation properties? Do you see changes in cyclic nucleotides following the incubation?

DR. CHERYL NAUSS-KAROL (Rutgers University): I've never studied the cyclic nucleotides, but the platelets do aggregate normally.

DR. HENRY YAMAMURA (University of Arizona): Dr. Wheler, did you happen to look at other brain lesions?

DR. TREVOR WHELER (National Institutes of Health): No, we were just concentrating on the cerebral cortex.

DR. HENRY YAMAMURA (University of Arizona): Aren't you concerned with the amount of leakage that occurs when you perform the taurine uptake and subcellular fractionation studies?

DR. TREVOR WHELER (National Institutes of Health): Yes, leakage is something that one must be concerned with. However, it is notable that the percentage of taurine recovery in the P2B fraction was greater than in the P2A and P2C fractions. Moreover, other workers have added taurine to cold homogenates and have found that there is no redistribution of it into the synaptosomal compartments; rather, it stays in the soluble supernatant.

DR. S. S. OJA (University of Tampere): Dr. Wheler, you made reference to work showing that beta-alanine influenced the uptake of both GABA and taurine. However, when both GABA and taurine are added to the same incubation medium, it appears that they are not strong competitive inhibitors of their respective uptake processes, indicating that they don't share the same transport system. In that case, is it justified to infer that beta-alanine influences the uptakes of taurine and GABA in the same manner?

DR. TREVOR WHELER (National Institutes of Health): I agree that GABA and taurine almost certainly do not share the same uptake systems. I was quoting work from Iversen's laboratory, in which it was shown by autoradiography that beta-alanine is taken up by cortex glial cells and that beta-alanine competes with taurine for uptake.

DR. SUSAN SCHMIDT (Harvard University): Dr. Wheler, you've made a strong case for the presence of taurine in synaptosomes. One of your supporting arguments is that taurine causes hyperpolarization in the cortex. Could you comment on what mechanisms underlie this change in polarization?

DR. TREVOR WHELER (National Institutes of Health): Mandel claims that taurine could have two different effects. It could cause hyperpolarization by either increasing chloride intake or altering potassium flux.

DR. DAVID RASSIN (Institute for Basic Research in Mental Retardation, New York): I would like to make a comment on the synaptic vesicle taurine study. I think there is no question that taurine is associated with vesicles. I found it; Dr. Harry Bradford found it; Dr. Lee found it; Dr. Lähdesmäke found it. However, I do not understand the role of taurine in synaptic vesicles, particularly if it cannot leave the vesicles. Both Drs. Bradford and myself failed to demonstrate release of taurine from vesicular preparations under conditions in which acetylcholine and norepinephrine are released. The only way we ever got taurine release from our vesicles was to add trichloroacetic acid.

DR. S. S. OJA (University of Tampere): I agree with you that taurine is not released from the vesicles. However, I would like to suggest that bound taurine can still serve an important function.

DR. FRANCIS HOSKIN (Illinois Institute of Technology): I am wondering if anyone knows the molar concentration of taurine and a transmitter such as acetylcholine inside the vesicles.

DR. DAVID RASSIN (Institute for Basic Research in Mental Retardation,

New York): I don't think anyone has ever determined the actual concentration of taurine, but it is much higher than the acetylcholine content.

DR. FRANCIS HOSKIN (Illinois Institute of Technology): Since that is the case, perhaps taurine is not a transmitter substance but rather is inside the vesicles to bind acetylcholine.

DR. TREVOR WHELER (National Institutes of Health): Dr. Bradford has suggested that taurine might possibly bind norepinephrine in synaptic vesicles. He also feels that transmitter candidate amino acids might be released from the cytoplasm rather than classically from vesicles.

DR. STEVEN BASKIN (Medical College of Pennsylvania): There is also some evidence that taurine binds with histamine.

DR. DAVID KLEIN (National Institutes of Health): Using the taurine depletion techniques of Drs. Kocsis and Huxtable, it is now possible to directly examine the role of taurine in synaptic vesicles. One needs to merely deplete the tissue of taurine and examine what effect this has on the transmitters.

PART III

Actions of Taurine in the Cardiovascular System

Introduction

This section is concerned with the possible roles of taurine in the heart. Although no definitive physiological role has been assigned, evidence continues to accumulate implicating taurine in myocardial function. It has been reported to exhibit antiarrhythmic activity, an effect thought to be related to its property of altering ion flux. Several studies have also linked it to various types of heart failure.

The first three chapters of this section examine the importance of the myocardial taurine pool. Although it has been assumed that changes in intracellular taurine content modify myocardial excitability and contractility, conclusive evidence supporting this hypothesis is presently unavailable. Chapters by Shaffer and Kocsis and by Huxtable et al. describe methodology to lower tissue taurine levels. They suggest that these methods can be employed to examine the function of the intracellular taurine pool. Bahl et al. examine the uptake and accumulation of taurine by isolated heart cells, a process important in maintaining the intracellular taurine pool.

The molecular basis for the physiological and pharmacological actions of taurine is discussed in the latter part of this section. Schaffer et al. propose that taurine regulates calcium homeostasis by interacting with specific proteins on the sarcolemmal membrane. Similarly, Khatter et al. describe the sarcolemmal effects of taurine. They also suggest that taurine may alter magnesium movements in the myocardium. Welty and Welty describe the possible role of calcium in the onset of heart failure in cardiomyopathic hamsters. A mechanism for the beneficial effect of taurine treatment is discussed. In the last chapter, Ito et al. provide evidence that taurine-mediated changes in blood pressure are related to presynaptic release of catecholamines.

Copyright © 1981, Spectrum Publications, Inc.
The Effects of Taurine on Excitable Tissues

CHAPTER 17

Methods of Reducing Tissue Taurine Levels

Joel E. Shaffer
James J. Kocsis

Although high levels of taurine occur in heart and other tissues of a wide variety of species (Jacobsen and Smith, 1968; Kocsis et al., 1976), the function(s) of taurine in these extrahepatic tissues is not well established. Development of methods to reduce tissue taurine levels should prove useful in determining the functional roles of tissue (intracellular) as opposed to plasma (extracellular) taurine. A basis for developing such a method is the use of structural analogues of taurine that inhibit taurine transport. Among analogues known to inhibit taurine transport *in vitro* are beta-alanine, the carboxylic acid analogue of taurine; γ-aminobutyric acid (GABA), the homologue of beta-alanine; and taurocyamine, the guanidino analogue of taurine (Goldman and Scriver, 1967; Gaut and Nauss, 1976; Grosso et al., 1978). Beta-alanine has been shown to greatly increase the urinary excretion of taurine in mice (Gilbert et al., 1960) and rats (Kocsis and Kostos, 1963; Shaffer and Kocsis, 1978). The hypertaurinuria induced by such taurine analogues measures their ability to mobilize tissue taurine, since taurine displayed from tissue into plasma is not reabsorbed as efficiently by the renal tubules as other amino acids (Goldman and Scriver, 1967).

The present report compares the ability of beta-alanine to reduce tissue taurine stores with that of taurocyamine and GABA in rats and other species.

MATERIALS AND METHODS

Taurocyamine was synthesized according to the procedure of Morrison et al. (1958). All other chemicals were obtained from regular commercial sources and were certified ACS grade or better. Male Swiss-Webster mice (20–25 g), male Sprague-Dawley rats (200–220 g), and male Hartley Strain guinea pigs (600–800 g) were obtained from Perfection Breeders, Douglass-ville, Pennsylvania.

Taurine Assay

After sacrifice, tissues were removed and blotted on saline-moistened filter paper, weighed, placed in screw-capped vials containing 5–10 ml 2%-perch-loric acid (PCA), and allowed to extract overnight at room temperature. Aliquots (2–5 ml) of the PCA extract or of diluted urine (5 ml) were passed through 9 × 40 −55 mm columns of 200–400 mesh Dowex-50 × 8 and chased with 2.0 ml deionized water. To an aliquot (0.05–0.40 ml) of the Dowex-50 eluate in a screw-capped 50-ml polypropylene centrifuge tube were added 0.1 ml dinitrofluorobenzene (DNFB), 0.1 ml N/1 NaOH, and 0.5 ml dimethylsulfoxide (DMSO). Reactants were stirred for 30 sec on a vortex mixer. DMSO markedly increased the rate of dinitrophenylation. Then 0.1 ml of 3 N HCl was added to adjust the solution to pH 1.5–2.0, and sufficient deionized water to make the final volume 5.0 ml. After 20 ml ethylacetate was added to the mixture, the capped tubes were shaken for 10 min and the ethyl acetate discarded. Optical density of the residual aqueous fraction containing dinitrophenyltaurine was then compared at 355 nm with replicates containing 5, 10, or 20 μg of taurine standard that had been treated with DNFB and processed as above.

RESULTS

Beta-alanine administration i.p., s.c., or p.o. increased urinary excretion of taurine (Fig. 17.1). Taurine excretion was always greatest on Day 1 following treatment but remained elevated over control levels throughout each of the treatment periods. It was found that although half as much beta-alanine was consumed by orally treated rats, these rats excreted as much taurine as parenterally treated rats and showed fewer side effects.

Rats treated i.p. or s.c. ate less food, drank more water, and produced more urine than they did during their respective 3-day control periods, and also lost weight. In addition, rats treated i.p. sometimes displayed piloerection, occasional diarrhea, and nose bleeds. By contrast, rats treated orally ate slightly less food and drank less water than they did during their control period, but they did not lose weight and showed no diuresis. Thus, oral adminis-

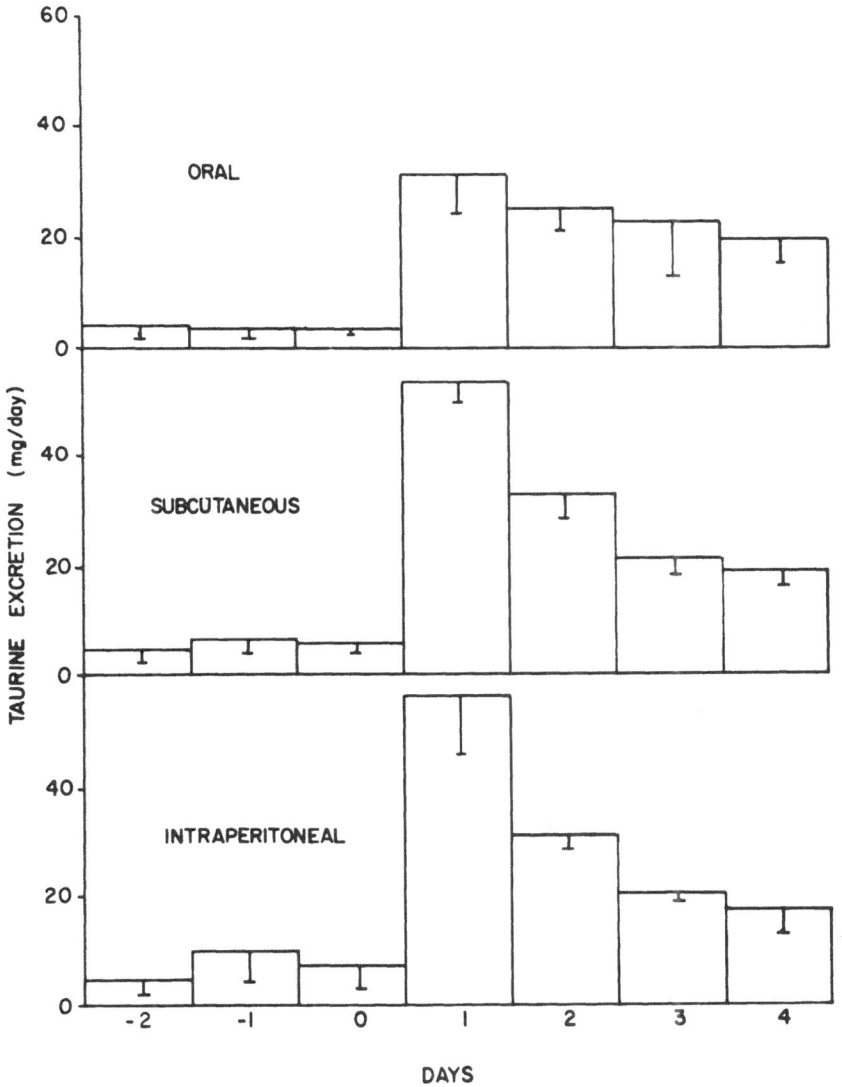

Fig.17.1.   The effect of different routes of administration of beta-alanine on the urinary excretion of taurine. Male Sprague-Dawley rats (200–220 g) were housed individually in metabolism cages and the urine collected daily with the aid of a deionized water rinse of the collecting funnel. After a 3-day control period, each of the three groups of four rats was treated for 4 days with beta-alanine either i.p. (3.6 g/kg, b.i.d. 1% b.w.), s.c. (3.6 g/kg, b.i.d., 1% b.w.), or p.o. (as a 3% w/v solution in the drinking water). Urinary taurine values are expressed as the means ± SD mg taurine/day/rat.

tration of beta-alanine proved to be more convenient and less toxic, and increased urinary taurine excretion as much as parenteral administration.

To determine whether beta-alanine-induced hypertaurinuria was accompanied by reduced tissue taurine levels, 3.6 g/kg beta-alanine was administered i.p. twice daily for 2, 4, or 7 days (Table 17.1). After 2 days of beta-alanine, only intestine, kidney, liver, and atria showed significant reductions of taurine; the greatest reduction of taurine, to 20% of control levels, occurred in the liver, while the least reduction, to 70% of control levels, was found in the atria. Four days of beta-alanine treatment reduced taurine levels further in all tissues examined except gastrocnemius and liver; liver taurine content actually increased over that seen after 2 days of beta-alanine. Seven days of beta-alanine resulted in reduced taurine levels in all tissues examined. In these experiments, heart tissues seemed to be more resistant than intestine or gastrocnemius to the taurine-depleting effects of beta-alanine.

When rats were treated orally with beta-alanine for a longer period of 28 days, no further depletion of taurine levels was observed in heart or gastrocnemius as compared with that seen after 7 days of beta-alanine (Fig. 17.2); however, taurine levels were found to be reduced in both cerebral cortex and thalamus of the 28-day-treated rats.

To see if other species responded to beta-alanine in a similar manner, weanling guinea pigs[1] were administered 3.6 g/kg beta-alanine s.c. once daily for 6 days. Like rats, guinea pigs treated in this fashion show increased urinary taurine excretion (data not shown). Table 17.2 shows that the greatest reduction in tissue taurine levels occurred in the gastrocnemius (59% of controls), followed by the intestine and left ventricle, each of which showed the same degree of reduction (22% of controls); liver taurine levels did not differ from control levels.

Mice were also administered beta-alanine and the homologue of beta-alanine, gamma-aminobutyric acid (GABA), as 1% and 3% solutions in the drinking water for 7 days. Although only beta-alanine was effective in reducing heart taurine levels, beta-alanine and GABA seemed to be equally effective in the gastrocnemius and liver of mice (Table 17.3). However, in another experiment (data not shown here) in which rats were treated with 3% GABA in the drinking water for 5 days, liver showed a reduction of tissue taurine levels to 60% of control levels, with no reduction in the ventricle, gastrocnemius, brain, or pancreas. Thus, GABA reduced tissue taurine levels more effectively in mice than in rats.

Experiments to determine proper concentrations of beta-alanine to use in

---

[1]Although weanling guinea pigs would not drink a 3% beta-alanine solution and would not tolerate two doses of 3.6 g/kg beta-alanine per day, adult (600–800 g) guinea pigs seemed to find 3% beta-alanine as potable as tap water.

Table 17.1 Effects of increasing duration of beta-alanine treatment on tissue taurine levels in rats[a]

	Left ventricle	Right ventricle	Atria	Gastroc-nemius	Intestine	Kidney	Liver
2 days							
Control	2.88 ± 0.27	2.36 ± 0.31	2.55 ± 0.24	1.52 ± 0.21	1.86 ± 0.25	1.20 ± 0.14	1.13 ± 0.34
Treated	2.86 ± 0.22	2.36 ± 0.31	1.84 ± 0.25[b]	1.44 ± 0.32	1.00 ± 0.17[b]	0.55 ± 0.10[b]	0.26 ± 0.17[b]
% of control	99%	100%	72%	95%	54%	46%	23%
4 days							
Control	3.02 ± 0.31	2.66 ± 0.25	2.82 ± 0.29	1.77 ± 0.22	2.32 ± 0.26	1.24 ± 0.14	1.15 ± 0.14
Treated	2.14 ± 0.14[b]	1.84 ± 0.18[b]	1.83 ± 0.21[b]	1.49 ± 0.26	1.20 ± 0.13[b]	0.74 ± 0.22[b]	0.49 ± 0.14[b]
% of control	71%	69%	65%	84%	52%	60%	43%
7 days							
Control	2.62 ± 0.21	2.56 ± 0.22	2.89 ± 0.40	2.35 ± 0.41	1.88 ± 0.24	1.26 ± 0.06	0.87 ± 0.24
Treated	1.82 ± 0.33[b]	1.74 ± 0.45[b]	1.62 ± 0.17[b]	1.73 ± 0.20[b]	0.78 ± 0.18[b]	0.51 ± 0.20*	0.22 ± 0.11[b]
% of control	69%	68%	56%	74%	41%	40%	25%

[a]Rats were housed individually and treated with beta-alanine (3.6 g/kg, i.p., b.i.d., 1% b.w.) for 2, 4, or 7 days. Control rats received 0.9% NaCl (i.p., b.i.d., 1% b.w.) for 2, 4, or 7 days. Rats were decapitated and the indicated tissues removed and analyzed for taurine. Data are expressed as the mean ± SD mg taurine/g tissue wet wt. of four to six rats.

[b]$p < 0.05$ compared with control.

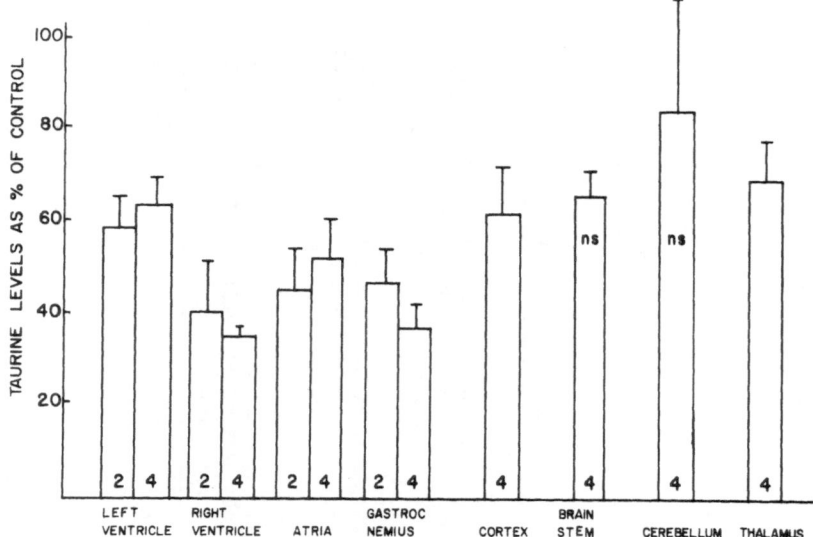

Fig. 17.2. The effect of 2 and 4 weeks of beta-alanine treatment on tissue taurine levels. Rats were housed in four groups of five rats each. Two groups were treated with 3% beta-alanine in the drinking water for 2 or 4 weeks (indicated by *2* or *4* on the graph), while the remaining two groups served as controls and were given tap water during the same period. Tissue taurine levels are expressed as a percent of those found in control rats. Control taurine levels for each tissue, expressed as the mean ± SD mg taurine/gm wet wt., for the 2- and 4-week control period were: left ventricle, 2-week, 2.99 ± 0.52 and 4-week, 2.48 ± 0.11; right ventricle, 2-week, 3.09 ± 0.35 and 4-week, 3.61 ± 0.66; atria, 2-week, 2.60 ± 0.39 and 4-week, 3.23 ± 0.35; gastrocnemius, 2-week, 2.41 ± 0.26 and 4-week, 2.14 ± 0.20; cerebral cortex, 4-week, 0.99 ± 0.17; brainstem, 4-week, 0.64 ± 0.29; cerebellum, 4-week, 1.00 ± 0.25; thalamus, 0.76 ± 0.16. Taurine levels in all tissues from beta-alanine-treated rats were significantly different from control levels ($p < 0.05$ using Student's t-test) except those marked *ns*.

Table 17.2   Effect of beta-alanine on tissue taurine levels in guinea pigs[a]

Left Ventricle	Gastrocnemius	Liver	Intestine
Control (4) (100%)			
1.73 ± 0.23	1.72 ± 0.19	0.28 ± 0.07	0.93 ± 0.07
Treated (4)			
1.25 ± 0.15[b]	1.02 ± 0.04[b]	0.29 ± 0.03	0.67 ± 0.10[b]
(72%)	(59%)	(103%)	(72%)

[a]Male Hartley strain guinea pigs (600–800 g) were injected subcutaneously with beta-alanine (3.6 g/kg, 1% b.w.) or with 0.9% NaCl (s.c., 1% b.w.) once daily for 6 days. After sacrifice, 24 hr after the last dose, the indicated tissues were removed and analyzed for taurine. Values are expressed as means ± SD mg taurine/g wet wt.

[b]$p < 0.025$ compared with control guinea pigs using Students t-test.

Table 17.3   Effect of beta-alanine and GABA on tissue taurine levels in mice[a]

Group	n	Heart	Gastrocnemius	Liver
Control	(7)	5.55 ± 0.33 (100%)	5.70 ± 0.36 (100%)	2.01 ± 0.3 (100%)
β-ala (1%)	(7)	4.98 ± 0.28 (90%)	5.44 ± 0.61 (95%)	1.61 ± 0.32 (91%)[b]
β-ala (3%)	(6)	4.66 ± 0.40 (84%)[b]	4.74 ± 0.22 (83%)[b]	0.94 ± 0.25 (47%)[b]
GABA (1%)	(6)	5.52 ± 0.24 (99%)	4.88 ± 0.63 (86%)[b]	1.28 ± 0.36 (64%)[b]

[a]Male Swiss Webster mice (20–25 g) were treated with either beta-alanine, 1% or 3% in the drinking water; GABA, 1% or 3% in the drinking water; or tap water or 7 days. Mice were then sacrificed and the heart, gastrocnemius, and liver removed and analyzed for taurine. Values are expressed as mean ± SD mg taurine/g wet wt. and as a percentage of control.

[b]$p < 0.025$ compared with control using Student's t-test.

drinking water for reducing tissue taurine levels showed that neither 5% nor 8% was more effective than 3% beta-alanine in rats or mice (data not shown). Also, as noted above, increasing the duration of treatment in rats from 7 to 14 or 28 days did not further reduce tissue taurine levels except for brain tissues. Thus, the use of 3% beta-alanine in the drinking water for 7 days was satisfactory for reducing taurine content in most tissues of rats or mice.

To compare the effectiveness of taurocyamine with that of beta-alanine in reducing tissue taurine levels, these agents were administered to rats in the drinking water for 6 days. Like beta-alanine, taurocyamine also increased the urinary excretion of taurine. This hypertaurinuria was not due to the conversion of taurocyamine to taurine, since our earlier studies had shown that ^{14}C-taurine was not found in urine after ^{14}C-taurocyamine administration to rats (unpublished data). Table 17.4 shows that while 3% beta-alanine seemed to be more effective in reducing taurine levels in the liver and gastrocnemius, a 1.5% taurocyamine solution was more effective in reducing brain taurine levels; the two agents were equally effective in lowering heart taurine levels.

DISCUSSION

The hypertaurinuria induced in rats by the transport inhibitor beta-alanine is probably due to inhibition of taurine uptake by tissues (Gaut and Nauss, 1976; Grosso et al., 1978) and to an inhibition of taurine reabsorption by the renal tubules (Goldman and Scriver, 1967; Scriver et al., 1966), since taurine and beta-alanine have a common transport system selective for beta amino acids (Goldman and Scriver, 1967). Certainly one source of the uri-

Table 17.4   Effect of beta-alanine and taurocyamine on tissue taurine levels in rats[a]

Ventricle	Gastrocnemius	Brain	Liver
Control (5)			
2.31 ± 0.35	1.62 ± 0.18	0.69 ± 0.06	0.81 ± 0.44
(100%)	(100%)	(100%)	(100%)
beta-Alanine (5)			
1.44 ± 0.28	1.07 ± 0.10[b]	0.61 ± 0.07	0.22 ± 0.03
(62%)	(66%)	(88%)	(27%)
Taurocyamine (5)			
1.31 ± 0.17[b]	1.34 ± 0.33	0.49 ± 0.06[b]	0.35 ± 0.13[b]
(57%)	(83%)	(72%)	(44%)

[a]Male Sprague-Dawley rats (200–220 g) were treated with either beta-alanine (3% in the drinking water), taurocyamine (1.5% in the drinking water), or tap water (controls) for 6 days. Rats were then decapitated and the indicated tissues removed and analyzed for taurine. Values are expressed as means ± SD mg taurine/g wet wt. and as a percentage of control.
[b]$p < 0.05$ compared with control using Student's t-test.

nary taurine is the diet, which contained 0.06% taurine in our experiments. However, since beta-alanine did reduce tissue taurine levels, the tissues are probably a major source of the extra urinary taurine.

The tissues that lost taurine first in these experiments—the liver, kidney, and intestine—were also the tissues that were repleted first (data not shown here). These tissues are known to turn over taurine relatively rapidly (Minato et al., 1969; Huxtable and Bressler, 1972; Spaeth and Schneider, 1974). On the other hand, the heart, gastrocnemius, and brain—tissues known to have slow turnover rates (Minato et al., 1969; Huxtable and Bressler, 1972; Spaeth and Schneider, 1974)—lost taurine later and were more difficult to deplete. Transport inhibitors like beta-alanine, taurocyamine, and GABA should reduce taurine levels in tissues with a rapid taurine turnover before those with slower turnover rates.

The fact that no tissues were completely depleted of taurine in these experiments may be due to several factors. It has been shown for example that tissue half-lives of taurine are increased when the sources of taurine are restricted. Tissues of cats placed on a taurine-free diet turned over taurine more slowly than tissues in control animals (Sturman et al., 1978), and tissues of rats on a B_6-deficient diet showed a similar prolongation of taurine half-lives (Sturman, 1973). In addition, rats in our experiments had access to taurine in the diet (0.06% taurine) and may also have increased their synthesis of taurine from precursor amino acids.

Although beta-alanine reduced tissue taurine levels in mice and guinea pigs, mouse tissues were more resistant to the taurine-depleting action of beta-alanine than those of rats or guinea pigs. This may be due to greater taurine synthesis in mice or to the fact that the mice consumed more food and hence more taurine/gram body weight than the rats or guinea pigs. In guinea pigs, tissue taurine levels seemed to be reduced as readily as those in rats, except for liver. Unlike most rodent species, guinea pigs do not conjugate bile acids with taurine (Vessey, 1978), but guinea pig liver probably synthesizes taurine from sulfur amino acids, like that of most species.

The fact that GABA is less effective than beta-alanine or taurocyamine as a taurine transport inhibitor (Gaut and Nauss, 1976) probably explains why GABA was the least effective of the three agents tested in reducing tissue taurine levels in our experiments; however, GABA injected i.p. did increase urinary taurine excretion in rats.

While it is uncertain to what extent treatment with either beta-alanine or taurocyamine may reproduce the various known biological effects of taurine, some information is available on the levels of both these agents during treatment. After oral administration in the drinking water of 3% beta-alanine for 7 days (our experiments) and of 1% taurocyamine (guanidinoethylsulfonate) for 9 days (Huxtable et al., 1981), beta-alanine levels in both heart and liver were 2 μmol/g, whereas taurocyamine levels in these tissues were approximately 15 μmol/g. Beta-alanine did not accumulate in these tissues, probably because it is readily converted to malonic semialdehyde and eliminated as CO_2 (Graff and Haberman, 1950; Pihl and Fritzson, 1955; Nutzenadel and Scriver, 1976). Taurocyamine, on the other hand, showed tissue accumulation because, like taurine, it does not appear to be metabolized.

Reducing myocardial taurine levels with beta-alanine has shown that intracellular taurine may act differently from extracellular taurine (Shaffer et al., 1979). A concentration of 10 mM taurine in the perfusate (100 times the plasma level) protected against the cardiac failure induced by calcium following a calcium-free perfusion period in isolated rat hearts ("calcium paradox") (Schaffer et al., 1980). Decreasing intracellular taurine levels in the heart with beta-alanine from 20 to 10 mM, however, did not enhance the development of cardiac failure. On the other hand, preliminary results in another model (Marmo, 1971) showed that rats with reduced heart taurine levels developed ventricular fibrillation sooner than control rats infused i.v. with $BaCl_2$.

Further work will show whether reducing tissue taurine levels with beta-alanine or taurocyamine will hasten development of pathological changes in cat retinal photoreceptor cells (Schmidt et al., 1976), in the stroke-prone, spontaneously hypertensive rat (Nara et al., 1978), in pancreatic islet tissue of streptozotocin-treated mice (Tokunaga et al., 1979), or in the brains of seizure-prone animals (Laird and Huxtable, 1978; Iwata et al., 1979).

SUMMARY

Oral administration of a 3% beta-alanine solution in the drinking water was as effective in increasing urinary taurine excretion and in decreasing tissue taurine levels in rats as subcutaneous or intraperitoneal administration. Beta-alanine was more effective in reducing tissue taurine levels in rats and guinea pigs than in mice. Other inhibitors of taurine transport were compared with beta-alanine for their ability to reduce tissue taurine levels. Taurocyamine, the guanidino analogue of taurine, was less effective in rats than beta-alanine in reducing taurine levels in gastrocnemius but more effective in reducing brain taurine levels, whereas the homologue of beta-alanine, GABA, was relatively ineffective in both mice and rats.

ACKNOWLEDGMENTS

These studies were supported by HL 20211. We wish to acknowledge the expert technical assistance of Ms. L. A. Semanko in these studies and to thank Dr. D. K. Reibel, Department of Physiology, Hershey Medical Center, Pennsylvania State University, Hershey, Pennsylvania, for the beta-alanine determinations.

REFERENCES

Gaut, Z. N.; and Nauss, C. B. Uptake of taurine by human blood platelets: A possible model for brain. In *Taurine,* Huxtable, R.; and Barbeau, A., eds. Raven Press, New York (1976), pp. 91–98.

Gilbert, J. B.; Ku, Y.; Rogers, L. L.; and Williams, R. J. The increase in urinary taurine after intraperitoneal administration of amino acids to the mouse. *J. Biol. Chem.,* 235, 1055–1060 (1960).

Goldman, H.; and Scriver, C. R. A transport system in mammalian kidney with preference for β-amino compounds. *Pediatr. Res.,* 1, 212–213 (1967).

Graff, J.; and Haberman, H.D. On the metabolism of β-alanine. *J. Biol. Chem.,* 186, 369–372 (1950).

Grosso, D. S.; Roeske, W. R.; and Bressler, R. Characterization of a carrier-mediated transport system for taurine in the fetal mouse heart *in vitro. J. Clin. Invest.,* 61, 944–952 (1978).

Huxtable, R.; and Bressler, R. Taurine and isethionic acid: Distribution and interconversion in the rat. *J. Nutr.,* 102, 805–814 (1972).

Huxtable R. J.; Laird, H. E.; and Lippicott, S. Rapid depletion of tissue taurine content by guanidinoethyl sulfonate. This volume, Chap. 18.

Iwata, H.; Yamagami, S.; Lee, E.; Matsuda, T.; and Baba, A. Increase of brain taurine contents of El mice by physiological stimulation. *Japan. J. Pharmacol.,* 29, 503–507 (1979).

Jacobsen, J. G.; and Smith, L. H., Jr. Biochemistry and physiology of taurine and taurine derivatives. *Physiol. Rev.,* 48, 424–511 (1968).

Kocsis, J. J.; and Kostos, V. J. Taurinuria and depletion of tissue taurine stores in rats treated with beta-alanine. *Pharmacologist*, 5, 268 (1963).

Kocsis, J. J.; Kostos, V. J.; and Baskin, S. I. Taurine levels in the heart tissues of various species. In *Taurine*, Huxtable, R.; and Barbeau, A., Raven Press, New York (1976), pp. 145–153.

Laird, H. E.; and Huxtable, R. J. Taurine and audiogenic epilepsy. In *Taurine and Neurological Disorders*, Barbeau, A.; and Huxtable, R., eds. Raven Press, New York (1978), pp. 339–357.

Marmo, E. Effects of different drugs with β-adrenolytic activity in experimental models of arrhythmias. *N.S. Arch. Pharmacol.*, 269, 231–247 (1971).

Minato, A.; Hirose, S.; Agiso, T.; Uda, K.; Takigawa, Y.; and Fujihira, E. Distribution of radioactivity after administration of taurine -^{35}S in rats. *Chem. Pharm. Bull.*, 17, 1498–1504 (1969).

Morrison, J. F.; Ennor, A. H.; and Griffiths, D. E. The preparation of barium monophosphotaurocyamine. *Biochem. J.*, 68, 447–452 (1958).

Nara, Y.; Yamori, Y.; and Lovenberg, W. Effect of dietary taurine on blood pressure in spontaneously hypertensive rats. *Biochem. Pharmacol.*, 27, 2689–2692 (1978).

Nutzenadel, W.; and Scriver, C. R. Uptake and metabolism of β-alanine and L-carnosine by rat tissues *in vitro:* Role in nutrition. *Am. J. Physiol.*, 230, 643–651 (1976).

Pihl, A.; and Fritzson, P. The catabolism of C^{14}-labeled β-alanine in the intact rat. *J. Biol. Chem.*, 215, 345–351 (1955).

Schaffer, S. W.; Chovan, J.; Kramer, J.; and Kulakowski, E. The role of taurine receptors in the heart. This volume, Chap. 20.

Schmidt, S. Y.; Berson, E. L.; and Hayes, K. C. Retinal degeneration in cats fed casein: I. Taurine deficiency. *J. Invest. Ophthalmol.*, 15, 47–52 (1976).

Scriver, C. R.; Pueschel, S.; and Davies, E. Hyper β-alaninemia associated with β-aminoaciduria and α-aminobutyricaciduria, somnolence and seizures. *N. Engl. J. Med.*, 274, 635–643 (1966).

Shaffer, J. E.; and Kocsis, J. J. Taurine mobilizing effects of beta alanine in rats. *Fed. Proc.*, 37 (3), 2963 (1978).

Shaffer, J. E.; Kramer, J.; Schaffer, S. W.; and Kocsis, J. J. Intracellular or extracellular effects of taurine? *Fed. Proc.*, 38, 695 (1979).

Spaeth, D. G.; and Schneider, D. L. Turnover of taurine in rat tissues. *J. Nutr.*, 104, 179–186 (1974).

Sturman, J. A. Taurine pool sizes in the rat: Effects of vitamin B-6 deficiency and high taurine diet. *J. Nutr.*, 103, 1566–1580 (1973).

Sturman, J. A.; Rassin, D. K.; Hayes, K. C.; and Gaull, G. E. Taurine deficiency in the kitten: Exchange and turnover of ^{35}S-taurine in brain, retina and other tissues. *J. Nutr.*, 108, 1462–1476 (1978).

Tokunaga, H.; Yoneda, Y.; and Kuriyama, K. Protective actions of taurine against streptozotocin-induced hyperglycemia. *Biochem. Pharmacol.*, 28, 2807–2811 (1979).

Vessey, D. A. The biochemical basis for the conjugation of bile acids with either glycine or taurine. *Biochem. J.*, 174, 621–626 (1978).

Copyright © 1981, Spectrum Publications, Inc.
The Effects of Taurine on Excitable Tissues

CHAPTER 18

Rapid Depletion of Tissue Taurine Content by Guanidinoethyl Sulfonate

Ryan J. Huxtable
Hugh E. Laird
Shirley Lippincott

TAURINE: APPROACHES TO A CARDIAC FUNCTION

Taurine is one of those enigmatic substances about which so much, and so little, is known. It is a simple substance, with a molecular weight of 125 and a molecular formula of $C_2H_7O_3NS$. Taurine occurs in high concentration in many mammalian tissues, comprising in excess of 60% of the total free amino acid pool in the rat heart. Furthermore, it is one of the most abundant amino acids in the brain. Taurine, which was first isolated from bull's urine in 1827 (Tiedemann and Gmelin, 1827), is a substance of high chemical stability and low metabolic reactivity. In mammals, only one potential pathway for the degradation of taurine has been proposed, that being the conversion to isethionic acid, but the evidence for this pathway is tenuous (Huxtable, 1978). Analysis of taurine can be readily achieved by a number of simple colorimetric tests in the laboratory, or it can be detected by an amino acid analyzer, on which it is one of the first substances eluted.

One might suppose that a substance exhibiting all these characteristics would have been well studied and well understood. However, such is not the case. Definitive studies on the function of taurine in the heart have yet to be done. It is often asked whether taurine has a function at all, or whether it is just an inert end product of sulfur amino acid metabolism. This chapter briefly reviews the broad lines of evidence for and against the mammalian functionality of taurine.

The major reason for thinking taurine important may be more intuitive than objective. This is the extremely high concentrations that occur in certain tissues. The sheer quantity of the material in the heart militates against it being nonfunctional. A more cogent reason is the extreme experimental difficulty in modifying concentrations of taurine either upward or downward. Recently, taurine deficiency has been produced in the cat, with the result that deficient animals become blind (Hayes et al., 1975; Schmidt et al., 1976). This is an important observation in that it links a deficiency in taurine directly to a pathological consequence. Congestive heart failure has been associated with elevations in taurine concentration in the affected ventricle, but which is cause and which is consequence, or whether the two events are related at all, has not been clearly elucidated (Peterson et al., 1973; Huxtable and Bressler, 1974).

Other indications of the functionality of taurine are the wide range of pharmacological phenomena associated with it. In both animals and humans, for example, taurine is a potent anticonvulsant. In the heart, it has antiarrhythmic and inotropic actions, and there are many indications that it modulates the movements of certain ions; in particular, calcium, and, to a lesser extent, potassium and chloride (Gruener et al., 1976). In addition, our laboratory has recently shown a connection between the actions of taurine and the sympathetic nervous system in the heart (Huxtable and Chubb, 1977). There are also indications of a relationship between the catecholamines and taurine in the central nervous system (Prous et al., 1978). These areas have been recently reviewed (Barbeau and Huxtable, 1978).

Probably the two most powerful arguments against the functionality of taurine are the species-to-species variation in taurine concentration that is found in the heart and the difficulty of showing pathological consequences, or loss of function, when taurine concentratons are altered.

There are good reasons as to why our knowledge of taurine is less than that of, say, GABA. GABA is chemically similar to taurine, in that both are α,ω amino acids and both have neuroinhibitory actions in the central nervous system. GABA, however, is restricted to the CNS and does not occur in the heart. Furthermore, it is metabolically unstable, and levels in tissue biopsy or autopsy samples can rise fairly rapidly with time (Perry et al., 1971). The analysis of GABA is more tedious, in that instead of being one of the earliest compounds eluted from an amino acid analyzer, it only elutes after several hours. Why, therefore, do we know more of the substance and have a clearer idea of its functionality than of taurine? Traditionally, functions for a substance are found as a result of modifying the concentration of that substance and observing physiopathological consequences. In practice, this often reduces to finding synthetic agonists or antagonists to the substance under study. These are available for GABA. Taurine levels are hard to modify experimentally, and no good taurine antagonist has been discovered. When one becomes available, we can expect an acceleration in our understanding of taurine. One line of investigation we have been pursuing over the last several years is that of the regulation of taurine in the heart, not only in the hope of getting indications of functionality, but in the expectation that if we understood the regulation, we would be able to control it.

REGULATION OF TAURINE IN THE HEART

The appropriate site at which to modify the taurine concentration in the rat heart is clearly that of its transport. We have shown that the bulk of taurine arises in the heart by an influx process from the serum, and not by endogenous biosynthesis. Cysteine sulfinic acid decarboxylase does not occur in the heart, so the pathway of taurine biosynthesis from cysteine via cysteine sulfinic acid cannot operate. Cysteamine dioxygenase, which converts cysteamine to hypotaurine, is present. However, it is not known how cysteamine arises in the heart. Summation of the work of a number of laboratories indicates that metabolic turnover of taurine in the heart is also unimportant; that is, that taurine disappears from the heart by an efflux process. It follows that the taurine levels in the heart are regulated primarily by transport (Huxtable, 1978).

TRANSPORT OF TAURINE INTO THE HEART

In the isolated, perfused rat heart, taurine is taken up by an active transport process, which is saturable at taurine concentrations of 200 μM or higher. The process exhibits a Km of 45 μM and a Vmax of 32 nmol/g dry weight/min (Huxtable and Chubb, 1977; Chubb and Huxtable, 1978a). The saturation kinetics, shown in Fig. 18.1, indicate that the heart cell membrane is impermeable to passive diffusion of taurine. This is to be expected, in view of the highly charged and hydrophilic nature of this amino acid. The pKa of the sulfonate group is 1.5, and, therefore, at a physiological pH of 7.2, the concentration ratio of the ionized sulfonate to the nonionized sulfonic acid moiety is 5×10^5:1. A Km of 45 μM classifies the cardiac transport system as one of medium affinity. It thus differs from the high affinity system we have found in rat brain synaptosomes, which exhibits a Km of 3.2 μM and a Vmax of 3.0 nmol/g protein/min (Hruska et al., 1978a). These values have been corrected for the contribution of a low affinity transport system, having a Km

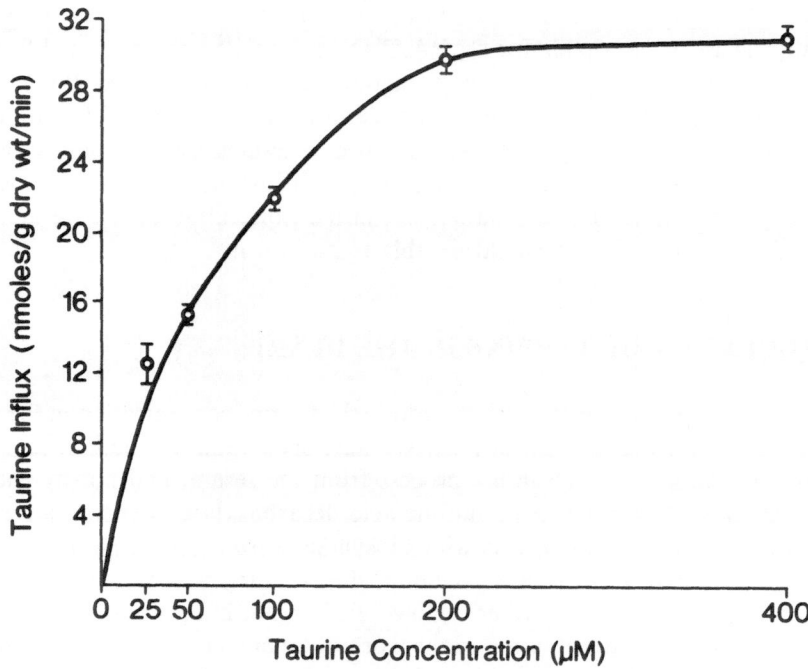

Fig. 18.1.   Uptake of [³H]taurine by the isolated perfused heart. Hearts from rats of 200–300 g body weight were isolated and perfused by the Langendorff technique with Krebs-Henseleit buffer at 37° in a nonrecirculating system. The buffer also contained 5.5 mM glucose and was equilibrated with 95% O_2 – 5% CO_2. After 5 min, [³H]taurine at the test concentration was added to the buffer. After an additional 6-min perfusion, the perfusion medium was changed back to Krebs-Henseleit buffer for 6 min to remove labeled taurine from vascular and interstitial space. The heart was then homogenized in 8 volumes of 5% TCA, and the homogenate centrifuged at 40,000 g for 15 min. Radioactivity in the supernatant was determined, and taurine uptake calculated from the known specific activity of the taurine used. Data is displayed as means ± SEM for four hearts per point.

of 3,300 µM and a Vmax of 700 nmol/g protein/min. The Km in the heart, however, is of the same order as the taurine concentration in rat serum, which is around 20 µM. Despite these differences in affinity, structural requirements for transport by synaptosomes and heart appear to be similar (Azari et al., 1978; Hruska et al., 1978b).

Awapara and Berg (1976) report similar findings for taurine transport in rat heart slices. They found transport to be saturable and to have a Km of 40 µM. However, in the fetal mouse heart, taurine transport is not saturated even at 2 mM taurine. The apparent Km of the sodium-dependent component of transport was 440 µM (Grosso et al., 1978).

INHIBITORS OF TAURINE TRANSPORT

The structure-activity requirements of the taurine-transporting system in the heart are quite precise. Compounds we have found to be effective inhibitors in the Langendorff heart preparation are shown in Table 18.1. The inhibitors hypotaurine and β-alanine both occur naturally in the rat, and both are straight-chain β-amino acids, as is taurine. Hypotaurine is the sulfinic acid analogue and β-alanine the carboxylic acid analogue of taurine. Guanidinoethylsulfonate and guanidinopropionate, the guanidylated derivatives of taurine and β-alanine, respectively, are also transport inhibitors of taurine.

Active transport of β-alanine occurs in the heart, apparently by the same system that transports taurine. However, as shown in Fig. 18.2, there is a nonsaturable component to transport. The pKa of β-alanine is 3.6. Thus, at a pH of 7.2, the proportion of β-alanine with a nonionized acid function is 125 times higher than for an equivalent concentration of taurine. The greater lipophilicity resulting from this is the probable reason for the observed passive diffusion of β-alanine into the heart. Lineweaver-Burk plots for β-alanine transport are shown in Fig. 18.3. At concentrations of β-alanine below 20 μM, the apparent Km is 18 μM, and the apparent Vmax is 10 nmol/g dry weight/min. These uncorrected kinetic constants increase at higher concentrations (Fig. 18.3).

Guanidinoethylsulfonate is transported in place of taurine by a saturable process (Fig. 18.4). The Lineweaver-Burk plot for transport is shown in Fig. 18.5. Over the concentration range 0–400 μM, transport occurs by a one-

Table 18.1 Inhibitors of taurine influx in the isolated perfused heart[a,b]

Compound	Concentration (μM)	Control		Treated		Percent change	p-value
Hypotaurine	150	(4)	10.6 ± 1.2	(4)	1.8 ± 1.0	−82.3	< 0.001
β-Alanine	500	(4)	21.4 ± 2.4	(4)	3.6 ± 1.6	−83.2	< 0.001
β-Guanidinoethyl sulfonate	150	(6)	18.4 ± 1.2	(8)	10.9 ± 1.0	−40.6	< 0.001
	250	(7)	10.2 ± 0.2	(7)	4.3 ± 0.1	−57.6	< 0.001
β-Guanidino- propionate	150	(9)	13.2 ± 1.3	(11)	10.4 ± 0.3	−21.3	< 0.05

[a]Values are given as means ± SD. Number of animals per group given in parentheses.

[b]Taurine uptake was determined as described for Fig. 18.1 (concentration used: 100 μM in presence of β-alanine, 25 μM for other experiments). Inhibitors were coperfused at the concentrations given.

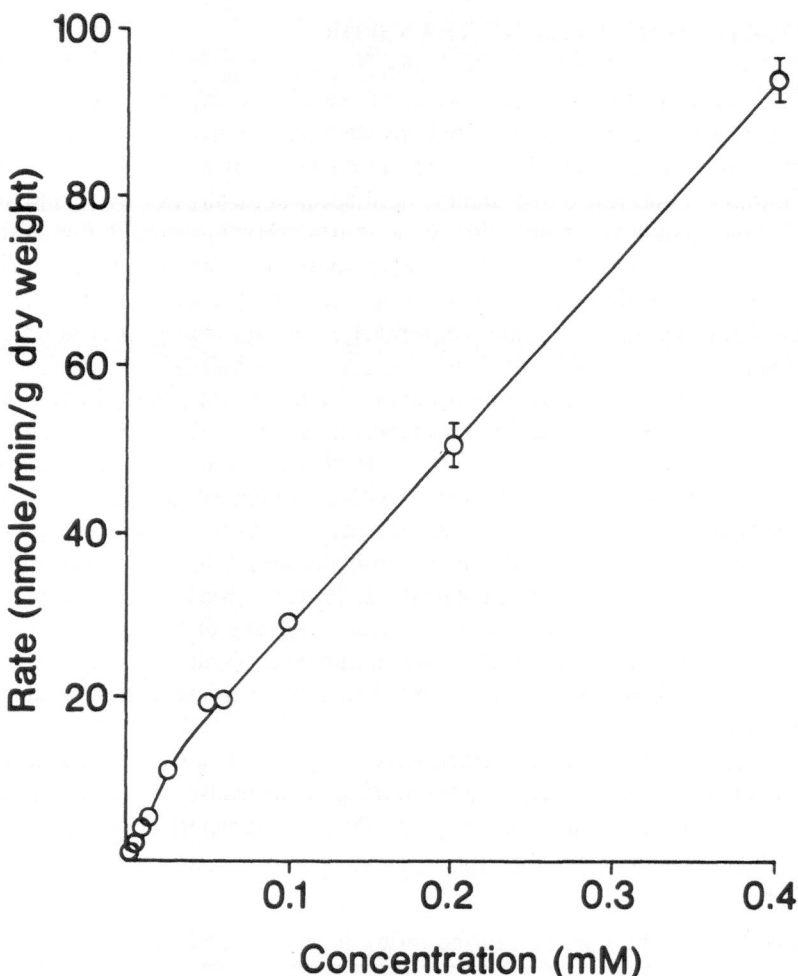

Fig. 18.2. Uptake of [³H]-β-alanine by the isolated perfused heart. Method used was as described for Fig. 18.1. Data displayed as means ± SEM for four to seven hearts per point.

component system, with an affinity (Km) of 153 μM and a Vmax of 65 nmol/g dry weight/min.

Can these inhibitors be used to decrease organ taurine content in the intact animal? Hypotaurine is of little practical use as a transport inhibitor, as it is readily oxidized to taurine both in the perfusion medium and after transport into the heart. This latter conversion may be nonenzymatic, as the search for hypotaurine oxidase activity in the heart has to date been unsuccessful.

β-Alanine has potential usefulness as a transport inhibitor. However, it has

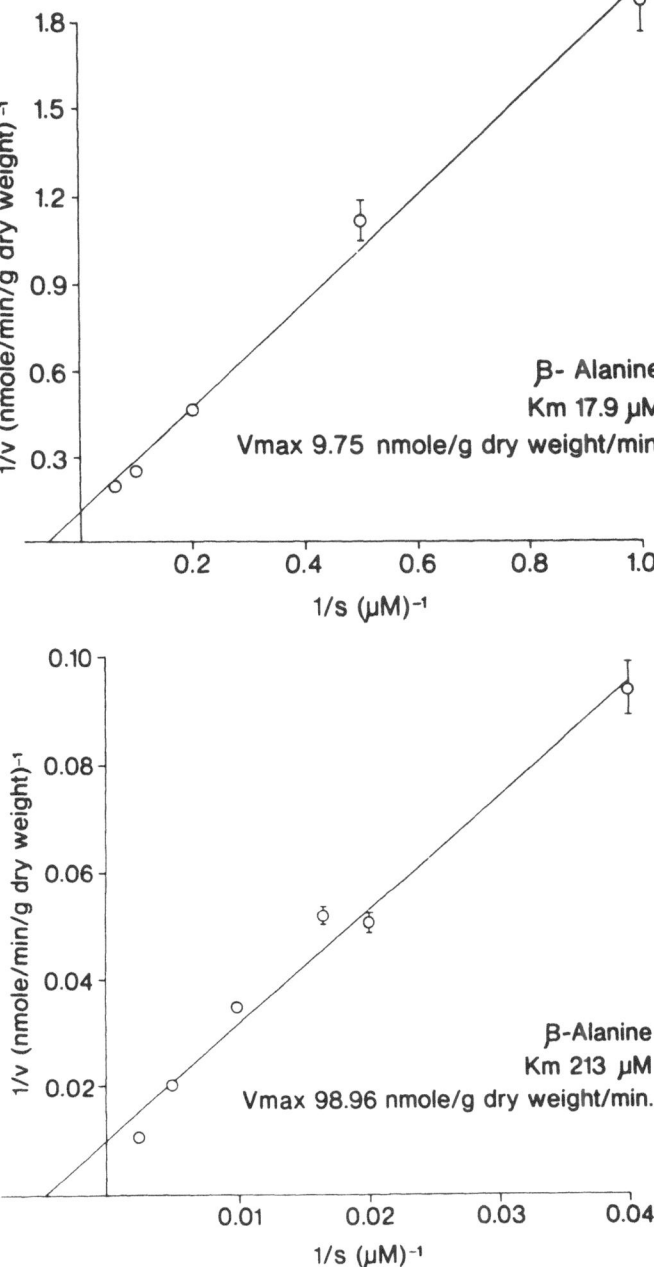

Fig. 18.3. Lineweaver-Burk plots for β-alanine influx in the heart over the concentration ranges: (*a*) 1–15 μM, and (*b*) 25–400 μM. Data shown as means ± SEM for four to seven hearts per point.

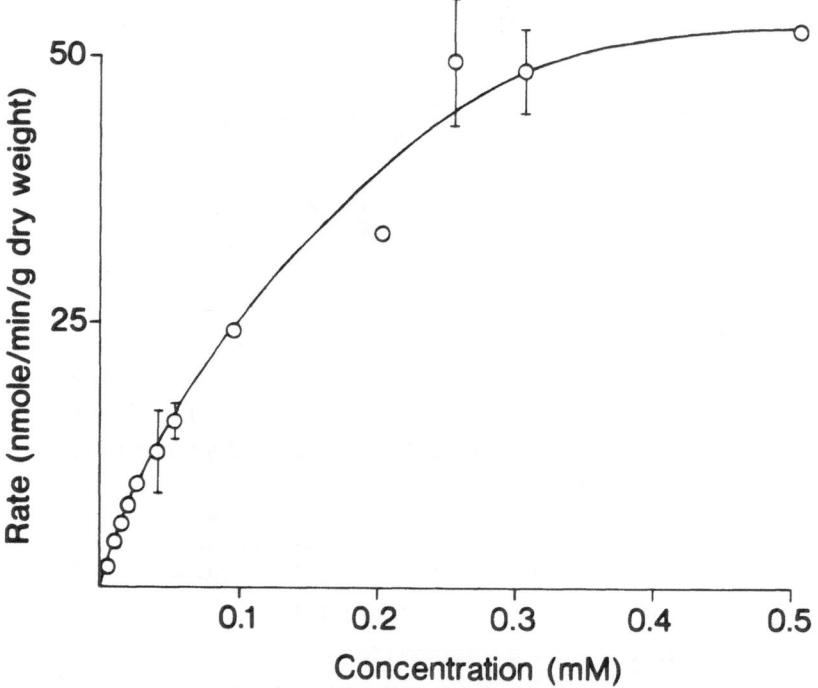

Fig. 18.4.   Uptake of [³H]guanidinoethylsulfonate by the isolated perfused heart. Method used was as described for Fig. 18.1. Data displayed as means ± SEM for three to four hearts per point.

physiological and pharmacological actions of its own, and there is a non-saturable component to its uptake by the heart. Guanidinoethylsulfonate, on the other hand, does not occur in mammalian heart (Guidotti and Costagli, 1970). Furthermore, its transport is saturable. We have therefore tested this compound *in vivo* for its effects on tissue taurine concentrations.

EFFECT OF DIETARY GUANIDINOETHYLSULFONATE ON TISSUE FREE AMINO ACID CONTENT

Two types of experiments were performed. In the first, animals were maintained ad libitum on Wayne rat chow containing 1.6 µmol taurine/g chow and distilled water containing 1% w/v guanidinoethylsulfonate (omitted with the control group). The second type of experiment was performed in a similar manner, except that both groups of animals were maintained on a tau-

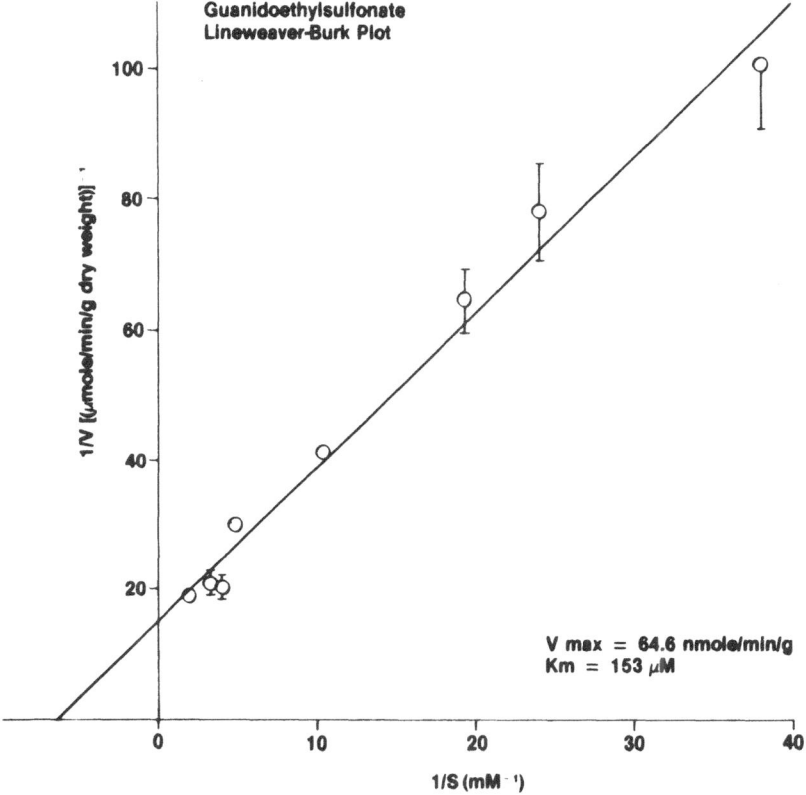

Fig. 18.5.   Lineweaver-Burk plot for guanidinoethylsulfonate influx in the heart.

rine-free diet (Bio-Mix 900 from Bio-Serv, Inc., Frenchtown, N.J. 08825). No difference in weight gain or in food or water consumption was observed between the two groups in either experimental procedure.

With the first procedure (animals on a normal dietary intake of taurine), after 4 weeks exposure to guanidinoethylsulfonate, the taurine concentration of various organs had fallen sharply, as shown in Table 18.2. Falls ranged from 54% of control in the kidney to 19% of control in the heart and muscle. All organs examined exhibited a decrease in taurine concentration. Decreases in taurine concentration were accompanied by an accumulation of guanidinoethylsulfonate (Table 18.2), ranging from 0.8 μmol/g wet weight of tissue in the frontal cortex to 16 μmol/g in the heart. Despite these dramatic modifications in taurine content, the only gross behavioral abnormality noted was hyperactivity. However, no specific behavioral tests were applied.

Table 18.2.   Effect of dietary guanidinoethyl sulfonate on organ taurine and guanidinoethyl sulfonate concentration (μmol/g tissue)[a]

| | Taurine | | | Guanidinoethyl sulfonate |
	Control	Treated	% control	Treated
Heart	23.38 ± 1.12	4.64 ± 0.48	20	16.01 ± 1.83
Lung	6.65 ± 0.47	1.98 ± 0.42	30	2.57 ± 0.24
Kidney	4.63 ± 0.96	2.50 ± 0.31	54	3.64 ± 0.18
Liver	4.46 ± 1.51	1.09 ± 0.33	24	10.84 ± 1.04
Spleen	4.19 ± 1.06	1.19 ± 0.14	28	2.75 ± 0.58
Intestine	5.55 ± 1.18	2.04 ± 0.36	37	2.30 ± 0.78
Muscle: leg	2.71 ± 0.25	0.34 ± 0.08	19	2.57 ± 1.00
neck	1.76 ± 0.25	0.34 ± 0.08	19	2.57 ± 1.00
Brain: cerebellum	4.90 ± 0.38	1.63 ± 0.31	33	1.53 ± 0.21
medulla pons	1.98 ± 0.37	0.90 ± 0.08	46	0.97 ± 0.09
frontal cortex	5.57 ± 1.30	2.81 ± 0.41	50	0.84 ± 0.04

[a]Data are means ± SEM of five samples. The decrease in taurine was statistically significant in all areas except the frontal cortex.

[b]Male Sprague-Dawley rats (initial weight 117 g, final weight 308 g) were maintained on tap water (controls) or tap water containing 1% guanidinoethyl sulfonate for the 4 weeks prior to sacrifice. Rats were maintained ad libitum on Wayne rat chow containing 11.6 μmol taurine/g. Tissues were homogenized in 4 volumes of 3.5% sulfosalicylic acid and centrifuged at 40,000 g for 20 min. Supernatant was analyzed for taurine (amino acid analyzer) and guanidinoethyl sulfonate (method of Guidotti and Costagli, 1970).

Free amino acid concentrations in three brain areas were determined (Table 18.3), the frontal cortex, medulla-pons, and cerebellum. Apart from decreased taurine content (statistically not significant at the $p < 0.05$ level in the frontal cortex), no alterations in free amino acid content were seen.

With the second procedure (animals receiving no dietary taurine), guanidinoethylsulfonate concentrations increased more markedly (Table 18.4). Furthermore, all tissues so far examined, except for two, were equilibrated with guanidinoethylsulfonate by the ninth day of the experiment, and no further increases in guanidinoethylsulfonate content occurred. The exceptions are the heart, in which equilibration occurred within 20 days, and muscle (30 days). Relative to the first procedure, tissues showing the most marked increase in guanidinoethylsulfonate at 30 days are leg muscle (15 X) and intestine (7 X), with other tissues falling between 1 X and 3 X. This suggests a particular sensitivity of skeletal muscle to the guanidinoethylsulfonate:taurine ratio in the serum.

Table 18.3. Effect of guanidinoethyl sulfonate on brain free amino acid concentrations (μmol/g tissue)[a,b]

	Frontal cortex		Cerebellum		Medulla pons	
	Control	Experimental	Control	Experimental	Control	Experimental
PE	ND	1.55 ± 0.27	ND	ND	0.87 ± 0.10	1.20 ± 0.54
Tau	5.57 ± 1.30	2.81 ± 0.41	4.90 ± 0.38	1.63 ± 0.31[c]	1.98 ± 0.36	0.90 ± 0.08[b]
Asp	2.89 ± 0.43	3.15 ± 0.41	3.07 ± 0.16	1.64 ± 0.20	3.36 ± 0.16	2.59 ± 0.46
Thr	0.64 ± 0.10	0.90 ± 0.20	0.92 ± 0.05	0.78 ± 0.06	1.02 ± 0.07	0.83 ± 0.15
Ser	1.29 ± 0.22	1.59 ± 0.22	1.09 ± 0.13	1.04 ± 0.10	1.42 ± 0.21	1.27 ± 0.12
Glu NH$_2$	2.12 ± 0.33	3.49 ± 0.66	3.27 ± 0.18	3.11 ± 0.26	1.68 ± 0.19	1.60 ± 0.20
Glu	10.13 ± 1.45	11.69 ± 1.73	10.90 ± 0.61	9.89 ± 0.55	5.57 ± 0.63	4.70 ± 0.91
Gly	1.40 ± 0.21	1.71 ± 0.25	1.51 ± 0.09	1.40 ± 0.04	3.45 ± 0.30	2.95 ± 0.45
Ala	1.19 ± 0.16	1.47 ± 0.17	1.48 ± 0.06	1.30 ± 0.08	1.32 ± 0.14	1.18 ± 0.14
GABA	4.38 ± 0.57	4.89 ± 0.56	3.50 ± 0.31	3.17 ± 0.14	2.75 ± 0.42	1.75 ± 0.38

[a]Data shown as means ± SEM of five samples. ND = not detected or nonmeasurable.

[b]Experimental conditions as for Table 18.2.

[c]$p < 0.001$.

[d]$p < 0.025$.

Table 18.4.   Tissue guanidinoethyl sulfonate accumulation (μmol/g tissue)[a,b]

	Days of treatment			
	9	20	30	40
Heart	15.93 ± 0.67	20.29 ± 0.55	20.95 ± 2.62	24.70 ± 0.61
Lung	9.24 ± 0.62	6.37 ± 0.44	6.45 ± 0.40	6.13 ± 0.45
Kidney	6.83 ± 0.28	5.85 ± 0.66	7.18 ± 0.65	8.93 ± 0.44
Liver	14.81 ± 1.21	13.37 ± 1.50	16.40± 0.67	14.96 ± 1.28
Spleen	7.75 ± 0.26	8.26 ± 0.94	8.66 ± 0.28	8.00 ± 0.40
Intestine	13.14 ± 1.58	8.51 ± 0.97	16.04 ± 0.91	10.98 ± 0.46
Muscle (leg)	22.10 ± 2.61	33.97 ± 1.61	41.24 ± 1.22	39.47 ± 2.73

[a]Data are means ± SEM of three animals per point.

[b]Male Sprague-Dawley rats (initial weight 125 g) were maintained on water containing 1% guanidinoethyl sulfonate and on taurine-free chow (Biomix 900, Bio-Serv, Frenchtown, New Jersey). After the indicated number of days, animals were sacrificed and guanidinoethyl sulfonate levels determined as described in Table 18.2.

Cardiac free amino acid concentrations during the course of this experiment are shown in Table 18.5. The only significant change observed is a sharp drop in taurine to 31% of control by Day 9, and to 24% of control by Day 20. A number of conclusions follow from these experiments: (1) A marked decrease in cardiac taurine content may be achieved within a few days in animals kept on a taurine-free diet and drinking water containing 1% guanidinoethylsulfonate. (2) Cardiac taurine levels cannot be decreased below 20–30% of normal regardless of the length of time guanidinoethylsulfonate is administered. This irreducible component of the taurine content may represent the proportion of taurine made available by biosynthesis rather than influx. (3) Taurine levels are modified without effects on other free amino acid levels.

IS GUANIDINOETHYLSULFONATE A TAURINE AGONIST?

With compounds in the guanidino series, we have achieved one of our objectives: that of finding a substance that would block taurine transport and lower its cellular concentrations. Guanidinoethylsulfonate lowers cardiac taurine to one-quarter or one-fifth of control within a period of a few days. However, this does not result in any dramatic effect on the heart or on the

Table 18.5. Effect of guanidinoethyl sulfonate on heart free amino acid concentrations (μmol/g tissue)[a,b]

	Days of treatment			
	9	20	30	40
Control				
Tau	30.24 ± 2.84	26.93 ± 2.00	32.67 ± 1.69	32.18 ± 0.36
Asp	2.05 ± 0.24	1.65 ± 0.40	2.01 ± 0.14	1.52 ± 0.34
Thr	0.88 ± 0.05	0.62 ± 0.10	0.97 ± 0.09	0.51 ± 0.17
Ser	0.93	0.88 ± 0.08	0.84 ± 0.19	0.30
Glu NH$_2$	5.91 ± 0.70	5.51 ± 0.29	6.68 ± 0.17	6.60 ± 0.96
Glu	9.10 ± 0.52	6.72 ± 0.55	8.04 ± 0.47	5.22 ± 0.50
Gly	0.69 ± 0.09	0.68 ± 0.10	1.26 ± 0.04	0.77 ± 0.16
Ala	4.35 ± 0.32	4.02 ± 0.03	4.60 ± 0.11	3.59 ± 0.04
Treated				
Tau	9.42 ± 1.20	6.43 ± 0.52	7.82 ± 0.72	9.65 ± 0.64
Asp	2.25 ± 0.20	2.14 ± 0.18	2.04 ± 0.38	2.38 ± 0.09
Thr	0.86 ± 0.11	0.59 ± 0.04	0.71 ± 0.10	0.42 ± 0.07
Ser	0.79 ± 0.10	0.71 ± 0.00	0.85 ± 0.34	0.70 ± 0.09
Glu NH$_2$	6.11 ± 0.31	5.87 ± 0.64	6.13 ± 1.08	7.52 ± 0.12
Glu	9.05 ± 0.56	7.39 ± 0.41	7.40 ± 0.98	7.78 ± 0.24
Gly	0.70 ± 0.00	0.73 ± 0.05	0.88 ± 0.12	0.67 ± 0.12
Ala	5.63 ± 0.10	4.17 ± 0.14	4.72 ± 1.25	4.53 ± 0.27

[a]Data are means ± SEM of two animals per group for controls and three per group for treatment groups (four for 40-day group).

[b]Experiment was as described for Table 18.4, except control animals were not given guanidinoethyl sulfonate.

animal. That is to say, the animals stay alive, are reasonably normal in their gross behavior, and have no immediately obvious pathology of their organs. Do these findings say anything about the functionality of taurine? Three possibilities present themselves. First, that taurine is nonfunctional, and therefore that altering the levels of something that has no effect will produce no effect. Second, of the total taurine, only a small pool is functional, and this is unaltered by the guanidinoethylsulfonate treatment. Or, third, that guanidinoethylsulfonate is a taurine agonist, and that the guanidinoethylsulfonate has functionally replaced the taurine that was depleted.

The transport of taurine into the heart is stimulated by β-adrenergic activation. This stimulation is associated with an increased Michaelis constant for transportation and an increased Vmax (Huxtable and Chubb, 1977;

Chubb and Huxtable, 1978a). We have suggested that the β-adrenergic system regulates taurine influx by a cyclic AMP-dependent phosphorylation of a membrane transport site that increases that rate of breakdown of the taurine:transport site complex (Huxtable and Chubb, 1977; Chubb and Huxtable, 1978a). Taurine increases the calcium flux across the heart and increases the size of the intracellular calcium pool (Dolara et al., 1973; Chubb and Huxtable, 1978b). These actions are those of a regulatory substance, one modulating the delivery of calcium to the contractile elements. Interference with this regulation (e.g., by depletion of taurine content) would be expected to produce a pathological consequence only if an appropriate stress is placed on the heart. There is evidence indicating that changes in calcium flux are achieved by increasing the binding affinity of calcium for a membrane site (Chovan et al., 1979). We propose that the magnitude of this effect is regulated by the flux of taurine across the membrane; that is, in terms of modifications in calcium movement, the total taurine content of the heart is less important than the membrane flux of taurine. We would also propose that any substance transported by the same carrier system that transports taurine will act as a taurine agonist. Thus, β-alanine, hypotaurine, and guanidinoethylsulfonate would be predicted to have the same actions on calcium movement as does taurine.

A NOTE ON GUANIDINOETHYLSULFONATE AND GUANIDINOPROPIONATE

Guanidinoethylsulfonate has also been referred to in the literature as guanidinotaurine, guanidyltaurine, and taurocyamine. The first two names are chemically incorrect. If it is desired to name it as a taurine derivative, then N-amidinotaurine is acceptable. However, we prefer the systematic name 2-guanidinoethylsulfonate.

Numerous studies have been performed on the biochemistry of guanidinoethylsulfonate in invertebrates such as polychaetes (reviewed in Jacobsen and Smith, 1968). The phosphorylated derivative appears to function as a phosphagen in these animals (Ennor and Morrison, 1958). The mammalian biochemistry of guanidinoethylsulfonate, however, is less well understood and has been poorly studied. It occurs in low amounts in human and rat urine, and it has been shown to be present in rat liver, kidney, and muscle (Guidotti and Costagli, 1970). The source of it is unknown. We injected rats i.p. with [^{3}H]taurine, and at various times homogenized tissues with a guanidinoethylsulfonate solution from which this compound was subsequently reisolated. We also did the reverse experiment; that is to say, rats were injected i.p. with [^{3}H]guanidinoethylsulfonate and tissues were subsequently extracted with taurine, which was then recrystallized to constant activity. We found no indications that [³H]taurine was converted to guanidinoethylsulfonate or that [³H]guanidinoethylsulfonate was converted to taurine.

Guanidinopropropionate has been reported to be transported in place of creatine and also to serve as a substrate for creatine phosphokinase (Fitch et al., 1974; Shields and Whitehair, 1973). This compound, therefore, depletes tissues of creatinine, but phosphoguanidinopropionate may be used to regenerate ATP.

SUMMARY

Further advances in our understanding of taurine in the heart await the development of pharmacological antagonists to taurine, and the finding of methods to modify taurine concentration in the heart. The balance of evidence indicates that influx is quantitatively of greater importance than biosynthesis in the regulation of cardiac taurine content, and that taurine is removed from the heart by efflux rather than metabolism. Influx occurs by an active transport process, and a transport inhibitor holds promise of lowering organ taurine levels, in light of the considerations above. Guanidinoethylsulfonate is an effective transport inhibitor in vitro. We find that administration of guanidinoethylsulfonate to rats as a 1% aqueous solution leads to a rapid and sharp drop in taurine concentration of all tissues examined. Other amino acids are not affected. In place of taurine, tissues accumulate guanidinoethylsulfonate. As animals survive up to an 80% decrease in taurine content of the heart (and other organs) with no gross acute abnormalities, this may indicate that guanidinoethylsulfonate is acting as a taurine agonist.

ACKNOWLEDGMENTS

This research was supported by USPHS grants HL 19394 and NS 14405. We thank R. Ronstadt for assistance with perfusion experiments and M. Buechel for assistance with guanidinoethylsulfonate determinations.

REFERENCES

Awapara, J.; and Berg, M. Uptake of taurine by slices of rat heart and kidney. In *Taurine*, R. Huxtable and A. Barbeau, eds. Raven Press, New York (1976), pp. 135–143.
Azari, J.; Bahl, J.; and Huxtable, R. Guanidinoethylsulfonate and other inhibitors of the taurine transporting system in the heart. *Proc. Western Pharm. Soc.*, 22, 389–393 (1979).

Barbeau, A.; and Huxtable, R., eds. *Taurine and Neurological Disorders*. Raven Press, New York (1978).

Chovan, J. P.; Kulakowski, E. C.; Benson, B. W.; and Schaffer, S. W. Taurine enhancement of calcium binding to rat heart sarcolemma. *Biochim. Biophys. Acta*, 551, 129–136 (1979).

Chubb, R.; and Huxtable, R. Isoproterenol-stimulated taurine influx in the perfused rat heart. *Eur. J. Pharmacol.*, 48 369–376 (1978a).

Chubb, J.; and Huxtable, R. Transport and biosynthesis of taurine in the stressed heart. In *Taurine and Neurological Disorders*, A. Barbeau and R. Huxtable, eds. Raven Press, New York (1978b), pp. 161–178.

Dolara, P.; Agresti, A.; Giotti, A.; and Pasquini, G. Effect of taurine on calcium kinetics of guinea pig heart. *Eur. J. Pharmacol.*, 24, 352–358 (1973).

Ennor, A. H.; and Morrison, J. F. Biochemistry of the phosphagens and related guanidines. *Physiol. Rev.*, 38, 631–674 (1958).

Fitch, C. D.; Jellinek, M.; and Mueller, E. J. Experimental depletion of creatine and phosphocreatine from skeletal muscle. *J. Biol. Chem.*, 249, 1060–1063 (1974).

Grosso, D. S.; Roseke, W. R.; and Bressler, R. Characterization of a carrier-mediated transport system for taurine in the fetal mouse heart *in vitro. J. Clin. Invest.*, 61, 944–952 (1978).

Gruener, R.; Bryant, H.; Markovitz, D.; Huxtable, R.; and Bressler, R. Ionic actions of taurine on nerve and muscle membranes: Electrophysiological studies. In *Taurine*, R. Huxtable and A. Barbeau, eds. Raven Press, New York (1976), pp. 225–242.

Guidotti, A.; and Costagli, P. F. Occurrence of guanidotaurine in mammals: Variation of urinary and tissue concentration after guanidotaurine administration. *Pharmacol. Res. Comm.*, 2, 341–354 (1970).

Hayes, K. C.; Carey, R. E.; and Schmidt, S. Y. Retinal degeneration associated with taurine deficiency in the cat. *Science*, 188, 949–951 (1975).

Hruska, R. E.; Huxtable, R. J.; and Yamamura, H. I. High affinity, temperature-sensitive and sodium-dependent transport of taurine in rat brain. In *Taurine and Neurological Disorders*, A. Barbeau and R. Huxtable, eds. Raven Press, New York (1978a), pp. 109–117.

Hruska, R. E.; Padjen, A.; Bressler, R.; and Yamamura, H. I. Taurine: Sodium-dependent, high-affinity transport into rat brain synaptosomes. *Mol. Pharmacol.*, 14, 77–85 (1978b).

Huxtable, R. J. Regulation of taurine in the heart. In *Taurine and Neurological Disorders*, A. Barbeau and R. Huxtable, eds. Raven Press, New York (1978), pp. 5–17.

Huxtable, R.; and Bressler, R. Taurine concentrations in congestive heart failure. *Science*, 184, 1187–1188 (1974).

Huxtable, R.; and Chubb, J. Adrenergic stimulation of taurine transport by the heart. *Science*, 198, 409–411 (1977).

Jacobsen, J. G.; and Smith, L. H., Jr. Biochemistry and physiology of taurine and taurine derivatives. *Physiol. Rev.*, 48, 424–511 (1968).

Perry, T. L.; Hansen, S.; Berry, K.; Mok, C.; and Lesk, D. Free amino acids and related compounds in biopsies of human brain. *J. Neurochem.*, 18, 521–528 (1971).

Peterson, M. B.; Mead, R. J.; and Welty, J. D. Free amino acids in congestive heart failure. *J. Mol. Cell. Cardiol.*, 5, 139–147 (1973).

Prous, J. G. de Y.; Carlsson, A.; and Gomez, M. A. M. The effect of taurine on motor behavior, body temperature and monoamine metabolism in rat brain. *Naunyn-Schmiedeberg's Arch. Pharmacol.*, 304, 95–99 (1978).

Schmidt, S. Y.; Berson, E. L.; and Hayes, K. C. Retinal degeneration in cats fed casein: 1. Taurine deficiency. *Invest. Ophthalmol.*, 15, 47–52 (1976).

Shields, R. P.; and Whitehair, C. K. Muscle creatine: *In vivo* depletion by feeding β-guanidinopropionic acid. *Can. J. Biochem.*, 51, 1046–1049 (1973).

Tiedemann, F.; and Gmelin, L. Einige neue Bestandtheile der Galle des Ochsen. *Ann. Physik. Chem.*, 9, 326–337 (1827).

Copyright © 1981, Spectrum Publications, Inc.
The Effects of Taurine on Excitable Tissues

CHAPTER 19

Accumulation of Taurine by Isolated Rat Heart Cells and Rat Heart Slices

J. Bahl
C. J. Frangakis
B. Larsen
S. Chang
D. Grosso
R. Bressler

INTRODUCTION

Taurine, which has no known myocardial physiologic function, is maintained in the heart at levels that are regulated within narrow limits and have proven difficult to alter by dietary manipulation (Huxtable, 1978). Studies on the system(s) responsible for maintaining cardiac taurine levels could provide insight into the increases in taurine levels associated with cardiac hypertrophy (Huxtable and Bressler, 1974a,b). Patients with congestive heart failure had twice the taurine per gram wet weight of ventricle than patients having died of other noncardiac causes (Huxtable and Bressler, 1974a). To gain insight into the role of taurine in normal and hypertrophied hearts, we studied binding of taurine in isolated adult rat heart myocytes. Transport of taurine into rat heart slices serves as a model that might provide information about the system(s) that maintains cardiac taurine levels in the heart.

HEART CELLS

Heart cells were prepared by the method of Grosso et al. (1977). This preparation provides a high yield of myocytes, 90–95% of which exclude the dye trypan blue, and 80% of these cells can be observed to beat spontaneously. The cells oxidize the substrates glucose, pyruvate, citrate, and palmitate linearly for at least 2 hours. Increased levels of cyclic-AMP have been measured in these cells following stimulation by the hormones epinephrine and glucagon, and after incubation with the phosphodiesterase inhibitor l-methyl-3-isobutylxanthine (MIX) (Grosso et al., 1977; Frangakis, 1978).

HPLC Taurine Assay

Taurine was quantified by the following HPLC method. Tissue (1–5 mg) was homogenized in 200 μliter of double-distilled water and 200 μliter of a 2% aqueous picric acid solution. The resulting solution was passed through a 0.5 $\times$ 5 cm column (2.5 cm Dowex 1 $\times$ 8 Cl$^-$ form over 2.5 cm Dowex 50 $\times$ 4H$^+$ form equilibrated with water) and the eluant lyophilized. The residue was dissolved in 100 μliter water and an aliquot mixed with an equal volume of o-phthalaldehyde (OPA) reagent (Roth, 1971) for 1.0 min prior to injection for HPLC analysis. The chromatography was carried out using a Bondapak alkylphenyl column (Waters) eluted isocratically at 2 ml/min with 43% of a 0.05 M NaH$_2$PO$_4$, pH 5.3, buffer solution and 57% 0.05 M NaH$_2$PO$_4$ in 25% water/75% methanol. Elution time for the taurine standard was consistently 4.0 min. Fluorescence detection (Ex., 390 nm; Em., 475 nm) was made using a Gilson Spectra/Glo Fluorometer with filters specifically for OPA derivatives. Quantitation was accomplished by comparing peak heights of the standard solutions (prepared as described above) with those resulting from the test samples.

The intact heart contains 220 nmol taurine per mg cellular protein, whereas freshly isolated heart cells have 55 nmol per mg protein. The taurine content in the cells fell rapidly to 3.7 nmol per mg protein, which remained as a constant value even after several hours of incubation and repeated buffer changes. This suggests that the cells were highly permeable to taurine. The small amount of taurine that was held by the cells may represent a pool of tightly sequestered material or may be a nondiffusible cellular constituent that is converted to free taurine during the analysis procedure.

Taurine Uptake by Myocytes

Taurine enters the isolated heart myocytes by an unsaturable process.

Cells, incubated with a constant quantity of labeled taurine and varying concentrations of unlabeled taurine and varying concentrations of unlabeled taurine (1 μM to 30 mM), were separated from the incubation medium by being centrifuged through a mixture of phthalate esters. No saturation was observed (Fig. 19.1). The volume of ^3H_2O associated with the pellet was equal to the volume of the incubating medium necessary to contain the DPMs of taurine found with the pellet.

Taurine Binding by Myocytes

When cells were pipetted onto Nuclepore filters and washed twice with 4 ml of ice-cold buffer, binding was observed. As shown in Fig. 19.2, this binding was temperature and sodium dependent. A dose-dependence rela-

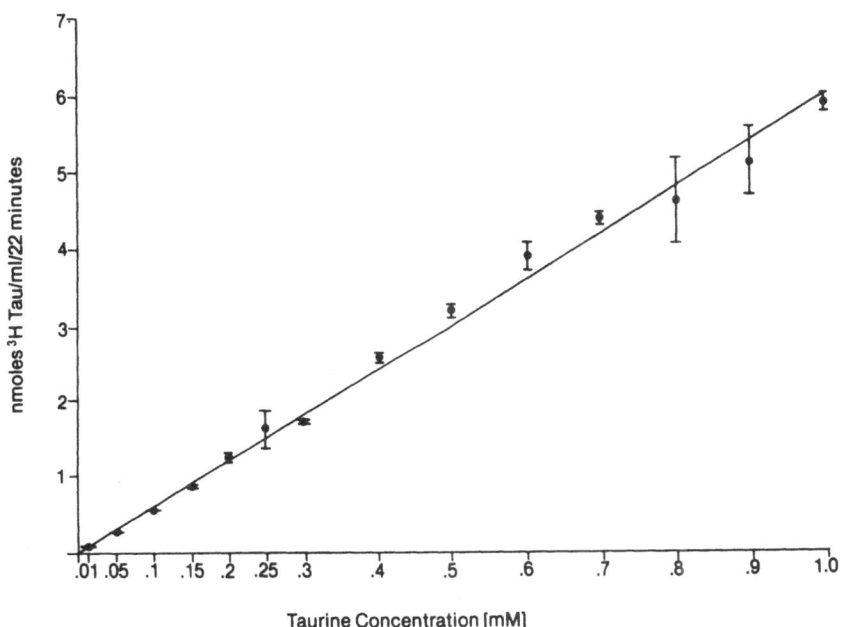

Taurine Concentration [mM]

Fig.19.1. ^3H-Taurine associated with centrifuged cells. Cells (1 mg protein in 1 ml), after incubation with ^3H-taurine and varying concentrations of unlabeled taurine for 22 min, were layered on a 1:1 mixture of dioctyl phthalate and dibutyl phthalate. Centrifugation for 30 sec with a Beckman microfuge pelleted the cells under the organic layer. The upper aqueous layer and most of the organic layer were removed by aspiration, and the tip of the tube cut. The pelleted cells were dissolved with NaOH and neutralized with HCl prior to counting. ^3H-taurine found in the pellet did not saturate between 0.01 mM and 1.0 mM taurine (shown here) nor in the range of 1.0 to 30 mM taurine (data not shown). Data reported as means ± SEM, $n = 5$.

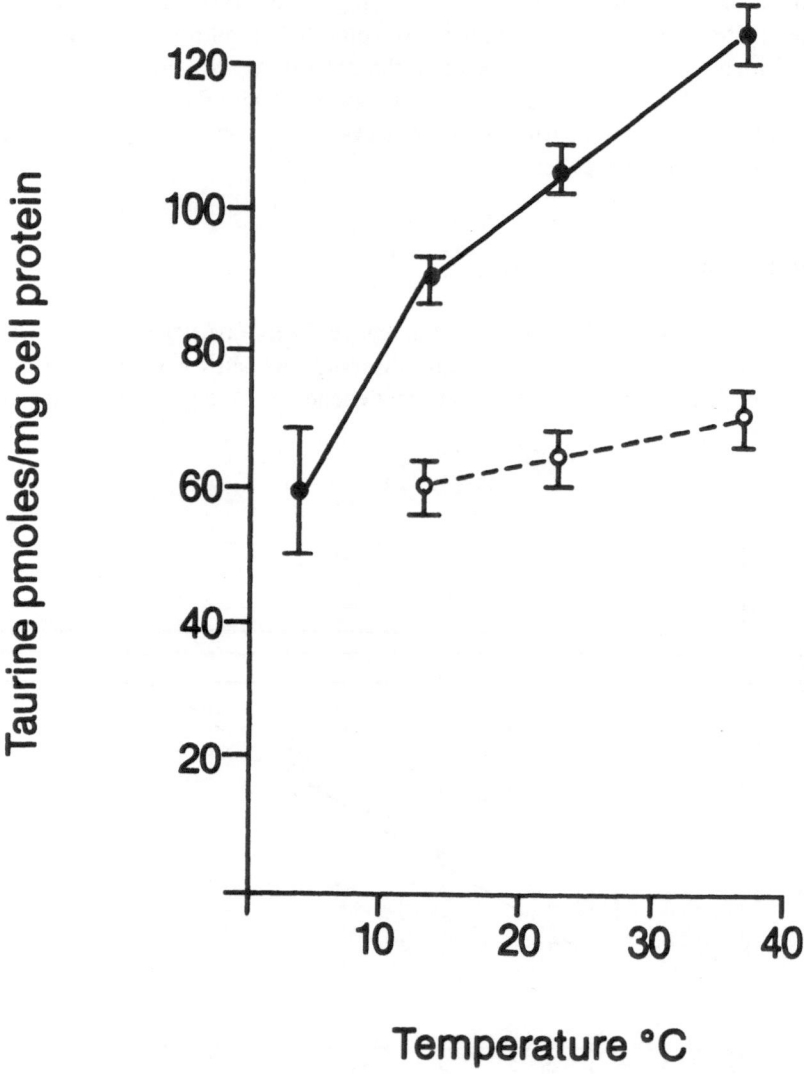

Fig. 19.2.   Effect of temperature and low sodium on taurine uptake. Cells (1 mg cellular protein) were incubated with 100 μM ^{3}H-taurine in buffer containing 141 mM sodium (●) or 20 mM sodium (O) at various temperatures for 2 min in a volume of 1 ml. Osmolarity was held constant with LiCl. Cells were pipetted onto (10 μm) Nuclepore filters and rinsed. Cells on filters were dissolved in fluor and counted by standard techniques. Data reported as mean ± SEM, n = 3.

tionship existed between sodium present in the medium and taurine binding. Binding at 4°C was used as a measure of sodium independent binding and was subtracted from the total binding at 37° to yield the sodium-dependent temperature-sensitive binding of taurine.

Calcium and Taurine Binding

Physiological concentrations of calcium produce a paradoxical toxic response in isolated heart cells (Clark et al., 1978). Cells can exist for a few minutes at calcium concentrations of 0.1 mM but cannot survive 1 mM calcium for more than a few seconds. As shown in Fig. 19.3, the addition of calcium to the medium increased taurine binding until the toxic 1 mM concentration of calcium was reached. Due to this toxicity, binding was measured after 2 min in the presence of 100 μM calcium.

Taurine Binding by Myocytes from Spontaneously Hypertensive Rats

The binding of taurine showed saturable kinetics (Fig. 19.4), and Scatchard analysis of binding yielded a high affinity system with an apparent Kd of 133 μM and a Vmax of 93 pmol mg^{-1}. In preliminary experiments with 6-month-old male, spontaneously hypertensive rats (SHR) (systolic b.p., 220 mm Hg; control, 130 mm Hg), no difference in taurine binding from control was observed. Binding was increased 50% by 10^{-6} M norepinephrine and could be blocked by pretreatment with propranolol. Dibutyrylcyclic AMP (10^{-8} M) stimulated binding 45%, and this increase was not blocked by propranolol.

Inhibition of Taurine Binding

The binding of ^{3}H-taurine was inhibited by the presence of taurine analogues but not by neutral alpha amino acids (Table 19.1). Hypotaurine was more effective than unlabeled taurine at decreasing the binding of ^{3}H-taurine. Both β-alanine and the guanidino derivative of taurine, guanidinoethylsulfonic acid, effectively inhibited binding, and, to a lesser extent, so did 3-amino-propylphosphonic acid.

Figure 19.5 indicates that incubation of cells in 30 mM taurine increase the amount of ^{45}Ca associated with the cells over control levels. Thus, taurine binding is increased with increasing calcium, and the presence of high taurine increases the ^{45}Ca associated with the cells.

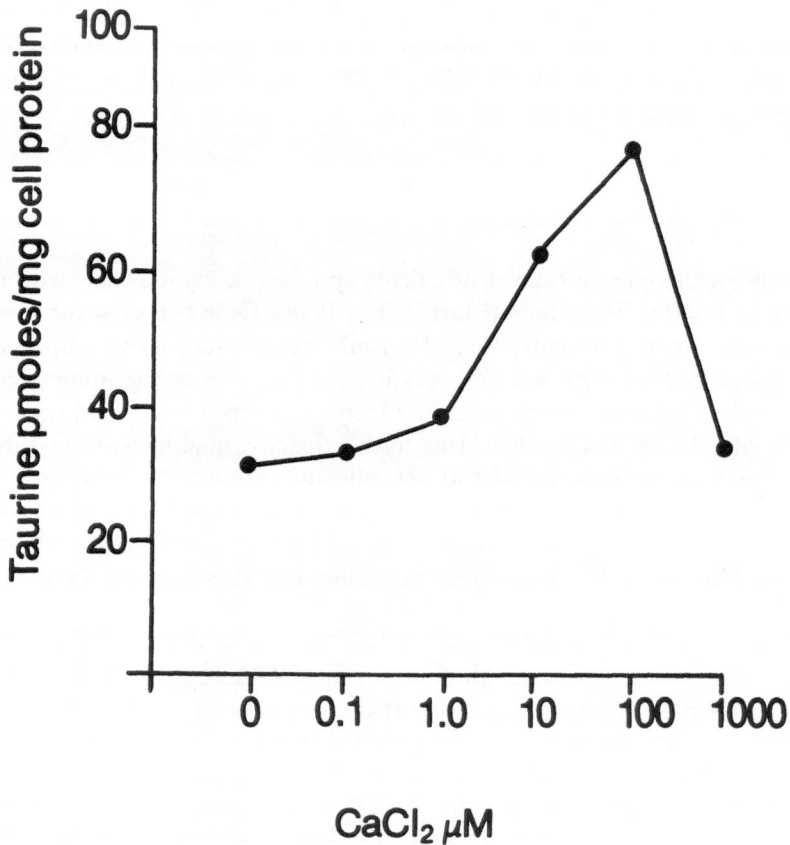

Fig. 19.3. Stimulation of taurine binding by calcium. Rat myocytes were incubated with 100 μM ^3H-taurine and several concentrations of $CaCl_2$ for 5 min. The zero calcium value of taurine binding was measured in cells incubated with 13 μM EGTA. Values are the means of four points.

HEART SLICES

Heart slices were prepared by the method of Awapara and Berg (1976) in phosphate buffer (8.95 mM NaH_2PO_4, 2 mM Na_2HPO_4, 121.7 mM NaCl, 5.4 mM KCl, 1 mM $CaCl_2$, 1 mM $MgCl_2$) with 25 mM glucose. Slices were preincubated in two changes of fresh buffer at 37° for 15 min and were then incubated in scintillation vials containing 1 ml of incubation medium per slice. After incubation, slices were lightly blotted prior to weighing. Aliquots of the incubation medium, taken both before and after incubation with the slices, were removed and counted for ^3H and ^{14}C. Before counting, the tissue was dissolved in NaOH and neutralized with HCl. The amount of taurine

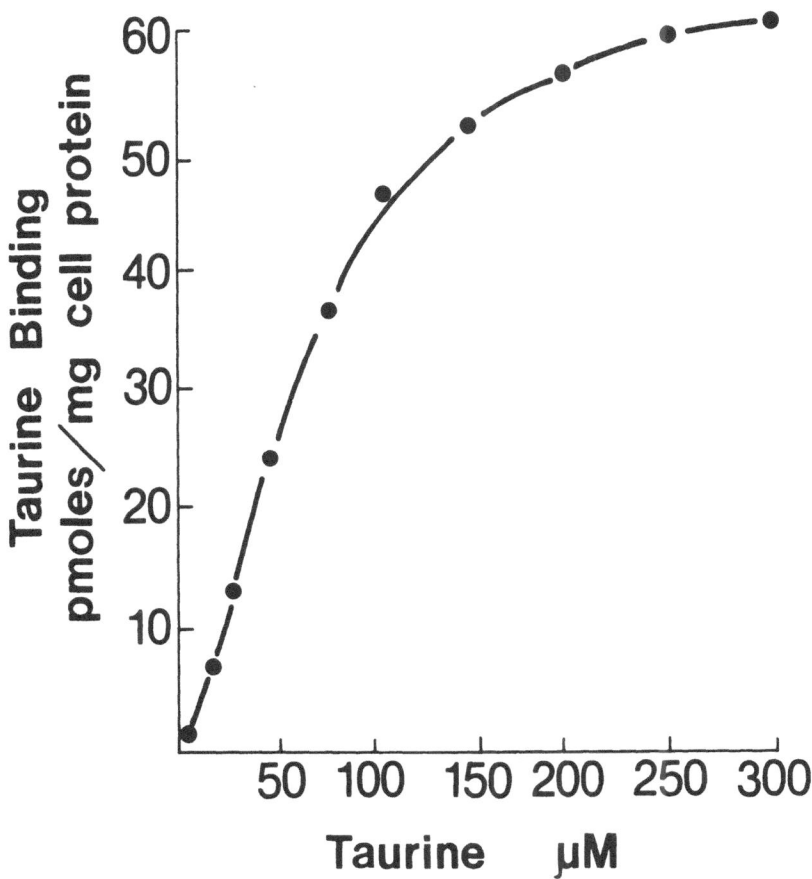

Fig. 19.4. Concentration dependence of taurine binding. Cells were incubated with ^{3}H-taurine for a period of 2 min. Saturation curve of net binding shown between range of 5 and 300 μM. Points are the means of three experiments, with $n = 4$ for each concentration tested.

taken up was calculated by subtracting the DPM of ^{3}H-taurine found in the extracellular space, as measured by ^{14}C-sorbitol space, from the total DPMs of ^{3}H-taurine associated with an individual slice and expressed as amount per milligram wet weight.

Taurine Uptake by Heart Slices

The time course of uptake is shown in Fig. 19.6. After a short period of

Table 19.1   Inhibition of binding in isolated heart cells[a]

Analogue	pmol/mg protein	Percent inhibition
Control	20.0 ± 2.9	—
Taurine (1 mM)	4.4 ± 2.0	80[b]
β-Alanine	6.6 ± 1.9	67[b]
Hypotaurine	2.8 ± 0.3	86[b]
Guanidinoethylsulfonic acid	6.2 ± 2.9	69[b]
3-Aminopropylphosphonic acid	8.0 ± 2.0	60[b]
α-Alanine	22.0 ± 5.0	0
α-Aminoisobutyric acid	23.0 ± 3.0	0
β-Aminoisobutyric acid	17.2 ± 10.0	14
α-Amino-n-butyric acid	19.0 ± 2.0	5
γ-Amino-n-butyric acid (GABA)	23.0 ± 7.0	0
Glycine	14.0 ± 8.0	30
Leucine	17.8 ± 4.0	11
Threonine	24.0 ± 3.1	0

[a]Cells were incubated for 2 min with ^{3}H-taurine (10 μM) and several analogues (1 mM). Data are expressed as mean ± SEM of three experiments.

[b]$p < 0.05$.

equilibration, the volume measure for extracellular space remained constant across time. Subtracting the taurine in the extracellular space from the total taurine yields the net amount transported into the slice, which increased linearly across time.

In Fig. 19.7, the kinetic data obtained from normotensive and spontaneously hypertensive animals are plotted. Lineweaver-Burk analysis reveals a high affinity system with an apparent Km of 62.5 μM, with Vmax of 63.7 pmol/mg^{-1} wet weight/hr^{-1} for the normotensive animals, and a Km of 45.5 μM and a Vmax of 59.6 pmol/mg^{-1} wet weight/hr^{-1} for the hypertensive animals.

Table 19.2 shows that the compounds which were potent inhibitors of taurine binding in isolated cells were also able to inhibit transport of taurine into heart slices.

DISCUSSION

Taurine entry into the isolated rat heart myocyte is by a nonsaturable process (diffusion) and does not appear to be concentrated against a gradient.

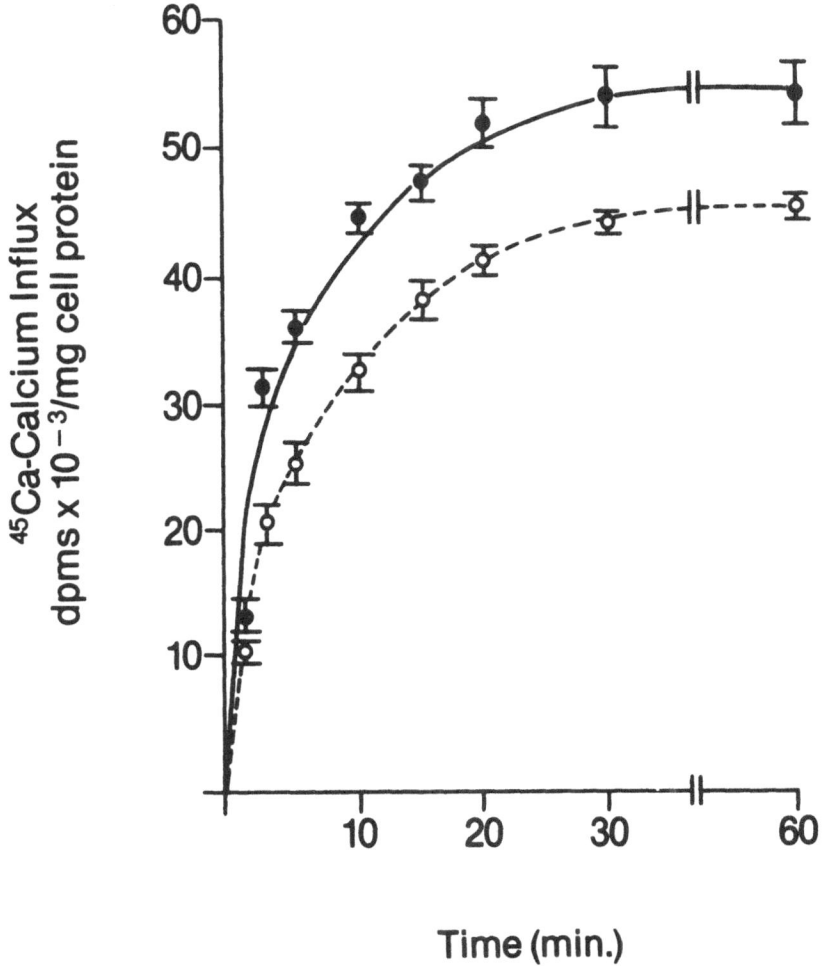

Fig.19.5. Influx of [45]Ca-calcium into isolated adult rat heart myocytes in the presence or absence of 30 μM taurine. Cells were filtered after various incubation time with 10 μM [45]Ca-calcium in the presence (●) or absence (O) of 30 mM taurine. Osmolarity was maintained by addition of LiCl. Values represent mean ± SEM of 4 replicate samples.

Taurine binds to the isolated myocyte in a saturable process that is temperature sensitive and requires both sodium and calcium ions. It is possible that the calcium dependence may not have been easily observed in other models where intracellular calcium levels are more tightly controlled (Awapara and

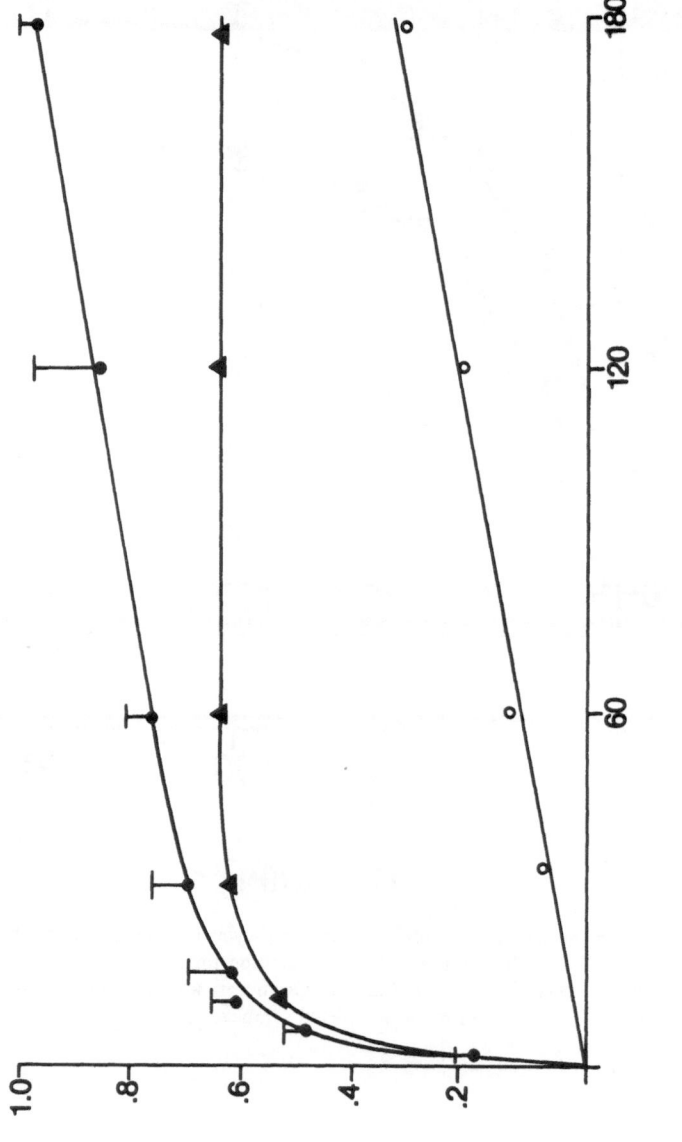

Fig. 19.6. Time course of taurine uptake into rat heart slices. Rat heart slices were incubated in buffer containing 1.0 μC/ml ^{3}H-taurine and 0.1 μC/ml ^{14}C-sorbitol for varying periods of time. The top curve is the total ^{3}H-taurine associated with the slice per milligram wet weight. The middle curve is ^{3}H-taurine appearing in the extracellular space as measured by ^{14}C-sorbitol. The difference between the two curves is the net transport of taurine into the intact cells of the slice. Each value is the mean of eight slices.

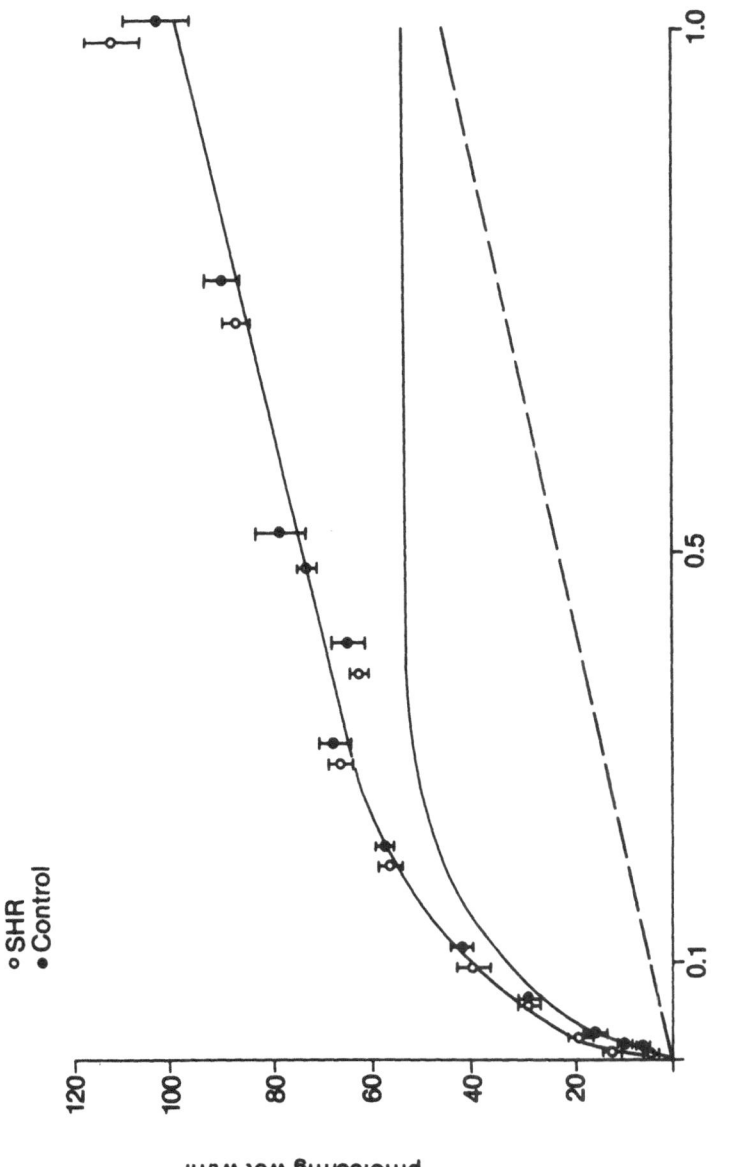

Fig. 19.7. Concentration dependence of taurine uptake in rat heart slices in normotensive and spontaneously hypertensive rats. Slices were incubated for 1 hr at 37° with ^{3}H-taurine and ^{14}C-sorbitol. Transport of taurine was measured at various concentrations of taurine up to 1 mM. Broken line, contribution to taurine uptake by low affinity transport system. Solid line, remaining uptake by high affinity transport system. Each value is the mean of between 20 and 24 slices, ± SEM.

Table 19.2   Inhibition of taurine uptake in rat heart slices[a]

	10 μM taurine		1 mM taurine	
	pmol/mg protein	Percent inhibition	pmol/mg protein	Percent inhibition
Control	6.8 ± 0.99	—	127.8 ± 11.0	—
2 mM β-alanine	1.9 ± 0.23	72.0	92.5 ± 5.0	27.6
2 mM guanidinoethylsulfonic acid	1.9 ± 0.15	71.8	110.0 ± 16.7	13.9
2 mM hypotaurine	0.74 ± 0.024	89.2	75.6 ± 2.5	39.2
2 mM α-Aminoisobutyric acid	6.8 ± 0.13	1.1	146.2 ± 8.9	−14.4

[a]Slices were incubated for 1 hr with either 10 μM or 1 mM ^{3}H-taurine. Values are the means ± SEM, $n = 6$.

Berg, 1976). Neutral alpha amino acids did not significantly reduce uptake, but analogues of taurine with similar charge and structure did inhibit binding. Altered levels of taurine seen in SHRs do not result from changes in binding under the conditions assayed.

The taurine uptake system had been previously characterized in rat heart slices (Awapara and Berg, 1976). We were unable to use this model to provide an explanation for the doubling of taurine seen in the SHR since control and SHR kinetic parmeters for high affinity transport are well within the range of experimental error and should be regarded as equivalent. Currently, the low affinity transport system and possible differences in the rate of efflux are being investigated to explain the differences in the amount of taurine observed between the two conditions.

The previous characterization of the taurine uptake system did not show decreased uptake when β-alanine was used as an inhibitor (Awapara and Berg, 1976). We found β-alanine a potent inhibitor of transport, as were hypotaurine and guanidinoethylsulfonic acid; but AIB, chosen to represent the neutral alpha amino acids, did not inhibit uptake significantly. The Km previously reported for slices (40 μM) is in good agreement with the 62.5 μM observed here for normal rat hearts and the Km of 45.5 μM for SHR. Thus, whatever differences that exist between the normotensive and hypertensive animal, resulting in the twofold increase in taurine myocardial concentration, do not appear to result from changes in the high affinity transport system. Current investigations center on the low affinity transport system and the efflux of taurine.

ACKNOWLEDGMENTS

The authors wish to thank Ms. Connie Sontag for her expert technical assistance. This work was supported by National Institutes of Health Grants HL 13636 and HL 20984.

REFERENCES

Awapara, J.; and Berg, M. Uptake of taurine by slices of rat heart and kidney. In *Taurine,* Huxtable, R.; and Barbeau, A., eds. Raven Press, New York (1976), pp. 135–143.

Clark, M. G.; Gannon, B. J.; Bodkin, N.; Patten, G. S.; and Berry, M. N. An improved procedure for the high yield preparation of intact beating cells from the adult rat. *J. Mol. Cell. Cardiol.,* 10, 1101–1121 (1978).

Frangakis, C. F. Dissertion: Uptake and disposition of taurine by isolated adult rat myocytes. Department of Pharmacology, University of Arizona (1978).

Grosso, D. A.; Frangakis, C. F.; Carlson E.; and Bressler, R. Isolation and characterization of myocytes from the adult rat heart. *Prep. Biochem.,* 715, 383–407 (1977).

Huxtable, R. J. Regulation of taurine in the heart. In *Taurine and Neurological Disorders,* Barbeau, A.; and Huxtable, R. J., eds. Raven Press, New York (1978), pp. 5–17.

Huxtable, R.; and Bressler, R. Elevation of taurine in human congestive heart failure. *Life Science,* 14, 1153–1159 (1974a).

Huxtable, R.; and Bressler, R. Taurine concentration in congestive heart failure. *Science,* 184, 1187–1188 (1974b).

Roth, M. Fluorescence reaction for amino acids. *Anal. Chem.,* 43, 880–882 (1971).

Copyright © 1981, Spectrum Publications, Inc.
The Effects of Taurine on Excitable Tissues

CHAPTER 20

The Role of Taurine Receptors in the Heart

S. W. Schaffer
J. Chovan
J. Kramer
E. Kulakowski

INTRODUCTION

It is well established that calcium serves as the coupler of excitation and contraction in muscle (Dhalla et al., 1977; Katz, 1977; Langer, 1976; and Williamson and Schaffer, 1976). The calcium necessary for maximal contraction appears to be derived from two calcium pools. Pool I calcium is supplied from the extracellular fluid and enters the cell during the plateau phase of the action potential. This pool is thought to be located on the sarcolemmal membrane (Langer, 1978; Limas, 1977; and Williamson et al., 1975). Pool II calcium appears to be separate from the sarcolemma since the calcium in this pool does not rapidly exchange with calcium from the extracellular space unless the muscle is stimulated. Most investigators feel that this latter pool is associated with the terminal cisternae of the sarcoplasmic reticulum.

In skeletal muscle, depolarization of the sarcolemma leads to both a rapid entry of calcium into the cell and a release of calcium from the sarcoplasmic reticulum (Fabiato and Fabiato, 1977; Kasai and Miyamoto, 1976). A similar mechanism has been proposed for excitation-contraction coupling in the heart. However, the sarcoplasmic reticulum (Pool II) is not as extensively developed in the heart as in skeletal muscle. This has led Langer (1976) to suggest that Pool II calcium is of minor importance in the heart, and that sufficient calcium can be supplied from Pool I to achieve maximal contraction. His proposal is based upon the premise that calcium entry into the cell proceeds by two separate processes, an electrogenic movement of calcium through pores and a nonelectrogenic, carrier-mediated exchange of calcium for sodium. Most of the coupling calcium would be derived from the carrier. Both models have one common feature; namely, Pool I calcium is required for myocardial contraction. Thus, it is generally accepted that the sarcolemma plays a central role in maintaining normal contraction.

Most of the wide range of effects of taurine on the mammalian heart are closely linked to calcium transport. Guidotti and Giotti (1970) found that taurine had no effect on normal contractility of isolated guinea pig atria, but antagonized the negative inotropic effect of a low-calcium medium. In a similar study, Dolara et al. (1973) found that taurine delayed the loss of tissue calcium and cardiac contractility when guinea pig hearts were perfused with buffer lacking calcium. These results led Dolara et al. (1973) to suggest that taurine facilitated the retention of calcium by some cellular component. Electrophysiological studies provided support for this view (Dolara et al., 1978). However, in apparent contradiction to this hypothesis, McBroom and Welty (1977) recently reported that treatment of cardiomyopathic hamsters with taurine led to a reduction in tissue calcium levels.

In this chapter we provide evidence clarifying these apparently contradictory results. Our data suggest that taurine serves a regulatory function to maintain myocardial calcium homeostasis. This action of taurine appears to be mediated by its interaction with the sarcolemmal membrane.

METHODS

Heart Perfusion

Hearts from 240–280-g male Wistar rats were perfused within 45 sec following decapitation. The standard working heart apparatus used in these studies was a modified version of that described by Neely et al. (1967). Basically, the coronary system of the heart was perfused from a reservoir placed 100 cm above the aortic cannula, while the left atrium received fluid from an atrial reservoir maintained at a presure head of 13 cm water. When the left atrial cannula was open, there was a net ejection of fluid against the 100-cm pressure head. Cardiac output was determined by a flow meter placed above the aortic cannula and by collecting the coronary effluent. Left ventricular pressure was measured with a Statham P23Gb pressure transducer by inserting a 22-gauge needle through the ventricle wall. Pressure work was calculated according to Neely et al. (1967).

Two parallel circuits with separate aortic and atrial reservoirs were used to permit rapid interconversion of perfusion fluid with a minimum of dead space. One circuit contained Krebs-Henseleit buffer supplemented with 5 mM glucose and 2.5 units/liter of insulin, while the other circuit contained the same buffer supplemented with 10 mM taurine in addition to 5 mM glucose and 2.5 units/liter of insulin. Hearts were regulated by means of an external stimulator at a rate of 300 beats/min.

Titration studies were done using Krebs-Henseleit buffer in a constant recirculation volume of 500 ml. Calcium levels were gradually increased by the addition of $CaCl_2$. After each addition the hearts were allowed to stabilize to their new steady-state levels of aortic pressure and output before continuing the experiment. Using the two parallel circuits, the control curve and the effect of the desired substance could be studied on the same heart.

The protocol for the calcium paradox studies consisted of an initial 15-min control perfusion, followed by varying lengths of calcium-free perfusion and 15 min of reperfusion with buffer containing 2.5 mM Ca^+. Changes in cardiac work were monitored throughout the procedure, and the extent of recovery is based upon a comparison between initial cardiac work and the work generated by the heart following 15 min of reperfusion. Those hearts which were examined for Ca^{2+} binding activity were decannulated immediately following the reperfusion period and studied as described below.

Sarcolemma Preparation

The plasma membrane was isolated according to the method of Sulakhe et al. (1973). Membrane to be used in calcium-binding studies was suspended in 20 mM tris-maleate buffer, pH 7.4, containing 5 mM KCl, 10 mM $MgCl_2$, and 100 mM NaCl at a final concentration of 10–20 μg/0.2 ml. The binding assay was performed using a Millipore filtration system described by Chovan et al. (1979).

Taurine-binding isotherms were obtained from membrane isolated from swine heart according to the method of Sulakhe et al. (1973). The membranes were suspended in 10 mM Tris HCl, pH 8.0, at a concentration of 5–10 mg protein/ml buffer and assayed using the Millipore filtration method of Kulakowski et al. (1978).

Solubilization

The taurine-binding proteins were extracted from a standard pig heart sarcolemma preparation by solubilization with the nonionic detergent Ammonyx-Lo (2% w/v). The solution was stirred for 90 min at 24°C and then centrifuged at 150,000 g for 70 min at 4°C. The resultant supernatant is referred to as the solubilized fraction. This solubilized fraction was dialyzed for 12 hr at 4°C against 10 mM Tris HCl, pH8.0, to reduce the detergent concentration to 0.02%. Following dialysis, the solubilized extract was stored as 2 ml fractions at −20°C for use in subsequent studies. This preparation has been shown to be stable at −20°C for up to 6 months.

The solubilized taurine-binding components were assayed by a Concanavalin A Sepharose method developed by us. The solubilized membrane fraction (100–150 μg solubilized protein/100 μliter buffer) was incubated at 24°C for 1 hr with an equal volume of Concanavalin A Sepharose suspension, prepared according to the method of March et al. (1974). Using this procedure, 30–40 μg of soluble membrane protein was bound to Con A Sepharose containing 12 mg Con A/ml gel solution. After incubation, the Con A Sepharose ligand complex was pelleted by centrifugation at 600 g for 10 min. The supernatant was removed and replaced by either 10 mM Tris HCl, pH 8.0, or buffer containing a large excess of unlabeled taurine. Following 10 min of preincubation, radiolabeled taurine was added to the Con A Sepharose suspension and incubated for an additional 20 min at 24°C. The reaction was terminated by the addition of 3.0 ml of ice-cold buffer and the filtration of the Con A Sepharose complex. The radioactivity of the complex was measured and corrected for nonspecific binding.

RESULTS

Effect of Taurine on Contractility of the Hypodynamic Heart

A link between extracellular taurine and calcium is illustrated by the titration studies shown in Fig. 20.1a. The results shown are from a representative heart subjected to various concentrations of extracellular calcium in the presence and absence of 10 mM taurine. Sigmoidal curves were obtained, and in this particular heart the maximal work output achieved in either the presence or absence of taurine was about 0.6 kg-m/g dry wt./min. It is seen that the major effect of taurine is to displace the work curve to lower values of calcium, indicating that taurine increases the sensitivity of the heart to extracellular calcium. It is also noted that taurine has very little effect on the maximal work output generated by the heart.

Taurine-induced shifts in work were also obtained in hearts made hypodynamic by addition of Verapamil, a pharmacological agent believed to exert its negative inotropic effect by inhibiting the slow inward movement of calcium into the myocardium during excitation (Fleckenstein, 1977). Figure 20.1b shows the contrasting effects of Verapamil and taurine on the heart. Addition of 12.8 μg/liter of Verapamil increased the calcium concentration

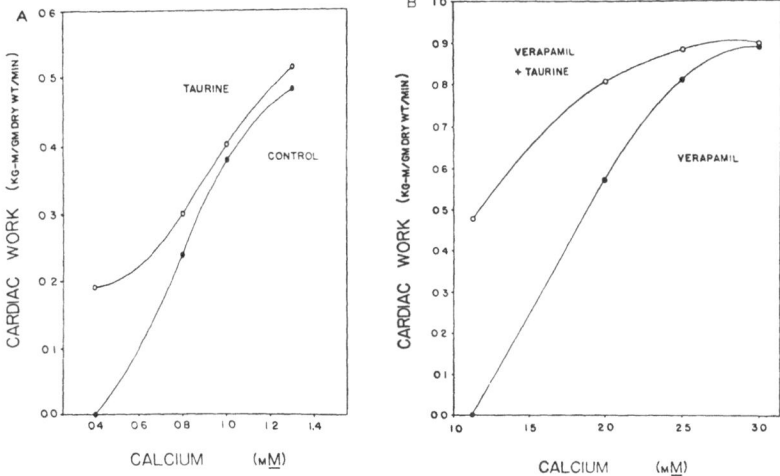

Fig. 20.1.   Effect of taurine on cardiac work of the hypodynamic heart. After an initial sta-bilizing period of approximately 15 min with Krebs Henseleit buffer containing either (*a*) 0.4 mM Ca^{2+} or (*b*) 1.2 mM Ca^{2+} and 12.8 µg/liter Verapamil, rat hearts were titrated with in-creasing concentrations of calcium in the presence or absence of 10 mM taurine. Cardiac work was calculated according to the method of Neely et al. (1967), and the data represent values from a typical heart.

required for half-maximal work output from 0.9 to 2.0 mM. The inhibitory effect of Verapamil was partially overcome by taurine, as noted by the dis-placement of the work curve to lower calcium values. This suggests that tau-rine is facilitating calcium transport into the hypodynamic myocardium.

Effect of Taurine on Mechanical Performance of the Calcium-overloaded Myocardium

A common model employed to produce calcium overload in the heart is the "calcium paradox." To induce this condition, rat hearts were initially per-fused with Krebs-Henseleit buffer containing 2.5 mM calcium for a 15-min stabilization period. The hearts were then subjected to several minutes of cal-cium-free perfusion, followed by reperfusion with buffer containing 2.5 mM calcium. This protocol led to irreversible mechanical failure, as shown in Fig. 20.2. In agreement with Yates and Dhalla (1975), we found that the degree of irreversible mechanical failure was dependent upon the length of calcium-free perfusion. In hearts perfused with buffer lacking taurine, failure was ob-

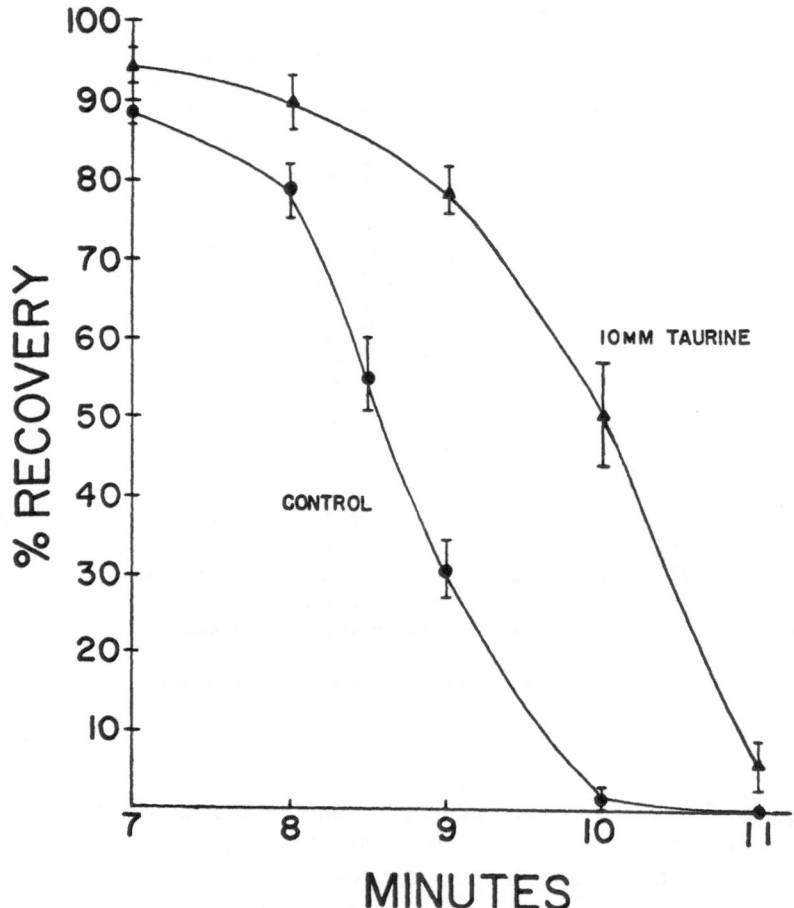

Fig. 20.2. Protective effect of taurine against heart failure resulting from the Ca²⁺ paradox. Rat hearts were initially perfused for a 15-min stabilization period with Krebs-Henseleit buffer containing 5 mM glucose and 2.5 U/liter of insulin. They were then subjected to several minutes of Ca²⁺-free perfusion, followed by 15 min of reperfusion with buffer containing 2.5 mM Ca²⁺. Taurine, when present, was included in all buffers at a concentration of 10 mM. The percent recovery is a comparison of cardiac work during stabilization and after 15 min of reperfusion. Each time point represents the mean ± SEM of four to ten hearts.

served when the period of calcium-free perfusion was extended beyond 7 min (Fig. 20.2). However, addition of taurine to the buffer delayed the onset of failure by about 1 min.

Dolara et al. (1973) have previously shown that taurine delays the loss of calcium from the heart during calcium-free perfusion. To test if the effect of taurine on the paradox was simply due to this delay in calcium efflux from the heart, we compared hearts that had been treated with taurine during the calcium-free perfusion period with hearts exposed to taurine only during the 15-min reperfusion period. It was found that the protective effect was only observed when hearts were treated with taurine during the reperfusion period. This observation suggests that taurine is preventing calcium overload during the reperfusion period, and that its effect on calcium efflux during the calcium-free perfusion period plays a minor role in the onset of irreversible damage.

Effect of Taurine on Calcium Binding to the Cell Membrane

It is widely believed that the irreversible damage resulting from the calcium paradox is associated with changes in the cell membrane that render it hyperpermeable to small ions (Crevey et al., 1978). During reperfusion this leaky membrane allows excessive accumulation of calcium by the cell. It has been argued that the resulting calcium overload leads to cellular damage and heart failure.

To test if the protective effect of taurine is related to changes in the membrane, we examined the calcium-binding pattern of isolated cell membrane prepared from hearts subjected to 9 min of calcium-free perfusion in the presence and absence of taurine. As seen in Fig. 20.3a, membranes obtained from hearts exposed to 9 min of calcium-free perfusion exhibit a greater capacity for calcium binding than membranes prepared from hearts that were not subjected to the calcium paradox. A Scatchard plot of these data (Fig. 20.3b) reveals that the increase in calcium binding during the paradox results from an increase in the number of both high and low affinity calcium-binding sites. This change may be related to the separation of the external lamina from the rest of the sarcolemma (Crevey et al., 1978). Treatment of hearts with 10 mM taurine was found to prevent this opening up of new sites during the paradox (Figs. 20.3a and b). Assuming that these sites are related to calcium transport, it would be attractive to suggest that taurine delays the onset of calcium overload by preventing changes in the membrane.

While treatment of the myocardium with taurine during the "calcium paradox" reduces the calcium-binding capacity of the cell membrane, incubation of isolated membrane from normal hearts with taurine enhances calcium binding (Fig. 20.4). A Scatchard plot of the data reveals that taurine primarily enhances calcium binding to the low affinity sites by increasing the affinity of calcium for these sites. It has been proposed that these sites represent Pool I calcium (Limas, 1977; Williamson et al., 1975). Thus, in the hypodynamic myocardium, taurine could enhance calcium binding to this pool,

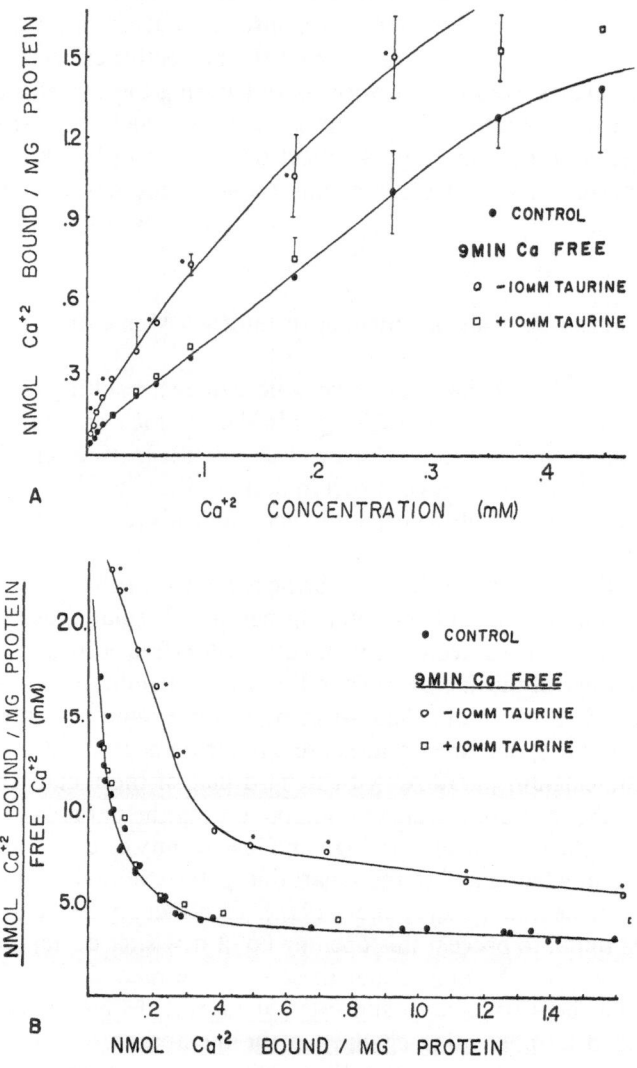

Fig. 20.3. Effect of Ca²⁺ paradox on sarcolemmal Ca²⁺ binding. (*a*) Sarcolemma were prepared from hearts subjected to 9 min of Ca²⁺-free perfusion and 15 min of reperfusion in the presence (□) and absence (○) of 10 mM taurine. The Ca²⁺-binding pattern of these two conditions were compared with Ca²⁺ binding (●) to sarcolemma prepared from hearts perfused for 45 min with normal buffer. All curves were obtained from representative hearts. An asterisk denotes a significant difference from the control (*p* < 0.01). (*b*) Scatchard plot of the data in (*a*).

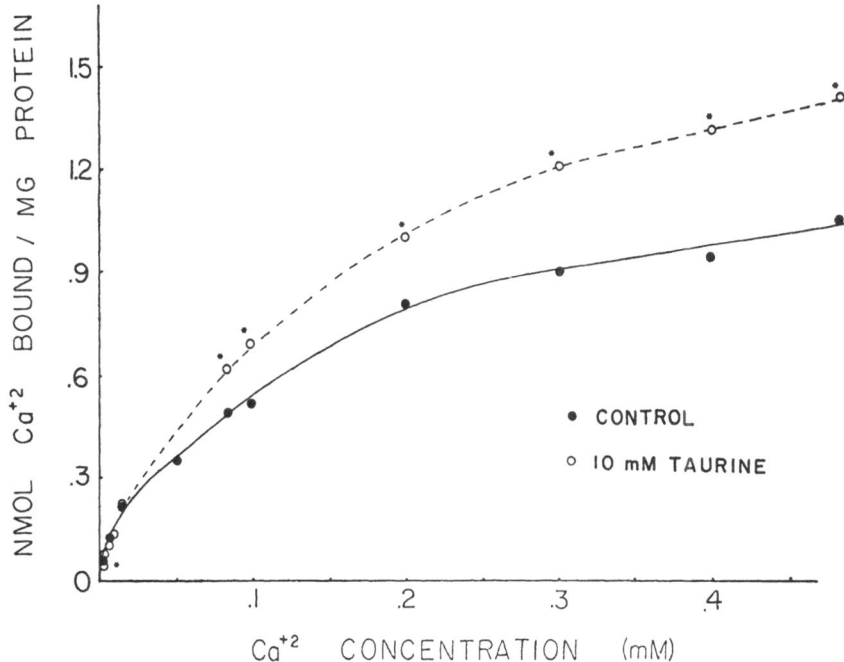

Fig. 20.4.   Taurine stimulation of Ca^{2+} binding to rat heart sarcolemma. Sarcolemma protein (70–90 μg/200 μliter), suspended in Tris-maleate, pH 7.4, containing 10 mM $MgCl_2$, 100 mM KCl, and 5 mM NaCl was preincubated for 10 min in the presence and absence of 10 mM taurine. $^{45}CaCl_2$ was added to a final concentration of 1.0–500 μM and incubated for an additional 10 min. Calcium-binding assays were carried out according to the method of Chovan et al. (1979). An asterisk denotes a significant difference from the control ($p < 0.01$).

thereby accelerating the flux of calcium into the cell and improving the contractile state of the heart.

Taurine-binding Proteins Located on the Myocardial Cell Membrane

The mechanism by which taurine mediates these changes in calcium binding appears to be closely linked to the binding of taurine to the membrane. Figure 20.5 shows the binding isotherms for taurine interaction with isolated pig heart sarcolemma. In agreement with previous data from rat heart sarcolemma, the isotherms reveal the existence of two sets of taurine-binding sites, both of which exhibit positive cooperativity. The $K_{0.5}$ determined for the high and low affinity sites are 0.2 and 3.0 mM, respectively.

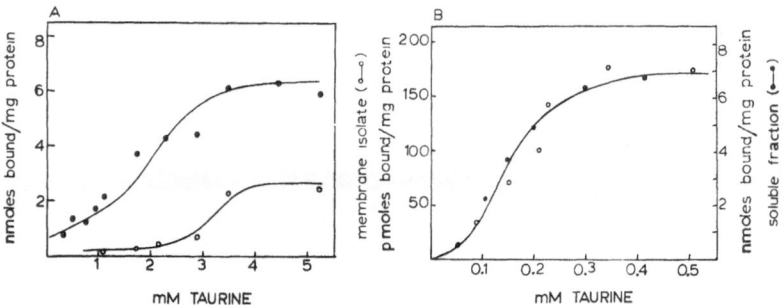

Fig. 20.5. Taurine binding isotherm of membrane isolate and detergent extract. Sarcolemma (120 μg/200 μliter) prepared from swine heart were suspended in Tris-maleate buffer, pH 7.4, containing 1.2 mM KH_2PO_4, 1.2 mM $MgSO_4$, 4.8 mM KCl, 1.25 mM $CaCl_2$, and 120 mM NaCl. The preparation was incubated at 25°C with ^{14}C-taurine (*a*, ○) or ^{3}H-taurine (*b*, ○) over the concentration ranges of 0.35–5.25 mM and 12.6–500 μM, respectively. The binding assays were determined using the method of Kulakowski et al. (1978). To assay the detergent-solubilized extract, it was passed over a Concanavalin-A sepharose column and the glycoprotein fraction was eluted with 0.1 M sodium phosphate buffer containing 0.1% Triton X-100 and 0.3% α-methyl mannoside. Following removal of α-methyl mannoside, the extract was incubated with the appropriate concentrations of ^{14}C-taurine (*a*, ●) and ^{3}H-taurine (*b*, ●), and bound taurine was determined by the standard polyethylene glycol method. Data were corrected for non-specific interactions.

Solubilization of the proteins with the nonionic detergent Ammonyx-Lo did not appreciably influence the binding profile of either protein. In Fig. 20.5 it is seen that the solubilized preparations retain the characteristic sigmoidal binding patterns and half-saturation values of the membrane isolate. The major effect of solubilization is to increase specific binding (amount bound/mg protein), an effect due to the greater purity of the solubilized preparation. Since the binding characteristics of the solubilized and membrane preparations are similar, it appears that the detergent adequately substitutes for the natural phospholipid environment.

Chromatographic Separation of the Taurine-binding Proteins

The two taurine-binding proteins can be distinguished on the basis of their specificity and affinity for taurine. In addition, they also differ in molecular size, as seen in Fig. 20.6. The chromatographic pattern in Fig. 20.6 was obtained by incubating the solubilized fraction with radioactive taurine prior to its application to a Sepharose 6B column. The elution pattern obtained was

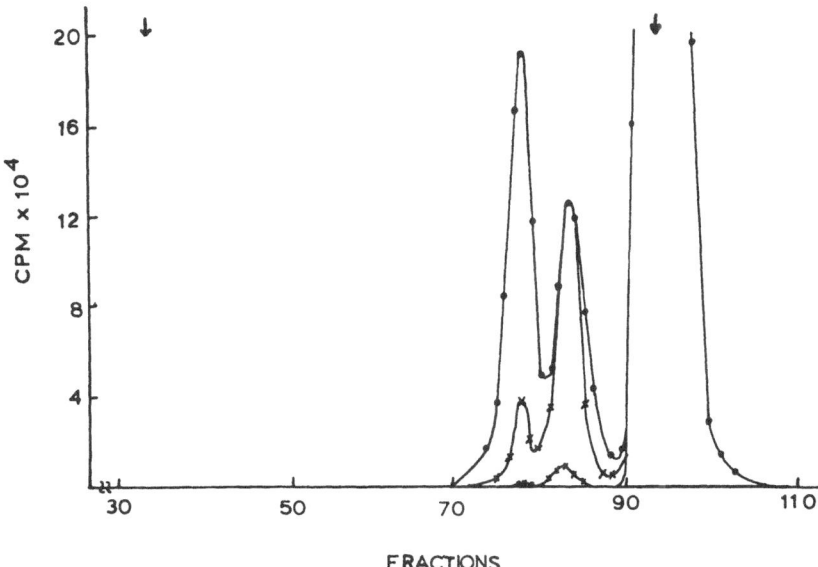

Fig. 20.6. Separation of taurine binding proteins on Sepharose-6B. Detergent-solubilized sarcolemmal extract was preincubated for 40 min with either 50 μM (★), 0.6 mM (X), or 4.2 mM (●) radiolabeled taurine and placed on ice for 10 min prior to application on a Sepharose-6B Column. All fractions (2.2 ml) were collected, counted, and corrected for specific activity. The initial arrow represents the void volume; and the second arrow, free taurine.

dependent upon the concentration of taurine used in the initial incubation. When the concentration employed was 50 μM, two radioactive peaks were observed, but the radioactive label appeared primarily in the second peak. As the concentration of taurine increased, the second peak became saturated, while the label in the first peak progressively increased. Since the second peak became saturated at a lower taurine concentration, we conclude that it must be the high affinity site. Using the method described by Cuatrecasas (1972), the molecular weights of the high and low affinity proteins were calculated to be 107,000 and 132,000, respectively.

Role of Taurine-binding Proteins

Three lines of evidence suggest that the high affinity protein is associated with the β-transport system. First, the Kd of taurine binding to this protein is similar to the Km of taurine uptake by perfused rat heart or rat myocytes

(Chubb and Huxtable, 1978; Grosso et al., 1978). Second, the specificities of taurine binding and uptake are similar. It is well accepted that hypotaurine, taurocyamine, and β-alanine are potent inhibitors of taurine uptake (Chubb and Huxtable, 1978; Grosso et al., 1978). As seen in Table 20.1, they are also potent inhibitors of taurine binding to the high affinity protein. Finally, pharmacological agents that alter taurine uptake also alter binding.

The role of the low affinity protein has not been established. Since all of the effects of taurine are manifest in the millimolar range, it is possible that the low affinity site may be linked to these actions. To test this hypothesis, several taurine analogues were examined to determine if they affect sarcolemmal calcium binding. It was found that all of the analogues that bound tightly to the low affinity protein also altered calcium binding. Moreover, a parallel existed between the affinity of the analogues for the binding protein and the concentration of the analogue required to mediate changes in calcium binding. As seen in Table 20.1, hypotaurine is a very potent inhibitor of taurine binding to the low affinity protein, while β-alanine and isethionic acid are quite ineffective. Their corresponding effects on calcium binding are seen in Fig. 20.7. In agreement with the binding data, hypotaurine enhances calcium binding while β-alanine and isethionic acid do not influence the binding pattern. Moreover, the effects of hypotaurine on calcium binding are elicited

Table 20.1   Effect of taurine analogues on taurine binding to the low and high affinity components of crude and solubilized pig heart sarcolemma[a]

	Percent taurine binding			
	Membrane isolate		Solubilized membrane	
Taurine analogue	High affinity	Low affinity	High affinity	Low affinity
Control	100	100	100	100
beta-Alanine	56	114	31	93
Hypotaurine	21	0	0	7
Isethionic acid	44	59	15	69
Homotaurine	58	48	43	60
Taurocyamine	28	55	0	29
Cysteine sulfinate	19	49	6	—

[a]All values represent the amount of taurine bound to either the membrane isolate or the solubilized extract following incubation with either 50 μM ^{3}H-taurine (high affinity) or 3.5 mM ^{14}C-taurine (low affinity) and an equal concentration of the analogue. Assays were performed as described in Fig. 20.5.

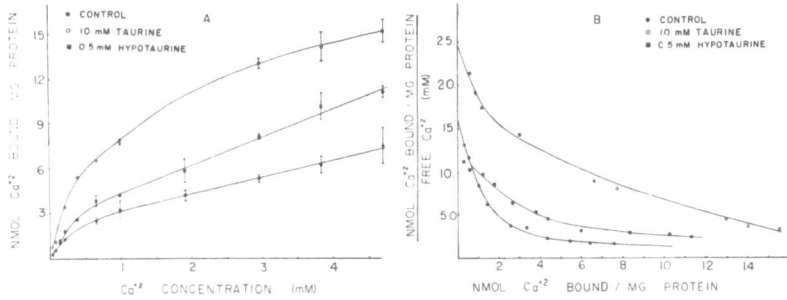

Fig. 20.7. Calcium binding to isolated rat heart sarcolemma in the presence of taurine or hypotaurine. Ventricular sarcolemma were isolated and assayed using the method of Chovan et al. (1979). The membrane was incubated in a medium containing 100 mM KCl, 5 mM NaCl, 10 mM $MgCl_2$, 20 mM Tris-maleate, and $^{45}CaCl_2$ at a concentration varying from 3.0 to 500 μM. The concentrations of taurine and hypotaurine employed were 10.0 and 0.5 mM, respectively. Values shown are the means ± SEM of four assays.

at a lower concentration than those changes mediated by taurine, in line with its affinity for the binding protein.

To further examine the role of the low affinity protein, we studied the effect of 10^{-9} and 10^{-10} M Verapamil on the changes in calcium binding mediated by taurine. In this concentration range, Verapamil had no noticeable influence on calcium binding to the membrane, but it inhibited taurine binding. The effect of Verapamil on taurine binding is shown in Fig. 20.8. A hyperbolic decrease in taurine binding accompanied the progressive increase in Verapamil concentration. Also associated with the rise in Verapamil concentration was a reduction in the effects of taurine on calcium binding. This is clearly demonstrated by the double reciprocal Scatchard plot of those calcium-binding sites that were affected by taurine (Fig. 20.9). In this plot, a change in the y intercept indicates a modification in the number of calcium-binding sites, while a change in the x intercept reveals an alteration in the Ka of calcium for its sites. Fig. 20.9 reveals that taurine increases Ka without affecting the number of binding sites, while increasing levels of Verapamil progressively reverse these effects of taurine. Of particular significance is the observation that Verapamil in the concentration range of 10^{-9} to 10^{-10} M inhibited the taurine effect although it exerted no direct effect on calcium binding. These results suggest that Verapamil restores the normal calcium-binding pattern, in part, by inhibiting taurine binding.

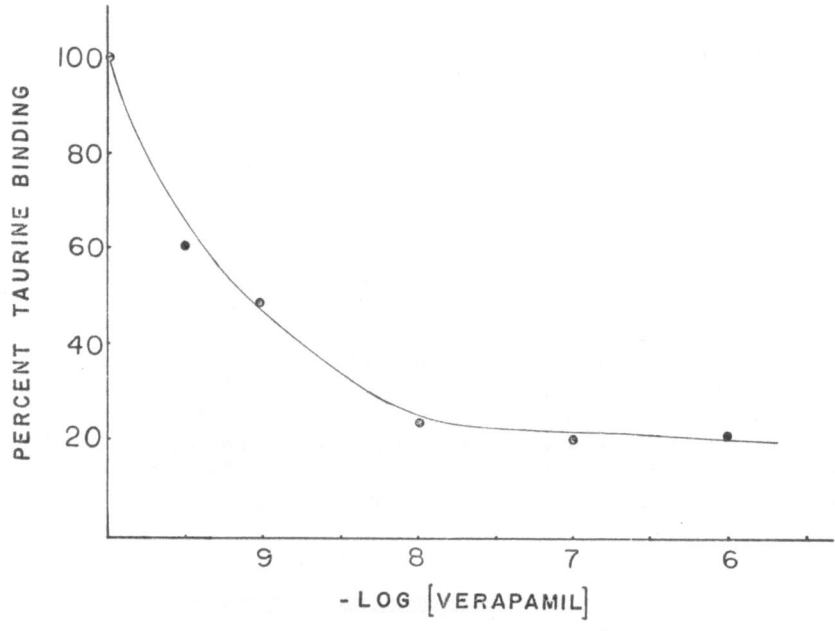

Fig. 20.8.   Inhibition of taurine binding to the low affinity protein by Verapamil. Sarcolemma were incubated with 3.5 mM ^{14}C-taurine and verapamil (10^{-10}–10^{-6} M) according to the procedures described in Fig. 20.5. Values shown are the means of three assays.

DISCUSSION

Maintenance of calcium homeostasis in the heart is crucial for normal function. A reduction in tissue calcium can lead to impaired mechanical performance, while calcium overload is associated with heart failure. In this report we have shown that taurine functions to regulate calcium homeostasis. In hearts made hypodynamic by either a reduction in perfusate calcium levels or by addition of verapamil, taurine enhances myocardial contractility. On the other hand, taurine delays the onset of heart failure resulting from experimentally induced calcium overload.

Our results indicate that taurine regulates myocardial performance by modifying the size of Pool I calcium. In the hypodynamic heart perfused with reduced calcium levels, taurine enhances Pool I calcium levels by increasing the affinity of the sarcolemmal sites for calcium. Similarly, Pool I calcium levels rise in the Verapamil-treated myocardium following addition of taurine. However, in the latter preparation the taurine effect appears to result in

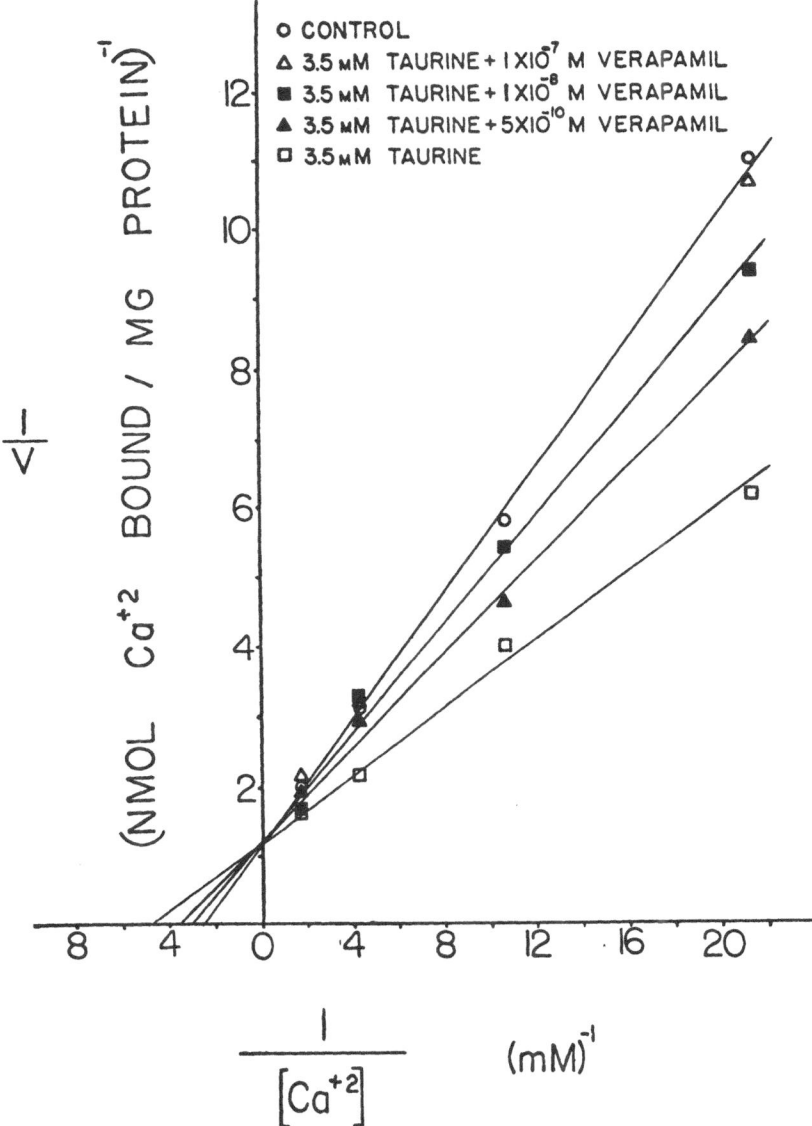

Fig. 20.9.   Effect of Verapamil on taurine enhancement of Ca^{2+} binding. Stimulation of sarcolemma Ca^{2+} binding by 3.5 mM taurine in the presence and absence of Verapamil (5.5×10^{-10} to 10^{-7} M) was determined using the method described in Fig. 20.4. Under all conditions, Ca^{2+} concentration was varied from 0.04 to 0.8 mM. The data are expressed as a double-reciprocal Scatchard plot. Values shown are the means of four assays.

part from a competition between taurine and Verapamil, leading to a partial displacement of Verapamil from the membrane.

Based upon electron microscopic evidence, Langer (1978) has proposed that separation of the external lamina from the rest of the membrane during calcium-free perfusion is the cause of membrane hyperpermeability. Our data imply that during the paradox, damage to the membrane leads to an increase in the size of Pool I calcium. Taurine delays the expansion of this pool, thereby protecting the heart against calcium overload. This observation is in agreement with Huxtable (1976), who suggested that taurine serves to stabilize membranes.

Mechanism of Taurine Action

Several models have been proposed to account for the effect of taurine on ion flux. Read and Welty (1965) originally suggested that the accumulation of cations by taurine-treated cells resulted from the metabolism of taurine to isethionic acid. However, the deamination of taurine proceeds too slowly, and the concentration of isethionic acid is too low to be compatible with this model (Fellman et al., 1978; Schaffer et al., 1978).

Barbeau and Donaldson (1974) proposed that taurine may have a stimulatory effect on Na^+-K^+ ATPase by removing inhibiting levels of Zn^{2+} through chelation. However, chelation of other ions, such as Ca^{2+}, by taurine has been shown to be weak (Dolara et al., 1978). Moreover, taurine was shown to have no effect on the activity of a partially purified preparation of rat brain Na - K^+ ATPase (Akera et al., 1976).

Dolara et al. (1978) have suggested that taurine salts of calcium could be formed intracellularly and could act as neutral calcium carriers. However, the interaction between taurine and calcium is so much weaker than the interaction between calcium and other cellular components (Calsequestrin and ATP) that it is unlikely that taurine salt formation is physiologically important.

Huxtable (1976) proposed that taurine affects ion flux by interacting directly with the membrane. We feel that this hypothesis is most consistent with our data. Several lines of evidence suggest that the actions of taurine are mediated by its interaction with sarcolemmal proteins. First, taurine binding to this protein occurs over the same concentration range in which taurine-mediated changes in both myocardial contraction and calcium binding to the membrane are observed. Second, there is a parallel between the affinity of various substances for the low affinity protein and the concentration required to mediate changes in calcium binding. Third, Verapamil inhibits both taurine binding and the taurine-mediated calcium-binding effects at a concentration where it has no effect on calcium binding to the sarcolemma. Finally, the effects of taurine on the "calcium paradox" appear to be mediated by changes in extracellular taurine rather than intracellular taurine (Shaffer et al., 1979; Sturman et al., 1978).

Role of Intracellular Taurine

The interaction of the low affinity protein with Con A Sepharose suggests that the binding protein is a glycoprotein located on the extracellular side of the membrane (Langer, 1978). This poses a central question concerning the role of taurine found within the cell. If the protein that mediates the actions of taurine is situated exclusively on the exterior of the membrane and is inaccessible to the intracellular space, then extracellular taurine would be responsible for the observed effects. On the other hand, a protein that spanned the membrane could conceivably bind taurine from either intracellular or extracellular pools. Very few studies have been concerned with this problem (Shaffer et al., 1979), and the issue will only be resolved following further experimentation.

ACKNOWLEDGMENTS

This work was supported by NIH grant HL-18554. Figures 7-9 were taken from Chovan et al. (1980) and reproduced with the permission of *Molecular Pharmacology*.

REFERENCES

Akera, T.; Ku, D.; and Brody, T. M. Alterations of ion movements as a mechanism of drug-induced arrhythmias and inotropic responses. In *Taurine*, R. Huxtable and A. Barbeau, eds. Raven Press, New York, (1976), pp. 121-134.

Barbeau, A.; and Donaldson, J. Zinc, taurine and epilepsy. *Arch. Neurol.*, 30, 52-58 (1974).

Chovan, J. P.; Kulakowski, E. C.; Benson, B. W.; and Schaffer, S. W. Taurine enhancement of calcium binding to rat heart sarcolemma. *Biochim. Biophys. Acta*, 551, 129-136 (1979).

Chovan, J. P.; Kulakowski, E. C.; Sheakowski, S.; and Schaffer, S. W. Calcium regulation by the low-affinity taurine binding sites of cardiac sarcolemma. *Mol. Pharmacol.*, 17, 295-300 (1980).

Chubb, J.; and Huxtable, R. Isoproterenol-stimulated taurine influx in the perfused rat heart. *Eur. J. Pharmacol.*, 48, 369-376 (1978).

Crevey, B. J.; Langer, G. A.; and Frank, J. S. Role of Ca^{++} in maintenance of rabbit myocardial cell membrane structural and functional integrity. *J. Mol. Cell. Cardiol.*, 10, 1081-1100 (1978).

Cuatrecasas, P. Properties of the insulin receptor isolated from liver and fat cell membranes. *J. Biol. Chem.*, 247, 1980–1991 (1972).

Dhalla, N. S.; Ziegelhoffer, A.; and Harrow, J. A. C. Regulatory role of membrane systems in heart function. *Can. J. Physiol. Pharmacol.*, 55, 1211–1234 (1977).

Dolara, P.; Agresti, A.; Giotti, A.; and Pasquini, G. Effect of taurine on calcium kinetics of guinea-pig heart. *Eur. J. Pharmacol.*, 24, 352–358 (1973).

Dolara, P.; Ledda, F.; Mugelli, A.; Mantelli, L.; Zilletti, L.; Franconi, F.; and Giotti, A. Effect of taurine on calcium, inotropism, and electrical activity of the heart. In *Taurine and Neurological Disorders,* A. Barbeau and R. J. Huxtable, eds. Raven Press, New York, (1978) pp. 151–159.

Fabiato, A.; and Fabiato, F. Calcium release from the sarcoplasmic reticulum. *Circ. Res.*, 40, 119–129 (1977).

Fellman, J. H.; Roth, E. S.; and Fujita, T. S. Taurine is not metabolized to isethionate in mammalian tissue. In *Taurine and Neurological Disorders,* A. Barbeau and R. J. Huxtable, eds. Raven Press, New York, (1978) pp. 19–24.

Fleckenstein, A. Specific pharmacology of calcium in myocardium, cardiac pacemakers, and vascular smooth muscle. *Ann. Rev. Pharmacol. Toxicol.*, 17, 149–166 (1977).

Grosso, D. S.; Roeske, W. R.; and Bressler, R. Characterization of a carrier-mediated transport system for taurine in the fetal mouse heart in vitro. *J. Clin. Invest.*, 61, 944–952 (1978).

Guidotti, A.; and Giotti, A. Taurina e sistema cardiovascolare. *Rec. Prog. Med.*, 49, 61–97 (1970).

Huxtable, R. J. Metabolism and function of taurine in the heart. In *Taurine,* R. Huxtable and A. Barbeau, eds. Raven Press, New York, (1976), pp. 99–119.

Kasai, M.; and Miyamoto, H. Depolarization induced calcium release from sarcoplasmic reticulum fragments. *J. Biochem.*, 79, 1053–1066 (1976).

Katz, A. M. Excitation-contraction coupling. In *Physiology of the Heart,* A. M. Katz, ed. Raven Press, New York, (1977), pp. 137–159.

Kulakowski, E. C.; Maturo, J.; and Schaffer, S. W. The identification of taurine receptors from rat heart sarcolemma. *Biochem. Biophys. Res. Comm.*, 80, 936–941 (1978).

Langer, G. A. Events at the cardiac sarcolemma: Localization and movement of contractile-dependent calcium. *Fed. Proc.*, 35, 1274–1278 (1976).

Langer, G. A. The structure and function of the myocardial cell surface. *Am. J. Physiol.*, H461–H468 (1978).

Limas, C. J. Calcium binding sites in rat myocardial sarcolemma. *Arch. Biochem. Biophys.*, 179, 302–309 (1977).

McBroom, M. J.; and Welty, J. D. Effects of taurine on heart calcium in the cardiomyopathic hamster. *J. Mol. Cell. Cardiol.*, 9, 853–858 (1977).

March, S. C.; Parikh, I.; and Cuatrecasas, P. A simplified method for cyanogen bromide activation of agarose for affinity chromatography. *Anal. Biochem.*, 60, 149–152 (1974).

Neely, J. R.; Liebermeister, H.; Battersby, E. J.; and Morgan, H. E. Effect of pressure development on oxygen consumption by isolated rat heart. *Am. J. Physiol.*, 212, 804–814 (1967).

Read, W. O.; and Welty, J. D. Taurine as a regulator of cell potassium in the heart. In *Electrolytes and Cardiovascular Diseases,* E. Bajusz, ed. S. Karger, New York, (1965), pp. 70–85.

Schaffer, S. W.; Sevem, A.; and Chovan, J. New Chromatographic procedure for measuring isethionic acid levels in tissue. In *Taurine and Neurological Disorders,* A. Barbeau and R. J. Huxtable, eds. Raven Press, New York, (1978), pp. 25–28.

Shaffer, J.; Kramer, J.; Schaffer, S.; and Kocsis, J. Intracellular or extracellular effects of taurine? *Fed. Proc.*, 38, 2465 (1979).

Sturman, J. A.; Rassin, D. K.; Hayes, K. C.; and Gaull, G. E. Taurine deficiency in the kitten: Exchange and turnover of [^{35}S]taurine in brain, retina and other tissues. *J. Nutr.*, 108, 1462–1476 (1978).

Sulakhe, P. V.; Drummond, G. I.; and Ng, D. C. Calcium binding by skeletal muscle sarcolemma. *J. Biol. Chem.*, 248, 4150–4157 (1973).

Williamson, J. R.; and Schaffer, S. Epinephrine, cyclic AMP, calcium, and myocardial contractility. In *Recent advances in Studies on Cardiac Structure and Metabolism, The sarcolemma, Vol. 9*, P. E. Roy and N. E. Dhalla, eds. University Park Press, Baltimore, (1976), pp. 205–223.

Williamson, J. R.; Woodrow, M. L.; and Scarpa, A. Calcium binding to cardiac sarcolemma. In *Recent advances in Studies on Cardiac Structure and Metabolism, Basic Functions of Cations in Myocardial Activity, Vol. 5*, A. Fleckenstein and N. S. Dhalla, eds. University Park Press, Baltimore, (1975), pp. 61–71.

Yates, J. C.; and Dhalla, N. S. Structural and functional changes associated with failure and recovery of hearts after perfusion with Ca^{++}-free medium. *J. Mol. Cell. Cardiol.*, 7, 91–103 (1975).

Copyright © 1981, Spectrum Publications, Inc.
The Effects of Taurine on Excitable Tissues

CHAPTER 21

Subcellular Effects of Taurine on Guinea Pig Heart

J. C. Khatter
P. L. Soni
R. J. Hoeschen
L. E. Alto
N. S. Dhalla

INTRODUCTION

Although taurine has been shown to exert a positive inotropic effect on the myocardium (Dietrich and Diacono, 1971; Guidotti et al., 1971; Iwata and Fujimoto, 1976; Huxtable 1976), the mechanism of its action is not clearly understood. According to a current concept of the excitation-contraction and relaxation process in the cardiac cell, calcium and ATP are believed to play key roles in determining heart performance under a wide variety of experimental conditions (Dhalla et al., 1977a, 1978a; Katz, 1977; Langer, 1978; Nayler et al., 1975). Excitation of myocardium is considered to raise intracellular concentration of free calcium by increasing calcium influx through sarcolemma as well as by releasing calcium from its storage sites at sarcolemma, sarcoplasmic reticulum, and, possibly, mitochondria; and this leads to the development of contraction. On the other hand, relaxation of the myocardium is a result of lowering the concentration of intracellular calcium by various mechanisms located at different membrane systems. Thus cardiac contraction and relaxation can be conceived as reflections of raising and lowering the cytoplasmic concentration of free calcium. It is possible that taurine may exert its influence on myocardium through changes in the movements of calcium at sarcolemma, sarcoplasmic reticulum, and mitochondria. Since the available techniques do not permit the assessment of calcium fluxes at different membrane systems in the intact cell, the effects of different agents on calcium accumulating activities of these organelles are usually studied in the isolated fractions under *in vitro* conditions. It was, therefore, the purpose of this study to examine the effects of various concentrations of taurine on sarcolemma, sarcoplasmic reticular (microsomal), and mitochondrial calcium binding and accumulating activities.

It is now widely accepted that ATP is mainly generated in mitochondria by oxidative metabolism of substrates. The assessment of the mitochondrial oxidative ability can be considered to provide some information concerning the ATP-producing system in myocardium. On the other hand, myofibrillar ATPases, by hydrolyzing ATP, are believed to provide energy for contractile work. In addition, different ATPases of sarcolemma, sarcoplasmic reticulum, and mitochondria hydrolyze ATP for the proper function of various "cationic pumps" located in these membrane organelles. Sarcolemmal adenylate cyclase converts ATP to cyclic-AMP, which serves as an important regulator of cellular permeability and metabolism. Thus the activities of myofibrillar and membrane ATPases as well as sarcolemmal adenylate cyclase can reflect the activities of different ATP-utilizing systems. The possibility that taurine may exert its cardiac effects through alterations in the ATP-producing and ATP-utilizing mechanisms has not been explained previously. It was, therefore, decided to investigate the actions of taurine on mitochondrial phosphorylation; sarcolemmal ATPase and adenylate cyclase; and mitochondrial, microsomal, and myofibrillar ATPase activities.

METHODS

Guinea pigs weighing 250–300 g were decapitated. The hearts were quickly removed. In a series of experiments, hearts were perfused by Langendoroff techniques by using Krebs-Henseleit solution containing 1.25 mM Ca^{2+} according to the procedure described by Yates and Dhalla (1975). The contractile force was recorded on the polygraph by using a force displacement transducer. The effects of taurine on this preparation were studied by switching the heart to perfusion medium containing different concentrations of taurine after a stabilization period of 15 min. In some experiments, the concentration of calcium or magnesium in the experimental medium was varied, and the effects of this intervention were observed in the absence and presence of 10 mM taurine. In another series of experiments, the unperfused hearts were employed for the isolation of different subcellular fractions. Sarcolemmal fraction was isolated by the hypotonic shock-LiBr treatment method as described previously (McNamara et al., 1974; Dhalla et al., 1976). Total ATP-ase activity of sarcolemma was determined by incubating 50–100 μg of membrane protein in a medium containing 50 mM Tris-HCl (pH 7.4), 4 mM $MgCl_2$, 100 mM NaCl, and 10 mM KCl. The membrane fraction was prein-cubated for 3 min at 37°C. The reaction was then started by the addition of 4 mM ATP and stopped 10 min later by the addition of 12% cold trichloro-acetic acid. The Pi released in the clear supernatant was measured by the method of Taussky and Shorr (1953). The Ca^{2+} ATPase and Mg^{2+} ATPase activities were determined in a similar manner except that the incubation medium contained 50 mM Tris HCl (pH 7.4), with or without 4 mM Ca^{2+} or Mg^{2+}. The difference between the total ATPase and Mg^{2+} ATPase was taken to be due to the Na^+-K^+-stimulated ATPase activity. The protein concentra-tion was determined by the method of Lowry et al. (1951).

Mitochondrial and sarcoplasmic reticular (microsomal) fractions were iso-lated by the procedures described by Harrow and Dhalla (1976), and the myofibrillar fraction was isolated according to the method of Solaro et al. (1971). The Mg^{2+} ATPase and total ATPase activities of the mitochondrial and microsomal fractions (0.1–0.2 mg/ml) were determined in a medium containing 100 mM KCl, 20 mM Tris-HCl (pH 6.8), 2 mM $MgCl_2$, and 2 mM ATP, in the absence and presence of 0.1 mM $CaCl_2$ at 38°C (Harrow and Dhalla, 1976). For the myofibrillar fraction, the Mg^{2+} ATPase and total ATPase activities were determined in a medium containing 10 mM histidine (pH 6.8), 5 mM sodium azide, 60 mM KCl, 3 mM $MgCl_2$, and 0.1 mM $CaCl_2$. The difference between total and Mg^{2+} ATPase activities was taken to be due to Ca^{2+}-stimulated, Mg^{2+}-dependent ATPase activities of these frac-tions. It should be pointed out that the assay conditions for the determina-tion of ATPase activities in subcellular fractions were optimal (Tomlinson et al., 1976).

ATP-independent calcium binding was studied by suspending 0.2 mg of sarcolemmal membrane protein/ml in a medium containing 100 mM Tris-HCl (pH 7.4) and incubating for 5 min in the presence of 0.1 mM Ca^{2+} at 37°C (Dhalla et al., 1976). ATP-dependent calcium binding and uptake ac-tivities by mitochondrial and microsomal fractions were studied by proce-dures described elsewhere (Tomlinson et al., 1976). The method for determin-ing the adenylate cyclase activity has been described previously (McNamara et al., 1974). The oxidative phosphorylation activity of mitochondria was studied by using pyruvate-malate as a substrate according to the method de-scribed by Tomlinson et al. (1976). Taurine was prepared fresh each time, and its effects were studied by incubating the cellular fractions in the absence and the presence of different concentrations for 3 min before starting the re-action. The results were analyzed statistically by the paired t-test.

RESULTS

Contractile Force of the Isolated Heart

In one set of experiments, the effect of different concentrations of taurine on the ability of isolated perfused hearts to generate contractile force was examined. Perfusion of the hearts with a medium containing taurine was observed to increase contractile force within 1 min, and this effect reached its maximum at about 3 min. The results shown in Fig. 21.1 indicate that the maximal positive inotropic action was apparent at 10 mM concentration of taurine.

The effects of 10 mM taurine on contractile force of the isolated hearts perfused with media containing different concentrations of Ca^{2+} or Mg^{2+} were also studied. The results described in Fig. 21.2 reveal that increasing the concentration of Ca^{2+} from 1.25 mM to 2 mM increased the contractile force,

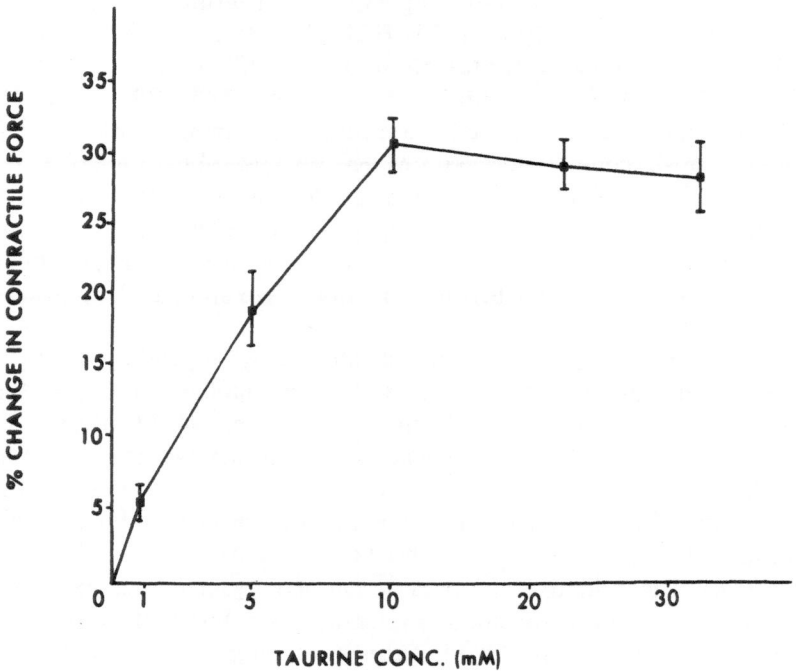

Fig. 21.1. Effects of different concentrations of taurine on the contractile force of isolated guinea pig heart. The percent increases in contractile force after 3 min of perfusion with taurine were calculated on the basis of initial values for contractile force. The concentration of calcium in the perfusion medium was 1.25 mM. Each value is a mean ± SE of four experiments.

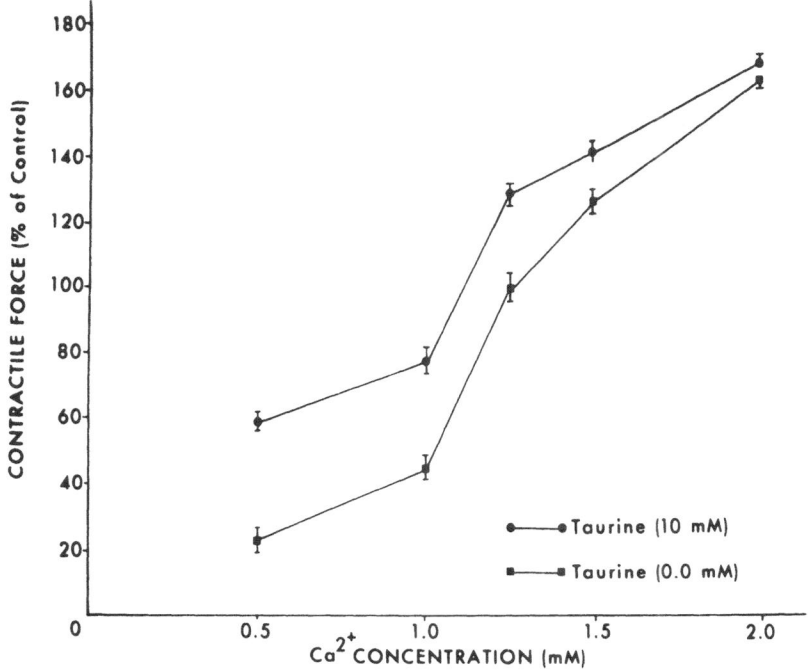

Fig. 21.2. Effects of 10 mM taurine in the presence of different concentrations of calcium on the contractile force of isolated guinea pig heart. After a stabilization period of 15 min with normal medium containing 1.25 mM calcium, the hearts were perfused with media containing different concentrations of calcium in the absence or presence of 10 mM taurine. The results are expressed as percent of the initial contractile force during perfusion with normal medium. Each value is a mean ± SE of four experiments.

but the positive inotropic effect of taurine was markedly reduced. On the other hand, when the concentration of Ca^{2+} in the medium was reduced from 1.25 mM to 0.5 mM, the contractile force was decreased but the positive inotropic action of taurine was increased (Fig. 21.2). Decreasing the concentration of Mg^{2+} from 4 mM to 1, 0.5, or 0.25 mM resulted in 10–20% increase in contractile force, but the positive inotropic effect of taurine was reduced by 50–75% of the contractile value at 4 mM Mg^{2+} (results not shown).

Calcium-accumulating Activities of Heart Membranes

In order to test if the cardiac effects of taurine are due to alterations in calcium movements at different membrane systems, the actions of taurine on calcium-accumulating activities of sarcolemma, microsomes, and mitochondria were studied under in vitro conditions. Some of the results from this study are shown in Table 21.1. It can be seen that ATP-independent Ca^{2+} binding to the sarcolemmal fraction was increased upon increasing the concentration of taurine; however, a statistically significant ($p < 0.05$) increase was observed only at 15–30 mM concentration of taurine. Guinea pig heart sarcolemmal fraction was also found to bind 5 ± 1.2 nmol Ca^{2+}/mg protein/5 min in the presence of 1–4 mM ATP. This ATP-dependent Ca^{2+} binding was not significantly ($p > 0.05$) altered by 1–30 mM concentration of taurine.

ATP-dependent Ca^{2+} binding and Ca^{2+} uptake activities of the guinea pig heart microsomal fraction in the absence or presence of 4 mM oxalate were found to be 38 ± 2.4 and $1,230 \pm 201$ nmol Ca^{2+}/mg protein/5 min, respectively. The results shown in Table 21.1 indicate that ATP-dependent calcium binding with microsomes was not significantly affected by taurine ($p > 0.05$). Likewise, taurine in 1–40 mM concentration was found to have no action ($p \leq 0.05$) on the calcium uptake activity (results not shown).

Mitochondrial fraction obtained from guinea pig heart was found to bind and accumulate 77 ± 2.1 and 284 ± 7.0 nmol Ca^{2+}/mg protein/5 min in the absence and presence of Pi and succinate, respectively. It was interesting to find that mitochondrial Ca^{2+} binding, but not Ca^{2+} uptake, was significantly ($p > 0.05$) increased in the presence of 15 mM or higher concentrations of taurine (Table 21.1). The increase in mitochondrial calcium binding by 15

Table 21.1   Effect of taurine on calcium-accumulating activities of heart sarcolemmal, microsomal, and mitochondrial fractions[a]

| | Calcium-accumulating activities (nmol/mg protein/5 min) | | | |
| | Sarcolemmal Ca^{2+} binding[b] | Microsomal Ca^{2+} binding | Mitochondrial activities | |
Concentration of taurine (mM)			Ca^{2+} binding	Ca^{2+} uptake
Control	18 ± 1.7	38 ± 2.4	77 ± 2.1	284 ± 7.0
5.0	19 ± 2.3	37 ± 1.7	81 ± 1.7	280 ± 3.9
15.0	22 ± 1.7^c	38 ± 2.1	97 ± 1.1^c	283 ± 3.1
30.0	24 ± 1.8^c	38 ± 1.8	107 ± 1.7^c	293 ± 7.3
40.0	—	39 ± 2.7	119 ± 1.5^c	303 ± 8.2

[a]Each value is a mean $\pm$ SE of four to six experiments.
[b]Ca^{2+}-binding activities were measured in the absence of ATP.
[c]Significantly different from controls ($p < 0.05$).

mM taurine was also observed when different concentrations of calcium (5–100 μM) were employed in the assay medium. It should also be pointed out that mitochondrial calcium binding, unlike sarcolemmal and microsomal calcium binding, was sensitive to 5 mM sodium azide. Furthermore, marker enzyme studies and electron microscopic examination (Anand et al., 1977) of the sarcolemmal, mitochondrial, and microsomal fractions revealed minimal cross-contamination.

ATP-producing and Hydrolyzing Activities

In another set of experiments, the effects of taurine on mitochondrial oxidative phosphorylation activities were studied. The control values for RCI, ADP:O ratio, and State 3 respiration of the guinea pig heart mitochondria were 8.2 $\pm$ 0.5, 3.1 $\pm$ 0.2, and 230 n-atoms O/mg protein/min, respectively. These activities were not altered significantly ($p > 0.05$) by 1–40 mM taurine.

From the data in Table 21.2, it is evident that mitochondrial Mg^{2+} ATPase activity was significantly increased by 15 mM or higher concentrations of taurine. The increase in mitochondrial ATPase activity was also seen at 0.5–10 mM concentrations of Mg^{2+} in the assay medium (results not shown). It should be pointed out that the mitochondrial fraction, unlike the microsomal fraction, does not exhibit Ca^{2+}-stimulated, Mg^{2+}-dependent ATPase activity. At any rate, in contrast to mitochondrial Mg^{2+} ATPase, micro-

Table 21.2   Effects of taurine on ATPase activities of heart microsomal, and mitochondrial, and myofibrillar ATPase activities[a]

Concentration of taurine (mM)	Mito-chondrial Mg^{2+} ATPase	Microsomal activities		Myofibrillar activities	
		Mg^{2+} ATPase	Ca^{2+} stimulated ATPase	Mg^{2+} ATPase	Ca^{2+} stimulated ATPase
		ATPase activities (μmol/mg protein/hr)			
Control	25 $\pm$ 1.70	38 $\pm$ 2.10	18 $\pm$ 3.70	11.52 $\pm$ 1.12	18.61 $\pm$ 1.06
5.0	26 $\pm$ 1.60	31 $\pm$ 1.27[b]	20 $\pm$ 2.81	10.86 $\pm$ 0.48	18.00 $\pm$ 1.18
15.0	31 $\pm$ 1.21[b]	27 $\pm$ 1.38[b]	21 $\pm$ 3.20	9.00 $\pm$ 0.61[b]	18.78 $\pm$ 1.12
30.0	34 $\pm$ 1.70[b]	26 $\pm$ 1.27[b]	21 $\pm$ 3.52	8.41 $\pm$ 0.78[b]	18.36 $\pm$ 1.06
40.0	35 $\pm$ 1.32[b]	25 $\pm$ 1.48[b]	21 $\pm$ 4.81	8.34 $\pm$ 0.81[b]	18.54 $\pm$ 1.18

[a]Each value is a mean $\pm$ SE of five to six experiments.
[b]Significantly different from controls ($p < 0.05$).

somal Mg^{2+} ATPase activity was found to decrease significantly in the presence of 5–40 mM taurine (Table 21.2). The microsomal Ca^{2+}-stimulated ATPase activity was not affected by 5–40 mM taurine. The data in Table 21.2 also indicate that myofibrillar Mg^{2+} ATPase, unlike myofibrillar Ca^{2+}-stimulated ATPase, was decreased ($p < 0.05$) by 15–40 mM taurine. The influence of taurine on sarcolemmal ATPase and adenylate cyclase was also examined, and the results are shown in Table 21.3.

Table 21.3   Effects of taurine on heart sarcolemmal ATPase and adenylate cyclase activities[a]

Concentration of taurine (mM)	ATPase activities (μmol/mg protein/hr)			Adenylate cyclase activity (pmol cyclic-AMP/mg protein/min)
	Ca^{2+} ATPase	Mg^{2+} ATPase	Na^+-K^+ ATPase	
Control	16.56 ± 2.10	12.64 ± 1.37	8.86 ± 1.30	250 ± 7.30
5 mM	17.59 ± 1.90	16.25 ± 1.16[b]	8.36 ± 0.43	278 ± 9.13
15 mM	18.36 ± 1.61	17.84 ± 1.28[b]	8.64 ± 1.20	256 ± 5.81
30 mM	17.17 ± 1.81	18.20 ± 1.61[b]	8.89 ± 1.21	277 ± 8.90
40 mM	17.63 ± 2.12	18.34 ± 1.37[b]	9.34 ± 1.37	—

[a]Each value is a mean ± SE of four to six experiments.
[b]Significantly different from control ($p < 0.05$).

The sarcolemmal Ca^{2+} ATPase was not affected significantly ($p > 0.05$), whereas Mg^{2+} ATPase activity was increased by 5–40 mM concentrations of taurine. The sarcolemmal Mg^{2+} ATPase activity was increased by 5 mM taurine at different concentrations (0.5–5 mM) of Mg^{2+} in the incubation medium. Guinea pig heart sarcolemmal preparation showed only a negligible amount of Ca^{2+}-stimulated Mg^{2+}-dependent ATPase activity. The activities of sarcolemmal Na^+-K^+ ATPase or adenylate cyclase were not significantly altered ($p > 0.05$) in the presence of 5–40 mM taurine.

DISCUSSION

We have shown that taurine increased ATP-independent calcium binding by sarcolemmal fraction obtained from guinea pig heart by the hypotonic shock-LiBr treatment method. Our results with sarcolemmal calcium binding confirm the findings of Chovan et al. (1979), who have employed sarcolemmal fraction isolated from rat heart by using the sucrose density gradient method. However, it is pointed out that the values for calcium-binding activity reported by Chovan et al. (1979) for a comparable concentration of calcium were markedly lower than those obtained in the present study. This is most probably due to the presence of a high concentration of Mg^{2+} in the assay medium, since this cation has been shown to inhibit ATP-independent calcium-binding activity in rat heart sarcolemma (Dhalla et al., 1976). An increase in membrane calcium-binding activities has been suggested to increase the sarcolemmal calcium stores that are available for release upon excitation of the myocardium (Dhalla et al., 1977a), and this may partly explain the positive inotropic action of taurine. This view is consistent with the observations that ATP-independent calcium binding with sarcolemma has been shown to decrease in the presence of cardiodepressant agents such as propranolol and divalent cations (Dhalla et al.; 1977b; Harrow et al., 1978). Furthermore, an increase in calcium content of guinea pig heart by taurine has been observed by Dolara et al. (1973). The positive inotropic effect of taurine was observed to be markedly reduced by increasing the concentration of calcium in the perfusion medium. On the other hand, it is pointed out that quinidine, which is known to decrease myocardial contractile force, was found to increase ATP-independent calcium binding to the sarcolemma (Dhalla et al., 1978b). This raises a question regarding the significance of the observed increase in sarcolemmal calcium binding by taurine in terms of its positive inotropic action. However, taurine has been shown to antagonize arrhythmias induced by different interventions, and it may well be that taurine exerts its stabilization action by increasing sarcolemmal calcium in a manner similar to quinidine. At any rate, taurine was found to have no effect on sarcolemmal ATP-dependent calcium binding, which has been suggested to reflect the calcium efflux mechanism at the myocardial cell membrane (Dhalla et al., 1977a). On the basis of their calcium wash-out studies in guinea pig heart, Dolara et al. (1973) were able to show changes in three kinetically defined calcium compartments by taurine. Although taurine has been shown to increase calcium binding by guinea pig heart microsomal fraction when assayed by dialysis techniques (Dolara et al., 1976), the results described in this study obtained by millipore filtration techniques, as well as those reported by others with guinea pig heart microsomes (Remtulla et al., 1978), reveal no such increase in calcium binding or uptake activities. Similarly, no effect of taurine on rat or dog heart microsomal calcium transport activities was observed (Entman et al., 1977; Chubb and Huxtable, 1978). The inability of taurine to alter calcium transport properties of sarcoplasmic reticulum is further evident from the negative effect of this agent on microsomal Ca^{2+}-stimulated, Mg^{2+}-dependent ATPase, which is considered to serve as a "calcium pump" mechanism at this membrane site. On the other hand, we have shown that cardiac mitochondrial calcium-binding activity, but not calcium uptake, was markedly increased by taurine. An increase in rat liver mitochondrial calcium-binding activity by taurine has also been reported by other investigators (Dolara et al., 1973). It therefore seems likely that the taurine-induced increase in calcium-binding activity of microsomal fraction observed by Dolara et al. (1976) may be due to contaminating mitochondria in their preparation. In view of our results and the data reported in the literature, we believe that the cardiac effects of taurine may not be mediated through its influence on calcium stores in the sarcoplasmic reticulum. Furthermore, the observed changes in calcium compartments in the hearts perfused with taurine (Dolara et al., 1973) may represent changes in calcium stores associated with sarcolemmal and mitochondrial membranes. Although the significance of mitochondrial calcium binding in the cardiac contraction and relaxation process is not unequivocally established at present, several cardiodepressent agents have been shown to decrease mitochondrial calcium-binding activity (Dhalla et al., 1977b). The observed increase in mitochondrial calcium binding by taurine may increase myocardial calcium stores, which become available for release upon excitation, and this may then contribute toward increasing the cardiac contractile force. Thus it appears that changes in mitochondrial and sarcolemmal calcium stores may be involved in cardiac effects of taurine.

Since taurine had no effect on the oxidative phosphorylation activity of mitochondria, it is unlikely that increased production of ATP is a mechanism for the taurine effect on the myocardium. Likewise, the activity of myofibrillar calcium-stimulated, Mg^{2+}-dependent ATPase, which is associated with the ability of heart to generate contractile force (Dhalla et al., 1977a), was not altered by taurine. This rules out the possibility of its effect on the major ATP-utilizing mechanism. In this regard, it should also be noted that other ATP-utilizing systems such as microsomal Ca^{2+}-stimulated ATPase, sarcolemmal adenylate cyclase, and Na^+-K^+ ATPase, which have been implicated in different cellular functions (Dhalla et al., 1977a), were not affected by taurine. Akera et al. (1976) were also unable to detect the influence of taurine on cardiac Na^+-K^+ ATPase preparations. From these results we are tempted to conclude that cardiac effects of taurine are not related to changes in major mechanisms for ATP production and utilization in the myocardium. However, it was interesting to observe that sarcolemmal and mitochondrial Mg^{2+} ATPase activities were increased and that microsomal and myofibrillar Mg^{2+} ATPase activities were decreased by taurine. Since the significance of Mg^{2+} ATPase activities of different cell components is not clear at present, the implications of these effects of taurine in terms of changes in contractile force cannot be stated with certainty. However, if it is assumed that Mg^{2+} ATPase activity is related to Mg^{2+}-binding sites located at the membrane and contractile protein, it is then conceivable that taurine may influence the distribution of Mg^{2+} in the myocardium by increasing the affinity of Mg^{2+} for sarcolemmal and mitochondrial membranes as well as by decreasing the affinity of Mg^{2+} for sarcoplasmic reticulum and myofibrils. The role of Mg^{2+} movements in myocardial function has recently been emphasized (Polimeni and Page, 1973), and our view is supported by our observations that the positive inotropic effect of taurine is markedly reduced by decreasing the concentration of Mg^{2+} in the perfusion medium.

SUMMARY

Taurine was found to produce a positive inotropic effect in isolated perfused guinea pig heart in a dose-dependent manner. This increase in the contractile force by taurine was depressed by increasing the concentration of Ca^{2+} or by decreasing the concentration of Mg^{2+} in the perfusion medium. ATP-independent sarcolemmal Ca^{2+}-binding and ATP-dependent mitochondrial Ca^{2+}-binding activities were increased by taurine, whereas sarcoplasmic reticular (microsomal) Ca^{2+} transport activities were not altered by this agent. The sarcolemmal Na^+-K^+ ATPase; adenylate cyclase and Ca^{2+} ATPase; the microsomal Ca^{2+}-stimulated ATPase and myofibrillar Ca^{2+}-stimulated ATPase activities; and the mitochondrial oxidative phosphorylation activities were not influenced by taurine. On the other hand, this agent was found to increase sarcolemmal and mitochondrial Mg^{2+} ATPase and decrease microsomal and myofibrillar Mg^{2+} ATPase activities. These results suggest that taurine may not alter major mechanisms for ATP production and utilization. The positive inotropic effect of taurine may partly be explained due to the changes in sarcolemmal and mitochondrial calcium stores as well as alterations in Mg^{2+} movements at various sites in the myocardium.

ACKNOWLEDGMENT

This work was supported by a grant from the Manitoba Heart Foundation.

REFERENCES

Akera, T.; Ku, D.; and Brody, T. M. Alterations of ion movements as a mechanism of drug-induced arrhythmias and inotropic responses. In *Taurine*, R. Huxtable and A. Barbeau, eds. Raven Press, New York (1976), pp. 121–133.

Anand, M. B.; Chauhan, M. S.; and Dhalla, N. S. Ca^{2+}/Mg^{2+} ATPase activities of heart sarcolemma, microsomes and mitochondria. *J. Biochem.*, 82, 1732–1739 (1977).

Chovan, J. P.; Kulakowski, E. C., Benson, B. W., and Schaffer, S. W. Taurine enhancement of calcium binding to rat heart sarcolemma. *Biochim. Biophys. Acta,* 551, 129–136 (1979).

Chubb, J.; and Huxtable, R. Transport and biosynthesis of taurine in the stressed heart. In *Taurine and Neurological Disorders,* A. Barbeau and R. J. Huxtable, eds. Raven Press, New York (1978), pp. 161–177.

Dhalla, N. S.; Anand, M. B.; and Harrow, J. A. C. Calcium binding and ATPase activities of heart sarcolemma. *J. Biochem.* 79, 1345–1350 (1976).

Dhalla, N. S.; Ziegelhoffer, A.; and Harrow, J. A. C. Regulatory role of membrane systems in heart function. *Can. J. Physiol. Pharmacol.,* 55, 1211–1234 (1977a).

Dhalla, N. S.; Lee, S. L.; Anand, M. B.; and Chauhan, M. S. Effects of acebutolol, practolol and propranolol on the rat heart sarcolemma. *Biochem. Pharmacol.,* 26, 2055–2060 (1977b).

Dhalla, N. S.; Das, P. K.; and Sharma, G. P. Subcellular basis of cardiac contractile failure. *J. Mol. Cell. Cardiol.,* 10, 363–385 (1978a).

Dhalla, N. S.; Harrow, J. A. C.; and Anand, M. B. Actions of some antiarrhythmic agents on heart sarcolemma. *Biochem. Pharmacol.,* 27, 1281–1283 (1978b).

Dietrich, J.; and Diacona, J. Comparison between ouabain and taurine effects on isolated rat and guinea pig hearts in low calcium medium. *Life Sci.,* 10, 499–507 (1971).

Dolara, P.; Agresti, A.; Giotti, A.; and Pasquini, G. Effects of taurine on calcium kinetics of guinea pig heart. *Eur. J. Pharmacol.,* 24, 352–358 (1973).

Dolara, P.; Agresti, A.; Giotti, A.; and Sorace, B. The effect of taurine on calcium exchange of sarcoplasmic reticulum of guinea pig heart studied by means of dialysis kinetics. *Can. J. Physiol. Pharmacol.,* 54, 529–533 (1976).

Entman, M. L.; Bornet, B. P.; and Bressler, R. The effect of taurine on cardiac sarcoplasmic reticulum. *Life Sci.,* 21, 543–550 (1977).

Guidotti, A.; Badiani, G.; and Giotti, A. Potentiation by taurine of inotropic effect of strophanthin K on guinea pig auricles. *Pharmacol. Res. Commun.,* 3, 29–38 (1971).

Harrow, J. A. C.; and Dhalla, N. S. Effects of quinidine on calcium transport activities of the rabbit heart mitochondria and sarcotubular vesicles. *Biochem. Pharmacol.,* 25, 897–902 (1976).

Harrow, J. A. C.; Das, P. K., and Dhalla, N. S. Influence of some divalent cations on heart sarcolemmal bound enzymes and calcium binding. *Biochem. Pharmacol.,* 27, 2605–2609 (1978).

Huxtable, R. Metabolism and function of taurine in the heart. In *Taurine,* R. Huxtable and A. Barbeau, eds. Raven Press, New York (1976), pp. 99–119.

Iwata, H.; and Fujimoto, S. Potentiation by taurine of the inotropic effect of ouabain and the content of intracellular Ca^{2+} and taurine in the heart. *Experientia,* 32, 1559–1560 (1976).

Katz, A. M. Excitation-contraction coupling. In *Physiology of the Heart,* A. M. Katz, ed. Raven Press, New York (1977), pp. 137–159.

Langer, G. A. The structure and function of the myocardial cell surface. *Am. J. Physiol.,* 235(5), H461–H468 (1978).

Lowry, O. H.; Rosebrough, N. J.; Farr, A. L.; and Randall, R. L. Protein measurement with the folin phenol reagent. *J. Biol. Chem.,* 193, 265–275 (1951).

McNamara, D. B.; Sulakhe, P. V.; Singh, J. N., and Dhalla, N. S. Properties of heart sarcolemmal Na^+-K^+ ATPase. *J. Biochem.* (Tokyo), 75, 795–803 (1974).

Nayler, W. G.; Dunnet, J.; and Barry, D. The calcium accumulating activity of subcellular fractions isolated from rat and guinea pig heart muscle. *J. Mol. Cell. Cardiol.,* 7, 275–288 (1975).

Polimeni, P. I.; and Page, E. Magnesium in heart muscle. *Circ. Res.,* 33, 367–374 (1973).

Remtulla, M. A.; Katz, S.; and Applegarth, D. A. Effect of taurine on ATP-dependent calcium transport in guinea pig cardiac muscle. *Life Sci.,* 23, 383–390 (1978).

Solaro, R. J.; Pang, D. C.; and Briggs, F. N. Purification of cardiac myofibrils with Triton X-100. *Biochim. Biophys. Acta,* 245, 259–262 (1971).

Taussky, H. H.; and Shorr, E. A microcolorimetric method for the determination of inorganic phosphorus. *J. Biol. Chem.*, 202, 675–685 (1953).

Tomlinson, C. W.; Lee, S. L.; and Dhalla, N. S. Abnormalities in heart membranes and myofibrils during bacterial infective cardiomyopathy in the rabbit. *Circ. Res.*, 39, 82–92 (1976).

Yates, J. C.; and Dhalla, N. S. Structural and functional changes associated with failure and recovery of hearts after perfusion with Ca^{2+}-free medium. *J. Mol. Cell. Cardiol.*, 7, 91–103 (1975).

Copyright © 1981, Spectrum Publications, Inc.
The Effects of Taurine on Excitable Tissues

Effects of Taurine on Subcellular Calcium Dynamics in the Normal and Cardiomyopathic Hamster Heart

Joseph D. Welty
M. Cathie Welty

INTRODUCTION

It has been considered for some time that the molecular basis of heart failure may be broadly categorized into defects in energy production or defects in energy utilization (Olson, 1959). Even though defects in energy production have been demonstrated in some instances, energy stores in general have been reported to be normal in many models of heart failure (Wollenberger, 1949). The excitation-contraction coupling component of energy utilization has long been suspected to have a role in the failing heart mechanism (Nayler, 1963). Calcium has been considered to play a central role in this excitation-contraction process, and recently Dhalla (1976) has suggested that both calcium deficiency and calcium overload within the cardiac cells may be involved in the mechanism of heart failure.

In the present study we have selected the cardiomyopathic hamster (Bio 14.6) as a model to investigate the possible role of calcium in the mechanism of heart failure. McBroom and Welty (1977) have observed a 12-fold increase in total heart calcium in this hamster model at 60 days of age, the maximal lesioning period. These authors have further demonstrated that administration of taurine (2-aminoethanesulfonic acid) will decrease the magnitude of this elevated heart calcium. Taurine has long been implicated in modulating cardiac activity by affecting ion movement (Read and Welty, 1963; Welty and Read, 1964; Welty et al., 1976), and recently, taurine receptors have been identified on rat heart sarcolemma (Kulakowski et al., 1978). In addition, taurine has been shown by many investigators to be elevated in various states of natural and experimentally induced heart failure (Peterson et al., 1973; Huxtable and Bressler, 1974). Therefore, a role for taurine in the energy-utilization mechanism has certainly been implicated.

The purpose of the present investigation is to study the subcellular calcium-modulating mechanism in the cardiomyopathic (Bio 14.6) hamster in an attempt to identify both the biochemical lesion(s) responsible for the observed elevation of total heart calcium and the nature of the progressive heart failure state. In addition, the effect of taurine on these calcium-modulating mechanisms in the cardiomyopathic as well as in the normal (RB) Syrian golden hamster is discussed.

METHODS AND MATERIALS

Sixty-day-old cardiomyopathic hamsters of the Bio 14.6 strain (Bio-Research Consultants, Cambridge, Mass.) served as the experimental model for this study, while random bred (RB) Syrian golden hamsters were used as controls. Animals of the Bio 14.6 strain develop cardiomyopathic lesions at 60 days of age (Bajusz et al., 1969) without signs and symptoms of heart failure (Bajusz, 1969). An elevated total heart calcium is also observed at this age (McBroom and Welty, 1977). These animals then progress through a cardiac compensatory period followed by overt cardiac failure by 300 days of age. For the mitochondrial calcium determinations, groups of Bio 14.6 and RB hamsters were given 0.1 M taurine in tap water *ad libitum* following weaning (25 days of age), until the time of the experiment. All animals were fasted 12 hr prior to experimentation and were sacrificed by cervical dislocation.

The subcellular fractions from five pooled hearts were obtained by the following procedures. The sarcolemmal fraction was obtained by a slight modification of the method of McNamara et al. (1974). The post-1,000 g supernatant was used for the isolation of the mitochondrial and sarcoplasmic reticular fractions by the method of Sulakhe and Dhalla (1971a). While it is not suggested that these fractions are pure, on the basis of putative marker enzymes, cross-contaminations of the subcellular fractions were considered negligible (St. Louis and Sulakhe, 1976). Electron microscopic evidence was obtained for the absence of intact mitochondria in the sarcolemmal and sarcoplasmic reticular fractions. Protein was measured by the method of Lowry et al. (1951).

Assays of the Sarcolemmal Fraction

The Na^+-K^+-stimulated ATPase activity was measured using 50 μg of membrane protein incubated in a total volume of 1 ml containing 100 mM NaCl, 10 mM KCl, 4 mM $MgCl_2$ and 50 mM Tris-HCl (pH 7.4), at 37°C in the presence or absence of 2 mM ouabain and/or 5 mM taurine. Following a 5-min preincubation period, the reaction was started by the addition of ATP at a final concentration of 4 mM. The reaction was allowed to continue for 10 min, stopped by the addition of 1 ml of 12% cold trichloroacetic acid, centrifuged, and the amount of Pi liberated was determined by the method of Taussky and Shorr (1953).

The Ca^{2+} ATPase activity was measured as described above for the Na^+-K^+ ATPase activity except that the incubation buffer contained 4 mM $CaCl_2$ and 50 mM Tris-HCl (pH 7.4), in the presence or absence of 5 mM taurine. The Mg^{2+} ATPase activity was measured as described previously for the Na^+-K^+ ATPase activity except that the incubation buffer contained 4 mM $MgCl_2$ and 50 mM Tris-HCl (pH 7.4), in the presence or absence of 5 mM taurine. The ATP hydrolysis that occurred in the absence of the cations was subtracted in order to calculate the cation-stimulated ATPase activity (McNamara et al., 1974).

The calcium-binding and uptake studies were performed on 300 μg of membrane protein at 30°C in a medium containing 25 mM Tris-maleate (pH 6.7), 5 mM $MgCl_2$, 100 mM KCl, $^{45}CaCl_2$ (10 mCi/nmol) in the absence and presence of 5 mM potassium oxalate, respectively. Here we use the term *uptake* to describe the accumulation of calcium in the presence of ATP and a precipitating anion. The term *binding* is used to imply only the presence of ATP. The taurine concentration, when present, was 5 mM, and the free $CaCl_2$ concentration was either 10^{-4} or 10^{-6} M, buffered according to the procedure of Katz et al. (1970). The reaction was started by the addition of ATP and terminated by Millipore filtration (0.45 μM/25 mm) at the desired intervals of 30 sec, 2 min, 4 min, and 6 min (Sulakhe et al., 1976).

Assays of the Mitochondrial Fraction

The Ca^{2+} ATPase activity of this fraction was determined by incubating 300 μg of the mitochondrial fraction protein in a medium containing 100 mM KCl, 1 mM $MgCl_2$, 4 mM ATP, 20 mM Tris-HCl (pH 6.8, 37°C), and 4 mM

CaCl$_2$, according to the procedure of Anand et al. (1977). Taurine, when present, was at a concentration of 5 mM. The mitochondria, in which the calcium content was determined, were isolated as previously described or by the method of Tomlinson et al. (1974).

The total calcium content was determined by atomic absorption spectrophotometry following extraction with 0.5 N HCl, sonication, and the addition of LaCl$_3$ (1%) (Reynafarje and Lehninger, 1969; Peng et al., 1977).

The calcium binding and sodium-induced calcium release of this fraction was determined on 300 μg/ml of protein with either 10^{-6} or 10^{-4}M calcium (Katz et al., 1970) in a medium containing 210 mM mannitol, 70 mM sucrose, 10 mM Tris-Cl buffer (pH 7.4), ^{45}CaCl$_2$ (5×10^5 cpm/ml), and 10 mM Tris-succinate at 25°C in a final volume of 3 ml, in the presence or absence of 5 mM taurine. Following 2 min of incubation, calcium release was induced by the addition, first of ruthenium red (2 nmol/mg protein), and then of 10 mM NaCl. Control preparations (Ca^{2+} free) were of equal ionic strength. Fifty-microliter aliquots were removed and examined for ^{45}Ca at 30 sec, 2 min, 4 min, and 6 min after induction of calcium release. The Millipore filtration technique of Carafoli et al. (1974) was employed to assay the aliquots.

Assays of the Sarcoplasmic Reticular Fraction

The Ca^{2+} ATPase activity of this fraction was determined on 200 μg of protein incubated in a medium containing 100 mM KCl, 10 mM MgCl$_2$, 4 mM ATP, and 20 mM Tris-HCl (pH 6.8) at 37°C, in the presence or absence of 5 mM taurine.

The calcium uptake and binding by this fraction was measured by the Millipore filtration technique in a medium containing 100 mM KCl, 20 mM Tris-HCl (pH 6.8), ^{45}CaCl$_2$ (10 mCi/mM), and 4 mM ATP, in the presence and absence of 5 mM potassium oxalate, respectively. The calcium concentration was maintained at 10^{-6} or 10^{-4}M (Katz et al., 1970) in the presence or absence of 5 mM taurine. The total volume of the medium was 3 ml. The reaction was started by the addition of ATP and terminated by Millipore filtration.

Statistics

Data were analyzed using Student's t-test for intergroup comparisons, and Duncan's New Multiple Range Test, as modified by Kramer (1956), was used for intragroup comparisons. Significance was set at the $p < 0.05$ level. All data are presented as mean $\pm$ standard deviation.

RESULTS

The 60-day-old Bio 14.6 cardiomyopathic hamsters at time of sacrifice displayed no gross signs of failure. The heart weights (0.35 ± 0.01 g) and heart weight:body weight ratios (3.9 ± 0.1 g $\times$ 10^3) were not different from the RB control animals. These results are consistent with our previous observations (Appelt et al., 1976). The hearts of the Bio 14.6 hamsters, on gross examination, displayed acute lytic lesions with minimal inflammation.

Sarcolemmal Fraction

The Na^+-K^+ ATPase activity of the Bio 14.6 sarcolemmal fraction was reduced by 48% when compared to the RB (Table 22.1). In the Bio 14.6, the addition of 5 mM taurine increased the Na^+-K^+ ATPase activity by 46% but was without effect in the RB preparations.

The Mg^{2+} ATPase activity of the Bio 14.6 sarcolemmal fraction was 53% greater than the RB value (Table 22.1). Taurine was without effect on this activity in either strain of hamster.

The Ca^{2+} ATPase activity of the Bio 14.6 sarcolemmal fraction was 35% greater than the RB value (Table 22.1). Taurine, as in the case of the Mg^{2+} ATPase activity, was without effect on the Ca^{2+} ATPase activity in either strain of hamster.

Table 22.1 ATPase activity of the subcellular fractions from cardiomyopathic (Bio 14.6) and control (RB) hamster hearts

	RB	RB plus taurine	Bio 14.6	Bio 14.6 plus taurine
Sarcolemma				
Na^+-K^+ ATPase (8)[a]	14.39 ± 1.05[b,d]	14.60 ± 0.74	7.46 ± 1.73[d,e]	10.90 ± 1.90[e]
Mg^{2+} ATPase (8)	22.80 ± 1.09[b,d]	23.24 ± 0.97	34.82 ± 1.76[d]	35.59 ± 2.03
Ca^{2+} ATPase (8)	21.82 ± 1.05[b,d]	22.59 ± 1.47	30.76 ± 1.86[d]	31.04 ± 1.03
Mitochondria				
Ca^{2+} ATPase (8)	8.56 ± 1.50[d,e]	5.51 ± 1.08[e]	6.61 ± 1.41[d,e]	4.32 ± 1.01[e]
Sarcoplasmic Reticulum				
Ca^{2+} ATPase (8)	5.17 ± 1.93[c]	5.27 ± 1.76	5.55 ± 0.89[e]	6.79 ± 1.35[e]

[a]Number of preparations.

[b]μmol Pi/mg protein/hr.

[c]μmol Pi/mg protein/5 min.

[d]$p < 0.05$ interstrain comparisons.

[e]$p < 0.05$ intrastrain comparisons.

Calcium binding of the sarcolemmal fraction from RB hamsters was essentially complete by 30 sec; it displayed an ATP saturation at a concentration of 250 μM, an optimal Mg^{2+} concentration at 5 mM, and a KCl optimal concentration at 100 mM, consistent with the results previously reported for the guinea pig by St. Louis and Sulakhe (1976). The sarcolemmal fraction from the Bio 14.6 hamster did not vary from these results except that Ca^{2+} binding (Table 22.2) was still incomplete after 6 min (Fig. 22.1). Using a calcium concentration of 10^{-6}M, the Bio 14.6 displayed greater Ca^{2+} binding than the RB control. At a concentration of 10^{-4}M calcium, no difference in initial binding was observed between the strains; however, by 30 sec the RB had reached a steady-state calcium binding, while the Bio 14.6 had not by 6 min. When taurine (5 mM) was present in the medium at a 10^{-6}M calcium concentration, no differences were observed between the strains; however, the Ca^{2+} binding in the RB was increased, while the Ca^{2+} binding in the Bio 14.6 was decreased from their respective controls. At a calcium concentration of 10^{-4}M, taurine (5 mM) increased (Fig. 22.1) the calcium binding only in the RB and only after 4 and 6 min of incubation.

The effect of varying the taurine concentration on calcium binding to the

Table 22.2 Sarcolemmal Ca^{2+} binding from cardiomyopathic (Bio 14.6) and control (RB) hearts

	30 sec	2 min	4 min	6 min
Ca^{2+} 10^{-6} M				
RB (8)[a]	0.29 ± 0.05[b,f]	0.27 ± 0.03[f]	0.34 ± 0.07[f]	0.32 ± 0.05[f]
Bio 14.6 (8)	0.56 ± 0.17[d,e,g]	0.64 ± 0.11[d,g]	0.70 ± 0.08[d,g]	0.78 ± 0.11[d,e,g]
Taurine[c]				
RB (8)	0.42 ± 0.08[e,f]	0.53 ± 0.11[f]	0.62 ± 0.18[f]	0.66 ± 0.18[e,f]
Bio 14.6 (8)	0.40 ± 0.02[e,g]	0.54 ± 0.12[g]	0.60 ± 0.18[g]	$0.60 \pm \pm 0.06$[e,g]
Ca^{2+} 10^{-4} M				
RB (8)	21.3 ± 2.9	23.5 ± 6.3	25.9 ± 4.9[f]	24.8 ± 6.9[f]
Bio 14.6 (8)	22.1 ± 1.1[e]	24.9 ± 4.7	29.0 ± 7.6	32.9 ± 9.8[e]
Taurine				
RB (8)	24.2 ± 3.7[e]	24.7 ± 4.8	30.7 ± 7.0[f]	30.1 ± 1.0[e,f]
Bio 14.6 (8)	23.5 ± 2.6[e]	26.5 ± 8.6	29.2 ± 1.2	33.5 ± 1.6[e]

[a]Number of preparations.

[b]nmol Ca^{2+}/mg protein.

[c]Taurine added to the incubation medium at a final concentration of 5 mM.

[d]$p < 0.01$ interstrain comparisons.

[e]$p < 0.05$ intrastrain comparisons.

[f]$p < 0.05$ intrastrain comparisons between treatments (RB).

[g]$p < 0.05$ intrastrain comparisons between treatments (Bio 14.6).

sarcolemma in the presence of $10^{-6}M$ calcium is shown in Fig. 22.2. At all taurine concentrations examined, the initial fast binding was increased from

SARCOLEMMAL Ca^{+2} BINDING

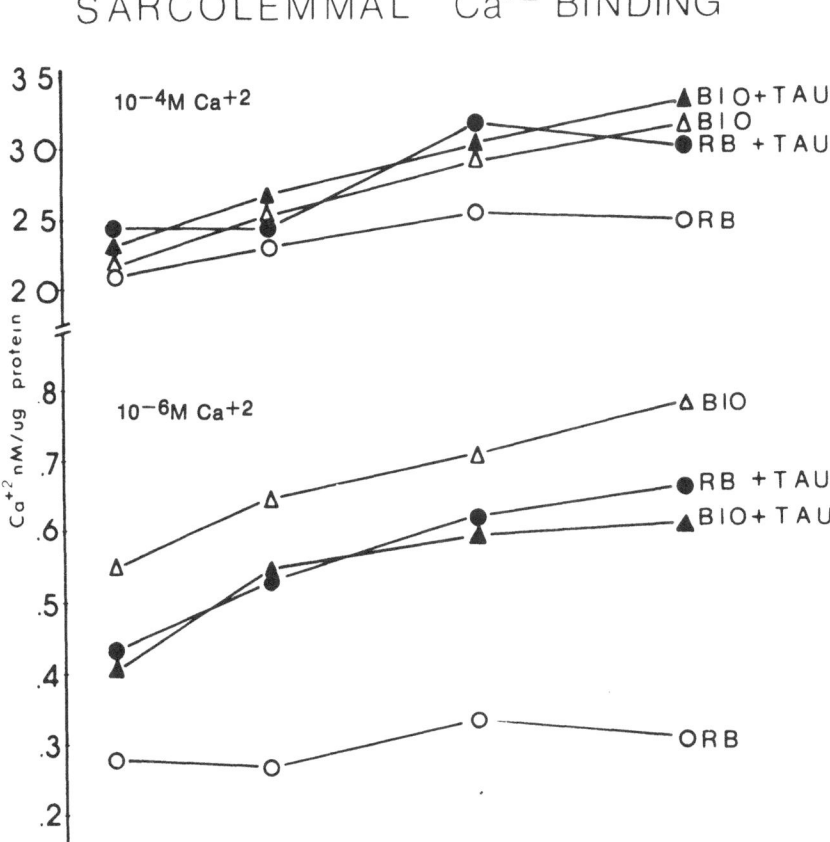

Fig. 22.1. Calcium binding by cardiac sarcolemmal membranes (300 μg protein) incubated at 30°C in a medium (3 ml) containing: 25 mM Tris-maleate (pH 6.7), 5 mM MgCl$_2$, 100 mM KCl, ^{45}CaCl$_2$ (10 mCi/nmol), 2.5 mM ATP, with either 1 μM or 100 μM CaCl$_2$. At desired intervals, 50 μliter aliquots of incubation mixture were filtered through Millipore disks. Key: O, Random Bred (RB); Δ, Cardiomyopathic (BIO); ●, Random Bred plus 5 mM taurine (RB + TAU); ▲, Cardiomyopathic plus 5 mM taurine (BIO + TAU). Data shown as nanomoles Ca^{2+}bound/μg protein.

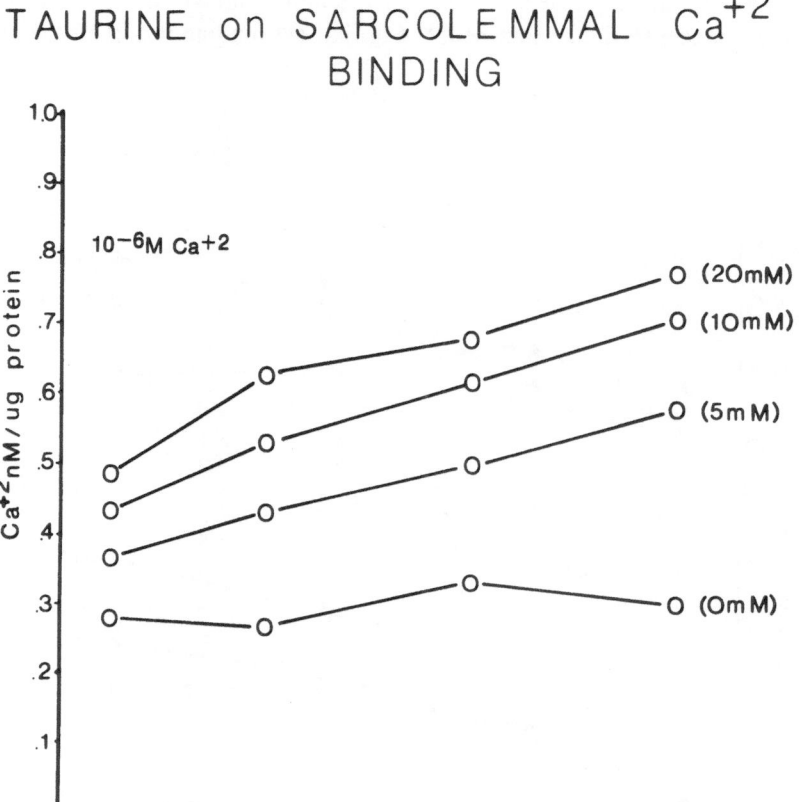

Fig. 22.2. Calcium binding by cardiac sarcolemmal membranes (300 μg protein) from normal (RB) hamsters in the presence of various concentrations of taurine. The incubation was at 30°C in a medium (3 ml) containing: 25 mM Tris-maleate (pH 6.7), 5 mM $MgCl_2$, 100 mM KCl, $^{45}CaCl_2$ (10 mCi/nmol), 1 μM $CaCl_2$, 2.5 mM ATP, and taurine (0 mM, 5 mM, 10 mM, 20 mM). At desired intervals, 50 μliter aliquots of incubation mixture were filtered through Millipore disks. Data shown as nanomoles Ca^{2+} bound/μg protein.

the control level. The calcium-binding pattern at 10 mM and 20 mM taurine, however, was not different. The slow secondary phase of calcium binding established by 5 mM taurine was not affected by further increases in taurine concentration.

Sarcolemmal calcium uptake in the presence of $10^{-6}M$ calcium appeared complete by 30 sec in both the Bio 14.6 and RB; however, in the presence of taurine, both strains displayed a slow secondary calcium uptake, which did not reach a steady state by 6 min (Fig. 22.3). Calcium uptake, either in the

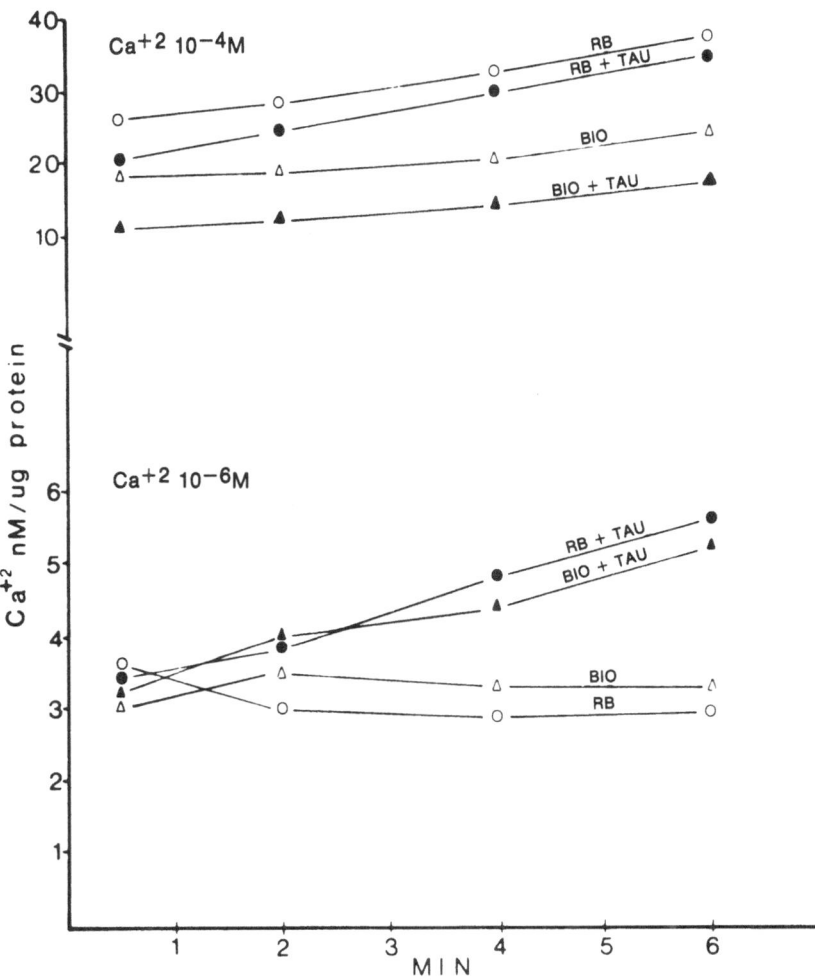

Fig. 22.3. Calcium uptake by cardiac sarcolemmal membranes (300 μg protein) incubated at 30°C in a medium (3 ml) containing: 25 mM Tris-maleate (pH 6.7), 5 mM potassium oxalate, 5 mM MgCl$_2$, 100 mM KCl, ^{45}CaCl$_2$ (10 mCi/nmol), 2.5 mM ATP, with either 1 μM or 100 μM CaCl$_2$. At desired intervals 50 μliter aliquots of incubation mixture were filtered through Millipore disks. Data shown as nanomoles Ca^{2+} bound/μg protein.

presence or absence of taurine, was identical in the two strains (Table 22.3). At a calcium concentration of 10^{-4}M, the Bio 14.6 uptake was depressed (26% at 30 sec, 30% at 2 min, 36% at 4 min, and 35% at 6 min) from the RB control values. Taurine was without effect on either strain at this calcium concentration.

Mitochondrial Fraction

The Ca^{2+} ATPase activity of the mitochondrial fraction may well represent Racker's F_1 ATPase (Kagawa and Racker, 1971) and as such is only an indirect measure of Ca^{2+} transport. This activity, however, in Bio 14.6 hearts was depressed 23% (Table 22.1) from the RB control hearts. Taurine was shown to reduce the activity in the RB by 36% and to further reduce the activity in the Bio 14.6 by 35%.

Calcium binding of the mitochondrial fraction from Bio 14.6 was dramatically reduced in the 10^{-6}M calcium medium from the RB, but no differences between the strains were observed at 10^{-4}M calcium. Taurine was without effect on calcium binding in either strain at either calcium concentration (Table 22.4). The addition of NaCl (10 mM final concentration) to

Table 22.3 Sarcolemmal uptake of calcium in cardiomyopathic (Bio 14.6) and control (RB) hamster hearts

	30 sec	2 min	4 min	6 min
$Ca^{2+} 10^{-6}$ M				
RB (8)[a]	3.6 ± 0.5[b]	3.0 ± 0.8	2.8 ± 0.7	3.0 ± 0.9
Bio 14.6 (8)	3.0 ± 0.2	3.5 ± 0.1	3.3 ± 0.8	3.2 ± 0.8
Taurine[c]				
RB (8)	3.5 ± 0.9[e]	3.8 ± 0.8	4.8 ± 0.9	5.6 ± 1.0[e]
Bio 14.6 (8)	3.3 ± 0.8[e]	4.0 ± 0.9	4.4 ± 0.9	5.4 ± 1.0[e]
$Ca^{2+} 10^{-4}$ M				
RB (8)	26.0 ± 2.5[e]	27.8 ± 0.6	33.7 ± 1.1	38.0 ± 4.7[e]
Bio 14.6 (8)	19.3 ± 1.8[d,e]	19.6 ± 1.3[d]	21.6 ± 1.5[d]	24.9 ± 4.5[d,e]
Taurine				
RB (8)	20.6 ± 5.5[e]	25.1 ± 4.2	30.6 ± 4.6	34.8 ± 4.3[e]
Bio 14.6 (8)	11.2 ± 2.3[d,e]	13.8 ± 5.5[d]	14.8 ± 5.3[d]	17.3 ± 6.5[d,e]

[a]Number of preparations.

[b]nmol Ca^{2+}/mg protein.

[c]Taurine added to the incubation medium at a final concentration of 5 mM.

[d]$p < 0.01$ interstrain comparisons.

[e]$p < 0.05$ intrastrain comparisons.

Table 22.4 Calcium binding of the mitochondrial fraction from cardiomyopathic (Bio 14.6) and control (RB) hamster hearts

	30 sec	2 min	4 min	6 min
$Ca^{2+}10^{-6}$ M				
RB (8)[a]	1.82 ± 0.33[b]	1.22 ± 0.31	1.21 ± 0.48	1.12 ± 0.49
Bio 14.6 (8)	0.29 ± 0.04[d]	0.25 ± 0.04[d]	0.27 ± 0.08[d]	0.26 ± 0.08[d]
Taurine[c]				
RB (8)	2.11 ± 0.51	1.36 ± 0.34	1.47 ± 0.62	1.43 ± 0.65
Bio 14.6 (8)	0.31 ± 0.05[d]	0.26 ± 0.05[d]	0.27 ± 0.09[d]	0.26 ± 0.08[d]
$Ca^{2+}10^{-4}$ M				
RB (8)	25.9 ± 2.5	20.5 ± 1.2	22.8 ± 1.6	19.8 ± 2.4
Bio 14.6 (8)	26.8 ± 3.6	25.3 ± 4.3	25.0 ± 6.2	25.9 ± 7.1
Taurine				
RB (8)	25.3 ± 2.5	22.6 ± 2.3	25.9 ± 4.2	23.5 ± 3.1
Bio 14.6 (8)	26.4 ± 3.6	25.4 ± 3.5	28.1 ± 6.6[d]	23.9 ± 5.3

[a]Number of preparations.

[b]nmol Ca^{2+}/mg protein.

[c]Taurine added to the incubation medium at a final concentration of 5 mM.

[d]$p < 0.01$ interstrain comparisons.

the mitochondrial fraction at 2 min incubation induced a significant release of calcium (Fig. 22.4) in both Bio 14.6 and RB hamsters in the presence of either 10^{-6} or 10^{-4}M calcium (Carafoli et al., 1974). At the 10^{-6}M calcium concentration, sodium induced a greater release of calcium in the RB than in the Bio 14.6 preparations. Taurine was without effect on calcium release in either strain. In contrast, at the 10^{-4}M calcium concentration, the addition of sodium induced an equivalent release of calcium from both the RB and Bio 14.6 mitochondrial fractions. In the presence of taurine, a greater release of calcium was observed in both the RB and Bio 14.6 compared to the nontaurine controls. Both strains showed an equivalent calcium release, which was greater at 4 min than at 6 min.

One group each of RB and Bio 14.6 hamsters was given 0.1 M taurine in the drinking water from weaning until time of sacrifice. The mitochondrial calcium content was determined from the hearts of these animals as well as from non-taurine-fed animals. In the non-taurine-fed group the Bio 14.6 mitochondrial calcium content (30.8 ± 2.6 nmol/mg protein) was threefold greater than the RB (9.0 ± 1.5 nmol/mg protein) calcium content. The feeding of taurine had no effect on the RB mitochondrial Ca^{2+} content values (7.8 ± 1.9 nmol/mg protein), but a decrease of some twofold was observed in the Bio 14.6 (12.4 ± 1.3 nmol/mg protein).

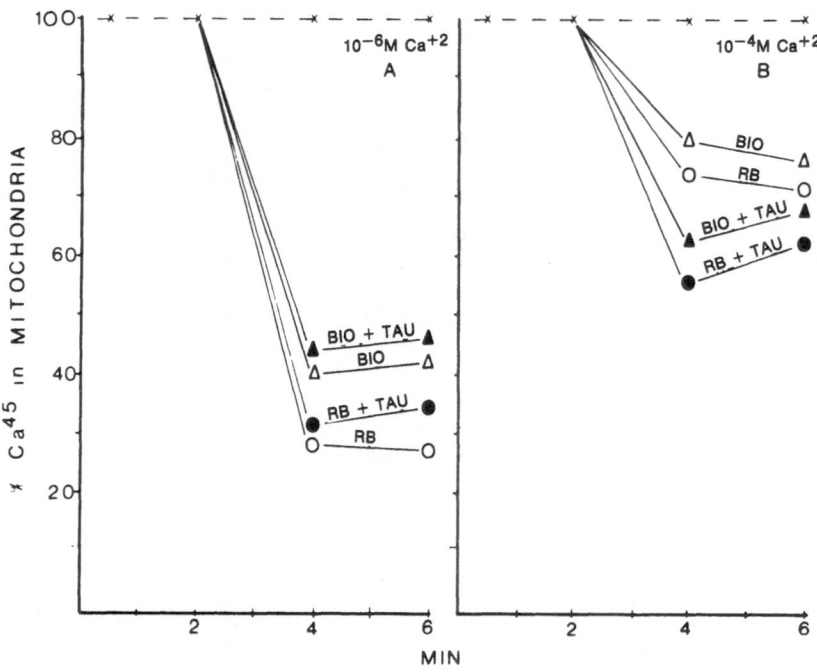

TAURINE on Na⁺ INDUCED Ca⁺² RELEASE in MITOCHONDRIA

Fig. 22.4. Ability of NaCl to release Ca²⁺ from heart mitochondria. Mitochondria (300 μg protein) were incubated in a medium (3 ml) at 25°C containing: 210 mM mannitol, 70 mM sucrose, 10 mM Tris-HCl (pH 7.4), ⁴⁵CaCl₂ (5 × 10⁵ cpm/ml of reaction), 10 mM Tris-succinate, in the presence or absence of 5 mM taurine, with a Ca² concentration of: (a) 1 μM CaCl₂, (b) 100 μM CaCl₂. Following 2 min incubation, ruthenium red (2 nmol/mg protein) was added, followed by 10 mM NaCl. Key: O, Random Bred (RB); △, Cardiomyopathic (BIO); ●, Random Bred plus 5 mM taurine (RB + TAU); ▲, Cardiomyopathic plus 5 mM taurine (BIO + TAU).

Sarcoplasmic Reticular Fraction

The calcium-dependent ATPase activities from the RB and the Bio 14.6 were not different (Table 22.1). Taurine was without effect on the RB, but a slight stimulation in activity was seen in the Bio 14.6 preparations.

The calcium binding and uptake in 10⁻⁶ or 10⁻⁴M calcium were not different when comparing RB and Bio 14.6 preparations. Taurine was without effect on the parameters studied (Table 22.5).

Table 22.5   Sarcoplasmic reticular calcium binding and uptake

	Binding	Uptake	Binding plus taurine[a]	Uptake plus taurine
10^{-6} M calcium				
RB (8)[b]	4.3 ± 0.7[c]	31.6 ± 5.8	5.0 ± 0.8	33.2 ± 6.0
Bio 14.6 (8)	4.0 ± 1.2	29.5 ± 7.2	4.5 ± 1.7	31.6 ± 9.4
10^{-4} M calcium				
RB (8)	16.8 ± 5.2	304 ± 16	21.2 ± 5.9	315 ± 12
Bio 14.6 (8)	14.1 ± 3.6	300 ± 22	19.0 ± 5.4	291 ± 15

[a]Taurine final concentration 5 mM.

[b]Number of preparations.

[c]nmol calcium/mg protein/2 min.

DISCUSSION

The present study was designed to study the calcium-regulating mechanisms in the cardiomyopathic hamster during the lesioning period, but prior to the development of failure (Bajusz, 1969). By this approach a biochemical lesion may well be identified that could explain the elevation of total heart calcium as well as the nature of the progressive heart failure state. The possible ameliorating effects of taurine could also be studied by this approach.

Several laboratories have studied various subcellular ion-regulating mechanisms in the cardiomyopathic hamster in varying degrees of cardiac failure. The results reported have often been conflicting (Dhalla, 1974). A consistent observation, however, has been an elevation of cardiac cellular calcium.

Our results suggest several possible explanations for the elevation of total heart calcium as well as the elevation of mitochondrial calcium in the cardiomyopathic hamster. We propose that defects reside in the sarcolemma (i.e., depressed Na^+-K^+ ATPase activity, increased sarcolemmal calcium storage, and increased Ca^{2+} ATPase activity), which favor calcium influx, as well as a defect in the mitochondria (i.e., depressed Ca^{2+} releasability), which also favors calcium influx.

The observed depression of Na^+-K^+ ATPase activity in this study may lead to an increase in total Ca^{2+} through the action of a Na^+-Ca^{2+} exchange mechanism proposed by Reuter (1974) and Katz (1975). According to this mechanism, the Na^+ gradient would serve as an energy source for extrusion of Ca^{2+}. An electroneutral exchange of two Na^+ for one Ca^{2+} would lead to a distribution ratio, that is: $[Ca]i/[Ca]o = [Na]_i^2/[Na]_o^2$. An increase in intracellular sodium would therefore lead to an increase in intracellular calcium.

Should the Na^+-K^+ ATPase activity be depressed, then an elevation of intracellular sodium would occur, with a resultant increase in cellular calcium. Katz (1975) has proposed a similar mechanism to explain the positive inotropic effect of cardiac glycosides. The observed stimulation of Na^+-K^+ ATPase activity in the Bio 14.6 and absence of an effect in the random bred hamster by taurine would then be consistent with the reported results of McBroom and Welty (1977), in which the feeding of taurine reduced the observed elevation of heart calcium in the Bio 14.6 but was without effect in the random bred hamsters.

The increased calcium storage capacity of the sarcolemma would also be consistent with the suggestion of myocardial calcium overload in this model of heart failure. The sarcolemmal calcium-binding data, in the low-calcium medium, suggest a fast initial binding pool in the random bred animals that is essentially filled by about 30 sec. The Bio 14.6 demonstrated greater initial binding plus a slower secondary binding pool that was not complete by 6 min. This increase in binding in the Bio 14.6 suggests a greater sarcolemmal-bound calcium pool, which could enter the cytoplasm with the passage of an action potential. In the presence of taurine the fast binding is decreased, but no effect on the slower calcium binding is observed. This again would suggest a role for taurine in reducing heart calcium in this cardiomyopathic animal model.

The recent suggestion that two Ca^{2+} ATPases reside in the sarcolemma adds another factor to be considered in the Ca^{2+} balance of cardiac cells. Sperelakis and Schneider (1976) have suggested that calcium influx in the myocardium is an ATP-dependent mechanism operating through the opening of specific calcium channels. Several investigators (Sulakhe and Dhalla, 1971b; Hui et al., 1976; McNamara et al., 1974) demonstrated the presence of a Ca^{2+} ATPase in the cardiac sarcolemma, which is considered to be directly involved in stimulus-induced calcium influx (Dhalla et al., 1978). Sonnenblick and Stam (1969) have postulated a Ca^{2+}stimulated, Mg^{2+}-dependent ATPase in cardiac sarcolemma, a "calcium pump" (Carafoli and Cromptom, 1978a), which is involved in lowering the intracellular calcium concentration. If, indeed, there are two separate ATPases in the sarcolemma, it is tempting to interpret our finding of an elevated Ca^{2+} ATPase in the Bio 14.6 as another probable cause of the elevated myocardial calcium concentration. Since this increase in Ca^{2+} ATPase activity has also been observed in early, moderate, and late stages of failure in the Bio 14.6 (Dhalla et al., 1976), our finding would implicate this defect in the sarcolemma in the mechanism of heart failure in this model. However, demonstration of the absence of this lesion in the newborn would be required to confirm this hypothesis.

Consistent with the hypothesis of a sarcolemmal defect leading to calcium overload is the finding of an elevated sarcolemmal Mg^{2+} ATPase activity.

Bajusz (1969), Bajusz and Lossnitzer (1968), and Nadkarni et al. (1972) have observed decreased myocardial levels of magnesium and a greater susceptibility to dietary magnesium deficiency in the young Bio 14.6 hamsters. These authors have suggested that an early magnesium deficiency may play a role in the cardiomyopathic process. Carafoli and Cromptom (1978b) have demonstrated that Mg^{2+} is a natural inhibitor of Ca^{2+} uptake in cardiac mitochondria. It would be tempting to speculate that, if the sarcolemmal Mg^{2+} ATPase represents a separate enzyme system controlling magnesium flux (McNamara et al., 1974; Anand et al., 1977), then the elevated activity found could help to explain the decreased Mg^{2+} content, which would decrease *in vivo* inhibition of the mitochondrial Ca^{2+} uptake process.

Evidence for the electrophoretic mitochondrial influx of calcium was first presented by Selwyn et al. (1970); however, many authors (Anand et al., 1977; Carafoli, 1974) have used the measurement of ATP hydrolysis as an indirect assessment of calcium movement. In our studies, the Ca^{2+} ATPase of the mitochondria was azide inhibitable, suggesting that the activity being measured is actually from Racker's F_1 ATPase (Racker et al., 1975). The ATPase activity from the Bio 14.6 animals, which was depressed, could either represent a dysfunction within the system or reflect the already large calcium load within the mitochondria. The observation that in the presence of taurine the activity is further depressed, which would tend to reduce the calcium load, is supported by the finding that the mitochondrial calcium content from taurine-fed Bio 14.6 animals was less than that from the non-taurine-fed counterparts.

In the presence of low Ca^{2+} concentrations, the depressed Ca^{2+} binding of the mitochondria is consistent with the interpretation of a depressed "Ca^{2+} ATPase." Decreased calcium binding is also a consistent observation in early, moderate, and late stages of heart failure in this model (Dhalla et al., 1978). In light of the demonstrated elevation of mitochondrial calcium content in this study, a decrease in the efflux of mitochondrial calcium must, therefore, exist in this model. This conclusion is supported in part by the observation that in a low-calcium concentration, the Na^+-induced Ca^{2+} release from the mitochondria of Bio 14.6 animals is depressed from that of the control animals. Taurine, in the high-calcium medium, enhanced the Na^+-induced release of Ca^{2+}, this observation being consistent with the finding of reduced mitochondria calcium content in the taurine-fed Bio 14.6 animals.

We have thus identified biochemical lesions in cardiac sarcolemma of the Bio 14.6 cardiomyopathic hamster that could explain the calcium overload observed. In addition, it is suggested that the decreased calcium binding observed in the mitochondria may be a compensatory adaptive mechanism, while the actual defect may reside in the releasability of mitochondrial calcium.

In the normal animal the effects of taurine appear to be associated with calcium modulation in the sarcolemma and mitochondria. The increase in sarcolemmal Ca^{2+} binding in the presence of taurine is consistent with the observation of Dolara et al. (1973) that taurine increases the affinity of some cell structures for calcium. Dolara et al. (1973) also reported that the perfusion of guinea pig hearts with an 8 mM taurine solution increased the heart calcium content. This is inconsistent with our previous report (McBroom and Welty, 1971) that the feeding of 0.1 M taurine in drinking water was without effect on the calcium content of the normal hamster heart. The differences in taurine concentrations as well as the species difference may explain the divergence of findings.

The finding that taurine enhances the ATP-dependent sarcolemmal Ca^{2+} binding and uptake may offer some insight into the understanding of the observation that taurine exerts a positive inotropic effect on isolated guinea pig auricles (Dietrich and Diacono, 1971) as well as potentiates the effect of cardiac glycosides (Guidotti et al., 1971). Since inotropic effects are thought to be modulated via calcium availability, taurine may provide additional calcium via the sarcolemma for this effect. Interpretation of what role this may play in the overall inotropic effects must be weighed against the calculations of Carafoli and Crompton (1978b) that the sarcolemma constitutes only 1% of the total area of Ca^{2+}-transporting membranes in the heart. However, if a mechanism could be established to link these findings with the observation that taurine enhances the releasability of Ca^{2+} from the mitochondria, a sufficient area (87%) of Ca^{2+}-transporting membrane would be involved to induce the positive inotropic response.

In the present study we have demonstrated several defects in the cardiac sarcolemma and mitochondria of the Bio 14.6 hamster that could explain elevated heart calcium content as well as lead to a failure state. We have suggested mechanisms whereby taurine could correct these defects. In addition, we have demonstrated a role for taurine in calcium modulation in the normal hamster heart.

REFERENCES

Anand, M. B.; Chauhan, M. S.; and Dhalla, N. S. Ca^{+2}/Mg^{+2} ATPase activities of heart sarcolemma, microsomes, and mitochondria. *J. Biochem.*, 82, 1731–1739 (1977).

Appelt, A. W.; Welty, J. D.; and Peterson, M. B. Changes in sarcolemmal and sarcoplasmic reticulum ATPase activities with age in the cardiomyopathic Syrian hamster. *J. Mol. Cell. Cardiol*, 8, 901–907 (1976).

Bajusz, E. Dystrophic calcification of myocardium as conditioning factor in genesis of congestive heart failure: An experimental study. *Am. Heart J.*, 78, 202–210 (1969).

Bajusz, E.; and Lossnitzer, K. A new disease model of chronic congestive heart failure: Studies on its pathogenesis. *Trans. N.Y. Acad. Sci.*, 30 (Sec. II), 939–948 (1968).

Bajusz, E.; Baker, J. R.; Nixon, C. W.; and Homburger, F. Spontaneous hereditary myocardial degeneration and congestive heart failure in the Syrian hamster. *Ann. N.Y. Acad. Sci.*, 156, 105–129 (1969).

Carafoli, E. Mitochondria in the contraction and relaxation of heart. In *Myocardial Biology*, Vol. 4, N. S. Dhalla, ed. University Park Press, Baltimore (1974), pp. 393–406.

Carafoli, E.; and Crompton, M. The regulation of intracellular calcium. In *Current Topics in Membranes and Transport*, F. Bronner and A. Kleinzeller, eds. Academic Press, New York (1978a), pp. 151–216.

Carafoli, E.; and Crompton, M. The regulation of intracellular calcium by mitochondria. *Ann. N.Y. Acad. Sci.*, 307, 269–284 (1978b).

Carafoli, E.; Tiozzo, R.; Crovetti, L. F.; and Kratzing, C. The release of calcium from heart mitochondria by sodium. *J. Mol. Cell. Cardiol.*, 6, 361–371 (1974).

Dhalla, N. S. Defects in calcium regulatory mechanisms in heart failure. In *Myocardial Biology*, Vol. 4, N. S. Dhalla, ed. University Park Press, Baltimore (1974), pp. 521–534.

Dhalla, N. S. Involvement of membrane systems in heart failure due to intracellular calcium overload and deficiency. *J. Mol. Cell. Cardiol.*, 8, 661–667 (1976).

Dhalla, N. S.; Tomlinson, C. W.; Singh, J. N.; Lee, S. L.; McNamara, D. B.; Harrow, J. A. C.; and Yates, J. C. Role of sarcolemmal changes in cardiac pathophysiology. In *The Sarcolemma*, P. Roy and N. S. Dhalla, eds. University Park Press, Baltimore (1976), pp. 377–394.

Dhalla, N. S.; Das, P. K.; and Sharma, G. P. Subcellular basis of cardiac failure. *J. Mol. Cell. Cardiol.*, 10, 363–385 (1978).

Dietrich, J.; and Diacono, J. Comparison between ouabain and taurine effects on isolated rat and guinea pig hearts in low calcium medium. *Life Sci.*, 10, 499–508 (1971).

Dolara, P.; Agresti, A.; Giotti, A.; and Pasquini, G. Effect of taurine on calcium kinetics of guinea-pig heart. *Eur. J. Pharm.*, 24, 352–358 (1973).

Guidotti, A.; Bandiani, G.; and Giotti, A. Potentiation by taurine of inotropic effect of strophanthin-K on guinea-pig isolated auricles. *Pharm. Res. Comm.*, 3, 29–38 (1971).

Hui, C. W.; Drummond, M.; and Drummond, G. I. Calcium accumulation and cyclic AMP stimulated phosphorylation in plasma membrane-enriched preparations of myocardium. *Arch. Biochem. Biophys.*, 173, 415–427 (1976).

Huxtable, R.; and Bressler, R. Taurine concentrations in congestive heart failure. *Science*, 184, 1187–1188 (1974).

Kagawa, Y.; and Racker, E. Partial resolution of the enzyme catalyzing oxidative phosphorylation. *J. Biol. Chem.*, 246, 5477–5490 (1971).

Katz, A. M. Congestive heart failure: Role of altered myocardial cellular control. *N. Engl. J. Med.*, 293, 1184–1191 (1975).

Katz, A. M.; Repke, D. I.; Upshaw, J. E.; and Polascik, M. A. Use of zonal centrifugation to fractionate fragmental sarcoplasmic reticulum, (Na^+-K^+)-activated ATPase and mitochondrial fragments. *Biochim. Biophys. Acta*, 205, 473–490 (1970).

Kramer, C. Y. Extension of multiple range tests to group means with unequal numbers of replication. *Biometrics*, 12, 307–310 (1956).

Kulakowski, E. C.; Maturo, J.; and Schaffer, S. W. The identification of taurine receptors from rat heart sarcolemma. *Biochem. Biophys. Res. Comm.*, 80, 936–941 (1978).

Lowry, O. H.; Rosebrough, N. J.; Farr, A. L.; and Randall, R. J. Protein measurement with the Folin phenol reagent. *J. Biol. Chem.*, 193, 265–275 (1951).

McBroom, M. J.; and Welty, J. D. Effects of taurine on heart calcium in the cardiomyopathic hamster. *J. Mol. Cell. Cardiol.*, 9, 853–858 (1977).

McNamara, D. B.; Sulakhe, P. V.; Singh, J. N.; and Dhalla, N. S. Properties of heart sarco-lemmal Na+-K+ ATPase. *J. Biochem.*, 75, 795–803 (1974).

Nadkarni, B. B.; Hunt, B.; and Heggtveit, H. A. Early ultrastructure and biochemical changes in myopathic hamster heart. In *Myocardiology*, Vol. 1, E. Bajusz and G. Rona, eds. University Park Press, Baltimore (1972), pp. 251–261.

Nayler, W. G. Significance of calcium ions in cardiac excitation and contraction. *Am. Heart J.*, 65, 404–411 (1963).

Olson, R. E. Myocardial metabolism in congestive heart failure. *J. Chronic Diseases*, 9, 442–464 (1959).

Peng, C. F.; Kane, J. J.; Murphy, M. L.; and Straub, K. D. Abnormal mitochondrial oxidative phosphorylation of ischemic myocardium reversed by Ca^{+2} chelating agents. *J. Mol. Cell. Cardiol.*, 9, 897–908 (1977).

Peterson, M. B.; Mead, R. J.; and Welty, J. D. Free amino acids in congestive heart failure. *J. Mol. Cell. Cardiol.*, 5, 139–147 (1973).

Racker, E.; Knowles, A. F.; and Eytan, E. Resolution and reconstitution of ion transport systems. *Ann. N.Y. Acad. Sci.*, 264, 17–33 (1975).

Read, W. O.; and Welty, J. D. Effect of taurine on epinephrine and digoxin induced irregularities of the dog heart. *J. Pharm. Exptl. Ther.*, 139, 283–289 (1963).

Reuter, H. Exchange of calcium ions in the mammalian myocardium: Mechanisms and physiological significance. *Circ. Res.*, 34, 599–605 (1974).

Reynafarje, B.; and Lehninger, A. L. High affinity and low affinity binding of calcium by rat liver mitochondria. *J. Biol. Chem.*, 244, 584–593 (1969).

St. Louis, P. J.; and Sulakhe, P. V. Adenosine triphosphate-dependent calcium binding and accumulation by guinea pig cardiac sarcolemma. *Can. J. Biochem.*, 54, 946–956 (1976).

Selwyn, H. J.; Dawson, A. P.; and Dunnett, S. J. Calcium transport in mitochondria. *FEBS Letters*, 10, 1–5 (1970).

Sonnenblick, E. H.; and Stam, A. C., Jr. Cardiac muscle: Activation and contraction. *Ann. Rev. Physiol.*, 31, 647–674 (1969).

Sperelakis, N.; and Schneider, J. A. A metabolic control mechanism for calcium ion influx that may protect the ventricular myocardial cell. *Am. J. Cardiol.*, 37, 1079–1085 (1976).

Sulakhe, P. V.; and Dhalla, N. A. Excitation-contraction coupling in heart: VII. Calcium accumulation in subcellular particles in congestive heart failure. *J. Clin. Invest.*, 50, 1019–1027 (1971a).

Sulakhe, P. V.; and Dhalla, N. S. VI. Demonstration of calcium activated ATPase in the dog heart sarcolemma. *Life Sci.*, 10, 185–191 (1971b).

Sulakhe, P. V.; Leung, N. L.; and St. Louis, P. J. Stimulation of calcium accumulation in cardiac sarcolemma by protein kinase. *Can. J. Biochem.*, 54, 438–445 (1976).

Taussky, H. H.; and Shorr, E. A microcolorimetric method for the determination of inorganic phosphorus. *J. Biol. Chem.*, 202, 675–685 (1953).

Tomlinson, C. W.; Yates, J. C.; and Dhalla, N. S. Relationship among changes in intracellular calcium stores, ultrastructure and contractility of myocardium. In *Myocardial Biology*, Vol. 4, N. S., Dhalla, ed. University Park Press, Baltimore (1974), pp. 331–345.

Welty, J. D.; and Read, W. O. Studies on some cardiac effects of taurine. *J. Pharm. Exptl. Ther.*, 144, 110–155 (1964).

Welty, J. D.; McBroom, M. J.; Appelt, A. W.; Peterson M. B.; and Read, W. O. Effect of taurine on heart and brain electrolyte imbalances. In *Taurine*, R. Huxtable and A. Barbeau, eds. Raven Press, New York (1976), pp. 155–163.

Wollenberger, A. The energy metabolism of the failing heart and the metabolic action of the cardiac glycosides. *J. Pharm. Exptl. Ther.*, 97, 311–352 (1949).

Copyright © 1981, Spectrum Publications, Inc.
The Effects of Taurine on Excitable Tissues

CHAPTER 23

Cardiovascular Actions of Taurine, γ-Aminobutyric Acid (GABA), and γ-Amino-β-hydroxybutyric Acid (GABOB) after Chemical Denervation

Ryuta Ito
Toshimitsu Uchiyama
Shuko Yoda
Noriaki Homma
Katsuo Furukawa

Studies of the mechanisms underlying the fall in blood pressure seen after administration of γ-aminobutyric acid (GABA) or γ-amino-β-hydroxybutyric acid (GABOB) showed that GABA or GABOB initially released catecholamines from sympathetic presynaptic sites. The catecholamines in turn elicited a release of histamine, causing a fall in blood pressure that was further modified by the ganglionic-stimulating and blocking actions of GABA (Ito et al., 1976, 1977a, b, 1978: johnson and O'Brien, 1976; Kawamoto, 1978). The GABA-induced hypotension was not blocked by picrotoxin. In our previous report describing the interactions of taurine with autonomic drugs (Ito et al., 1970b), taurine was shown to: (1) potentiate or inhibit epinephrine-induced blood pressure changes, (2) reduce epinephrine-induced arrhythmias, and (3) decrease cardiac conductivity. In the work reported here, the cardiovascular actions of taurine, GABA, and GABOB were studied after chemical adrenergic denervation in order to evaluate the sympathetic pre- and postsynaptic actions of these agents.

METHODS

Blood Pressure and Heart Rate (Rats)

Femoral arterial pressure was measured with a membrane manometer and transducer in male Wistar rats (200–350 g) anesthetized with 1 g/kg urethane. Heart rate was recorded from lead II of the EKG. Chemical adrenergic denervation was produced in rats according to the method of Johnson and O'Brien (1976) by i.p. administration of 40 mg/kg guanethidine once daily for 5 weeks or by i.v. administration of a single dose of 100 mg/kg 6-hydroxydopamine (6-OHDA) 2 days before the experiment. Changes in blood pressure and heart rate of chemically denervated and nondenervated control rats following i.v. administration of taurine, GABA, or GABOB were determined and compared with those produced by i.v. administration of either epinephrine, a postsynaptically active agent, or tyramine, a presynaptic releasing agent.

Vagus-amine test (Guinea Pigs)

Our version of the vagus-amine test conducted in guinea pigs (Ito, et al., 1970a) is a modified form of the original method developed by Roberts and Baer (1961). Guinea pigs whose vagus nerves are electrically stimulated at a rate that slows the heart just short of the point at which ectopic beats (arrhythmias) result are very sensitive test subjects for measuring the release of endogenous catecholamines. The released catecholamines produce arrhythmias, and the capacity of taurine, GABA, or GABOB to facilitate or block such release may be evaluated by administering them before an intravenous challenge with tyramine or 1-propanolol (presynaptic releasing agents) and epinephrine or isoproterenol (postsynaptically active agents), followed by vagal stimulation as described in the figure legends for each experiment (see Figs. 23.7–23.9).

RESULTS

Effects on Blood Pressure (Rats)

Chemical denervation by chronic guanethidine or 6-hydroxydopamine (6-

OHDA) administration was confirmed by comparison of blood pressures in the treated groups of rats with those of a control (nondenervated) group. In the control group, blood pressure was 92.4 ± 2.7 mm Hg (mean $\pm$ SE); in the 6-OHDA-treated group, 83.4 ± 3.3 mm Hg ($p < 0.10$); and in the guanethidine-treated group, 81.7 ± 1.7 mm Hg ($p < 0.05$). In the chemically denervated groups, it was also noted that while the epinephrine-induced rise in blood pressure was potentiated ($p < 0.005$), the subsequent reduction in blood pressure was not as great as in the controls ($p < 0.10$). Tyramine responses were also largely diminished in the chemically denervated animals.

The effects of taurine (Figs. 23.1 and 23.2) were studied both before and after chemical denervation (CD). Before CD, a 100 mg/kg i.v. dosage of taurine induced, in many cases, a decrease in blood pressure followed by a gradual increase. Both the fall and rise in blood pressure were roughly proportional to the dose of taurine given (between 1 and 100 mg/kg). Following CD, the fall in blood pressure seen in the nondenervated control rats disappeared, but the rise in blood pressure increased. A dose-response relationship was again observed between the dosage of taurine given and the increase in bood pressure.

As with taurine, the effects of GABA and GABOB were also studied both before and after CD (Fig. 23.3). Prior to CD, an intravenous administration of 1–10 mg/kg of GABA or GABOB reduced blood pressure; this was sometimes followed by a slight rise. At 100 mg/kg, for example, GABA induced a slight increase in blood pressure. Following CD, however, GABA and GABOB administration only produced a slight rise in blood pressure.

Effects on Heart Rate (Rats)

The effects of CD on heart rate were illustrated by comparison of the denervated and nondenervated control groups (Fig. 23.4). In the nondenervated control groups, the heart rate was 359.0 ± 11.3 beats/min (mean $\pm$ SE), while in the CD group treated with guanethidine, it was 374.6 ± 2.8 beats/min, and in the CD group treated with 6-OHDA, it was 386 ± 11.1 beats/min.

In the CD groups, the epinephrine-induced tachycardia disappeared, and bradycardia was markedly intensified ($p < 0.005$). Tyramine-induced tachycardia was also markedly reduced ($p < 0.005$).

When the effects of taurine before and after CD were studied (Figs. 23.4 and 23.5), it was noted that before CD, taurine induced a tachycardia followed by a slight bradycardia, a dose-response relationship existing in both. After CD, the tachycardia disappeared, and the bradycardia became more pronounced and was dose dependent.

GABA and GABOB effects were also examined (Fig. 23.6). Before CD,

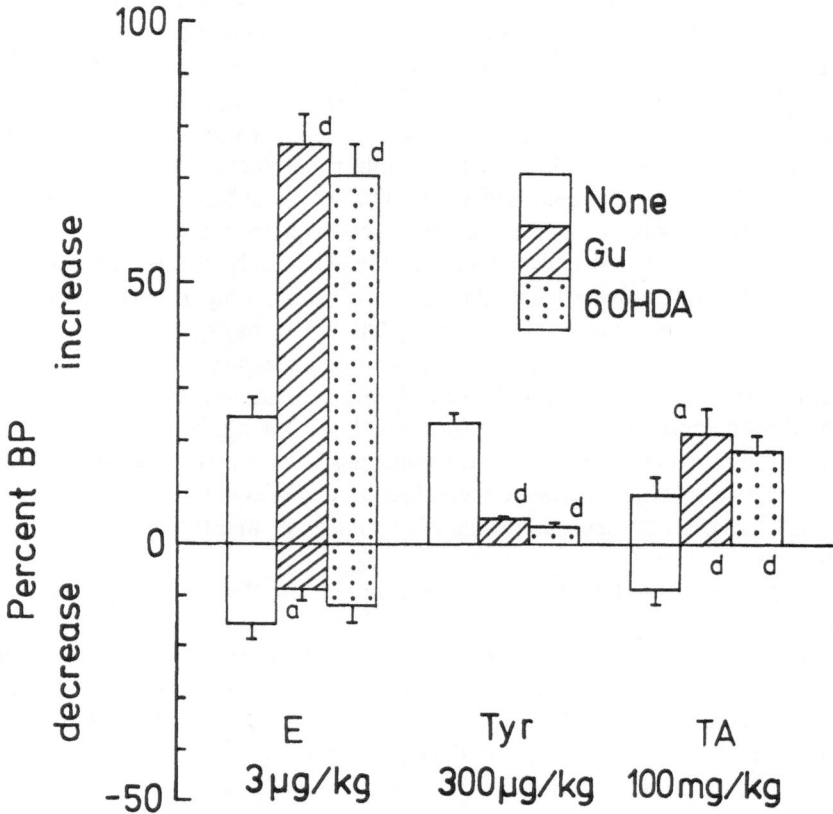

Fig. 23.1. Effect of taurine (TA) on blood pressure after chemical denervation in rats. Chemical denervation was induced in rats with either i.p. doses of 40 mg/kg guanethidine once daily for 5 weeks or with a single i.v. dose of 6-hydroxydopamine (6-OHDA). Blood pressure changes (as a percentage of that seen before i.v. injection) were measured after i.v. administration of the indicated doses of epinephrine (E), tyramine (Tyr), and taurine (TA). Before denervation, biphasic changes in blood pressure were seen after both E and TA, a rise followed by a fall in the case of E, and usually a fall followed by a rise in the case of TA; Tyr produced only a rise. After denervation, E produced a much larger rise and a smaller fall, Tyr produced a smaller rise than before denervation, and TA produced only a rise in blood pressure. Significance of differences from control (nondenervated) rats: a($p < 0.10$); d($p < 0.005$). Bars showing variation of the mean in Figs. 23.1–23.9 represent the standard error.

these drugs elicited a slight slowing, ranging from 3% to 20%. After CD, no obvious changes were noted.

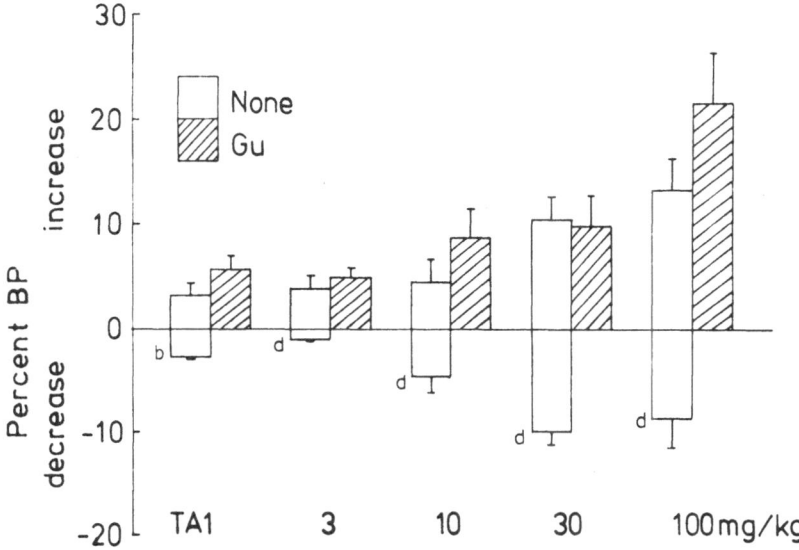

Fig. 23.2. Effect of various doses of taurine on blood pressure after guanethidine-induced denervation in rats. The indicated doses of taurine were injected i.v. into control (nondenervated) rats or into rats denervated by i.p. doses of 40 mg/kg guanethidine (Gu) once daily for 5 weeks. After denervation, no fall in blood pressure was seen, only a rise. Significance of differences from Gu-treated group: b($p < 0.05$); d($p < 0.005$).

Vagus-amine Test (Guinea Pigs)

Our modification in guinea pigs of the vagus-amine test (Ito et al., 1970b) originally developed by Roberts and Baer (1961) provides a very sensitive test system for evaluating the release of endogenous catecholamines. The vagus is electrically stimulated at a rate that slows the heart just short of the point at which ectopic beats (arrhythmia) are produced. In this system, i.v. injection of epinephrine or the release of catecholamines from sympathetic presynaptic sites by tyramine will each produce arrhythmia. The effectiveness of taurine, GABA, or GABOB in inhibiting the release of endogenous catecholamines was tested in guinea pigs in three ways: (1) Taurine, GABA, or GABOB was injected i.v. rapidly in various doses 10 sec before the vagus was stimulated for 20 sec at a rate previously determined to be just short of producing ectopic beats; (2) taurine, GABA, or GABOB was injected i.v. slowly 10 min before i.v. injections of tyramine or epinephrine; and (3) taurine (100 mg/kg) was injected daily i.p. for 10 days prior to i.v. injection of presynap-

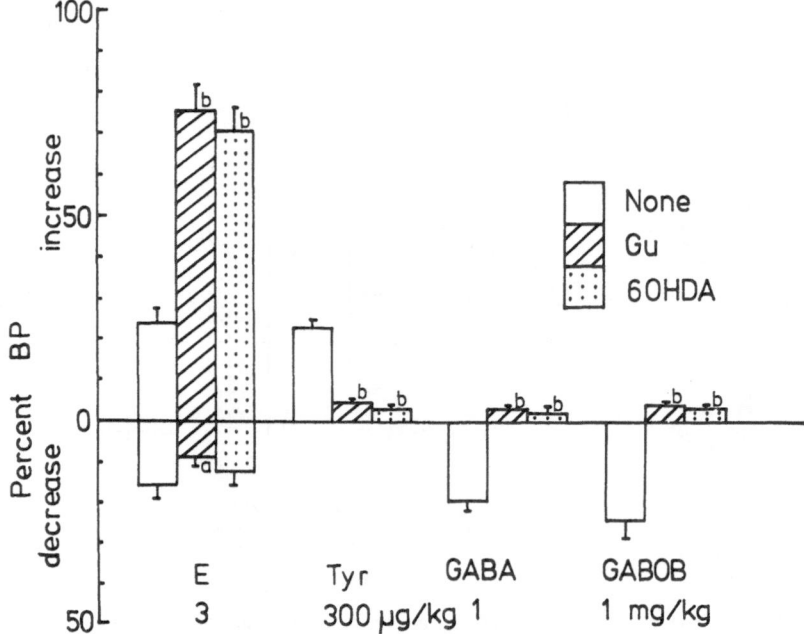

Fig. 23.3. Effect of GABA and γ-amino-β-hydroxybutyric acid (GABOB) on blood pressure after chemical denervation in rats. Chemical denervation was induced in rats with either guanethidine (Gu) 40 mg/kg i.p., once daily for 5 weeks, or with a single i.v. dose of 100 mg/kg 6-hydroxydopamine (6-OHDA). Blood pressure changes were measured after i.v. administration of the indicated doses of epinephrine (E), tyramine (Tyr), GABA, or GABOB. Before denervation, GABA and GABOB induced a fall in blood pressure; after denervation, only a small rise was noted. Responses to E and Tyr are described in the legend to Fig. 23.1. Significance of differences from control (nondenervated) rats: a($p < 0.10$); b($p < 0.05$).

tic-releasing agents (tyramine and 1-propanolol) or of postsynaptically active agents (epinephrine and isoproterenol).

Rapid injection of taurine into guinea pigs seemed to induce arrhythmias (or catecholamine release) in 20% to 50% of the animals, although no dose-response relationship was observed (Fig. 23.7); rapid injection of a 100 mg/kg dose of taurine, however, did not seem to produce arrhythmia. It is of interest that following slow i.v. injection of taurine, the guinea pig became sensitized to epinephrine (Fig. 23.8).

As with taurine, rapid injection of GABA and GABOB induced arrhythmias or catecholamine release in 20% to 50% of the animals. No dose-response relationship was observed with these drugs either (Fig. 23.7).

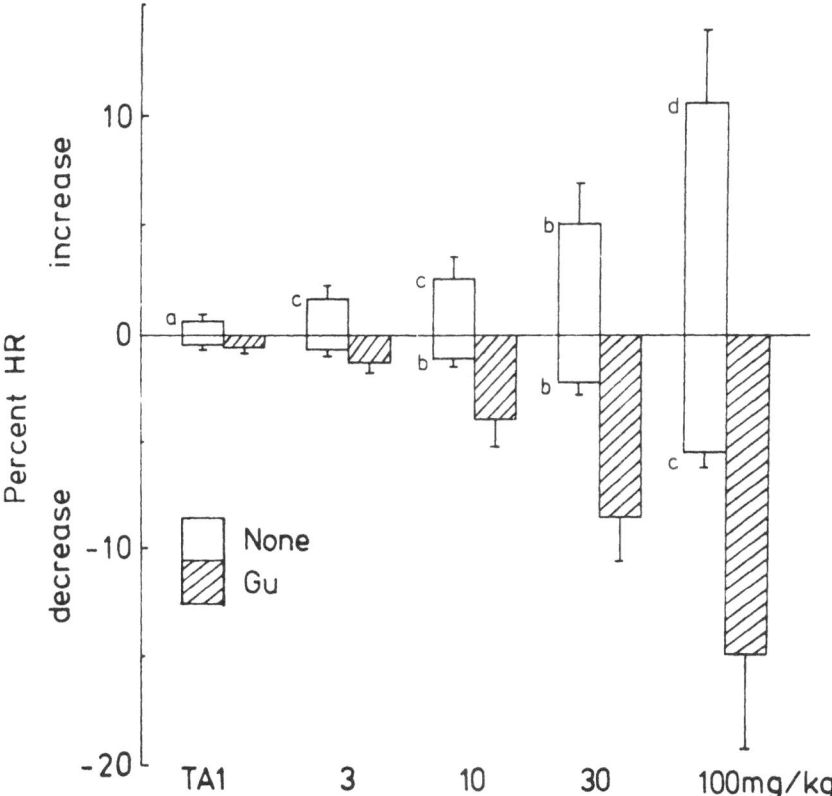

Fig. 23.4. Effect of taurine (TA) on heart rate after chemical denervation in rats. Changes in heart rate as a percentage of the heart rate seen before the i.v. injection were measured in the same rats as those shown in Fig. 23.1. Before denervation, i.v. administration of epinephrine (E) first decreased heart rates, then increased them, in a direction opposite to that of the blood pressure; in the case of TA, heart rates first increased, then decreased, but tyramine (Tyr) increased heart rates simultaneously with its inducing a rise in blood pressure. After denervation, E and TA produced only bradycardia. Significance of difference from control (nondenervated) rats: b(p < 0.05); d(p < 0.005).

Slow pretreatment with a single i.v. dose of taurine was found to completely block tyramine-induced catecholamine release. Thus, pretreatment with taurine 10 min before epinephrine injection reduced sensitivity to epinephrine in some animals while rendering other animals more sensitive (Fig. 23.8). Tyramine-induced CA release was also completely blocked by a single pretreatment with GABA or GABOB.

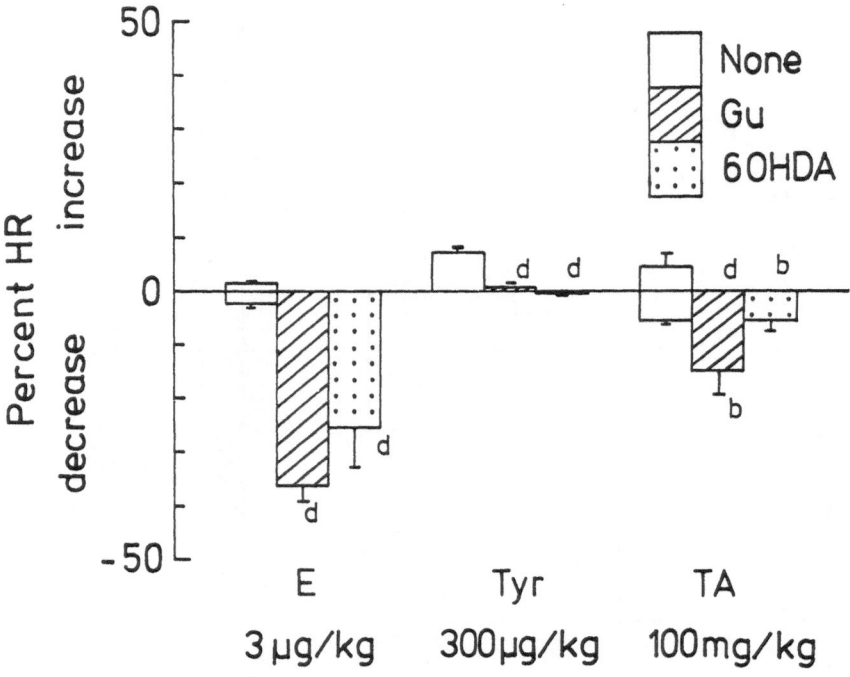

Fig. 23.5. Effect of increasing the doses of taurine on heart rate after guanethidine-induced denervation in rats. The changes in heart rate as a percentage of the heart rate before the i.v. injection of taurine were measured in the same rats as those shown in Fig. 23.2. In chemically denervated rats, only a bradycardia corresponding to the increase in blood pressure was seen, in contrast to the biphasic changes seen in the nondenervated control rats. Significance of differences from guanethidine-treated rats: a($p < 0.10$); b($p < 0.05$); c($p < 0.01$); d($p < 0.005$).

Chronic intraperitoneal pretreatment with 100 mg/kg of taurine (but no i.v. pretreatment with taurine just before the experiment) caused only reduced tyramine or 1-propranolol-induced CA release, and left sensitivities to epinephrine or isoproterenol unchanged (Fig. 23.9).

DISCUSSION

Differences in Actions between Taurine and GABA or GABOB

The actions of taurine on blood pressure and heart rate were different from those of GABA or GABOB both before and after chemical denervation (Table 23.1). Blood pressure in nondenervated rats generally decreased immedi-

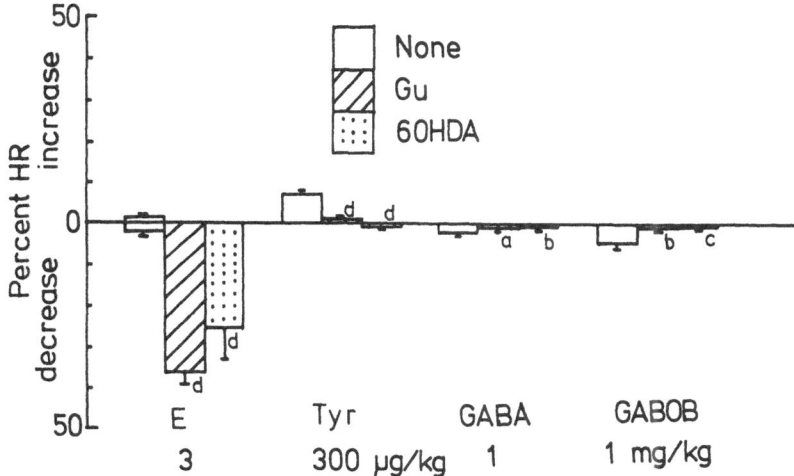

Fig. 23.6. Effect of GABA and γ-amino-β-hydroxybutyric acid (GABOB) on heart rate after chemical denervation in rats. Changes in heart rate were measured as a percentage of the heart rate before the i.v. injections of GABA or GABOB in the same rats as those shown in Fig. 23.3. In nondenervated control rats, GABA and GABOB each produced a bradycardia that corresponded to the fall in blood pressure seen in the same rats shown in Fig. 23.3. After denervation, a slight bradycardia was noted, corresponding to the slight rise in blood pressure seen in these rats (Fig. 23.3). Significance of differences from control rats: a($p < 0.10$); b($p < 0.05$); c($p < 0.01$); d($p < 0.005$).

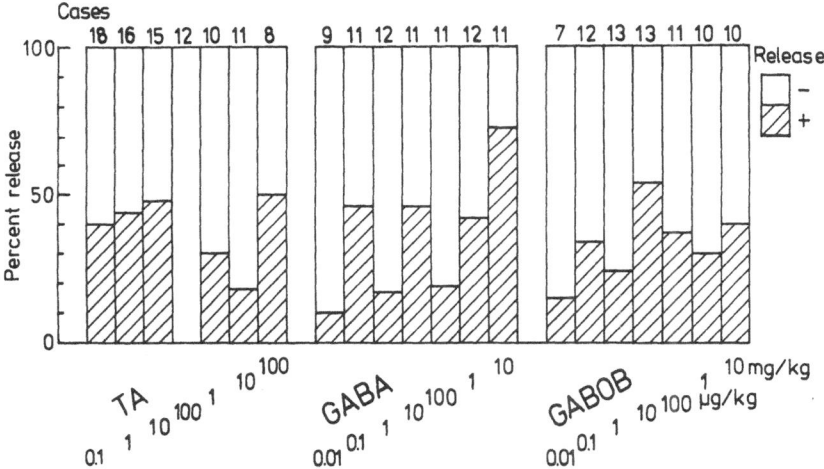

Fig. 23.7. Effects of various doses of rapidly injected taurine (TA), GABA, or γ-amino-β-hydroxybutyric acid (GABOB) on induction of ectopic beats in guinea pigs following controlled stimulation of the vagus nerve. Ten seconds after rapid i.v. injection of a dose of taurine (TA), GABA, or GABOB into guinea pigs, the vagus is electrically stimulated for 20 sec at a rate that had been previously established to maintain vagal slowing just short of producing ectopic beats in the EKG.

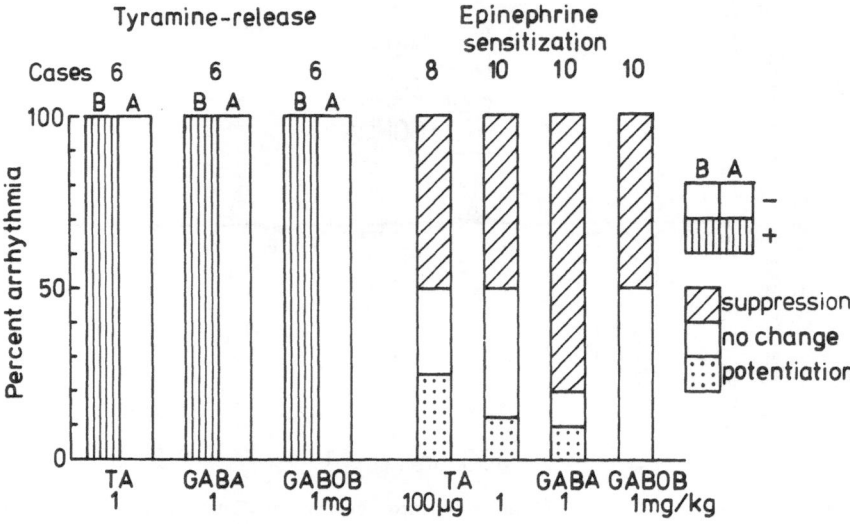

Fig. 23.8. Effect of a single dose of taurine (TA) or GABA (or GABOB) injected slowly i.v. prior to i.v. tyramine or epinephrine on induction of ectopic beats in guinea pigs. *Left:* After (*a*) slow i.v. injection of taurine (TA), GABA, or GABOB, a rapid i.v. injection of 0.1 μg/kg tyramine produced no arrhythmia (ectopic beats) in guinea pigs under vagal stimulation, as compared to 100% incidence of arrhythmia before (*b*) TA(1 or 10 mg/kg) GABA, or GABOB. Earlier evidence showed that the arrhythmias induced by i.v. tyramine injection were due to catecholamine release (Ito et al., 1978b). *Right:* The threshold dose of i.v. injected epinephrine needed to elicit arrhythmias in guinea pigs under vagal stimulation was first determined in preliminary experiments. After slow i.v. administration of the indicated doses of TA, GABA, or GABOB, the threshold dose of epinephrine was again injected to determine its ability to produce arrhythmias.

ately after taurine administration, then gradually increased above control levels; while after chemical denervation, no reduction occurred, but the rise became more pronounced. On the other hand, GABA induced only a reduction of blood pressure in nondenervated animals. This decrease disappeared and a rise was sometimes observed after chemical denervation. With respect to heart rate, taurine first induced an increase followed by a decrease in non-denervated animals, but after denervation, the increase disappeared and the decrease became more pronounced. A possible explanation for these changes may be related to the depletion of catecholamine that occurs in the presynaptic sympathetic nerve endings after denervation, since the fall in blood pressure and the tachycardia (the β-like actions) disappeared after denervation.

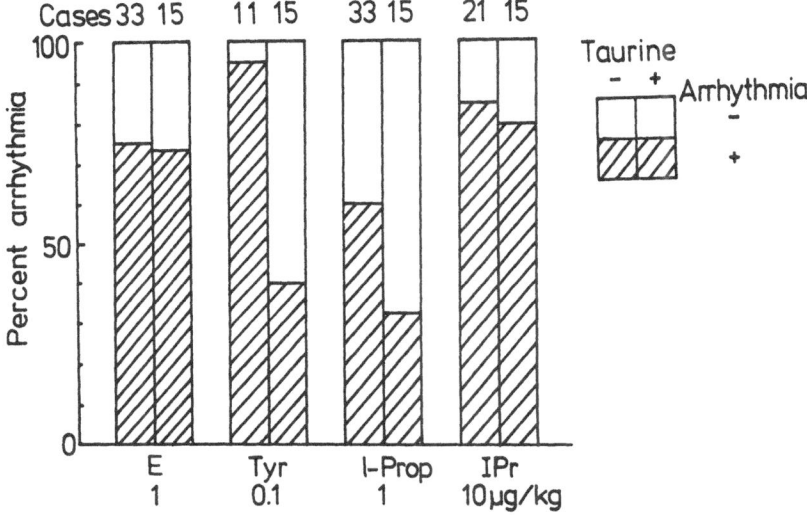

Fig. 23.9. Effect of pretreatment with taurine on arrhythmias elicited in guinea pigs by epinephrine (E), tyramine (Tyr), 1-propanolol (1-Prop), or isoproterenol (1Pr). Repeated daily i.p. doses of 100 mg/kg taurine were administered to guinea pigs for 10 days as pretreatment prior to acute i.v. injection of the four agents that elicited arrhythmia; control animals received no taurine and, taurine was not given i.v. to these guinea pigs, as pretreatment. Taurine reduced arrhythmias induced by the presynaptically acting agents (tyramine and 1-propanolol) but did not prevent that induced by the postsynaptically acting agents epinephrine and isoproterenol.

Table 23.1.   Difference between taurine and GABA or GABOB before and after chemical denervation

		TA		GABA or GABOB	
		Before	After	Before	After
Blood pressure[a,b]	α	c	cc	g	f
	β	d	e		e
Heart rate	c	c	e	e	e
	d	d	dd	d	e

[a]α-like action, blood pressure rise.
[b]β-like action, blood pressure fall.
[c]Increase.
[cc]Increased largely.
[d]Decrease.

[dd]Decreased largely.
[e]No change.
[f]Slight change.
[g]Slight or no change.

Presynaptic Actions

We have previously reported (Ito et al., 1976, 1977a,b, 1978a,b; Kamei, 1978; Kawamoto, 1978) that GABA and GABOB initially induced a release of catecholamines and then subsequently blocked the release. Dual opposing actions of taurine such as sensitization to epinephrine and membrane stabilization have been reported previously (Ito et al., 1970b).

In the guinea pig vagus-amine test, taurine, GABA, and GABOB each released catecholamine when used in very low doses, but no dose-response relationship was observed. It was found that catecholamine release decreased as the doses of taurine and GABA were increased and suppression of the changes became more pronounced. A possible explanation of these observations is that lower doses of these agents induced a nonspecific release of catecholamines but higher doses blocked the release, the block becoming stronger as the dose increased. Evidence for this blocking action may be seen in the suppression of the tyramine-induced arrhythmias (or catecholamine release). Blocking by GABA was complete but that by taurine was not. A balance between release and block seems to characterize the action of these agents.

A greater activity of taurine at presynaptic rather than postsynaptic sites after chronic pretreatment with taurine is suggested by its actions on epinephrine and isoproterenol-induced arrhythmias.

Postsynaptic Actions

The fact that taurine increases blood pressure suggests that taurine may act at postsynaptic sites in the vascular bed. Taurine may also increase Ca^{2+} release and binding as described for the heart sarcolemma membrane by Schaffer et al. (this volume, Chap. 20), Khatter et al. (this volume, Chap. 21), and Welty and Welty (this volume, Chap. 22). In accord with this concept is the increased sensitivity to epinephrine seen after denervation, which in turn would increase the response of receptors and the subcellular Ca^{2+} mechanism. A rise in blood pressure induced by GABA or taurine after denervation with guanethidine, 6-hydroxydopamine or reserpine is probably due to the same mechanism and requires further study.

Dual opposing actions, such as facilitating and blocking of catecholamine release by taurine, have been reported previously (Akera et al., 1976; Chubb and Huxtable, 1978; Dolara et al., 1978; Huxtable, 1976; Ito et al., 1970b). If the balance of positive and negative effects between the facilitating and blocking actions of an agent on catecholamine release were equal, there might be no observable change in arrhythmia. In other experiments, catecholamine release was also demonstrated after taurine administration (Fig. 23.8). This might also be due to the nonuniformity of sympathetic discharge (Gabriel et al., 1978; Lathers et al., 1978). Reduction in the tyramine-induced release of catecholamines after repeated pretreatment might occur if: (1) greater activity of taurine occurred at presynaptic sites than at postsynaptic sites, resulting in a blocking of catecholamine release; (2) taurine blocked the formation of catecholamine precursors, resulting in reduced catecholamine stores; (3) taurine possessed a cocaine-like action. Thus, early potentiation by taurine may be due to a block of catecholamine reuptake, and repeated treatment may reduce catecholamine stores at presynaptic sites or the sensitivity of the postsynaptic sites to epinephrine or isoproterenol. Thus, when catecholamine release occurs from presynaptic sites and reuptake is blocked, the increased catecholamine concentration at the postsynaptic sites may antagonize the blocking action of taurine, in addition to its possibly increasing Ca^{2+} binding.

The observations that taurine, GABA, or GABOB did not decrease the various pharmacological effects of epinephrine or isoproterenol indicate that postsynaptic stabilization by these amino acids was absent or rather weak. Differences between the postsynaptic actions of taurine and GABA became evident after denervation; that is, taurine seemed to exert a stronger Ca^{2+} binding action than GABA. Since our previous paper (Ito et al., 1970b) showed that the beta blocker sotalol also increased the bradycardia induced by taurine, the increase in taurine-induced bradycardia seen after denervation was probably due to absence of the released catecholamines in the denervated animals and their stimulation of the heart rate. Such changes by GABA were weaker. The above results indicate that taurine stimulates postsynaptic sites in the sympathetic system of the vascular bed. Interaction between taurine and norepinephrine in the pineal glands has been recently proposed by Wheler and Klein (this volume, Chap. 10), suggesting the possibility that this interaction may also occur in the cardiovascular system.

CONCLUSIONS

The present studies reveal the following with respect to the effects of chemical adrenergic denervation:

1. At presynaptic sites, low doses of taurine, GABA, and GABOB all initiate some nonspecific release of catecholamines, but higher doses block catecholamine release.

2. At postsynaptic sites, taurine partially constricts blood vessels. Taurine increases heart rate initially through a release of catecholamines but later blocks the catecholamine action. The postsynaptic blocking action of GABA on the conductive system is weaker than that of taurine.

ACKNOWLEDGMENT

Thanks are due to Taisho Pharmaceutical Company for their financial support and for their supply of taurine.

REFERENCES

Akera, T.; Ku, D.; and Brody, M. Alteration of ion movements as a mechanism of drug induced arrhythmias and inotropic responses. In *Taurine,* ed. by R. Huxtable and A. Barbeau. Raven Press, New York, (1976) pp. 121–134.

Chubb, J.; and Huxtable, R. J. Transport and biosynthesis of taurine in the stressed heart. In *Taurine and Neurological Disorders,* ed. by A. Barbeau and R. Huxtable. Raven Press, New York, (1978) pp. 161–178.

Dolara, P.; Ledda, F.; Mugelli, A.; Mantelli, L.; Zilletti, L.; Franconi, F.; and Giotti, A. Effect of taurine on calcium, inotropism, and electrical activity of the heart. In *Taurine and Neurological Disorders,* ed. by A. Barbeau and R. Huxtable. Raven Press, New York, (1978) pp. 151–159.

Gabriel, L. L.; Kelliher, G. J.; Roberts, J.; and Overton, W. R. Involvement of adrenergic nervous influences in ouabain-induced nonuniformity of ventricular repolarization. *J. Pharmacol. Exp. Ther.,* 207, 1–7 (1978).

Huxtable, R. J. Metabolism and function of taurine in the heart. In *Taurine,* ed. by R. Huxtable and A. Barbeau, Raven Press, New York, (1976) pp. 99–119.

Ito, R.; Shimura, M.; Usui, A.; Tsunoda, Y.; Makabe, T.; Kanno, Y.; Sekine, O.; Kato, Y.; and Honda, M. The application of Vagus-amine test, a method for evaluation of arrhythmogenic or antiarrhythmic activities in situ to rabbits, guinea-pigs and rats. *J. Med. Soc. Toho,* 17, 21–25 (1970a).

Ito, R.; Ohmoto, M.; Chang, H. S.; Honda, M.; Shimura, M.; and Ozeki, M. Influence of taurine and isethionic acid on cardiovascular action of autonomic drugs and atrioventricular node in situ. *J. Med. Soc. Toho,* 17, 205–212 (1970b).

Ito, R.; Uchiyama, T.; Kawamoto, S.; and Kamei, K. Interaction between autonomic drugs, autacoids, and GABA or α-amino-β-hydroxy-butyric acid. *Japan. J. Pharmacol.,* 26 (Suppl), 147 pp. (1976).

Ito, R.; Uchiyama, T.; Ichikawa, M.; Hamatani, J.; and Furukawa, K. Analysis of mechanism of GABA- and GABOB-induced blood-pressure fall (2). *Japan. J. Pharmacol.,* 27 (Suppl.), 101 pp (1977a).

Ito, R.; Uchiyama, T.; and Ichikawa, M. Mechanism of blood-pressure fall by exogenous GABA and GABOB. In *Proceedings of 18th International Congress of Neurovegetative Research (Tokyo),* pp. 134–135 (1977b).

Ito, R.; Uchiyama, T.; Ichikawa, M.; Furukawa, K.; Saito, K.; and Hamatani, J. Analysis of mechanism of GABA- and GABOB-induced blood-pressure fall (3). *Japan. J. Pharmacol.,* 28 (Suppl.), 106 pp. (1978a).

Ito, R.; Echiyama, T.; Ichikawa, M.; Saito, K.; Hamatani, J.; and Henomatsu, H. Blood-pressure fall mechanism of exogenous GABA and GABOB. *7th International Congress of Pharmacology,* Abstract 2067, p. 646 (1978b).

Johnson, E. M., Jr.; and O'Brien, F. Evaluation of the permanent sympathectomy produced by the administration of guanethidine to adult rats. *J. Pharmacol. Exp. Therap.,* 196, 53–61 (1976).

Kamei, K. Interactions between GABOB induced blood-pressure fall and autonomic drugs and autacoids in rabbits. *J. Med. Soc. Toho,* 25, 64–74 (1978).

Kawamoto, S. Interactions between GABA induced blood-pressure fall and autonomic drugs and autacoids in rabbits. *J. Med. Soc. Toho,* 25, 44–63 (1978).

Khatter, J. C.; Soni, P. L.; Hoeschen, R. J.; Alto, L. E.; and Dhalla, N. S. Subcellular effects of taurine on guinea pig heart. This volume, Chap. 21.

Lathers, C. M.; Kelliher, G. J.; Roberts, J.; and Beasley, A. B. Nonuniform cardiac sympathetic nerve discharge: Mechanism for coronary occlusion and digitalis-induced arrhythmia. *Circulation,* 57, 1058–1065 (1978).

Roberts, J.; and Baer, R. The method for the evaluation of subatrial rhythmic function in the heart of the intact animals. *J. Pharmacol. Exp. Therap.,* 129, 36–41 (1961).

Schaffer, S. W.; Chovan, J.; Kramer, J.; and Kulakowski, E. The role of taurine receptors in the heart. This volume, Chap. 20.

Welty, J. D.; and Welty, M. C. Effects of taurine on subcellular dynamics in the normal and cardiomyopathic hamster heart. This volume, Chap. 22.

Wheler, G. H. T.; and Klein, D. C. Function and regulation of taurine in the pineal gland. This volume, Chap. 10.

Discussion

DR. JOHN STURMAN (Institute for Basic Research in Mental Retardation, New York): Taurine-deprived cats lose about 90% of their myocardial taurine content without any apparent adverse effect to the heart. However, attempts to further reduce the taurine levels have failed, raising the possibility that the remaining taurine is physiologically important. Do you feel that this remaining taurine pool could be further reduced by treatment with taurocyamine?

DR. RYAN HUXTABLE (University of Arizona): If the remaining taurine is maintained by equilibrium with serum, a transport inhibitor could conceivably cause a further reduction in tissue taurine content. Concerning the role of taurine in the heart, we must remember that the heart is an autonomic organ. It can beat and pump blood in the absence of any outside regulation. It's very likely that taurine serves to regulate myocardial contraction without being directly involved in the contraction process itself.

DR. JAMES KOCSIS (Jefferson Medical School): I would like to ask Dr. Huxtable a question. What method did you use to measure taurocyamine tissue levels?

DR. RYAN HUXTABLE (University of Arizona): The analytical method we used is a very old one, involving the use of biacetyl.

DR. PAULA GOLDBERG (Medical College of Pennsylvania): I have two questions for Dr. Huxtable. First, did you determine the taurine content of the heart prior to your uptake studies to see if the content influenced the saturation kinetics? And second, is there any change in functional activity of the heart associated with taurine uptake?

DR. RYAN HUXTABLE (University of Arizona): Regarding the first question, taurine uptake into the heart is linear for at least 40 minutes. Although there was a slight increase in the total taurine content of the heart, it was very low compared to the total amount of tissue taurine. Concerning your second question, we have not carried out many physiological measurements; however, it was obvious that there was a stimulation of contractility upon exposure to taurine.

DR. HEITAROH IWATA (Osaka University): Dr. Huxtable, did you examine if the effect of isoproterenol or reserpine on taurine uptake is altered by alpha-blockers?

DR. RYAN HUXTABLE (University of Arizona): As far as I know, there are no alpha-receptors on myocytes. Therefore, I would not expect alpha-blockers to have any effect.

DR. HEITAROH IWATA (Osaka University): We found that alpha-blockers inhibited the increase in taurine influx mediated by starvation, reserpine, or glucocorticoids.

DR. RYAN HUXTABLE (University of Arizona): How are they doing it?

DR. HEITAROH IWATA (Osaka University): I have no idea. I would like to ask your opinion on the mode of reserpine action.

DR. RYAN HUXTABLE (University of Arizona): We used your protocol for those experiments; namely, the hearts were isolated from the animals 4 hours after being treated with reserpine. Since the heart is probably depleted of catecholamine stores after reserpine treatment, I believe the mechanism for increased taurine uptake in these animals is exposure to the catecholamines found in the blood.

DR. HEITAROH IWATA (Osaka University): Are repeated injections of isoproteronol required to mediate an accumulation of taurine in the heart?

DR. RYAN HUXTABLE (University of Arizona): No, there is some accumulation after a single injection either *in vivo* or *in vitro*.

DR. BARRY LOMBARDINI (Texas Tech University): Dr. Huxtable, the stimulation of taurine influx by beta-adrenergic agents *in vitro* was very large, but in animals chronically exposed to isoproterenol there did not appear to be a large increase in heart taurine levels. Thus, how important is this adrenergic effect *in vivo*?

DR. RYAN HUXTABLE (University of Arizona): If animals are treated with isoproterenol for a period of 10 days, the total taurine content of the heart rises substantially and then falls when isoproterenol stimulation is removed. This increase results from stimulation of taurine influx rather than its biosynthesis.

DR. WALTER LOVENBERG (National Institutes of Health): The relationship between taurine and the adrenergic system seems to differ in isolated heart cells and the pineal gland. In this regard, have the effects of taurine on cyclic-AMP accumulation by the heart cell been examined?

DR. JOSEPH BAHL (University of Arizona): We found taurine to slightly decrease cyclic-AMP levels, but I don't believe that the change was statistically significant.

DR. STEPHEN SCHAFFER (Hahnemann Medical College): We reported last year that perfusion of rat hearts with 10 mM of taurine led to a transient, twofold decrease in tissue cyclic-AMP content.

DR. STEVEN BASKIN (Medical College of Pennsylvania): We also examined this question and found that taurine seemed to stimulate retinal phosphodiesterase. However, we are not certain if our results are due to a variation in calcium.

DR. WALTER LOVENBERG (National Institutes of Health): Returning to my initial comment, dibutyryl cyclic-AMP appears to stimulate taurine uptake in the heart while stimulating its release from the pineal gland. In light of this apparent conflict, is it possible that dibutyryl cyclic-AMP merely enhances exchange of taurine across the cell rather than a net uptake or release?

DR. DAVID KLEIN (National Institutes of Health): I don't know. It is possible that taurine release is merely a flux phenomenon; however, we see approximately 15% of the intracellular taurine content released under normal conditions and an additional 15% when we add isoproterenol. This argues against a mechanism involving just exchange. It is possible that both the heart and the pineal gland contain a transport system which differs only in its orientation in the membrane. In the pineal gland it may be facing toward the outside of the cell and in the heart it may be facing inward.

DR. SUSAN SCHMIDT (Harvard University): When incubating retinas with normal buffer, there is a constant, slow rate of taurine efflux, which I believe is due to the disruption of some photoreceptors. The amount lost to the incubation medium is approximately 0.5% of the taurine in the retina per minute.

DR. WALTER LOVENBERG (National Institutes of Health): Is that efflux process stimulated by cyclic nucleotides?

DR. SUSAN SCHMIDT (Harvard University): It isn't stimulated, but rather it is lost as a result of cell disruption. I believe that this may be happening in other systems as well.

DR. RYAN HUXTABLE (University of Arizona): We have examined this question in the perfused rat heart preloaded with radioactive taurine. In this preparation, isoproterenol failed to stimulate the release of radioactivity. However, since the radioactivity may not have been equilibrated with all of the taurine pools, one cannot unequivocally conclude that taurine release was not increased.

DR. STEVEN BASKIN (Medical College of Pennsylvania): Dr. Klein, do you have any information on the rate of taurine release from the pineal gland?

DR. DAVID KLEIN (National Institutes of Health): Benson and Grosso found that the amount of taurine stays relatively constant during most of the day but decreases approximately 20% at dusk. The decrease could result from either inhibition of uptake or adrenergically induced release of taurine. Unfortunately, it is not known whether taurine is released first or if norepinephrine release occurs initially. Nevertheless, it supports the possibility that taurine is released, and this release can alter physiological levels of taurine.

DR. STEVEN BASKIN (Medical College of Pennsylvania): Dr. Dhalla found no effect of taurine on Na^+-K^+ ATPase activity while Drs. Voaden and Welty saw an effect. We have evidence that taurine may only affect Na^+-K^+ ATPase activity when calcium is present in the medium. Drs. Dhalla and Welty, do you feel that taurine could stimulate the enzyme by affecting its calcium-binding site?

DR. NARANJAN DHALLA (University of Manitoba): I think that calcium could account for the different observations.

DR. JOSEPH WELTY (University of South Dakota): We have measured the Na^+-K^+ ATPase activity of whole brain homogenates under various conditions. In the presence of taurine, the ATPase activity is normally depressed. However, after addition of EGTA, taurine appears to stimulate enzyme activity. This suggests that the taurine response is dependent upon the calcium concentration of the medium.

DR. MARY VOADEN (London University): We considered the possibility that taurine might interact with calcium and thereby increase the Na^+-K^+ ATPase activity. Does taurine, in fact, combine with calcium?

DR. STEPHEN SCHAFFER (Hahnemann Medical College): I think that the evidence Dr. Dolara published in *Biochemical Pharmacology* indicates that taurine is a very weak chelator of calcium. I don't think the effects of taurine can be accounted for by formation of calcium salts.

DR. STEVEN BASKIN (Medical College of Pennsylvania): We failed to observe any interaction between calcium and taurine using a calcium electrode. However, our failure to observe direct chelation does not mean that taurine cannot act at a calcium site on Na^+-K^+ ATPase.

DR. RYAN HUXTABLE (University of Arizona): It is true that the complex between calcium and taurine is very weak, but I don't think this should be dismissed out of hand. Dr. Dolara calculated that about 8% of the free calcium in the cell exists as a taurine salt.

DR. NARANJAN DHALLA (University of Manitoba): Taurine appears to prevent to some extent calcium overload in the myopathic hamster and in the calcium paradox model. Do you feel that taurine could be used clinically to treat against calcium overload?

DR. RYAN HUXTABLE (University of Arizona): I believe taurine could be used in the treatment of Friedreich's ataxia, a disease in which patients die at a very early age. They die, not from ataxia, but from an associated cardiomyopathy, which is characterized by calcium overload and is similar to the cardiomyopathic hamster model.

DR. J. KHATTER (University of Manitoba): Dr. Schaffer, does taurine prevent the ultrastructural changes which are associated with the calcium paradox?

DR. STEPHEN SCHAFFER (Hahnemann Medical College): We have not performed that experiment yet. However, it is a very attractive experiment to perform since the glycocalyx becomes separated from the rest of the membrane during the calcium paradox. Perhaps taurine prevents this splitting of the membrane.

DR. NARANJAN DHALLA (University of Manitoba): I would like to ask Drs. Schaffer and Welty if they determined the taurine concentration in hearts from either cardiomyopathic hamsters or hearts that were subjected to the calcium paradox. This information would be useful in determining whether taurine is acting intracellularly or extracellularly.

DR. JOSEPH WELTY (University of South Dakota): The taurine content of 6-day-old myopathic hamsters is slightly elevated. However, of more interest is the observation that myocardial taurine levels of the normal hamster, but not the myopathic hamster, are reduced 15–20% following isotonic intraperitoneal glucose injections.

DR. STEPHEN SCHAFFER (Hahnemann Medical College): In collaboration with Dr. Kocsis we have performed a couple of experiments to answer Dr. Dhalla's question. In one experiment, we compared the response of three sets of hearts to the calcium paradox. One group was treated with beta-alanine to reduce intracellular taurine levels by 10 mM; another group contained normal intracellular taurine levels but were perfused with buffer containing 10 mM of taurine; the last group was a control heart perfused with buffer lacking taurine. We found that, whereas a reduction in intracellular taurine levels had no effect, a 10 mM increase in extracellular taurine content protected the heart against failure resulting from the calcium paradox. This suggests that the protective effect of taurine results from a change in extracellular, rather than intracellular, taurine levels. We have also examined tissue taurine levels; preliminary data indicate that the taurine content of hearts subjected to the calcium paradox is low. Perhaps Dr. Kocsis would like to comment on the results.

DR. JAMES J. KOCSIS (Jefferson Medical College): In the last several years Dr. Rovetto and I have examined the taurine content of hearts subjected to numerous conditions, including hypoxia, ischemia, and calcium-free perfusion. We have only found changes upon perfusion with EDTA in the absence of calcium.

DR. GENE A. LENTINI (Philadelphia College of Osteopathy): Is the inotropic effect of taurine at a particular calcium concentration dose dependent, and would this effect be reversed by the analogue, beta-alanine?

DR. STEPHEN SCHAFFER (Hahnemann Medical College): The taurine effect is concentration dependent; however, I am not sure how beta-alanine would affect the taurine response.

DR. NARANJAN DHALLA (University of Manitoba): Dr. Huxtable, does taurine, or one of its metabolic intermediates, mediate the actions of taurine?

DR. RYAN HUXTABLE (University of Arizona): If I may indulge in a little bit of wild speculation, perhaps taurine can form a highly reactive, phosphorylated derivative analogous to ATP or acetylphosphate.

DR. JAMES KOCSIS (Jefferson Medical College): I think that it is possible that the activity of taurine is related to some metabolite. One candidate certainly is taurocyamine. Taurocyamine phosphate is known to be a high energy phosphate. Another candidate certainly is gamma-glutamyl taurine. Dr. Feuer mentioned that this new peptide has some taurine-like activity. In fact, it appears to be a more potent agent than taurine itself.

PART IV

Clinical Implications of Taurine

Introduction

The use of taurine as a therapeutic agent has only recently been recognized. Clinical trials have concentrated on its antiepileptic activity and revealed its promise in the treatment of epilepsy. In this section several other potential implications of taurine are discussed.

Taurine is known to alter ion transport and stabilize biological membranes. These effects appear to be the basis for the first two chapters in this section. In the first chapter, Kuriyama et al. describe the effect of taurine on streptozotocin-induced diabetes. They believe that its ability to antagonize the impairment of beta-cells by streptozotocin is related to its membrane-stabilizing activity. Durelli and Mutani find that taurine reduces potassium-induced muscle hyperexcitability in both normal volunteers and patients suffering from dystrophic myotonia. The data reveal the potential use of taurine in the treatment of myotonic disorders.

Normal tissue taurine content is maintained in part by dietary sources of the amino acid. Rassin and Gaull review the importance of taurine to human nutrition. They discuss the possibility that taurine deficiency may be associated with certain forms of cardiac arrhythmias, muscle or adrenal dysfunction, and retinal degeneration. Yamori et al. provide evidence that taurine may be involved in the pathogenesis of hypertension and atherosclerosis. They report that liver and serum taurine content is reduced in stroke-prone hypertensive rats. Treatment of these animals with taurine was found to retard the development of severe hypertension. Baskin and Finney review the factors that regulate tissue taurine content. They suggest that certain pharmacological agents may exert their action in part by altering taurine levels. Lombardini and Crass show that taurine is released from the occluded or chemically damaged heart. The release is relatively specific for taurine and can be detected in patients with clinically diagnosed myocardial infarction.

Copyright © 1981, Spectrum Publications, Inc.
The Effects of Taurine on Excitable Tissues

Taurine: A Sulfur-containing Amino Acid Possibly Important for Maintaining Cellular Integrity

Kinya Kuriyama
Seitaro Ohkuma
Makoto Muramatsu

The study of taurine (2-aminoethanesulfonic acid) was initiated in 1827, when Tiedman and Gmelin found this compound in ox bile. In spite of intensive studies on taurine by many investigators, its actions and effects and their significance in various mammalian organs have long been unknown.

Apart from being conjugated with several types of bile acids, taurine exists uncombined, and is distributed in almost all organs in large quantities. In the human, the highest amount of taurine has been found in cardiac and skeletal muscle. Although species and sex differences in the distribution of taurine exist, it is generally considered that muscular structures (heart, skeletal muscles), endocrine organs (pituitary gland, pancreas, ovary, adrenal glands, etc.), and certain parts of the brain contain a large amount of taurine (Iwata and Kuriyama, 1975). Accumulated evidence also has demonstrated that taurine indeed plays various physiological roles in these organs and tissues. For example, there is now evidence that taurine may play important roles as a neurotransmitter or neuromodulator in the central nervous system, a regulator of membrane excitability in the heart, and a possible modulator of secretory processes in various endocrine organs. Its wide distribution and numerous physiological actions are what make the study of this compound both a fascinating and sometimes complicated task.

One of the important strategies for the study of taurine may be to seek a basic or common function that is applicable to all cells containing taurine, in addition to the organ- and/or tissue-specific actions of this compound. In this chapter we present some of our recent data on physiological actions of taurine in the brain, heart, adrenal gland and pancreas, and we discuss whether or not such a concept is applicable to the study of physiological and pharmacological actions of taurine.

TAURINE AND NEUROTRANSMISSION

Although it has been well documented that taurine possesses depressant actions on the mammalian central nervous system (CNS), its mechanism of action is not fully defined. There is considerable evidence that taurine may exhibit this action by acting as a specific neurotransmitter in the mammalian CNS: Ca^{2+}-dependent release from cerebral cortical slices by electrical stimulation (Davison and Kaczmarek, 1971), high affinity uptake into synaptosomes (Schmid et al., 1975), synaptosomal enrichment of cysteine sulfinate decarboxylase (CSD) activity (one of the enzymes involved in the biosynthesis of taurine) (Agrawal et al., 1971), inhibitory action on spinal neurons (Curtis and Watkins, 1965) and its reversal by strychnine (Nicoll and Barker, 1973), and uneven distribution in the CNS (Schaw and Heine, 1965). In contrast to the idea that taurine acts as a neurotransmitter, the other hypothesis, that taurine plays a modulatory role for membrane excitability rather than acts as a neurotransmitter, has also been presented (van Gelder, 1972; Barbeau and Donaldson, 1974; Huxtable and Bressler, 1974; Kuriyama and Nakagawa, 1976). Since the proof of uneven distribution of the test compound at regional, cellular, and subcellular levels is considered to be one of the important criteria required for the identification of neurotransmitters, we have attempted to demonstrate the specific localization of taurine and CSD activity in the rat spinal cord, where one can easily differentiate the gray matter from the white matter, using our newly developed microassay method (Yoneda et al., 1977b). As shown in Fig. 24.1, taurine was, however, evenly distributed in the spinal cord independently of the density of neuronal synapses, and no areas having an especially high taurine content were detected, as compared with that of γ-aminobutyric acid (GABA). The GABA content in the gray matter was higher than that in the white matter, and it was also found that GABA content in tissue squares of the dorsal half of the spinal cord are significantly higher than those found in the ventral half. The highest content of GABA was observed in the dorsal area of the dorsal horn,

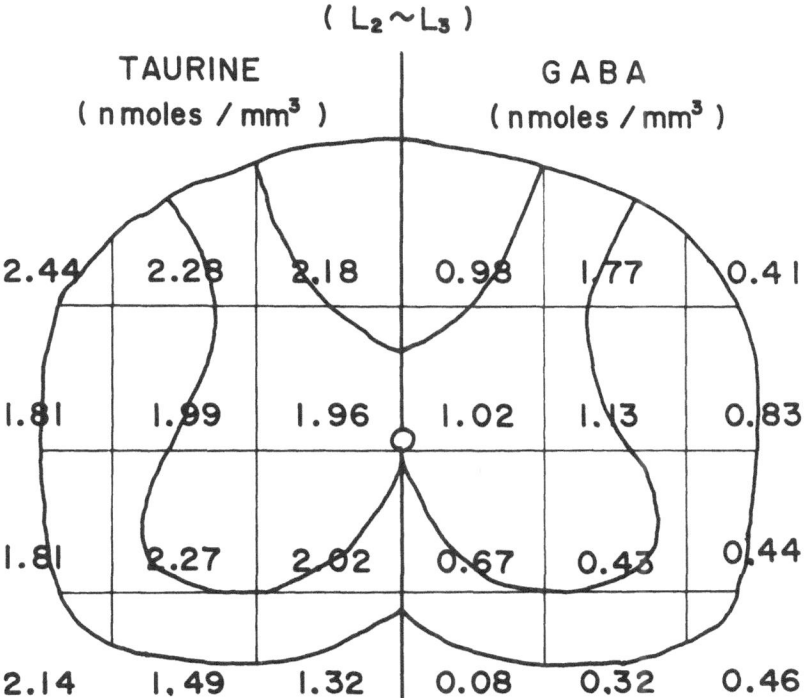

($L_2 \sim L_3$)

TAURINE
(nmoles / mm^3)

GABA
(nmoles / mm^3)

2.44	2.28	2.18	0.98	1.77	0.41
1.81	1.99	1.96	1.02	1.13	0.83
1.81	2.27	2.02	0.67	0.43	0.44
2.14	1.49	1.32	0.08	0.32	0.46

Fig. 24.1.   Distribution of taurine and GABA in rat spinal cord (L_2-L_3. Frozen lumbar segments of the rat spinal cord were cut horizontally in 480-μm sections, and each section was further dissected into 500 × 500 μm squares in a cold box ($-30°$C). Mean ± SEM obtained from six to eight separate experiments is shown. (From Yoneda and Kuriyama, 1978.

as reported by Otsuka and Konishi (1976). Considering these results, coupled with a similar even distribution profile of taurine found in the rat thalamus (Yoneda and Kuriyama, 1978a), it seems rather unlikely that taurine plays a neurotransmitter role, at least in the rat spinal cord and thalamus. In fact, much of the evidence against the possible role of taurine as a neurotransmitter—such as glial uptake (Ehinger, 1973), postnatal decline in cerebral content (Agrawal et al., 1971), and suppression of catecholamine release from the adrenal grand (Nakagawa and Kuriyama, 1975)—is also available in the literature. In addition, several pharmacological studies have demonstrated that cerebral taurine behaves differently in comparison with cerebral GABA, a major candidate for inhibitory neurotransmitter activity in the mammalian CNS. The administration of a convulsive dose of strychnine did not induce any detectable change in the spinal distribution of taurine (Kuriyama et al., 1978b). Furthermore, in spite of the possible involvement of GABA in morphine analgesia (Yoneda et al., 1976) and the significant alterations in the spinal and thalamic microdistribution of GABA by the administration of an analgesic dose of morphine (Yoneda et al., 1977a; Kuriyama and Yoneda, 1978), no significant changes of taurine content in these CNS structures were detected by the acute administration of this narcotic drug.

In vitro addition of taurine, however, significantly attenuated the Ca^{2+}-dependent, high KCl-evoked release of ^{14}C-ACh and ^3H-NE from cerebral cortical slices (Fig. 24.2) without affecting the unstimulated (spontaneous) release. Similarly, taurine administered orally for 3 days to animals (as well as added *in vitro*) inhibited the Ca^{2+}-dependent, KCl-evoked release of ^{14}C-ACh from superior cervical ganglion (SCG) (Kuriyama et al., 1978a). Since *in vitro* addition of leucine or methionine had no such suppressive action (Muramatsu et al., 1978), it is suggested that the effects of taurine described above may be specific for taurine. In addition to the well-known significance of intracellular free Ca^{2+} in regulating the excitability of neuronal tissues, our results also clearly indicated that taurine suppresses only Ca^{2+}-dependent and -stimulated release of various neurotransmitters. These results suggest that alterations in intracellular concentrations of Ca^{2+} induced by this amino acid may play a significant role in these suppressive actions. In fact, it has been demonstrated that *in vitro* addition of taurine significantly attenuates the re-

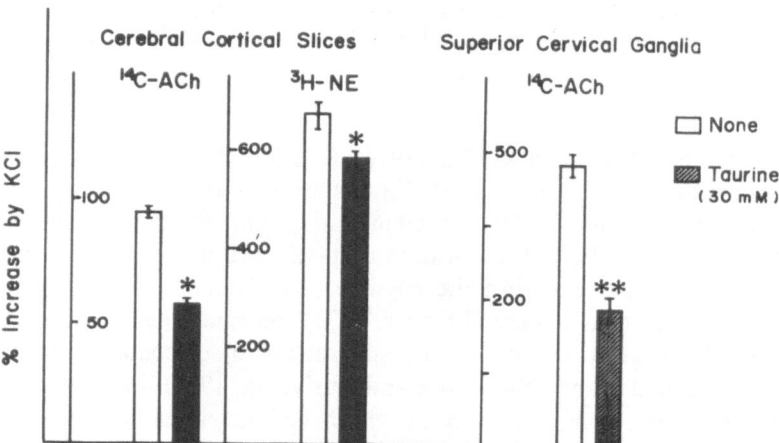

Fig. 24.2. Effect of *in vitro* addition of taurine on KCl-evoked release of ^{14}C-ACh and ^3H-NE from neuronal tissues. Sixty millimole of KCl was used to evoke the stimulated release of ^{14}C-ACh and ^3H-NE. Mean ± SEM obtained from four to six separate experiments is shown. *, $p < 0.05$; **, $p < 0.01$, compared with control values. (From Muramatsu et al., 1978.)

lease of ^{45}Ca from preloaded crude synaptosomal particulate (P_2) without affecting the uptake process (Kuriyama and Nakagawa, 1979). On the other hand, not only the uptake of ^{45}Ca by crude microsomal fraction (P_3) but also its release from this particulate fraction were not affected at all by the addition of taurine (Table 24.1).

Evidence presented herein strongly suggests that taurine may not act as a neurotransmitter but, as a modulator for membrane excitability by decreasing the concentration of intracellular free Ca^{2+} available for the release of other neurotransmitters in the mammalian CNS and probably in the peripheral nervous system. In fact, a definite interaction between taurine and Ca^{2+} in aqueous solution has been demonstrated recently (Dolara et al., 1978).

Since a relatively large amount of taurine is detected in the corpus striatum (CS) and cerebral cortex (CC) in the mammalian CNS (Yoneda et al.,

Table 24.1.   Effect of taurine on ^{45}Ca release from crude synaptosomal (P_2) and microsomal (P_3) fractions from rat brain[a,b]

Fraction		Taurine Concentration (mM)	^{45}Ca released (%)	
			(+) ATP	(−) ATP
P_2		0	7.3 ± 0.9	41.0 ± 2.0
		10	4.9 ± 0.5[c]	33.2 ± 2.5[c]
		0	31.6 ± 1.0	48.1 ± 0.6
	A	10	32.0 ± 3.5	47.2 ± 2.7
		30	35.4 ± 3.5	46.1 ± 2.7
P_3		0	31.9 ± 4.7	34.4 ± 2.1
	B	10	28.5 ± 3.9	38.7 ± 3.4
		30	28.7 ± 2.4	36.3 ± 3.3

[a]From Kuriyama and Nakagawa, 1979, and unpublished observations.

[b]P_2 fraction from the rat brain was preloaded by incubation at 30°C for 3 min with a medium containing 0.1 mM $^{45}CaCl_2$, 150 mM NaCl, 5 mM KCl, 3 mM $MgCl_2$, 1.24 mM KH_2PO_4, 10 mM glucose, 3 mM ATP, and 30 mM Tris-HCl (pH 7.4), and the release of ^{45}Ca was measured by incubating the preloaded particles at 30°C for 8 min in the same medium except Ca^{2+} concentration was altered to 0.01 mM. In the case of P_3 fraction, the same experimental conditions were employed except for the following points: Incubation times for the preloading and measurement of release were changed to 30 and 5 min, respectively. In addition, in experimental Series A, the concentration of NaCl was decreased to 125 mM, whereas in experimental Series B, NaCl was omitted and replaced by 125 mM KCl. Taurine was added in the medium used for the measurements of ^{45}Ca release. Mean ± SEM obtained from 4–5 separate experiments is shown.

[c]$p < 0.01$, compared with each non-taurine-added value.

1978a), we have further studied the functional roles of taurine in these CNS structures. It has been shown that the descending striato-nigral and ascending nigro-striatal neuronal fiber connections indeed exist in the extrapyramidal tract. In addition, the technique of intracerebral microinjection of kainic acid, a structural analogue of glutamic acid, has been developed recently to analyze the interconnection of various CNE neurons. Intracerebral injection of kainic acid is considered to cause a selective and irreversible degeneration of neurons near the injection site without affecting the axons terminating in or passing through the area (Coyle and Schwarcz, 1976). Unilateral intrastriatal injection of this agent induced neuronal degeneration at the injected area and significant reduction of the striatal, pallidal, and nigral GABA content and glutamate decarboxylase (GAD: a rate-limiting enzyme for GABA biosynthesis) activities in the injected side, as previously reported (Schwarcz and Coyle, 1977; Jessell et al., 1978; Kurihara et al., 1979). As shown in Fig. 24.3, microinjection of kainic acid into the left CS causes the degeneration of neuronal cell bodies in the injected area (CS-L) without altering the morphological features of contralateral CS (CS-R) as well as in both sides of the globus pallidus (GP) and substantia nigra (EN). These

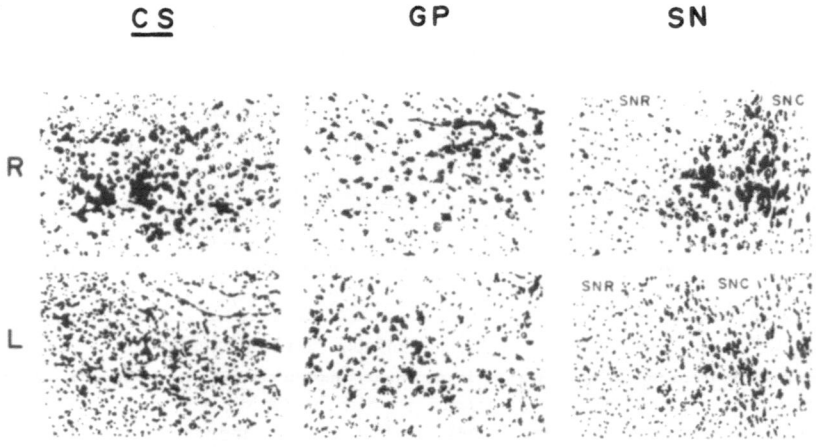

Fig. 24.3. Light micrographs of corpus striatum (CS), globus pallidus (GP), and substantia nigra (SN) at 7 days after the intrastriatal injection of kainic acid (2.0 μg/μliter). Key: L, left (injected side); R, right (noninjected side); SNR, pars reticulata of the SN; SNC, pars compacta of the SN. Each tissue section (10 μm) was stained with cresyl violet. Magnification: $\times$ 100. The kainic acid-treated striatum (CS-L) shows various histological abnormalities such as disappearance, necrosis and atrophy of neurons, and gliosis, compared with the noninjected side (CS-R). In the GP and SN, no such changes were noted, in both the treated and untreated sides. (From Kurihara et al., 1979.

results clearly indicate that neurochemical changes observed in the GP and SN following the intrastriatal injection of kainic acid, such as reductions in GABA content and GAD activity, are not induced by the diffusion of this agent from the CS. Unlike GABA, a significant reduction of taurine content was found only in the left striatum, into which kainic acid was injected (Fig. 24.4). On the other hand, CSD as well as GAD activities in the CS, GP, and SN of the injected side were significantly reduced by the unilateral microinjection of kainic acid into the left CS (Fig. 24.5). These results suggest that the descending striato-nigral or striato-pallidal taurine-containing neurons may not be present at all in these CNS structures in spite of the existence of GABA-ergic neurons therein. Considering the present findings, combined with our previous study (Yoneda and Kuriyama, 1978), it is conclusive that CSD activity is not a good tool for estimating the localization of endogenous taurine, and that this enzyme is unlikely to be the rate-limiting enzyme for taurine biosynthesis in the mammalian CNS, unlike the presence of cross-relationships between GABA and GAD (Roberts and Kuriyama, 1968).

In contrast to our proposal that taurine may play a modulatory role for membrane excitability in the mammalian CNS, there are also many reports to support the idea that taurine plays a possible role as a neurotransmitter.

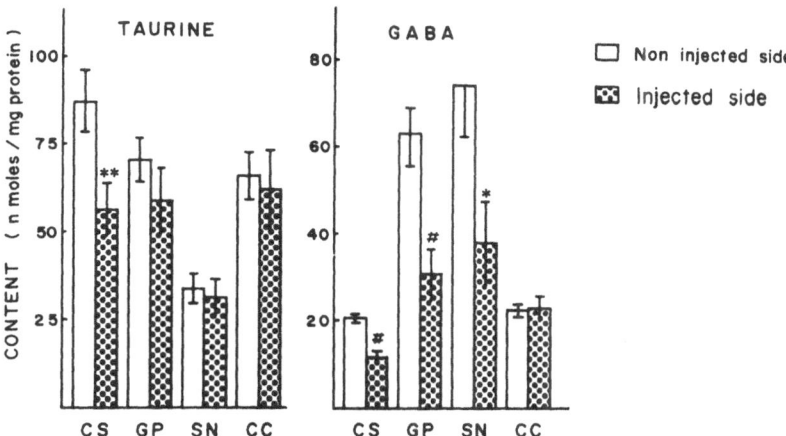

Fig. 24.4. Effect of striatal injection of kainic acid on taurine and GABA content in some extrapyramidal circuitry. Taurine and GABA content were determined at 7 days after the unilateral intrastriatal kainic acid (2 μg/1.0 μliter) injection. Key: CS, corpus striatum; GP, globus pallidus; SN, substantia nigra; CC, cerebral cortex. Mean ± SEM obtained from five to six separate experiments is shown. *, $p < 0.05$; **, $p < 0.02$; #, $p < 0.01$, compared with noninjected side. (From Kurihara et al., 1979 and unpublished observations.)

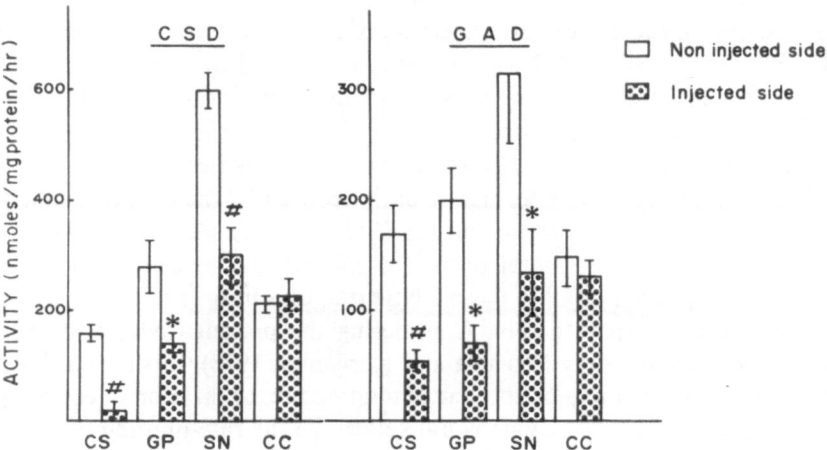

Fig. 24.5. Effect of striatal injection of kainic acid on CSD and GAD activities in some extra-pyramidal circuitry. CSD and GAD activities were determined at 7 days after the unilateral intrastriatal kainic acid ($2 \mu g/1.0 \mu liter$) injection. Mean ± SEM obtained from six separate experiments is shown. *, $p < 0.05$; #, $p < 0.01$, compared with noninjected side. (From Kurihara et al., 1979, and unpublished observations.)

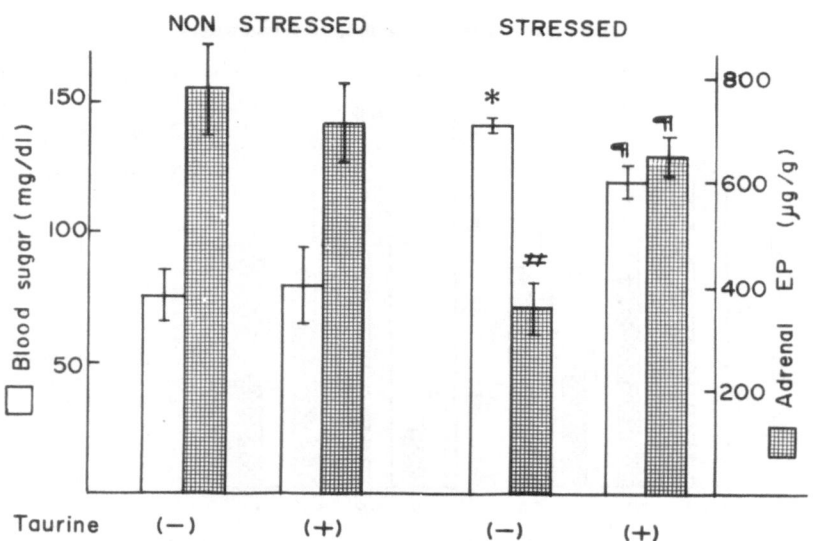

Fig. 24.6. Effect of oral administration of taurine on stress-induced alterations in blood sugar level and adrenal epinephrine (EP) content. Taurine was given orally, 4–7 g/kg/day for 3 days. Mean ± SEM obtained from four separate experiments is shown. *, $p < 0.05$; #, $p < 0.01$, compared with each nonstressed and non-taurine-treated value; and ¶, $p < 0.05$, compared with each stressed control and non-taurine-treated value. (From Nakagawa and Kuriyama, 1975.)

For example, the light-stimulated release of taurine from frog rod outer segments (Salceda et al., 1977), localization of this amino acid in retinal photoreceptor cells (Orr et al., 1976) and decrease in the content of taurine following the degeneration of this retinal component (Hayes et al., 1975), axonal transport of taurine in the goldfish visual system (Ingoglia et al., 1978), localization of taurine in the molecular layer of the rat cerebellum (Nadi et al., 1977), and selective depressant action of taurine on the Purkinje cell dendritic zone in the rat cerebellum (Frederickson et al., 1978) have been reported. If our view that taurine acts as a modulator of the excitability of neuronal membranes is correct, one must conclude that taurine may have a dual action as a modulator of membrane excitability and as a neurotransmitter in the mammalian nervous system.

TAURINE AND ENDOCRINE ORGANS

It has been demonstrated that considerably high levels of taurine are detected in various endocrine and reproductive organs such as the pineal and pituitary glands (Guidotti et al., 1972), generative organs (Lorincz and Kuttner, 1968), adrenal gland (Nakagawa and Kuriyama, 1975), and pancreas (Briel et al., 1972). Much evidence to support possible roles of taurine in regulating the secretory function of these organs has also been reported, such as the circadian rhythm in pituitary taurine content (Neuhoff and Tonge, 1973); decrease in uterine taurine content by estradiol (Kalman and Lombrozo, 1961); increase in taurine excretion (Pentz et al., 1959) as well as in muscular taurine content (Ryan and Carver, 1963) by adrenocortical hormone; rise in splenic, renal, and muscular content of taurine by hypophysectomy (Awapara, 1956); and enhancement of hypoglycemic action of insulin by taurine (Donadio and Fromageot, 1964). Although these facts strongly suggest that taurine may indeed have a regulatory role in the secretory processes in these endocrine organs, exact mechanisms underlying these regulatory functions are still not clearly defined. In this study, we have attempted to deal with this problem, using the pancreas as a model system.

Since it has been demonstrated that the stress-induced increase in blood sugar level, as well as the decrease in adrenal epinephrine content, is significantly inhibited by the oral administration of taurine for 3 days (Fig. 24.6), further studies on the physiological roles of taurine in the regulation of serum level of glucose were carried out using the pancreas, the other important organ having an ability to regulate the serum level of glucose.

Streptozotocin (STZ), one of the antibiotics developed for the therapy of leukemia, has been found to cause hyperglycemia by inducing selective damage of β-cells in the islets of Langerhans (Rakieten et al., 1963). Acute administration of this agent to male ddy mice induced a transient but significant hyperglycemia and a drastic fall in the pancreatic immunoreactive insulin (IRI) content, whereas the serum level of IRI was not altered significantly in spite of the occurrence of considerable hyperglycemia (Fig. 24.7), as reported by Junod et al. (1967). Oral administration of taurine for 3 days prior to the sacrifice of these STZ-administered animals showed a tendency to suppress the elevation of serum level of glucose, but this change was not statistically significant. The prolonged use of taurine (5–7 days), however, significantly inhibited the hyperglycemia induced by STZ (Tokunaga et al., 1979). In addition to this suppressive effect on hyperglycemia, taurine pretreatment for 7 days also significantly prevented the STZ-induced fall in pancreatic IRI content. Similar preventive effects were also observed when either methionine or cysteine (sulfur-containing amino acid and precursor for the biosynthesis of taurine, respectively) was orally administered, whereas isoleucine and valine had no such preventive properties (Fig. 24.8). Since the administration of these three amino acids, which have a suppressive effect on STZ-induced hyperglycemia, significantly increased the taurine content in the pancreas, it is suggested that the increase in pancreatic taurine may be intimately related to the protective effect against the pancreato-toxic action of STZ. It has been demonstrated, however, that a remarkable strain difference is present in the sensitivity to STZ in mice (Rossini et al., 1977), and ddy mouse is one of the resistant strains for this agent. Studies using CD-1 mice (one of the STZ-sensitive strains), however, showed that the administration of taurine also exhibits significant inhibitory effects on the hyperglycemia and the decrease in the serum and pancreatic content of IRI induced by acute administration of STZ (Fig. 24.9). These results indicate that the protective action of taurine on STZ-induced hyperglycemia occurs independently of strain difference or the susceptibility to STZ. Light microscopic observations of pancreatic sections from STZ-treated CD-1 mice stained with aldehyde-fuchsin revealed that granules stained with this dye, namely β-granules, completely disappeared in the pancreas of animals treated with STZ alone, whereas many granules stained with this dye still remained intact in the pancreas of animals pretreated with taurine (Fig. 24.10). These histological observations clearly indicate that the site for protective effect of taurine on STZ-induced hyperglycemia is indeed present in the pancreas per se.

Repeated administration of STZ was found to induce a continuous hyperglycemia, as previously reported by Like and Rossini (1976). Pretreatment and continuous administration of taurine not only significantly suppressed this continuous hyperglycemia but also attenuated the drastic fall in pancreatic IRI content (Fig. 24.11). Considering these results, we conclude that taurine has a protective action on both transient and continuous hyperglycemic conditions induced by STZ in CD-1 mice. In this regard it is note-

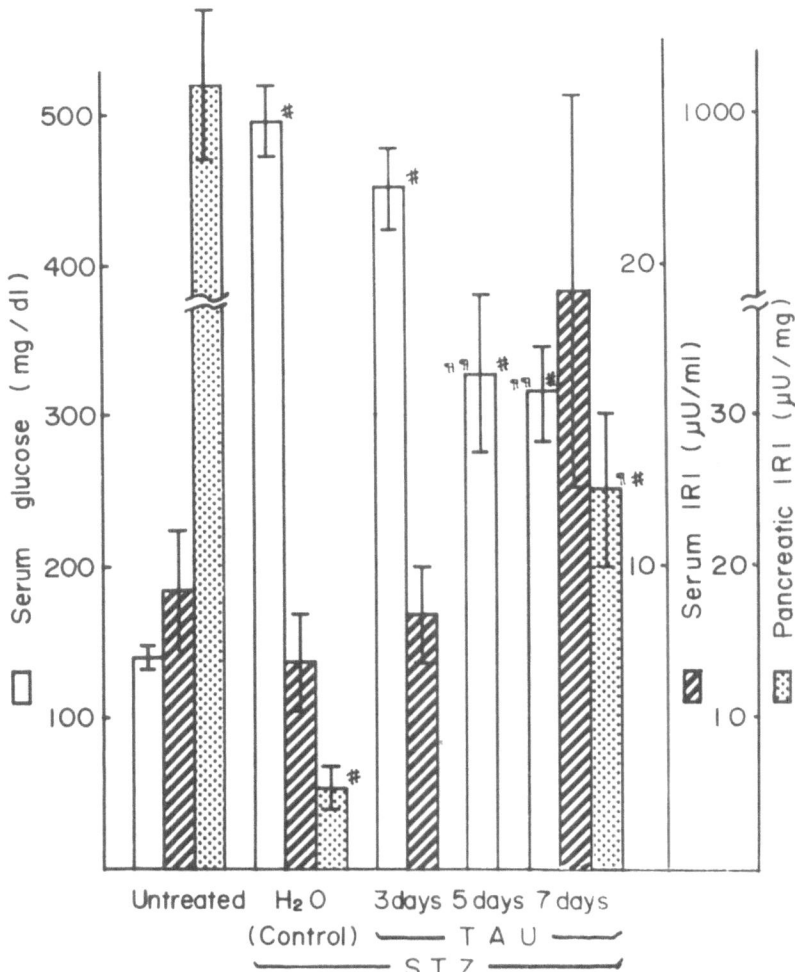

Fig. 24.7. Protective effect of taurine on STZ-induced alterations in serum levels of glucose and immunoreactive insulin (IRI), and pancreatic IRI content in ddy mice. Taurine was administered orally for 2, 4, or 6 days prior to the administration of STZ (250 mg/kg, i.v.), and taurine administration was continued until the sacrifice of animals (24 hr after the STZ administration). Each value represents the mean ± SEM obtained from 5 to 24 separate experiments. #, $p < 0.01$, compared with each untreated value; and ¶, $p < 0.02$; ¶¶, $p < 0.01$, compared with each H_2O (control) value. (From Tokunaga et al., 1979, *Biochem. Pharmacol.*, Vol. 28, used with permission.)

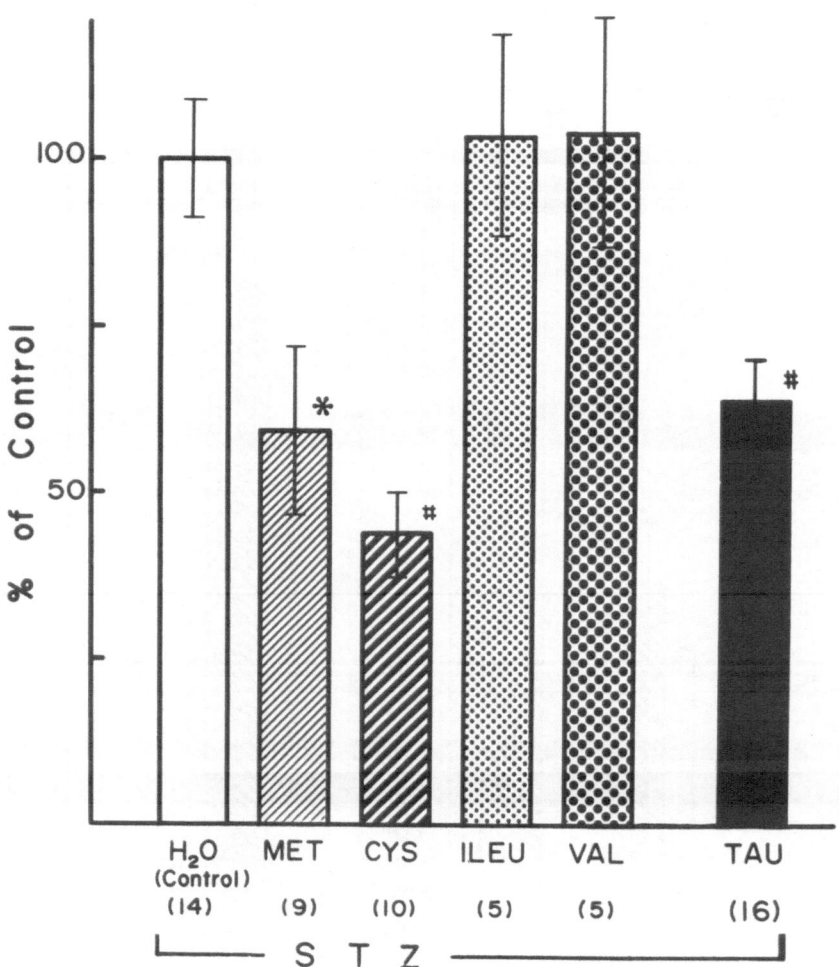

Fig. 24.8.   Effect of various amino acids on STZ-induced hyperglycemia in ddy mice. Each amino acid was administered orally for 7 days prior to the sacrifice of STZ-administered animals. Each value represents the mean ± SEM (numbers of separate experiments are indicated in parentheses). *, $p < 0.05$; #, $p < 0.01$, compared with the H$_2$O (control) value. (From Tokunaga et al., 1979, *Biochem. Pharmacol.*, Vol. 28, used with permission.)

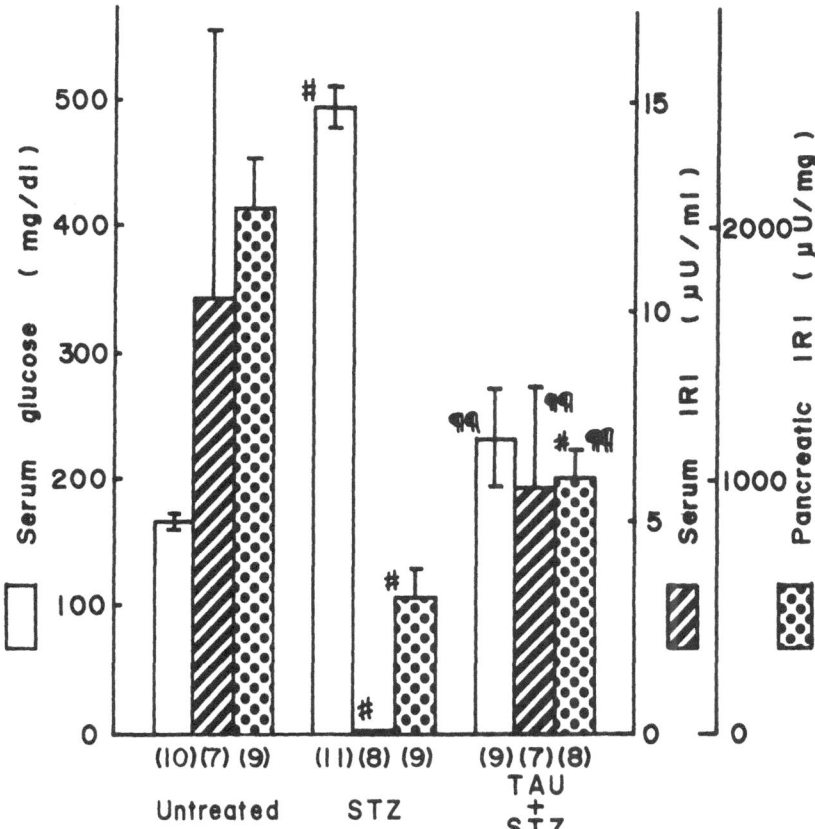

Fig. 24.9. Protective effect of taurine on hyperglycemic conditions induced by acute administration of STZ in CD-1 mice. Taurine was administered orally for 6 days prior to the administration of STZ (150 mg/kg, i.v.), and taurine administration was continued until the sacrifice of animals (24 hr after the STZ administration). Each value represents the mean ± SEM (numbers of separate experiments are indicated in parentheses). #, $p < 0.01$, compared with each untreated value; ¶¶ , $p < 0.02$, compared with each STZ value. (From Tokunaga et al., 1979, *Biochem. Pharmacol.*, Vol. 28, used with permission.)

worthy that *in vitro* addition of taurine indeed suppresses the Ca^{2+}-dependent release of IRI from the isolated islets of Langerhans (unpublished observations).

On the other hand, endogenous pancreatic taurine content was signifi-

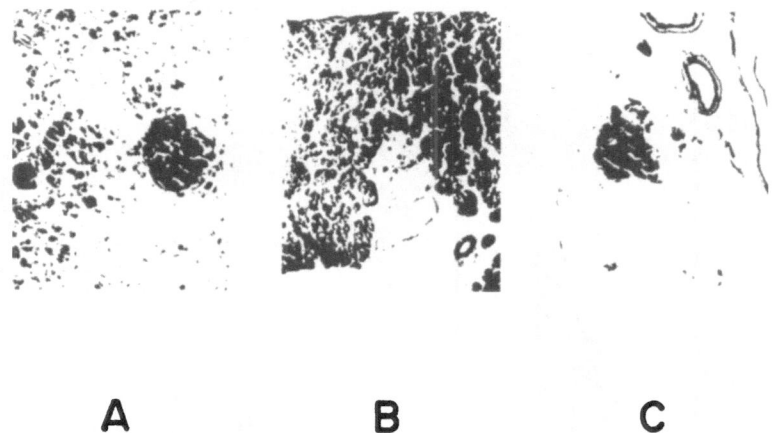

Fig. 24.10.   Light microscopic observations of pancreatic islets in CD-1 mice (aldehyde-fuchsin stain, × 400). Key: A, control; B, STZ alone; C, TAU + STZ. Animals were sacrificed 24 hr after the administration of vehicle or STZ. (From Tokunaga et al., 1979, *Biochem. Pharmacol.,* Vol. 28, used with permission.)

cantly increased in accordance with the rise in serum level of glucose in both ddy and CD-1 mice (Fig. 24.12). These results suggest that the pancreatic taurine content may be augmented consistently when the hyperglycemia and/or the damage of pancreatic β-cells occurs.

Although mechanisms underlying these protective actions of taurine against the damage of β-cells by STZ are not clear at present, it may possibly be related to the membrane-stabilizing action of this compound. In fact, there are many observations that support the idea that taurine acts as a membrane stabilizer, such as the antiarrhythmic action on epinephrine- or digoxin-induced arrhythmia (Read and Welty, 1963), anticonvulsant effect on experimentally induced epilepsy (van Gelder, 1972; Izumi et al., 1973), and protective effect on the loss of calcium from phospholipase C-treated sarcoplasmic reticulum (Huxtable and Bressler, 1973). Considering present results coupled with these previous findings, it is suggested that the elevation of taurine level in the pancreas may induce membrane-stabilizing actions on β-granules in the islets of Langerhans and thereby exhibit a protective action on STZ-induced damage in these pancreatic structures.

Evidence presented in this section also suggests that taurine may act as a stabilizer or modulator of membranous functions of cells, and that it may play an important physiological role in maintaining the function and/or in-

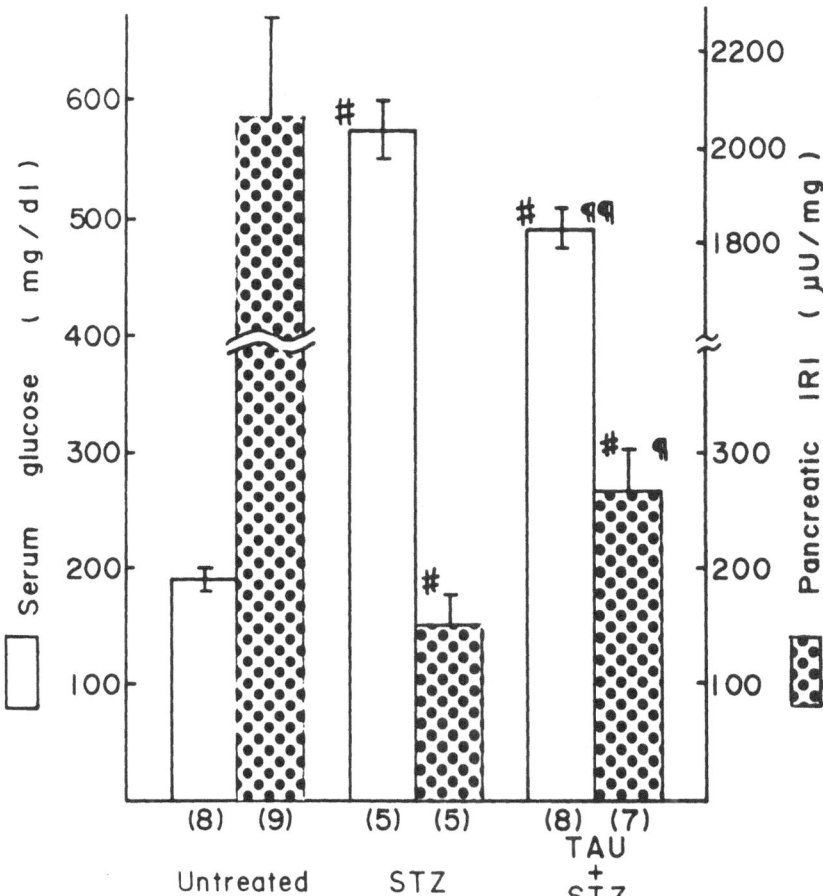

Fig. 24.11. Protective effect of taurine on continuous hyperglycemia induced by repeated administration of STZ in CD-1 mice. STZ was intraperitoneally administered (50 mg/kg) daily for 5 days, and animals were decapitated 2 weeks after the initiation of STZ administration. Taurine was orally administered from 7 days before the initiation of STZ administration to the day of sacrifice (for 21 days). Each value represents the mean ± SEM (numbers of separate experiments are indicated in parentheses). #, $p < 0.01$, compared with each untreated value; and ¶, $p < 0.05$; ¶¶, $p < 0.02$, compared with each STZ value. (From Tokunaga et al., 1979, *Biochem. Pharmacol.*, Vol. 28, used with permission.)

tegrity of chromaffin granules in the adrenal medulla as well as in the β-cells of the islets of Langerhans of the pancreas. Whether or not such a view of a physiological role of taurine is applicable to other endocrine and reproductive organs remains to be elucidated.

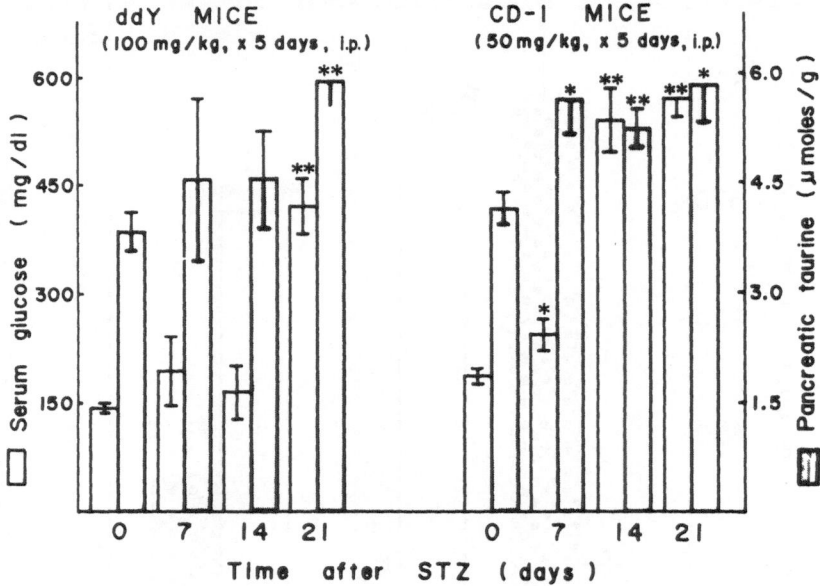

Fig. 24.12.   Effect of repeated administration of STZ on serum level of glucose and pancreatic taurine content. Mean ± SEM obtained from 9 to 16 separate experiments is shown. *, $p < 0.05$; **, $p < 0.02$, compared with each non-STZ-treated values. (From Yoneda et al., 1978b.)

FUNCTION OF TAURINE IN THE HEART

The heart is one of the organs that contains a large amount of taurine. It has been reported that taurine possesses antiarrhythmic effects on epinephrine- or digoxin-induced arrhythmia (Read and Welty, 1963) and a potentiating action on the inotropic effect of strophanthin-K (Guidotti et al., 1971). Furthermore, a significant elevation of cardiac taurine content in congestive heart failure, as well as stress-induced hypertension, was demonstrated (Huxtable and Bressler, 1974). Although these findings suggest the significance of this endogenous compound in the regulation and/or maintenance of cardiac function, molecular mechanisms underlying these phenomena remain to be elucidated.

The high concentration of cardiac taurine is considered to be maintained by the uptake of circulating taurine, rather than being derived from biosynthesis (Huxtable, 1976). In fact, slices from the mouse heart were found to significantly accumulate ^{14}C-taurine from the external medium against a concentration gradient (Hirai et al., 1979a), as previously reported in the rat heart (Awapara and Berg, 1976). This uptake system for taurine in mice was a saturable one consisting of a single component (apparent Km = 160 μM, Vmax = 8.85 nmol/g/min) and dependent on the concentration of Na$^+$ in the external medium. It was also found that taurine uptake by mouse cardiac slices is inhibited competitively by the addition of β-amino acids such as hypotaurine and β-alanine but unaffected by structurally related amino acids such as L-cysteine, L-cysteic acid, L-cysteine sulfinic acid, and L-glutamic acid. Repeated administration of isoproterenol (ISP), a β-adrenergic agonist, to male ddy mice caused an increase in heart mass, namely cardiac hypertrophy. Consistent with the occurrence of cardiac hypertrophy, it was found that significant increases, not only in the rate of taurine uptake but also in the taurine content, occur in mouse cardiac tissue, as previously demonstrated in the rat heart by Chubb and Huxtable (1978). These results suggest that the uptake system for taurine may be a main regulatory factor for the endogenous level of this compound, even in hypertrophic cardiac muscle, and that alterations in cardiac taurine may also be closely related to the occurrence and/or the maintenance of cardiac hypertrophy. This concept is further supported by the fact that a similar increase in the uptake and content of taurine is observed in right ventricular muscle of the rat heart that is made hypertrophic by the administration of the alkaloid monocrotaline (Hirai et al., 1979b). Monocrotaline is a naturally occurring pyrrolizidine alkaloid derived from the plant *Crotalaria spectabilis*. It has been demonstrated that the administration of this alkaloid to rats produces pulmonary lesions including the swelling of pulmonary capillary endothelial cells, thrombosis and lesions of the arterial media, and subsequent hypertrophy of the right cardiac ventricles (Hayashi and Lalich, 1967). Similar histological changes were also observed in this study. Subcutaneous administration of this agent to the rat induced a rise in the weight of right cardiac ventricular muscle as well as an increase in the uptake and content of taurine in this part of the heart without affecting other parts of the cardiac muscle such as the left ventricle and septum (Fig. 24.13).

These results, coupled with previously mentioned findings from various laboratories, suggest that taurine indeed plays an important role in regulating membrane excitability in the heart, and that the adaptive increase of this amino acid occurs invariably in hypertropic cardiac tissue, possibly by increasing the uptake process. Although the significance of these alterations in taurine level in the hypertrophic cardiac tissue is unclear at present, they could be related to various factors such as general changes in amino acid pool sizes and turnover, or they may be linked to specific cardiac actions of taurine, as suggested by Huxtable (1976). In view of the stabilizing or

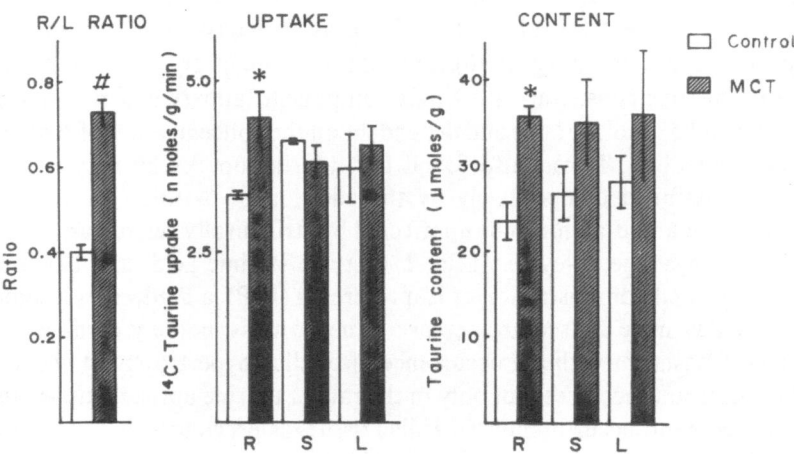

Fig. 24.13. Monocrotaline (MCT)- induced right ventricular hypertrophy and uptake of taurine. Male Wistar rats received 60 mg/kg of MCT subcutaneously and were sacrificed at 21 days after the administration. R/L ratio indicates the weight ratio of the right ventricle to left ventricle. Key: R, right ventricle; S, septum; L, left ventricle. Right and left ventricles, and septum were incubated with Krebs-Ringer Tris buffer (pH 7.4) containing ^{14}C-taurine (0.45 nmol/3 ml) for 10 min. Mean ± SEM obtained from three to five separate experiments is shown. *, $p < 0.05$, compared with non-MCT-treated values. (From Hirai et al., 1979.)

modulating effect of taurine on various membranous structures, as discussed in previous sections, one could also postulate that the increase of taurine in hypertrophic cardiac tissues may be related to the physiological function of the heart, the increased demand for maintaining morphological integrity of hypertrophic cardiac cells, or both.

COMMENT AND CONCLUSION

What does taurine actually do in various mammalian organs? In spite of several years of research on taurine and the involvement of many researchers in this research field, few definitive answers are given presently on these questions. An overview of the taurine field at the present time is like looking at a classical Japanese landscape painting: One knows that a unity exists if carefully examined, but most of the details are only vaguely perceived. Customarily, the appreciation of Japanese landscape painting follows a sequence that checks for a unity before going into the appreciation of the details. A similar type of approach may well be beneficial for future advancements in the study of physiological-pharmacological aspects of taurine. Trials to disclose general actions of taurine related to the maintenance of the essence of cell functions (unity) and organ- and/or tissue-specific actions of taurine (details) must be reconsidered. For example, there is much indirect evidence suggesting the possibility that taurine may be an inhibitory neurotransmitter in the mammalian CNS, while other evidence refutes this idea and suggests possible roles of this compound as a modulator or stabilizer of neuronal membranes. The latter concept is also applicable to other organs or tissues. These views are further supported by the fact that taurine is distributed universally in almost all tissues, whereas other putative neurotransmitters such as GABA and acetylcholine are only found in nervous tissues. Obviously, much more refined experimental data must be accumulated before establishing such a hypothesis; i.e., that taurine has dual roles in various mammalian organs.

REFERENCES

Agrawal, H. C.; Davison, A. N.; and Kaczmarek, L. K. Subcellular distribution of taurine and cysteinesulphinate decarboxylase in developing rat brain. *Biochem. J.,* 122, 759–763 (1971).

Awapara, J. The taurine concentration of organs from fed and fasted rats. *J. Biol. Chem.,* 218, 571–576 (1956).

Awapara, J.; and Berg, M. Uptake of taurine by slices of rat heart and kidney. In *Taurine,* R. Huxtable and A. Barbeau, eds. Raven Press, New York (1976), pp. 135–143.

Barbeau, A.; and Donaldson, J. Zinc, taurine and epilepsy. *Arch. Neurol.,* 30, 52–58 (1974).

Briel, G.; Gylfe, E.; Hellman, B.; and Neuhoff, V. Microdetermination of free amino acids in pancreatic islets isolated from obese-hyperglycemic mice. *Acta Physiol. Scand.,* 84, 247–253 (1972).

Chubb, J.; and Huxtable, R. J. Transport and biosynthesis of taurine in the stressed heart. In *Taurine and Neurological Disorders,* A. Barbeau and R. J. Huxtable, eds. Raven Press, New York (1978), pp. 161–178.

Coyle, J. T.; and Schwarcz, R. Lesion of striatal neurones with kainic acid provides a model for Huntington's chorea. *Nature,* 263, 244–246 (1976).

Curtis, D. R.; and Watkins, J. C. The pharmacology of amino acids related to gamma-aminobutyric acid. *Pharmacol. Rev.,* 17, 347–391 (1965).

Davison, A. N.; and Kaczmarek, L. K. Taurine—A possible neurotransmitter? *Nature,* 234, 107–108 (1971).

Dolara, P.; Franconi, F.; Giotti, A.; Basosi, R.; and Valensin, G. Taurine-calcium interaction measured by means of ^{13}C nuclear magnetic resonance. *Biochem. Pharmacol.,* 27, 803–804 (1978).

Donadio, G.; and Fromageot, P. Influence exercée par la taurine sur l'utilisation du glucose par le rat. *Bull. Soc. Chim. Biol.,* 46, 293–302 (1964).

Ehinger, B. Glial uptake of taurine in the rabbit retina. *Brain Res.*, 60, 512–516 (1973).

Frederickson, R. C. A.; Neuss, M.; Morzorati, S. L.; and McBride, W. J. A comparison of the inhibitory effects of taurine and GABA on identified Purkinje cells and other neurons in the cerebellar cortex of the rat. *Brain Res.*, 145, 117–126 (1978).

Guidotti, A.; Badiani, G.; and Giotti, A. Potentiation by taurine of inotropic effect of strophanthin-K on guinea-pig isolated auricles. *Pharmacol. Res. Comm.*, 3, 29–38 (1971).

Guidotti, A.; Badiani, G.; and Pepeu, G. Taurine distribution in cat brain. *J. Neurochem.*, 19, 431–435 (1972).

Hayashi, Y.; and Lalich, J. J. Renal and pulmonary alterations induced in rats by a single injection of monocrotaline. *Proc. Soc. Exp. Biol. Med.*, 124, 392–396 (1967).

Hayes, K. C.; Carey, R. E.; and Schmidt, S. Y. Retinal degeneration associated with taurine deficiency in the cat. *Science*, 188, 949–951 (1975).

Hirai, K.; Yoneda, Y.; and Kuriyama, K. Taurine uptake in mouse cardiac slices and modification by isoproterenol-induced cardiac hypertrophy. *Folia Pharmacol. Japon.*, 75, 15 pp. (1979a) (in Japanese).

Hirai, K.; Yoneda, Y.; and Kuriyama, K. Alterations in content and transport of taurine during experimentally induced cardiac hypertrophy. *Folia Pharmacol. Japon.*, in press (1979b) (in Japanese).

Huxtable, R. J. Metabolism and function of taurine in the heart. In *Taurine*, R. Huxtable and A. Barbeau, eds. Raven Press, New York (1976), pp. 99–119.

Huxtable, R. J.; and Bressler, R. Effect of taurine on a muscle intracellular membrane. *Biochim. Biophys. Acta*, 323, 573–583 (1973).

Huxtable, R. J.; and Bressler, R. Taurine concentration in congestive heart failure. *Science*, 184, 1187–1188 (1974).

Ingoglia, N. A.; Sturman, J. A.; Rassin, D. K.; and Lindquist, T. D. A comparison of the axonal transport of taurine and proteins in the goldfish visual system. *J. Neurochem.*, 31, 161–170 (1978).

Iwata, H.; and Kuriyama, K. *Taurine: Its Metabolism and Physiological-Pharmacological Actions*, Ishiyaku-Shuppan, Tokyo (1975), pp. 43–48 (in Japanese).

Izumi, K.; Donaldson, J.; Minnich, J.; and Barbeau, A. Ouabain-induced seizures in rats: Suppressive effects of taurine and GABA. *Canad. J. Physiol. Pharmacol.*, 51, 885–889 (1973).

Jessel, T. M.; Emson, P. C.; Paxinos, G.; and Cuello, A. C. Topographic projections of substance P and GABA pathways in the striato- and pallido-nigral system: A biochemical and immunohistochemical study. *Brain Res.*, 152, 487–498 (1978).

Junod, A.; Lambert, A. E.; Orci, L.; Pictet, R.; Gonet, A. E.; and Renold, A. E. Studies of the diabetogenic action of streptozotocin. *Proc. Soc. Exp. Biol. Med.*, 126, 201–205 (1967).

Kalman, S. M.; and Lombrozo, M. E. The effect of estradiol on the free amino acids of the rat uterus. *J. Pharmacol. Exp. Ther.*, 131, 265–269 (1961).

Kurihara, E.; Kuriyama, K.; and Yoneda, Y. Interconnection of GABA-ergic neurons in the rat extrapyramidal tract: Analysis using intracerebral microinjection of kainic acid. *Brain Res.*, Submitted for publication (1979).

Kuriyama, K.; and Nakagawa, K. Role of taurine in adrenal gland: A preventive effect on stress-induced release of catecholamines from chromaffin granules. In *Taurine*, R. Huxtable and A. Barbeau, eds. Raven Press, New York (1976), pp. 335–345.

Kuriyama, K.; and Nakagawa, K. Increased association of calcium with brain mitochondria in the presence of taurine. *Japan. J. Pharmacol.*, 29, 309–313 (1979).

Kuriyama, K.; and Yoneda, Y. Morphine induced altrations of γ-aminobutyric acid and taurine content and L-glutamate decarboxylase activity in rat spinal cord and thalamus: Possible correlates with analgesic action of morphine. *Brain Res.*, 148, 163–179 (1978).

Kuriyama, K.; Muramatsu, M.; Nakagawa, K.; and Kakita, K. Modulating role of taurine on release of neurotransmitters and calcium transport in excitable tissues. In *Taurine and Neurological Disorders*, A. Barbeau and R. J. Huxtable eds. Raven Press, New York (1978a), pp. 201–216.

Kuriyama, K.; Yoneda, Y.; and Kurihara, E. Microdistribution of taurine and cysteine sulfinate decarboxylase activity in rat spinal cord and thalamus: Comparison with γ-aminobutyric acid and L-glutamic acid decarboxylase. In *Taurine and Neurological Disorders*, A. Barbeau and R. J. Huxtable, eds. Raven Press, New York (1978b), pp. 35–48.

Like, A. A.; and Rossini, A. A. Streptozotocin-induced pancreatic insulitis: New model for diabetes mellitus. *Science*, 193, 415–417 (1976).

Lorincz, A. B.; and Kuttner, R. E. Comparative studies on free amino acids in female reproductive tissues. *Am. J. Obestet. Gynecol.*, 101, 462–472 (1968).

Muramatsu, M.; Kakita, K.; Nakagawa, K.; and Kuriyama, K. A modulating role of taurine on release of acetylcholine and norepinephrine from neuronal tissues. *Japan. J. Pharmacol.*, 28, 259–268 (1978).

Nadi, N. S.; McBride, W. J.; and Aprison, M. H. Distribution of several amino acids in regions of the cerebellum of the rat. *J. Neurochem.*, 28, 453–455 (1977).

Nakagawa, K.; and Kuriyama, K. Effect of taurine on alteration in adrenal functions induced by stress. *Japan. J. Pharmacol.*, 25, 737–746 (1975).

Neuhoff, V.; and Tonge, S. R. Some pharmacological indications of a role for taurine in the regulation of pituitary activity. *J. Pharm. Pharmacol.*, 25, 138p–139p (1973).

Nicoll, R. A.; and Barker, J. L. Effect of strychnine on dorsal root potentials and amino acid responses in frog spinal cord. *Nature New Biol.*, 246, 224–225 (1973).

Orr, H. T.; Cohen, A. I.; and Lowry, D. H. The distribution of taurine in the vertebrate retina. *J. Neurochem.*, 26, 609–611 (1976).

Otsuka, M.; and Konishi, S. GABA in the spinal cord. In *GABA in Nervous System Function*. E. Roberts, T. N. Chase, and D. B. Tower, eds. Raven Press, New York (1976), pp. 197–202.

Pentz, E. I.; Moss, W. T.; and Denko, C. W. Factors influencing taurine excretion in human subjects. *J. Clin. Endocr.*, 19, 1126–1133 (1959).

Rakieten, N.; Rakieten, M. L.; and Nadkarni, M. V. Studies on the diabetogenic actions of streptozotocin. *Cancer Chemother. Rep.*, 29, 91–98 (1963).

Read W. O.; and Welty, J. D. Effect of taurine on epinephrine and digoxin induced irregulalities of the dog heart. *J. Pharmacol. Exp. Ther.*, 139, 283–289 (1963).

Roberts, E.; and Kuriyama, K. Biochemical-physiological correlations in studies on the γ-aminobutyric acid system. *Brain Res.*, 8, 1–35 (1968).

Rossini, A. A.; Appel, M. C.; Williams, R. M.; and Like, A. A. Genetic influence of the streptozotocin-induced insulitis and hyperglycemia. *Diabetes*, 26, 916–920 (1977).

Ryan, W. L.; and Carver, M. J. Immediate and prolonged effects of hydrocortisone on the free amino acids of rat skeletal muscle. *Proc. Soc. Exp. Biol. Med.*, 114, 816–819 (1963).

Salceda, R.; López-Colomé, A. M.; and Pasantes-Morales, H. Light-stimulated release of [^{35}S]taurine from frog retinal rod outer segments. *Brain Res.*, 135, 186–191 (1977).

Schaw, R. K.; and Heine, J. D. Ninhydrin positive substances present in different areas of normal rat brain. *J. Neurochem.*, 12, 151–155 (1965).

Schmid, R.; Sieghart, W.; and Karobath, M. Taurine uptake in synaptosomal fraction of rat cerebral cortex. *J. Neurochem.*, 25, 5–9 (1975).

Schwarcz, R.; and Coyle, J. T. Neurochemical sequelae of kainate injections in corpus striatum and substantia nigra of the rat. *Life Sci.*, 20, 431–436 (1977).

Tokunaga, H.; Yoneda, Y.; and Kuriyama, K. Protective actions of taurine against streptozotocin-induced hyperglycemia. *Biochem. Pharmacol.*, 28, 2807–2811 (1979).

van Gelder, N. M. Antagonism by taurine of cobalt induced epilepsy in cat and mouse. *Brain Res.*, 47, 157–165 (1972).

Yoneda, Y.; and Kuriyama, K. A comparison of microdistributions of taurine and cysteine sulphinate decarboxylase activity with those of GABA and L-glutamate decarboxylase activity in rat spinal cord and thalamus. *J. Neurochem.*, 30, 821–825 (1978).

Yoneda, Y.; Takashima, S.; and Kuriyama, K. Possible involvement of GABA in morphine analgesia. *Biochem. Pharmacol.*, 25, 2669–2670 (1976).

Yoneda, Y.; Kuriyama, K.; and Kurihara, E. Morphine alters distribution of GABA in thalamus. *Brain Res.*, 124, 373–378 (1977a).

Yoneda, Y.; Takashima, S.; Hirai, K.; Kurihara, E.; Yukawa, Y.; Tokunaga, H.; and Kuriyama, K. Microassay methods for taurine and cysteine sulfinate decarboxylase activity. *Japan. J. Pharmacol.*, 27, 881–888 (1977b).

Yoneda, Y.; Tokunaga, H.; and Kuriyama, K. Microdistribution of taurine in rat central nervous system. *Neurochem Res.*, 3, 680 (1978a).

Yoneda, Y.; Tokunaga, H.; and Kuriyama, K. Inhibitory effects of taurine on streptozotocin-induced hyperglycemia. *Sulfur-containing Amino Acids*, 1, 159–174 (1978b) (in Japanese).

Copyright © 1981, Spectrum Publications, Inc.
The Effects of Taurine on Excitable Tissues

CHAPTER 25

Taurine Treatment of Human Myotonia: *In Vivo* Study of the Correlations between Taurine and Transmembrane Ion Fluxes

Luca Durelli
Roberto Mutani

The stabilizing action of taurine upon the membranes of excitable tissues is well established. Although taurine seems to influence the ionic conductance of the cellular membrane, it remains unclear which ions are preferentially affected. The amino acid possesses little effect on sodium permeability (Gruener et al., 1976; Nistri and Costanti, 1976) but increases potassium (Read and Welty, 1965; Huxtable, 1976), chloride (Hösli et al., 1975; Nicoll et al., 1976; Hue et al., 1978) or both potassium and chloride conductances (Gruener et al., 1976; Oja and Kontro, 1978). It has been suggested (Dolara et al., 1978; Izumi et al., 1978; Phillis, 1978) that taurine primarily increases the intracellular availability of free calcium, which then affects the ionic permeability of the cellular membrane.

Numerous studies have been devoted to clarifying the role of taurine in nervous, retinal, and cardiac tissue, but only a few communications have dealt with the effects of taurine on skeletal muscle (Baskin and Dagirmanjian, 1973; Gruener et al., 1976). In addition, no attempt to date has been made to correlate the therapeutic effect of taurine in man with a change in ionic permeabilities of the treated tissue. We therefore developed a method that allowed us to quantify the action of this amino acid upon the excitability of skeletal muscle, as well as to indirectly monitor transmembrane ionic flux.

The study was carried out on nine normal volunteers and nine patients suffering from dystrophic myotonia. Myotonia has been defined (Bryant, 1973) as the phenomenon of certain skeletal muscle fibers to produce after-discharges following a small depolarizing stimulus. The instability of the myotonic membrane, which persists after curarization or denervation (Brown and Harvey, 1939), can be explained by increased membrane resistance (Bryant, 1973; Lipiky and Bryant, 1973), resulting in part from a selective decrease in chloride conductance (Bretag, 1973; Lipiky and Bryant, 1973; Barchi, 1975; Furman and Barchi, 1978). Since taurine affects chloride permeability, the use of this amino acid as a therapeutic agent in the treatment of myotonia was examined.

POTASSIUM AND MUSCULAR EXCITABILITY

One of the major difficulties in evaluating clinical studies on myotonia is the lack of proper methodology defining muscular excitability. Most of the common methods suffer from inadequate reliability. The myotonic after-discharge is unevenly elicited by means of electrical stimulation (Desmedt, 1964), and the evaluation of "percussion" myotonia requires a perfectly reproducible tapping force. The most extensively used method is the measure of the relaxation time after maximal voluntary effort (Lewis, 1966). In this case the cooperation of the patient could change with time, and the end point of the decreasing myotonic after-potentials can be difficult to determine with certainty (Leyburn and Walton, 1959). For these reasons, several criteria of evaluation are frequently used together (Griggs et al., 1978).

In the method we utilized (Durelli and Mutani, 1979), the excitability changes of skeletal muscle were simply quantified in terms of the concentration of administered potassium, which produced: (a) in normal volunteers, the appearance of electrically induced muscular after-discharges or spontaneous electrical activity observed by EMG; (b) in myotonic patients, the appearance of spontaneous myotonic discharges. Statistical means and standard deviations could be easily calculated and objectively compared. It is known that the administration of potassium chloride (KCl) aggravates myotonia (Bryant, 1973) and increases the excitability of normal skeletal muscle (Brown, 1937). The action of KCl is partly presynaptic, by increasing cholinergic release (Gage and Quastel, 1965; Parson et al., 1965), but, since its effect remains unchanged after denervation (Brown, 1937), the primary site of action seems to be on the muscle membrane itself. The rise of extracellular KCl induces depolarization (Hodgkin and Horowicz, 1959) accompanied by an increase in membrane resistance, which is called anomalous rectification.

The conduction to sodium and potassium ions decreases, while chloride permeability is not appreciably affected (Boyle and Conway, 1941; Katz and Lou, 1947; Hodgkin and Horowicz, 1959; Falk and Landa, 1960; Adrian and Freygang, 1962a). In addition, the rise of extracellular KCl produces a net inward flux of the salt against an electrochemical gradient (Boyle and Conway, 1941). The cation is the first ion to move across the muscle membrane, promptly followed by chloride, whose transmembrane flux and intracellular accumulation are proportional to the extracellular increase in potassium (Conway, 1957).

Progressively increasing concentrations of KCl were given through the distal end of the radial artery by means of a constant-rate infusion pump. During each infusion, the spontaneous, or electrically induced, activity of the thenar eminence muscle was recorded with needle electrodes. Venous blood samples were also drawn from the origin of the cephalic vein, and actual potassium and chloride levels were compared with the expected levels computed from the dilution of a dye (Evans Blue) simultaneously given with KCl. Since the dye cannot leave the vascular space, data showing that the expected and actual values are statistically identical indicate that the administered electrolytes have not left the plasma. On the other hand, when the expected and actual values are dissimilar a movement of the electrolytes has occurred. Since the thenar eminence muscle constitutes most of the metabolically active tissue in the short arteriovenous circle studied, any movement could be ascribed to changes in the concentration of ions in these muscle cells. The direction of the transmembrane ion movement is expressed by the sign of the difference between the expected and actual values. Other authors (Grob et al., 1957) used the arteriovenous difference as an indirect index of electrolyte movement. However, this index was not utilized in the present study because it could not account for dynamic changes in arterial blood flow or plasma dilution due to venous return from nonperfused areas.

The results in Fig. 25.1 show the appearance of muscle hyperexcitability during KCl infusions in both normal and myotonic subjects. In normal volunteers (Fig. 25.1a), at a potassium infusion concentration of 113.9 ± 17.6 mEq/liter (corresponding to an actual venous level of 8.8 ± 1.3 mEq/liter), after-discharges similar to that displayed in Fig. 25.2b appeared following indirect stimulation of the muscle. In Fig. 25.2, muscle responses to supramaximal stimulation of the motor nerve are shown. In Fig. 25.2b, a polyphasic response is presented that cannot be interpreted in terms of the type of decreased conduction velocity observed in chronic potassium intoxication (Buchtal and Engbaek, 1957; Troni et al., 1978), since the duration of the response is not increased. The last negative peak of Fig. 25.2b can be explained by the limited number of hyperexcitable motor units developing after-discharge oscillations of membrane potential wide enough to reach the

A

B

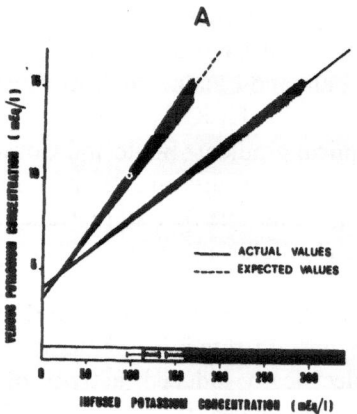 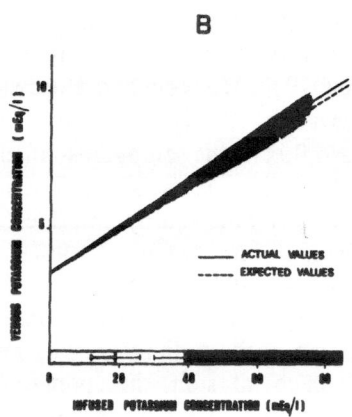

Fig. 25.1. Regression lines of the actual (solid lines) and expected (dashed lines) venous potassium levels before taurine treatment plotted against potassium intra-arterial infusion concentration. In this and the following figures, each line is represented within the standard error range of its regression coefficient (shaded area). The blank area above the abscissa in (*a*) corresponds to the absence of electrically induced muscular after-discharges; the dotted area, to the presence of such after-discharges; and the black area, to spontaneous discharges in normal volunteers. Similarly, the blank area above the abscissa in (*b*) corresponds to the absence of electrically induced myotonic after-discharges; the dotted area, to the presence of such discharges; and the black area, to spontaneous myotonic discharges in myotonic patients.

threshold for a second action potential. At a potassium infusion concentration of 161.9 ± 20.9 mEq/liter (corresponding to a venous level of 11.0 ± 1.5 mEq/liter), spontaneous electrical activity of two different patterns could be recorded from the studied muscle (Fig. 25.1a). In six subjects, motor unit activity discharging in multiple-spike trains was recorded, while in the other subjects, progressive high-frequency-spiking motor units were observed. The spontaneous activity appeared 10–40 sec after the beginning of the KCl infusion and ended 5–10 sec after the pump was stopped.

In myotonic patients, the response to KCl administration was biphasic (Fig. 25.1b). At a potassium infusion concentration of 19.0 ± 6.9 mEq/liter (corresponding to a venous level of 5.1 ± 0.6 mEq/liter), a striking improvement of the myotonic phenomenon occurred. The electrically induced after-discharges and, sometimes, even percussion myotonia disappeared. On the other hand, at a potassium infusion concentration of 38.9 ± 8.9 mEq/liter (corresponding to a venous level of 6.5 ± 0.7 mEq/liter), spontaneous myotonic discharges appeared. It should be pointed out that spontaneous discharges occurred in myotonic patients at KCl infusion concentrations significantly lower than in normal subjects ($p < 0.001$; Student t-test).

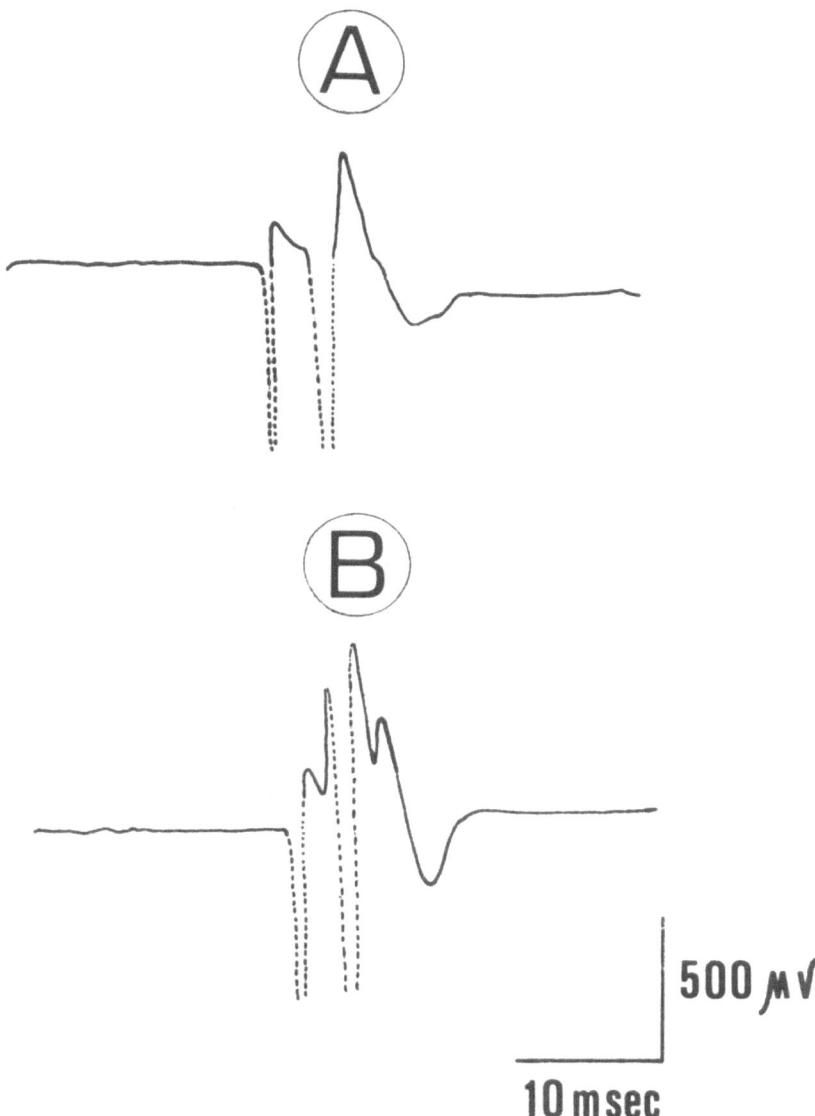

Fig. 25.2.   Motor responses recorded in a normal volunteer from the opponens pollicis muscle following maximal electrical stimulation of the median nerve at the wrist with single shocks of 0.2 msec duration, before (*a*) and during (*b*) intra-arterial KCl loading (potassium infusion concentration of 130 mEq/liter).

Significant regression lines ($p < 0.001$; Student t-test) were calculated for the actual and expected potassium venous levels as a function of the infused potassium concentration in both normal subjects and myotonic patients. (Fig. 25.1). In normal volunteers, the regression coefficient for the actual values is lower ($p < 0.01$; Student t-test) than the coefficient for the expected values (Fig. 25.1a). Significant regression lines ($p < 0.001$; Student t-test), plotting the differences between the expected and actual values of potassium and chloride venous levels against potassium infusion concentration, are shown in Fig. 25.3. These results are consistent with previous studies indicating that in normal muscle, potassium-induced hyperexcitability is associated with a net inward movement of KCl proportional to the external potassium concentration (Boyle and Conway, 1941; Grob et al., 1957). In contrast, no significant difference could be calculated between the line slope of the actual and expected venous values in myotonic patients (Fig. 25.1b). Likewise, no significant regression lines were observed when plotting the difference between the expected and actual values of potassium and chloride venous levels against potassium infusion concentration (Fig. 25.3). These results suggest that an abnormality in ion movement across the membrane of human myotonic muscle exists *in vivo*. Since the reduction of chloride conductance seems to be the most substantiated defect of the myotonic membrane (Bryant, 1973; Furman

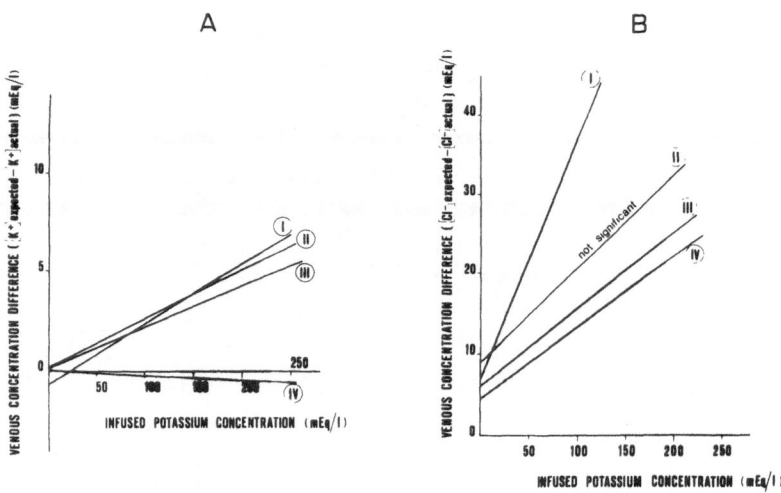

Fig. 25.3. Regression lines of the differences between the expected and actual potassium (*a*) or chloride (*b*) venous levels plotted against potassium intra-arterial infusion concentration. In (*a*), lines I and II are for normal volunteers, before and after taurine treatment, respectively; lines III and IV are for myotonic patients, after and before taurine treatment, respectively. In (*b*), lines I and II are for myotonic patients, after and before taurine administration, respectively; lines III and IV are for normal volunteers before and after taurine administration, respectively.

and Barchi, 1978), myotonia is similar to the *in vitro* muscle model in which
the external chloride is replaced by impermeable anions (Falk and Landa,
1960; Bretag, 1973). In this experimental situation, muscle fibers become
hyperexcitable and exhibit reduced chloride conductance. In addition, they
are much more sensitive than normal muscle cells to increased external potassium (Falk and Landa, 1960). When chloride permeability is affected, an increase of potassium at the external surface of the membrane will produce
depolarization and cathodal rectification (Falk and Landa, 1960), although
an intracellular accumulation of potassium will not occur since the cation
cannot move across the membrane with the simulaneous transport of chloride
(Conway, 1957).

Our results are therefore consistent with the hypothesis that chloride conductance in the myotonic membrane is impaired and that the difference in
sensitivity between the normal and myotonic muscle in response to KCl
results from their varying chloride permeability. The repetitive discharge of
the myotonic membrane has been likened to that of the cardiac pacemaker
muscle fibers (Bryant, 1973). In both cases a rhythmic inactivation of potassium permeability after the spike maintains the membrane potential sufficiently depolarized so that the threshold may be reached and a new action
potential triggered (Bryant, 1973). When the external potassium concentration is raised, the repetitive activity is enhanced since potassium conductance
is reduced both through the anomalous rectification mechanism (Katz and
Lou, 1947; Adrian and Freygang, 1962a) and by a block in delayed rectification (Adrian and Freygang, 1962b). In addition, elimination of the chloride
current during the negative after-potential, which occurs in myotonic muscle,
renders the muscle more excitable even with small increases in external potassium (Bryant, 1973; Adrian and Bryant, 1974; Barchi, 1975).

As seen in Fig. 25.1b, KCl infusion at low concentrations antagonizes the
myotonic phenomenon. This result can be interpreted as an effect of elevated
extracellular potassium on membrane conductance (Katz and Lou, 1947;
Adrian and Freygang, 1962a).

By raising external potassium, membrane conductance to cations initially
increases and then, at higher potassium concentrations, decreases. The initial increase presumably counteracts the closing of the potassium channels responsible for the pacemaker myotonic potentials, thus making repetitive activity less likely.

ACUTE TAURINE TREATMENT OF MUSCLE
HYPEREXCITABILITY

No significant difference in the basal plasma concentration of taurine was observed between normal volunteers and myotonic patients (Table 25.1). After taurine (O-DUE®, Falorni, Italy) parenteral treatment, each of the 18 subjects was given KCl infusions at progressively increasing concentrations. Six normal individuals and six myotonic patients received taurine intravenously at a dosage of 200 mg/kg. This treatment increased the plasma level of the amino acid more than 20-fold (Table 25.1) Mutani et al., 1978). The six remaining subjects were given intra-arterial injections of taurine at a dosage of 3–5 μmol/ml/min, which raised plasma taurine content approximately fourfold (Table 25.1). The results obtained by the two procedures were similar and were pooled for statistical analysis.

Both in normal volunteers and in myotonic patients, taurine greatly decreased potassium-induced muscle hyperexcitability. After taurine treatment no spontaneous discharges could be observed in normal subjects at the infusion concentrations that previously were effective. Moreover, electrically induced after-discharges were observed in only six out of the nine subjects, and only when the potassium infusion concentration was increased to 157.2 ± 13.4 mEq/liter (corresponding to a potassium venous level of 10.3 ± 1.1 mEq/liter). This is significantly higher than the concentration required before taurine treatment ($p < 0.01$; Student t-test). In myotonic patients, taurine administration had little effect on the initial improvement of the myotonic phenomenon that occurred at low KCl infusion concentrations. However, taurine increased the concentration of potassium (65.2 ± 10.4 mEq/liter, corresponding to a venous level of 7.6 ± 0.9 mEq/liter) required to elicit myotonic spontaneous discharges.

Statistical analysis of the electrolyte venous content of normal volunteers revealed the following: (a) significant regression lines ($p < 0.001$; Student t-

Table 25.1.   Venous plasma levels of taurine in normal volunteers and myotonic patients before taurine treatment, after intravenous administration, or during intra-arterial infusion of the amino acid[a]

		Normal Subjects (μmol/liter)	Myotonic patients (μmol/liter)
Basal values	(9)	54.4 ± 25.7	53.4 ± 22.9
Intravenous treatment[b]	(6)	1206.6 ± 227.6	1141.8 ± 150.8
Intra-arterial treatment[c]	(3)	234.0 ± 24.6	240.3 ± 34.3

[a]Values are expressed as means ± SD. Figures in parentheses represent the number of subjects.
[b]Dosage of 200 mg/kg.
[c]Dosage of 3–5 μmol/ml/min.

test) of the expected and actual venous levels of potassium plotted against potassium infusion concentrations (Fig. 25.4a) showed significant differences in the line slopes ($p < 0.01$; Student t-test); (b) regression lines of the differences between the expected and actual values of plasma potassium and chloride plotted against potassium infusion concentrations were significant ($p < 0.001$; Student t-test) (Fig. 25.3); (c) no significant difference between the regression coefficients of the lines before and after taurine treatment was observed. In contrast, the data obtained for myotonic patients were remarkably affected by taurine: (a) the slope of the regression line of the actual venous potassium level plotted against potassium infusion concentration (Fig. 25.4b) is significantly lower than the slope of the corresponding line calculated before taurine treatment ($p < 0.02$; Student t-test); (b) the regression lines plotting the differences between the expected and actual values of the venous potassium and chloride content against potassium infusion concentration (Fig. 25.3), which were not significant before taurine treatment became significant ($p < 0.01$; Student t-test) after it; (c) the line slope for the potassium concentration differences is similar to that of the corresponding line for normal volunteers (Fig. 25.3a).

The results suggest that in myotonic patients, taurine decreased potassium-induced muscle hyperexcitability and increased ion movement across the membrane, leading to an intracellular accumulation of KCl as observed in normal muscle. The stabilizing action of taurine upon excitable tissues has been associated with an increased membrane conductance for potassium

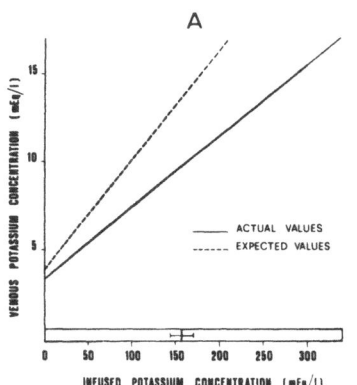
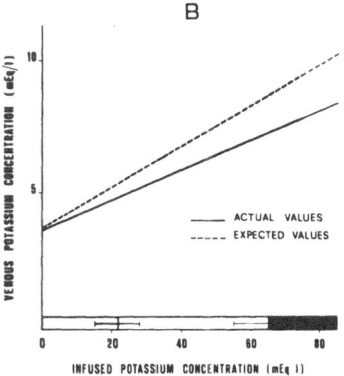

Fig. 25.4. Regression lines of the actual (solid lines) and expected (dashed lines) venous potassium levels after acute taurine parenteral treatment in normal volunteers (*a*) and myotonic patients (*b*). The lines are plotted against the potassium intra-arterial infusion concentration. The abscissas in (*a*) and (*b*) correspond to those described in the legend to Fig. 25.1.

(Read and Welty, 1965; Huxtable, 1976), for chloride (Hösli et al., 1975; Nicoll et al., 1976; Hue et al., 1978), or for both ions simultaneously (Gruener et al., 1976; Oja and Kontro, 1978). Our results are consistent with the hypothesis that *taurine acts by increasing chloride conductance alone or chloride and potassium conductances simultaneously*. We have studied a model of hyperexcitable muscle in which only the chloride conductance is decreased, and the potassium permeability is normal or enhanced (Bryant, 1973; Lipiky and Bryant, 1973). If the stabilizing effect of taurine treatment was associated with only an increase in potassium permeability, then the reduced chloride conductance would have inhibited the inward flux of potassium ions (Conway, 1957). Thus only an increase of chloride conductance can fully explain the observation that KCl accumulates intracellularly in the myotonic muscle after taurine treatment.

In normal volunteers no changes in the regression coefficients of the calculated lines were observed with taurine administration. This result also supports the hypothesis that taurine preferentially alters chloride permeability. Increases in ion conductance can only be detected by an indirect method (as the one used in the present study) if basal permeability is low (Shanes, 1958). Usually, chloride conductance represents at least 70% of total resting membrane conductance of the normal skeletal muscle (Bryant, 1973) and is high compared with potassium ion permeability (Falk and Landa, 1960). In addition, chloride conductance is not appreciably affected by increases in extracellular potassium concentration (Hodgkin and Horowicz, 1959), although this change decreases the membrane permeability of other cations (Boyle and Conway, 1941). Since ion movement across the membrane of the normal muscle is already pronounced before taurine treatment, the increase in chloride permeability produced by taurine would probably not affect ion flux sufficiently to change venous electrolyte concentration. However, in myotonia, where chloride conductance is impaired, ionic flux is abnormal. Thus, the increase in chloride conductance produced by taurine would change ion transmembrane movement and thereby affect venous electrolyte content.

CHRONIC TAURINE TREATMENT OF MYOTONIA

Myotonic patients were administered oral taurine at a dosage of 100–150 mg/kg for 6 months. Takahashi and Nakane (1978) have shown that a similar chronic treatment can produce a three- to fourfold increase in the basal plasma concentration of the amino acid, and the effectiveness of the intraarterial administration of taurine showed that such an increment in plasma content could decrease the myotonic phenomenon. At the end of the 6-month period, the patients were interviewed and submitted to both clinical and routine EMG examinations. In addition, the KCl intra-arterial loading test was repeated in six out of the nine patients.

The plasma concentration of taurine was determined to be 111.1 ± 62.3 μmol/liter, that is, about two times the basal value. All patients showed an improvement in the common myotonic symptoms; for example, they showed improvements in chewing, walking, or shaking hands. In most patients myotonic after-discharges were still elicited by the insertion of the EMG electrode or by percussion of their thenar eminence muscle.

A further improvement of the myotonic phenomenon (complete absence of electrically induced myotonic after-discharges and sometimes even of percussion myotonia) was achieved by intra-arterial administration of KCl to raise the potassium plasma level to 4.4 ± 0.3 mEq/liter, which is significantly lower ($p < 0.02$; Student t-test) than the corresponding value before treatment. Spontaneous myotonic discharges occurred at a potassium infusion concentration of 84.3 ± 9.1 mEq/liter (corresponding to a venous level of 8.9 ± 1.3 mEq/liter of potassium), which is significantly higher ($p < 0.05$; Student t-test) than the amount required to elicit the response after acute taurine administration. Thus, chronic taurine treatment appears to reduce the excitability of the myotonic muscle. It is also of interest that oral taurine administration extends the range of potassium venous concentrations over which an improvement in the myotonic phenomenon occurs (from 4.4 to 8.9 mEq/liter), such that improvement is observed at a physiological concentration of potassium. Although the relationship between plasma potassium concentration and the severity of myotonia is still controversial (Leyburn and Walton, 1960; Griggs et al., 1978), our results imply that during chronic taurine treatment a moderate increase in plasma potassium can produce some improvement in the symptoms. However, since taurine and potassium plasma concentrations were not monitored on a daily basis, it is impossible to be certain that the daily fluctuations in the severity of myotonia were associated with changes in potassium levels.

Significant regression lines ($p < 0.001$; Student t-test) were calculated for the expected and actual potassium venous levels and plotted against potassium infusion concentration (Fig. 25.5a). The regression coefficient of the line obtained from the actual values is significantly lower ($p < 0.001$; Student t-test) than that obtained from the expected values, suggesting an outward movement of potassium ions from the vascular space. This view is also consistent with the significant regression lines ($p < 0.01$; Student t-test) calculated for the plot of the differences between the expected and actual values of potassium and chloride against the potassium infusion concentrations (Figs. 25.5b and 25.5c). In addition, the regression coefficient of the line plotted for differences between the expected and actual values of potassium is larger ($p < 0.01$; Student t-test) than that of the corresponding line obtained

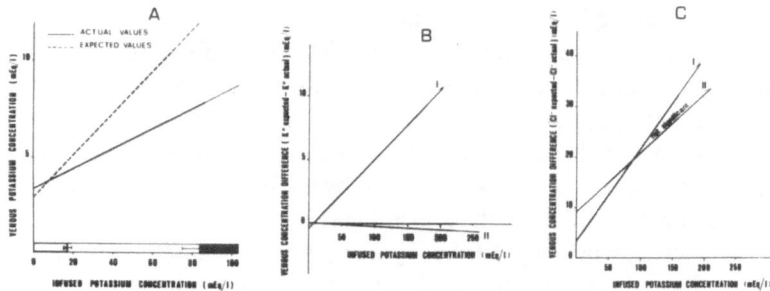

Fig. 25.5. Regression lines after chronic taurine oral treatment in myotonic patients: (*a*) actual (solid line) and expected (dashed line) venous potassium levels, and differences between expected and actual values for potassium (*b*) or chloride (*c*) plotted against potassium intra-arterial infusion concentration. The blocks above the abscissa in (*a*) are as explained for Fig. 25.1(*b*). In (*b*) and (*c*), the lines obtained after chronic treatment (line I) are compared with lines calculated before taurine (line II). Line II in (*b*) is not statistically significant.

after acute taurine treatment. These data support the hypothesis that movement of potassium and chloride ions into the intracellular space is increased more by chronic taurine administration than by acute treatment. Thus, the more pronounced the taurine-mediated reduction in muscle hyperexcitability, the larger the increase in transmembrane ion flux. This again confirms the view that the action of taurine upon hyperexcitable myotonic muscle is associated with an increase in membrane conductance.

In addition, the stabilizing action of taurine upon chronic administration raises the possibility of a dual site of action for the amino acid (Mutani et al., 1974; Van Gelder, 1978). The local binding to receptors at the membrane surface may account for its immediate effects seen in *in vitro* studies (Gruener et al., 1976; Izumi et al., 1978), in iontophoretic experiments (Hösli et al., 1975; Phillis, 1978), or in acute administration to humans or animals (Mutani et al., 1974, 1978). A delayed, long-lasting, and even more remarkable effect can be explained in terms of intracellular accumulation of taurine (Huxtable and Bressler, 1973), which is particularly pronounced in heart or skeletal muscle (Jacobsen and Smith, 1968). A modulating role of taurine on amino acid metabolism (Mutani et al., 1978; Van Gelder, 1978) and on the availability of intracellular calcium (Huxtable, 1976; Dolara et al., 1978) has been proposed.

In conclusion, though the real effectiveness of taurine in the treatment of myotonic disorders requires assessment by more extensive clinical experiments, the present study has extended the research on taurine to hyperexcitable tissues. Our results indicate that the action of taurine is associated with enhanced ion movement across the cell membrane in man. In particular, the results are consistent with the hypothesis that taurine acts on the muscle membrane by increasing the conductance of either chloride ion alone or both chloride and potassium simultaneously.

ACKNOWLEDGMENT

The authors gratefully acknowledge the financial help provided by Falorni S.p.A. (Florence, Italy) throughout these studies.

REFERENCES

Adrian, R. H.; and Bryant, S. H. On the repetitive discharge in myotonic muscle fibers. *J. Physiol.* (London), 240, 505–515 (1974).

Adrian, R. H.; and Freygang, W. H. The potassium and chloride conductance of frog muscle membrane. *J. Physiol.* (London), 163, 61–103 (1962a).

Adrian, R. H.; and Freygang, W. H. Potassium conductance of frog muscle membrane under controlled voltage. *J. Physiol.* (London), 163, 104–114 (1962b).

Barchi, R. L. Myotonia—An evaluation of the chloride hypothesis. *Arch. Neurol.* (Chicago), 32, 175–180 (1975).

Baskin, S. L.; and Dagirmanjian, R. Possible involvement of taurine in the genesis of muscular dystrophy. *Nature,* 245, 464–465 (1973).

Boyle, P.; and Conway, E. J. Potassium accumulation in muscle and associated changes. *J. Physiol.* (London), 100, 1–63 (1941).

Bretag, A. H. Mathematical modelling of the myotonic action potential. In *New Developments in Electromyography and Clinical Neurophysiology, Vol. 1,* J. E. Desmedt, ed. Karger, Basel (1973), pp. 464–482.

Brown, G. L. The action of potassium chloride on mammalian muscle. *J. Physiol.* (London), 91, 4P–5P (1937).

Brown, G. L.; and Harvey, A. M. Congenital myotonia in the goat. *Brain,* 62, 343–363 (1939).

Bryant, S. H. The electrophysiology of myotonia with a review of congenital myotonia of goats. In *New Developments in Electromyography and Clinical Neurophysiology, Vol. 1,* J. E. Desmedt, ed. Karger, Basel (1973), pp. 420–450.

Buchthal, F.; and Engbaek, L. Propagation velocity of intracellularly recorded action-potentials in striated frog muscle fibres. *Acta Physiol. Scand.,* 42, Suppl. 145 (1957).

Conway, E. J. Nature and significance of concentration relations of potassium and sodium ions in skeletal muscle. *Physiol. Rev.,* 37, 84–132 (1957).

Desmedt, J. E. Observations sur la réaction myotonique en stimulo-détection. *Revue Neurol.,* 110, 324–336 (1964).

Dolara, P.; Ledda, F.; Mugelli, A.; Mantelli, L.; Zilletti, L.; Franconi, F.; and Giotti, A. Effect of taurine on calcium, inotropism, and electrical activity of the heart. In *Taurine and Neurological Disorders,* A. Barbeau and R. J. Huxtable, eds. Raven Press, New York (1978), pp. 151–159.

Durelli, L.; and Mutani, R. Myotonia, potassium and taurine—A preliminary report. *J. Neurol. Sci.,* 42, 103–109 (1979).

Falk, G.; and Landa, J. F. Effects of potassium of frog skeletal muscle in a chloride-deficient medium. *Amer. J. Physiol.*, 198, 1225–1231 (1960).

Furman, R. E.; and Barchi, R. L. The pathophysiology of myotonia produced by aromatic carboxylic acids. *Ann. Neurol.*, 4, 357–365 (1978).

Gage, P. W.; and Quastel, D. M. J. Dual effect of potassium on transmitter release. *Nature*, 206, 625–626 (1965).

Griggs, R. C.; Moxley, R. T.; Riggs, J. E.; and Engel, W. K. Effects of acetazolamide on myotonia. *Ann. Neurol.*, 3, 531–537 (1978).

Grob, D.; Liljestrand, A., and Johns, R. J. Potassium movement in normal subjects: Effect on muscle function. *Amer. J. Med.*, 23, 340–355 (1957).

Gruener, R.; Bryant, H.; Markowitz, D.; Huxtable, R.; and Bressler, R. Ionic action of taurine on nerve and muscle membrane: Electrophysiologic studies. In *Taurine*, R. Huxtable and A. Barbeau, eds. Raven Press, New York (1976), pp. 225–242.

Hodgkin, A. L.; and Horowicz, P. The influence of potassium and chloride ions on the membrane potential of single muscle fibres. *J. Physiol.* (London), 148, 127–160 (1959).

Hösli, L.; Hösli, E.; Andres, P. F.; and Wolff, J. R. Amino acid transmitters—Action and uptake in neurons and glial cells of human and rat CNS tissue culture. In *Golgi Centennial Symposium*, M. Santini, ed. Raven Press, New York (1975), pp. 473–488.

Hue, B.; Pelhate, M.; and Chanelet, J. Sensitivity of postsynaptic neurons of the insect central nervous system to externally applied taurine. In *Taurine and Neurological Disorders*, A. Barbeau and R. J. Huxtable, eds. Raven Press, New York (1978), pp. 225–236.

Huxtable R. Metabolism and function of taurine in the heart. In *Taurine*, R. Huxtable and A. Barbeau, eds. Raven Press, New York (1976), pp. 99–119.

Huxtable, R.; and Bressler, R. Effect of taurine on a muscle intracellular membrane. *Biochim. Biophys. Acta*, 323, 573–583 (1973).

Izumi, K.; Ngo, T. T.; and Barbeau, A. Metabolic modulation in central nervous system by taurine. In *Taurine and Neurological Disorders*, A. Barbeau and R. J. Huxtable, eds. Raven Press, New York (1978), pp. 137–149.

Jacobsen, J. G.; and Smith, L. H., Jr. Biochemistry and physiology of taurine and taurine derivatives. *Physiol. Rev.*, 48, 424–511 (1968).

Katz, B.; and Lou, C. H. Electric rectification in frog's muscle. *J. Physiol.* (London), 106, 29P–30P (1947).

Lewis, I. Trial of diazepam in myotonia: A double blind single crossover study. *Neurology Minneapolis*, 16, 831–836 (1966).

Leyburn, P.; and Walton, J. N. The treatment of myotonia: A controlled clinical trial. *Brain*, 82, 81–91 (1959).

Leyburn, P.; and Walton, J. N. The effect of changes in serum potassium upon myotonia. *J. Neurol. Neurosurg. Psychiatry*, 23, 119–126 (1960).

Lipiky, R. J.; and Bryant, S. H. A biophysical study of the human myotonias. In *New Developments in Electromyography and Clinical Neurophysiology, Vol. 1*, J. E. Desmedt, ed. Karger, Basel (1973), pp. 451–463.

Mutani, R.; Bergamini, L.; Fariello, R.; and Delsedime, M. Effects of taurine on cortical acute epileptic foci. *Brain Res.*, 70, 170–173 (1974).

Mutani, R.; Bergamini, L.; and Durelli, L. Taurine in experimental and human epilepsy. In *Taurine and Neurological Disorders*, A. Barbeau and R. J. Huxtable, eds. Raven Press, New York (1978), pp. 359–373.

Nicoll, R. A.; Padjen, A.; and Barker, J. L. Analysis of amino acid responses on frog motoneurones. *Neuropharmacology*, 15, 45–53 (1976).

Nistri, A.; and Costanti, A. The action of taurine on the lobster muscle fibre and frog spinal cord. *Neuropharmacology*, 15, 635–641 (1976).

Oja, S. S.; and Kontro, P. Neurotransmitter actions of taurine in central nervous system. In *Taurine and Neurological Disorders*, A. Barbeau and R. J. Huxtable, eds. Raven Press, New York (1978), pp. 181-200.

Parson, R. L.; Hofmann, W. W.; and Feigen, G. A. Presynaptic effects of potassium ion on the mammalian neuromuscular junction. *Nature*, 208, 590-591 (1965).

Phillis, J. W. Overview of neurochemical and neurophysiological actions of taurine. In *Taurine and Neurological Disorders*, A. Barbeau and R. J. Huxtable, eds. Raven Press, New York (1978), pp. 289-303.

Read, W. O.; and Welty, J. D. Taurine as a regulator of cell potassium in the heart. In *Electrolytes and Cardiovascular Diseases*, E. Bajusz, ed. Karger, Basel and New York (1965), pp. 70-85.

Shanes, A. M. Electrochemical aspects of physiological and pharmacological action in excitable cells: 1. The resting cell and its alteration by extrinsic factors. *Pharmacol. Rev.*, 10, 59-164 (1958).

Takahashi, R.; and Nakane, Y. Clinical trial of taurine in epilepsy. In *Taurine and Neurological Disorders*, A. Barbeau and R. J. Huxtable, eds. Raven Press, New York (1978), pp. 375-385.

Troni, W.; Bruetto, S.; and Genero, O. Electromyographic study on experimental hyperkalaemia: 1. Electrophysiological alterations. *Acta Neurol.* (Napoli), 33, 381-389 (1978).

Van Gelder, N. M. Glutamic acid and epilepsy: The action of taurine. In *Taurine and Neurological Disorders*, A. Barbeau and R. J. Huxtable, eds. Raven Press, New York (1978), pp. 387-402.

Copyright © 1981, Spectrum Publications, Inc.
The Effects of Taurine on Excitable Tissues

CHAPTER 26

Taurine: Significance in Human Nutrition

David K. Rassin
Gerald E. Gaull

Considerable biochemical evidence exists for the importance of taurine during development in man and in other mammals. The only proven biochemical role for this compound in man, however, is its conjugation with bile acids to form bile salts, which are essential for fat absorption. It is not incorporated into proteins, but small amounts of taurine-containing peptides have been found in brain (Reichelt and Kvamme, 1973; Reichelt and Edminson, 1974). The possible importance of taurine to neurotransmission, retinal function, cardiac function, muscle function, and epilepsy has been extensively reviewed (Jacobsen and Smith, 1968; Huxtable and Barbeau, 1976; Barbeau and Huxtable, 1978). The role of taurine in development also has been reviewed (Sturman et al., 1977, 1978a; Gaull and Rassin, 1979).

The high concentrations of taurine in brain during fetal development of the rat, the monkey, the rabbit, and man at a time when the presumed major synthetic pathway (catalyzed by cysteinesulfinic acid decarboxylase) has little measurable activity implies that an exogenous source of taurine is necessary. In addition, man does not develop the high activity of hepatic cysteinesulfinic acid decarboxylase observed in the rat; therefore, he may always be dependent upon a dietary supply of taurine (Gaull et al., 1977). The growing evidence for the role of taurine during development and the dietary requirement for taurine in man have led us to concentrate our investigations on taurine nutrition in the neonate.

An interesting pattern of plasma and urine amino acid concentrations was observed in a study designed to investigate the effects on preterm infants of feeding formulas containing different protein quantity and quality, and comparing these infants with others fed human milk (Raiha et al., 1976; Gaull et al., 1977; Rassin et al., 1977a, 1977b). The amino acid concentrations in the plasma and the urine of formula-fed infants were generally either increased or unchanged when compared with those of infants fed banked human milk. There was one striking exception: From the first week of the study, the concentration of taurine in the urine of infants fed the formulas was lower than that of infants fed human milk. The plasma taurine concentration decreased steadily, and by the fourth week of the study, it was significantly lower in infants fed formulas than in infants fed human milk (Gaull et al., 1977) (Fig. 26.1).

Taurine was the only amino acid measured to have this pattern of apparent deficiency. Human milk has a considerable amount of taurine, whereas bovine milk, from which most formulas are prepared, contains only minimal amounts of taurine (Gaull et al., 1977; Rassin et al., 1978a) (Table 26.1). The formulas in this study of preterm infant feeding, thus, reflected the lack of taurine in the bovine milk from which the formulas were prepared. Indeed, we have analyzed a number of commercially available infant formulas and find that they all contain only small amounts of taurine (Table 26.1); soy-based formulas contain none. Other investigators also have documented but have not commented upon the low concentrations of taurine in plasma of human preterm infants (Dickinson et al., 1970) and in urine of human full-term infants (Jonxis, 1951; Souchon, 1952; Jagenburg, 1959) fed artificial formulas when compared with infants fed human milk.

We are currently investigating the metabolic effects of feeding artificial formulas to full-term human infants. Feeding the full-term infant, of course, represents a more general nutritional problem than that represented by the feeding of the preterm infant, who represents a new and special biological entity. Preliminary results indicate that plasma and urine concentrations of taurine in full-term infants fed taurine-deficient formulas are lower than those found in breast-fed human infants (Rassin et al., 1979) (Table 26.2). Furthermore, the concentrations of taurine in plasma and urine of breast-fed term infants are considerably higher than those of preterm infants fed pooled human milk at a fixed volume. The reason for this difference is not yet clear. In effect, however, the differences in plasma and urine taurine concentrations between infants fed formulas and those fed human milk are greater in term infants than they are in preterm infants.

Thus, the rapidly growing human infant appears to have a dietary requirement for taurine. No adverse clinical signs or symptoms have been identified in human infants fed commercial formulas, although no systematic investigation has yet been made. However, cats and kittens fed a synthetic diet containing partially purified casein as the source of protein become taurine deficient and develop retinal degeneration. This degeneration, which eventually

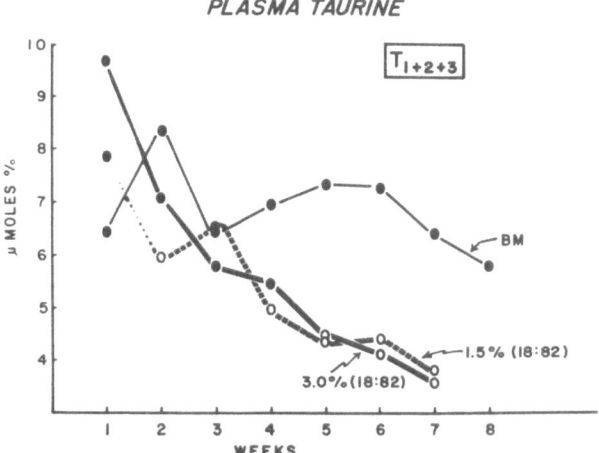

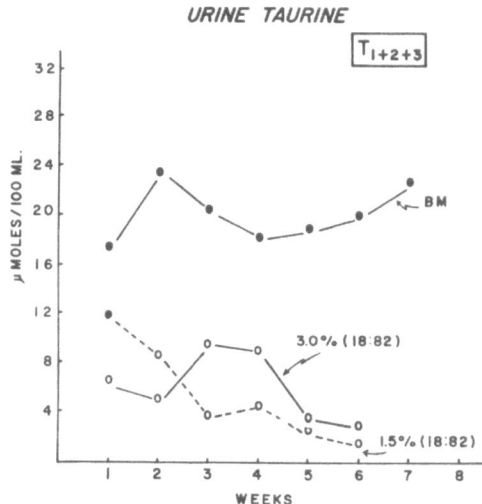

Fig. 26.1. Effect of a low-taurine diet on mean plasma (*a*) and mean urine (*b*) concentration of taurine in preterm infants as a function of age. BM = pooled, banked human milk; 1.5% (18:82) = 1.5 g% protein with 18 parts bovine whey and 82 parts bovine casein proteins; 3.0% (18:82) = 3.0 g% protein preparation of the same formula. (Adapted from Gaull et al., 1977.)

Table 26.1. Taurine content of milk

	Taurine (μmol %)
Species	
Cat	287
Rhesus monkey	56
Man: American[a]	34
Man: Pooled, banked Finnish[b]	26
Rat	15
Rabbit	14
Cow	1
Formulas[c]	
1.5 g% (60:40)	2
1.5 g% (60:40)	3
1.5 g% (18:82)	1
3.0 g% (18:82)	1
Formulas[d]	
Similac® (Ross Laboratories)	2
Enfamil® (Mead Johnson)	3
SMA® (Wyeth Laboratories)	3

[a]These data were obtained from the analysis of individually collected samples from women in the United States (Rassin et al., 1978a).

[b]These data were obtained from the analysis of samples of pooled, banked milk from Finnish women that was used as the control diet in the preterm infant feeding studies discussed in the text (Gaull et al., 1977).

[c]These data were obtained from the analysis of specially prepared formulas comparable to various commercially available formulas used as the experimental diets in the preterm infant feeding studies discussed in the text (Gaull et al., 1977).

[d]These data were obtained from the analysis of randomly selected cans of infant formula bought in a supermarket (Rassin and Gaull, unpublished observations).

results in blindness, can be prevented or reversed by feeding taurine, but not by feeding methionine, cysteine, or inorganic sulfate (Berson et al., 1976; Knopf et al., 1978). In addition, cats fed dog food, apparently prepared differently from cat food and lacking in taurine, also become taurine deficient and develop retinal lesions (Aguirre, 1978).

The retina of the cat is especially dependent upon taurine and actively resists depletion of this amino acid. Thus, when the taurine concentration in most organs, including the brain, has decreased 10-fold, that in the retina has decreased less than 50%. Retinal degeneration appears to begin when the taurine concentration has been reduced by more than 50% (Sturman et al., 1978b; Rassin et al., 1978b). These findings demonstrate that the structural

Table 26.2. Taurine in plasma and urine of full-term infants[a]

	Weeks after birth				
	1	2	4	8	12
Plasmas (μmol/dl $\pm$ SEM)					
Human milk	8.2 $\pm$ 0.6	9.8 $\pm$ 1.0	9.7 $\pm$ 0.5	9.0 $\pm$ 0.4	11.2 $\pm$ 1.1
1.5 g% (60:40)	7.8 $\pm$ 1.1	6.4 $\pm$ 1.9	6.7 $\pm$ 1.3[b]	6.6 $\pm$ 0.5[c]	6.3 $\pm$ 0.7[b]
1.5 g% (18:82)	8.2 $\pm$ 1.5	5.5 $\pm$ 1.3[b]	4.2 $\pm$ 0.4[c]	5.2 $\pm$ 0.7[c]	3.5 $\pm$ 0.6[c]
Urines (μmol/dl $\pm$ SEM)					
Human milk	49.6 $\pm$ 2.2	39.1 $\pm$ 8.0	39.1 $\pm$ 3.5	33.1 $\pm$ 3.5	42.0 $\pm$ 7.9
1.5 g% (60:40)	28.1 $\pm$ 6.5	5.1 $\pm$ 2.6[c]	2.5 $\pm$ 1.0[b]	3.2 $\pm$ 1.1[c]	3.6 $\pm$ 1.1[c]
1.5 g% (18:82)	12.2 $\pm$ 4.5[b]	1.9 $\pm$ 0.9[c]	1.0 $\pm$ 0.2[c]	0.6 $\pm$ 0.1[c]	1.3 $\pm$ 0.5[c]

[a]Preliminary results from infants fed human milk tested against those from infants fed 1.5 g% protein with either 60:40 or 18:82 ratio of whey to casein proteins (Rassin et al., 1979).
[b]$p < 0.05$.
[c]$p < 0.01$.

integrity of the retina of the cat, especially in the photoreceptor cells, is dependent upon a dietary supply of taurine. Thus, taurine is implicated in a structural function in this experimental model as well as in the potential neurotransmitter or neuromodulator role that also has been proposed. It is likely that the retina of the kitten would be even more susceptible to a deficiency of taurine during early development, prenatal or neonatal, rather than at maturity, when a large pool of taurine has already been acquired.

The effects of feeding a taurine-deficient diet (artificial formulas) to the human infant may be more moderate than those observed in the kitten, for a number of reasons: The human infant conjugates bile acids solely with taurine at birth, but later develops the capacity to conjugate with glycine as well (Encrantz and Sjovall, 1959; Sjovall, 1960; Poley et al., 1964; Challacombe et al., 1975). Human infants fed human milk remain predominantly taurine conjugators of bile acids, whereas those fed taurine-deficient formulas become predominantly glycine conjugators of bile acids (Brueton et al., 1978; Watkins et al., 1979). The cat conjugates bile acids with taurine only (Rabin et al., 1976). The ability to conjugate bile acids with both taurine and glycine may serve to ameliorate the effects of taurine deficiency in man. In addition, even those human infants fed the currently available taurine-deficient formulas usually have foods containing taurine added to their diet within a few weeks or months of birth. Even in the cat after 10 weeks of a taurine-deficient diet, the bile taurine pool is minimally affected, albeit at the expense of the liver pool of taurine (Sturman et al., 1978b). In the beagle puppy fed by total parenteral nutrition of amino acids (a diet deficient in taurine), the liver pool of taurine falls rapidly, while the bile taurocholate pool is much slower to reflect the taurine deficiency (Malloy et al., 1981). Indeed, in man in relatively short-term experiments (up to 9 days), the pool of taurocholate reflected the pool of taurine available in the liver and the diet, rather than that in plasma or muscle, indicating that the bile taurine pool in man cannot be predicted from plasma taurine measurements (Hardison, 1978).

The finding that formula-fed infants tend to become predominately glycine conjugators rather than predominatly taurine conjugators of bile acids (Brueton et al., 1978) suggests that the deficiency of taurine may not benefit the infant. Taurine-conjugated bile acids are more efficient bile acids than those conjugated with glycine (Schersten, 1970). Furthermore, recent work clearly establishes the importance of bile acid conjugates, especially taurocholate pool size, in the control of hepatic cholesterol synthesis and, therefore, of total body cholesterol pool size (Nervi and Dietschy, 1978). The plasma concentration of cholesterol in breast-fed infants is remarkably higher than that of infants fed synthetic formulas (Table 26.3). It has been suggested that the human infant may have a dietary requirement for preformed cholesterol (Fomon, 1974). It is possible also that in man taurine-conjugated bile acids are more effective in abetting the absorption of dietary cholesterol. The interaction of taurine and cholesterol may be important and is apparently species specific. Taurine reduces serum, liver, and aorta cholesterol in the rat (an animal that primarily conjugates bile acids with taurine) but does not reduce serum, liver, or cholesterol content in the rabbit (an animal that primarily conjugates bile acids with glycine) (Herrmann, 1959).

The lower absorption of lipids by infants fed some formulas may be a result, in part, of the failure of these formulas to provide enough cholesterol and taurine as bile salt precursors. Supplementation of the formulas with those two compounds also might improve the absorption of the fat-soluble vitamins.

Table 26.3.   Serum cholesterol at 12 weeks of age[a]

	Cholesterol (mmol/liter)
Human milk	4.45 ± 0.12
1.5% (60:40)	3.08 ± 0.15[b]
1.5% (18:82)	2.80 ± 0.24[b]

[a]Abbreviations as in Table 26.2 (from Rassin et al., 1979).
[b]$p < 0.01$.

Some organs in the human infant may be adversely affected by taurine deficiency even when these organs are unaffected in the kitten. Taurine has also been implicated in the maintenance of muscle function (Huxtable and Bressler, 1973), cardiac function (Grosso and Bressler, 1976), and the structural integrity of adrenal medullary granules (Nakagawa and Kuriyama, 1975). In the taurine-deficient kitten, the concentration of taurine in heart, muscle, and adrenal gland is 10-fold less than that in the corresponding organs from control kittens (Sturman et al., 1978b). The possibility that a deficiency of taurine may be associated with some forms of cardiac arrhythmias, muscle dysfunction, or adrenal dysfunction should be considered and systematically explored. Relevant to these possible relationships between dietary taurine and organ function is the finding that taurine in the diet may reduce the development of hypertension in genetically susceptible rats (Nara et al., 1978). These authors suggest that taurine is not involved in normal regulation of blood pressure. In a genetically susceptible animal, however, the lack of dietary taurine interacts with genetic factors and results in the disease state (Nara et al., 1978). Another example of the interaction of a genetic liability and dietary taurine is found in mice susceptible to gallstone formation (Fujihara et al., 1978). A mouse strain has been isolated that is genetically prone to cholelithiasis when given an "atherogenic" diet containing cholesterol and cholic acid. The formation of gallstones in these animals is markedly reduced by including 2% taurine in the drinking water (Fujihara et al., 1978). Characterization of the role of taurine in infant nutrition by extrapolation from animal models is difficult, especially because of the species differences (such as the cat versus man) discussed above.

The high concentrations of taurine in brain during fetal life—coupled at the same time with the low activity of the enzyme responsible for catalyzing the synthesis of taurine, cysteinesulfinic acid decarboxylase—implies that the mother must transfer considerable taurine to the fetus. It might be expected that one reflection of this supply of taurine to the fetus is the concentration of taurine in the plasma of the mother during pregnancy. Different changes in plasma taurine during pregnancy, however, are seen in different species (Table 26.4). The pregnant Rhesus monkey has a large increase in plasma taurine concentration at 50 and 100 days gestation and returns to preconception concentrations by 150 days (Kohrs et al., 1978). The pregnant rat has a decreased plasma taurine concentration at 12 and 21 days gestation (Palou et al., 1977). Women have a virtually unchanged plasma taurine concentration during pregnancy (Armstrong and Yates, 1964; Schoengold et al., 1978; Rassin, Raiha and Gaull, unpublished observations) but do excrete significantly less taurine in the urine during pregnancy (Armstrong and Yates, 1964). Thus, man and the rat, two species with vastly different capacities to synthesize taurine, are similar at least in the direction of the change, whereas the

Table 26.4.   Taurine in plasma of different species during pregnancy

Man	0–12	13–28	>28	Postpartum		Non-pregnant
	Weeks of gestation					
Plasma (μmol/dl)[a]	16.2	16.4	19.1	19.7		—
Urine (μmol/mg creatinine)[a]	0.58	0.32	0.14	0.03 (0–8 wks)	0.44 (>8 wks)	—
Plasma (μmol/dl)[b]	6.96	5.06	4.17	5.98		9.93
Plasma (μmol/dl)[c]	4.6	2.6	3.4	3.0		4.9

Rhesus monkey	50	100	150	7	30	Pre-conception
	Days of gestation			Postpartum		
Serum (control diet) (μmol/dl)[d]	47.0	47.3	13.6	18.2	21.2	14.3
Serum (protein-deficient diet) (μmol/dl)[d]	34.7	28.3	11.8	18.2	20.9	13.3

Rat	12	21	Preconception
	Days of gestation		
Plasma (μmol/dl)[d]	38.9	27.1	47.9

[a]Adapted from Armstrong and Yates, 1964; one subject.

[b]Adapted from Schoengold et al., 1978; more than 7 subjects per group.

[c]Rassin, Raiha, and Gaull, unpublished observations; 6 subjects per group.

[d]Adapted from Kohrs et al., 1978; 7 or 8 animals per group.

[e]Adapted from Palou et al., 1977; 7 to 9 animals per group.

Rhesus monkey is considerably different. Of course, these changes may be more representative of maternal hormonal changes and dietary differences than of fetal requirements, but they do illustrate the puzzling species differences observed with regard to the biochemical properties of taurine.

A recent investigation of plasma amino acids in normal children and in children with protein-calorie malnutrition further emphasizes the importance of taurine in determining nutritional states (Ghisolfi et al., 1978). Five groups of children were studied: normal (Group I), well nourished but suffering from infection (Group II), well nourished but having fasted for 2–4 days due to infection (Group III), moderate protein-calorie malnutrition (Group IV), and severe protein-calorie maltrition (Group V). The diets of the control groups were not fully described. Nevertheless, plasma taurine concentrations were far lower in the malnourished children (Groups IV and V) (Table 26.5). These differences were more dramatic than those observed for a number of the classical indices of malnutrition based on various ratios of other plasma amino acids (Ghisolfi et al., 1978). Indeed, it is the conclusion of these authors that plasma taurine may be the best discriminant between the fasting and the malnourished state in children.

In conclusion, there is considerable evidence for the important relationship between diet and the supply of taurine during the early development of the human infant. The precise biological role of this nutrient is unknown apart from the important precursor relationship to the bile salts. Recent experimental data from the spiny lobster, *Panulirus argus,* may be relevant to a biological understanding of the importance of taurine in nutrition. The antennular system of this lobster contains specific taurine receptors; these receptors are insensitive to α-amino acids, which are constituents of seawater (Fuzzessery et al., 1978). The major sources of food for the lobster usually contain taurine as one of their five most abundant amino acids. It has been suggested that the antennular taurine receptors have specifically evolved to allow this animal to identify food sources against a background of other amino acids present in the lobster's environment (Fuzzessery et al., 1978). Thus, in the lobster, taurine may serve the most important nutritional role of all: identifying what is food and so ensuring that the animal survives.

Table 26.5.   Taurine in plasma and urine of man during malnutrition

	I Normal diet (controls)	II Normal diet (infection)	III 2–4 day fast (infection)	IV Moderate protein- calorie malnutrition	V Severe protein- calorie malnutrition
Plasma (μmol/dl)[a]	7.63	7.21	7.38	3.74	3.53
	Controls	Protein calorie malnutrition	After protein treatment		
Urine (μmol/24 hr)[b]	87.9	221.2	24.0		

[a]Adapted from Ghisolfi et al., 1978.
[b]Adapted from Bigwood, 1962.

REFERENCES

Aguirre, G. D. Retinal degeneration associated with the feeding of dog food to cats. *J. Am. Vet. Med. Assoc.*, 172, 791–796 (1978).

Armstrong, M. D.; and Yates, K. N. Amino acid excretion during pregnancy. *Am. J. Obst. and Gynec.*, 88, 381–390 (1964).

Barbeau, A.; and Huxtable, R. J., eds. *Taurine and Neurological Disorders.* Raven Press, New York (1978).

Berson, E. L.; Hayes, K. C.; Rabin, A. R.; Schmidt, S. Y.; and Watson, G. Retinal degeneration in cats fed casein: 2. Supplementation with methionine, cysteine or taurine. *Invest. Ophthalmol.*, 15, 52–58 (1976).

Bigwood, E. J. Protein deficiency and aminoaciduria in Kwashiorkor, in Central Africa. *Nutritio et Dieta*, 4, 17–50 (1962).

Brueton, M. J.; Berger, H. M.; Brown, G. A.; Ablitt, L.; Iyangkaran, N.; and Wharton, B. A. Duodenal bile acid conjugation patterns and dietary sulphur amino acids in the newborn. *Gut*, 19, 95–98 (1978).

Challacombe, D. N.; Edkins, S.; and Brown, G. A. Duodenal bile acids in infancy. *Arch. Dis. Child.*, 50, 837–843 (1975).

Dickinson, J. C.; Rosenblum, H.; and Hamilton, P. B. Ion exchange chromatography of the free amino acids in the plasma of infants under 2,500 gm at birth. *Pediatrics*, 45, 606–613 (1970).

Encrantz, J. D.; and Sjovall, J. On the bile acids in duodenal contents of infants and children. *Clin. Chim. Acta*, 4, 793–799 (1959).

Fomon, S. J. *Infant Nutrition*, 2nd ed. W.B. Saunders, Philadelphia (1974), pp. 172–174.

Fujihara, E.; Daneta, S.; and Oshima, T. Strain difference in mouse cholelithiasis and the effect of taurine on the gallstone formation in $C_{57}BL/C$ mice. *Biochem. Med.*, 19, 211–217 (1978).

Fuzzessery, Z. M.; Carr, W. E. S.; and Ache, B. W. Antennular chemosensitivity in the spiny lobster, *Panuliris argus:* Studies of taurine sensitive receptors. *Biol. Bull.*, 154, 226–240 (1978).

Gaull, G. E.; and Rassin, D. K. Taurine and brain development: Human and animal correlates. In *Neural Growth and Differentiation*, E. Meisami and M. A. B. Brazier, eds. Raven Press, New York (1979), pp. 461–477.

Gaull, G. E.; Rassin, D. K.; Raiha, N. C. R.; and Heinonen, K. Milk protein quantity and quality in low-birth-weight infants: 3. Effects on sulfur amino acids in plasma and urine. *J. Pediatr.*, 90, 348–355 (1977).

Ghisolfi, J.; Charlet, P.; Ser, N.; Salvayre, R.; Thouvenot, J. P., and Duole, C. Plasma free amino acids in normal children and in patients with protein-calorie malnutrition: Fasting and infection. *Pediat. Res.*, 12, 912–917 (1978).

Grosso, D. A.; and Bressler, R. Taurine and cardiac physiology. *Biochem. Pharmacol.*, 25, 2227–2232 (1976).

Hardison, W. G. M. Hepatic taurine concentration and dietary taurine as regulators of bile acid conjugation with taurine. *Gastroent.*, 75, 71–75 (1978).

Herrmann, R. G. Effect of taurine, glycine and β-sitosterols on serum and tissue cholesterol in the rat and rabbit. *Circ. Res.*, 7, 224–227 (1959).

Huxtable, R. J.; and Barbeau, A., eds. *Taurine.* Raven Press, New York (1976).

Huxtable, R. J., and Bressler, R. Effects of taurine on a muscle intracellular membrane. *Biochim. Biophys. Acta*, 323, 573–583 (1973).

Jacobsen, J. G.; and Smith, L. H., Jr. Biochemistry and physiology of taurine and taurine derivatives. *Physiol. Rev.*, 48, 424–511 (1968).

Jagenburg, O. R. The urinary excretion of free amino acids and other compounds by the human. *Scand. J. Clin. Lab. Invest.* [Suppl. 43], 11, 3–183 (1959).

Jonxis, J. H. P. The influences of differences in food on the amino acid excretion. *Arch. Dis. Child.*, 26, 272 (1951).

Knopf, K.; Sturman, J. A.; Armstrong, M.; and Hayes, K. C. Taurine: An essential nutrient for the cat. *J. Nutr.*, 108, 773–778 (1978).

Kohrs, M. B.; Kerr, G. R.; and Harper, A. E. Serum amino acids during gestation of rhesus monkeys fed two different levels of protein. *J. Nutr.*, 108, 525–534 (1978).

Malloy, M. H.; Rassin, D. K.; Gaull, G. E.; and Heird, W. C. Development of taurine metabolism in beagle pups: Effects of taurine-free total parenteral nutrition. *Biol. Neon.* (1981) in press.

Nakagawa, K.; and Kuriyama, K. Effect of taurine on alteration in adrenal functions induced by stress. *Japan. J. Pharmacol.*, 25, 737–746 (1975).

Nara, Y.; Yamori, Y.; and Lovenberg, W. Effect of dietary taurine on blood pressure in spontaneously hypertensive rats. *Biochem. Pharmacol.*, 27, 2689–2692 (1978).

Nervi, F. O.; and Dietschy, J. M. The mechanisms and the interrelationship between bile acid and chylomicron-mediated regulation of hepatic cholesterol synthesis in the liver of the rat. *J. Clin. Invest.*, 61, 895–909 (1978).

Palou, A.; Arola, L.; and Alemany, M. Plasma amino acid concentrations in pregnant rats and in 21-day foetuses. *Biochem. J.*, 166, 49–55 (1977).

Poley, J. R.; Dower, J. C.; Owen, C. A.; and Stickler, G. B. Bile acids in infants and children. *J. Lab. Clin. Med.*, 63, 838–846 (1964).

Rabin, B.; Nicolosi, R. J.; and Hayes, K. C. Dietary influence on bile acid conjugation in the cat. *J. Nutr.*, 106, 1241–1246 (1976).

Raiha, N. C. R.; Heinonen, K.; Rassin, D. K.; and Gaull, G. E. Milk protein quantity and quality in low-birth-weight infants: 1. Metabolic responses and effects on growth. *Pediatrics*, 57, 659–674 (1976).

Rassin, D. K.; Gaull, G. E.; Heinonen, K.; and Raiha, N. C. R. Milk protein quantity and quality in low-birth-weight infants: 2. Effects on selected aliphatic amino acids in plasma and urine. *Pediatrics*, 59, 407–422 (1977a).

Rassin, D. K.; Gaull, G. E.; Raiha, N. C. R.; and Heinonen, K. Milk protein quantity and quality in low-birth-weight infants: 4. Effects on tyrosine and phenylalanine in plasma and urine. *J. Pediatr.*, 90, 356–360 (1977b).

Rassin, D. K.; Sturman, J. A.; and Gaull, G. E. Taurine and other free amino acids in milk of man and other mammals. *Early Human Develop.*, 2, 1–13 (1978a).

Rassin, D. K.; Sturman, J. A.; Hayes, K. C.; and Gaull, G. E. Taurine deficiency in the kitten: Subcellular distribution of taurine and [^{35}S]taurine in brain. *Neurochem. Res.*, 3, 401–410 (1978b).

Rassin, D. K.; Jarvenpaa, A.-L.; Raiha, N. C. R.; and Gaull, G. E. Breast feeding versus formula feeding in full-term infants: Effects on taurine and cholesterol. *Pediat. Res.*, 13, 406 (1979).

Reichelt, K. L.; and Edminson, P. D. Biogenic amine specificity of cortical peptide synthesis in monkey brain. *FEBS Lett.*, 47, 185–189 (1974).

Reichelt, K. L.; and Kvamme, E. Histamine-dependent formation of N-acetylaspartyl peptides in mouse brain. *J. Neurochem.*, 21, 849–859 (1973).

Schersten, T. Bile acid conjugation. In *Metabolic Conjugation and Metabolic Hydrolysis*, Vol. 2, W. H. Fishman, ed. Academic Press, New York (1970), pp. 75–121.

Schoengold, D. M.; DeFiore, R. H.; and Partlett, R. C. Free amino acids in plasma throughout pregnancy. *Am. J. Obstet. and Gynecol.*, 131, 490–499 (1978).

Sjovall, J. Bile acids in man under normal and pathological conditions. *Clin. Chim. Acta*, 5, 33–41 (1960).

Souchon, F. Papierchromatographische Untersuchungen der freien amino sauren im Sauglingsham. *Z. Ges. Exp. Med.,* 118, 219–229 (1952).

Sturman, J. A.; Rassin, D. K.; and Gaull, G. E. Taurine in development. *Life Sci.,* 21, 1–22 (1977).

Sturman, J. A.; Rassin, D. K.; and Gaull, G. E. Taurine in the development of the central nervous system. In *Taurine and Neurological Disorders,* A. Barbeau and R. J. Huxtable, eds. Raven Press, New York (1978a), pp. 49–71.

Sturman, J. A.; Rassin, D. K., Hayes; K. C.; and Gaull, G. E. Taurine deficiency in the kitten: Exchange and turnover of [^{35}S]taurine in brain, retina, and other tissues. *J. Nutr.,* 108, 1462–1476 (1978b).

Watkins, J. B.; Jarvenpaa, A. L.; Raiha, N.; Szczepanik Van-Leween, P.; Klein, P. D.; Rassin, D. K.; and Gaull, G. Regulation of bile acid pool size: Role of taurine conjugates. *Pediat. Res.,* 13, 410 (1979).

Copyright ©1981, Spectrum Publications, Inc.
The Effects of Taurine on Excitable Tissues

CHAPTER 27

Pathophysiological Role of Taurine in Blood Pressure Regulation in Stroke-Prone Spontaneously Hypertensive Rats (SHR)

Yukio Yamori
Yasuo Nara
Ryoichi Horie
Akira Ooshima
Walter Lovenberg

INTRODUCTION

Stroke-prone spontaneously hypertensive rats (SHRSP), which were established by selective breeding (Yamori et al., 1974a; Okamoto et al., 1974), develop severe hypertension and cerebrovascular lesions in more than 90% of the population (Yamori et al., 1976). These SHRSP, as well as spontaneously hypertensive rats (SHR), are regarded as the best models to date for studies on the pathogenesis and prevention of the spontaneously developing hypertensive diseases observed in man (Yamori and Okamoto, 1974; Folkow et al., 1973).

One of the most interesting new findings obtained in SHRSP is the dietary prevention of hypertensive diseases; high protein diet reduces the incidence of stroke with or without concomitant reduction in blood pressure (Yamori et al., 1977a,b). Further analyses of the mechanisms underlying the effect of high protein diet have clarified the following (Yamori et al., 1979a): (1) High fish protein diets or other diets containing animal protein attenuate the development of severe hypertension and reduce the incidence of stroke, while soybean protein diets reduce the incidence of stroke without any effect on blood pressure. (2) The strength of arterial walls is well maintained in SHRSP fed a high protein diet for a long time. (3) Urinary sodium excretion is accelerated in SHRSP fed high protein diets, which therefore reduces the adverse effect of salt on hypertensive cardiovascular diseases.

DIETARY SULFOAMINO ACIDS AND STROKE

Since various diets containing different amounts and kinds of protein affect the blood pressure and incidence of stroke, amino acid contents of these diets were analyzed in relation to their effects on blood pressure and their ability to reduce the incidence of stroke. Six typical diets, which differed from each other in protein content and the source of protein but not in lipid and salt content, were used: normal diet containing 17% protein (Funahashi F_{II}); standard diet NIH); low protein diet containing 10% protein; high soybean or fish protein diets containing 50% soybean or fish protein, respectively; and a chemically defined diet consisting of amino acids corresponding to a 17% protein diet. Samples, 20 mg, of each diet were hydrolyzed with 2 ml of 6N HCl at 110°C for 24 hr, and their amino acid content was analyzed by an amino analyzer (Hitachi 835). Significant inverse correlations were noted between methionine content or total sulfoamino acid content (methionine, cystine, and taurine) and blood pressure recorded at the tail (Yamori et al., 1974b) in 15–30 rats of each group at the age of 80 days (Fig. 27.1). The significant inverse correlation was also noted between sulfoamino acid contents and the stroke incidence, which was determined in rats by autopsy following natural death. These data suggest that dietary sulfoamino acids attenuate the development of hypertension and therefore decrease the incidence of stroke.

CHRONIC EFFECT OF SULFOAMINO ACIDS ON BLOOD PRESSURE

SHRSP, 8 or 10 rats from each group, were fed on normal diet (Funahashi F_{II}) or the normal diet containing 1.5% methionine and 1.2% cysteine (equimolar to methionine) from the age of 1.5 months until natural death occurred. Taurine-treated SHRSP, stroke-resistant SHR (SHRSR), and normotensive Wistar-Kyoto (WK) rats (6 to 8 rats each) were fed standard

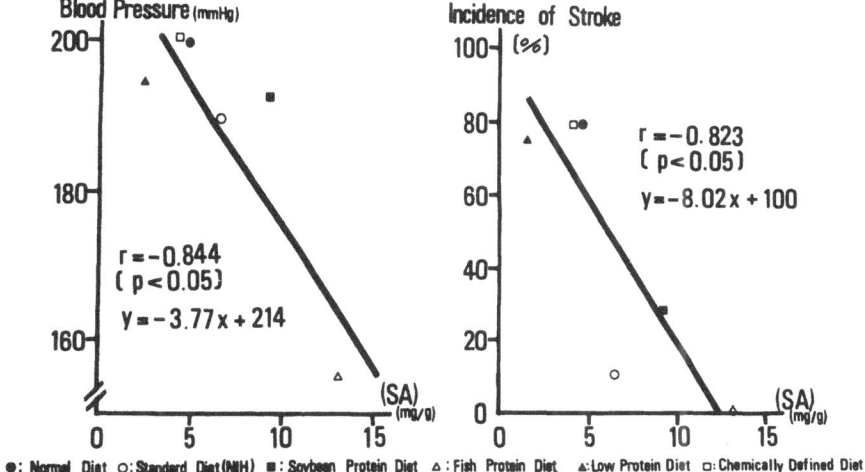

Fig. 27.1.   Relationship between dietary sulfoamino acid contents (SA) and blood pressure or stroke incidence in SHRSP.

diet (NIH) with tap water containing 3% taurine, while the corresponding control groups received normal tap water. Blood pressure was monitored during the experimental period from 32 to 107 days after birth.

Chronic cysteine administration had no effect on blood pressure. However, methionine treatment did decrease the degree of hypertension significantly in comparison with normal diet-fed SHRSP (Fig. 27.2). Consistently, the stroke incidence was significantly decreased in the methionine-treated group (43%), but no reduction was observed in the cysteine-treated group (88%) (Fig. 27.3). The average blood pressure in taurine-treated SHRSP was significantly lower than that in control SHRSP: 197 ± 5 (mean ± SE) vs. 210 ± 3 at the age of 70 days, and 185 ± 3 vs. 208 ± 2 at the age of 107 days. Taurine-treated SHRSR showed a slightly, but not significantly, lower blood pressure than nontreated control SHRSR: 179 ± 4 vs. 190 ± 6 at the age of 70 days, and 179 ± 4 vs. 187 ± 6 at the age of 107 days. No significant differences were observed in blood pressure between taurine-treated and nontreated WK: 125 ± 3 vs. 129 ± 4 at the age of 107 days.

These data indicate that the effect of sulfoamino acids on blood pressure are different. The difference in the depressor effects of methionine and cysteine suggests that the sulfoamino acids might be directly linked to neural

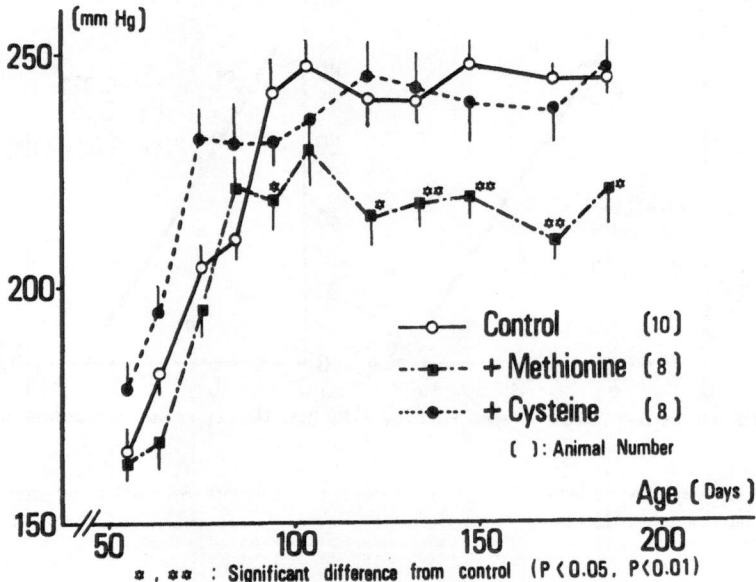

Fig. 27.2.   Chronic effect of methionine and cysteine on blood pressure in SHRSP.

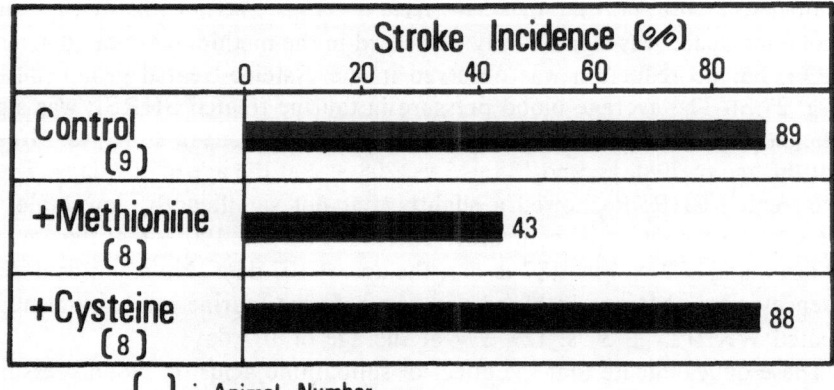

Fig. 27.3.   Effect of methionine and cysteine on stroke incidence.

regulation of blood pressure, since the brain is four or five times more permeable to methionine than to cysteine (Oldendorf, 1971). Although taurine, a metabolite of methionine and cysteine, is less easily taken up in the brain (Cutler and Coull, 1978), it reduces the blood pressure only in SHRSP that develop severe hypertension. This suggests that endogenous taurine metabolism or taurine transport into the brain might be different in SHRSP with severe hypertension.

ENDOGENOUS TAURINE IN SHR

Nine rats from each group (SHRSP, SHR, and WK) were sacrificed under sodium pentobarbital anesthesia (50 mg/kg) by cardiac puncture, and blood was collected in a syringe containing anticoagulant. The heart, liver, skeletal muscle (gastrocnemius), aorta, and brain were dissected out; and the brain was separated into four regions (telencephalon, cerebellum, diencephalon, and lower brainstem) by the method of Glowinski and Iversen (1966). Tissues were homogenized in 3 vol of 1% picric acid. The supernatant obtained by the centrifugation of the homogenate at 11,000 g for 20 min was applied to an ion-exchange column (AG 1 $\times$ 4 resin, 0.5 $\times$ 5 cm) to remove the picric acid; and the effluent, which was evaporated to dryness, was redissolved in 0.25 N citric acid, pH 2.1, to measure taurine by a Beckman 120C amino acid analyzer. Cysteic acid decarboxylase was assayed as described previously (Nara et al., 1978).

The taurine content in the liver was significantly reduced in SHRSP, the group that showed the largest blood pressure response to perorally administered taurine (Fig. 27.4). A significant reduction was also noted in the serum of SHRSP compared with that of WK (Fig. 27.5). The liver and serum SHRSR taurine content, however, was intermediate between SHRSP and WK. In other tissues and various brain regions, there were no significant differences in the taurine content of the three groups. In contrast to the findings reported by Huxtable and Bressler (1974), no significant change in taurine content of the heart was noted in SHR. Moreover, no differences were noted in cysteic acid decarboxylase activity among the three strains.

EFFECT OF TAURINE ADMINISTRATION ON THE ENDOGENOUS LEVELS

Taurine serum and tissue concentration was determined in SHRSP, SHRSR, and WK that were treated with taurine as described earlier.

There was no significant increase in serum taurine level in the treated

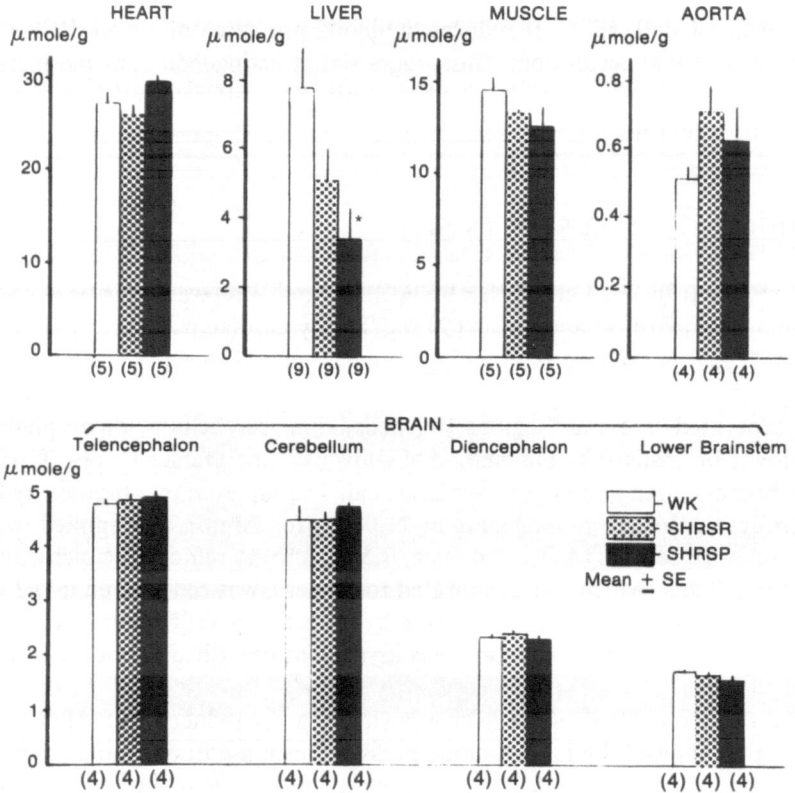

*Significant difference between WK and SHRSP (p< 0.01)   ( ): animal numbers

+The amount was measured as described in Methods and Materials. Aortas from three animals were used to measure the taurine amount. Animals were 3 month old rats. The blood pressure in WK, SHRSR and SHRSP was 129 ± 3, 189 ± 2 and 227 ± 6, respectively.

Fig. 27.4.   Taurine content of the brain, heart, liver, skeletal muscle, and aorta in SHRSP, SHRSR, and Wistar-Kyoto (WK) rats.

groups of these three strains (Fig. 27.5), but accumulation of taurine was noted in the brain tissues which showed a significant increase in the treated SHRSP except for the cerebellum (Fig. 27.6).

The turnover of taurine in the brain is reported to be very slow, with a half-

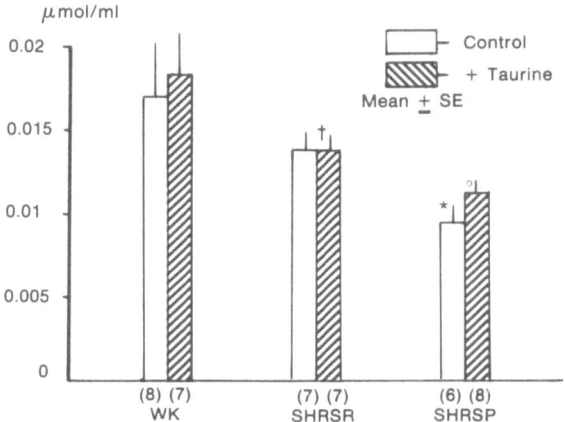

( ) animal numbers
*Significant difference between control WK and SHRSP (p< 0.025)
°Significant difference between taurine administrated WK and SHRSP (p< 0.025)
†Significant difference between taurine administrated SHRSR and SHRSP (p< 0.05)

Fig. 27.5. Serum concentration of taurine in SHRSP, SHRSR, and Wistar-Kyoto (WK) rats.

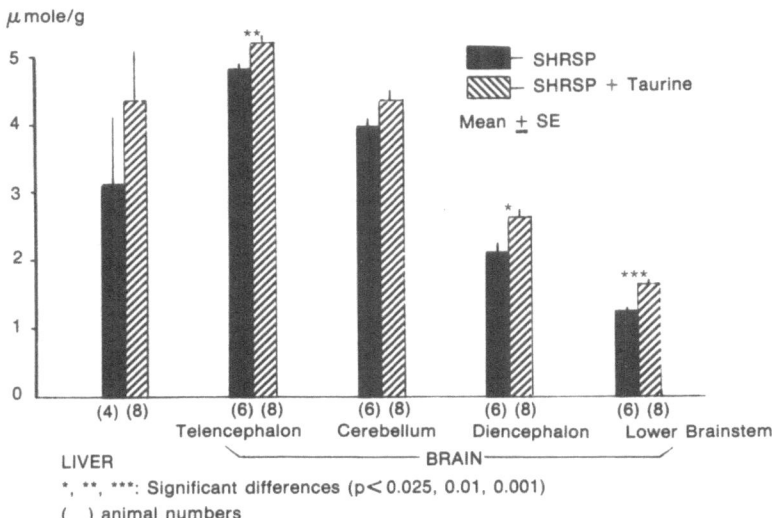

Fig. 27.6. Taurine content of the brain and liver in SHRSP after taurine administration.

life of 24 days (Spaeth and Schneider, 1974). However, it turns over more rapidly in the liver, with a half-life of less than 1 day. Such a great difference in the turnover may be the main reason why no increase in taurine content was noted in any tissues except for the brain of treated rats that had been deprived of food and water containing taurine for 1 day before sacrifice. The small but significant increment of taurine in the central nervous system, as well as the rather slow onset of the depressor effect, suggests that the possible site of action of taurine may be the regions of the brain closely related to autonomic blood pressure regulation.

DIRECT EFFECT OF SULFOAMINO ACIDS ON BLOOD PRESSURE

The aforementioned findings and accumulating evidence indicating the importance of the nervous system in the development of spontaneous hypertension (Yamori, 1976; Nagaoka and Lovenberg, 1977) led us to examine the direct effect of sulfoamino acids on blood pressure in SHRSP, SHR, and WK. Femoral arteries were cannulated 1 day prior to treatment, which consisted of a 10-μliter injection of taurine, methionine, and cysteine (30–600 μg) into the lateral ventricle of rats lightly anesthetized with α-chloralose (40 mg/kg, i.v.). Direct blood pressure was observed by a polygraph through an indwelling catheter for 10 hr after the injection.

The intraventricular injection of taurine (30–500 μg) quickly reduced blood pressure by 30–50 mmHg in rats; the sustained reduction was noted also in SHRSP (Fig. 27.7). The blood pressure of the treated rats did not recover to the initial level, and animals often died after the intraventricular injection of large doses of taurine. In contrast, acute blood pressure reduction was not noted after the intraventricular injection of methionine and cysteine, although a slow depressor effect (30–80 mmHg) was observed 2–4 hr after the intraventricular injection in SHRSP (Fig. 27.7). Such a slow onset of depressor effect by methionine and cysteine indicates that these sulfoamino acids may be metabolized to taurine, which directly reduces blood pressure. Although the physiological role of taurine in the central nervous system has not yet been clarified (Jacobsen and Smith, 1968), there is significant evidence indicating that taurine may be a neurotransmitter or neuromodulator (Davison, 1971; Kaczmarek and Davison, 1972; Oja and Lähdesmäki, 1974; Lombardini, 1976). Based on these findings, one might postulate a pathophysiological role of taurine in central blood pressure regulation, especially in SHRSP with severe hypertension accompanied by a reduction in the endogenous taurine level of liver and serum.

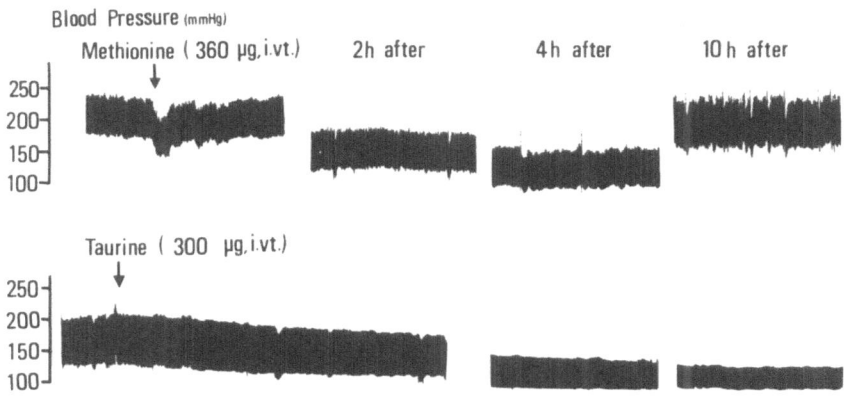

Fig. 27.7.   Central effect of sulfoamino acids on blood pressure in SHRSP.

URINARY TAURINE EXCRETION IN HYPERTENSIVE MODELS

Since taurine transport into the kidney may affect serum taurine levels (Chesney et al., 1978), 24-hr urine samples were collected from 3-month-old SHRSP, SHR, and WK (5 in each group), individually kept in metabolic cages, and analyzed by an amino acid analyzer (Hitachi 835).

Significant increases in urinary taurine excretion were noted in SHRSP and SHR compared with WK (Fig. 27.8). Other amino acids that were significantly different between the three groups were proline (significantly increased in SHRSP and SHR) and tyrosine and isoleucine (both significantly decreased in SHRSP and SHR). Although detailed mechanisms of these differences in urinary amino acid excretion in hypertensive models are completely unknown, increases in taurine and proline excretion cannot be simply explained by amino aciduria secondary to hypertension itself, since urinary excretion of other amino acids was not increased while decreases in tyrosine and isoleucine excretion were noted. These findings raise the possibility that abnormal taurine metabolism occurs in hypertensive rats, particularly in SHRSP which respond to exogenous taurine administration through a reduction in severe hypertension.

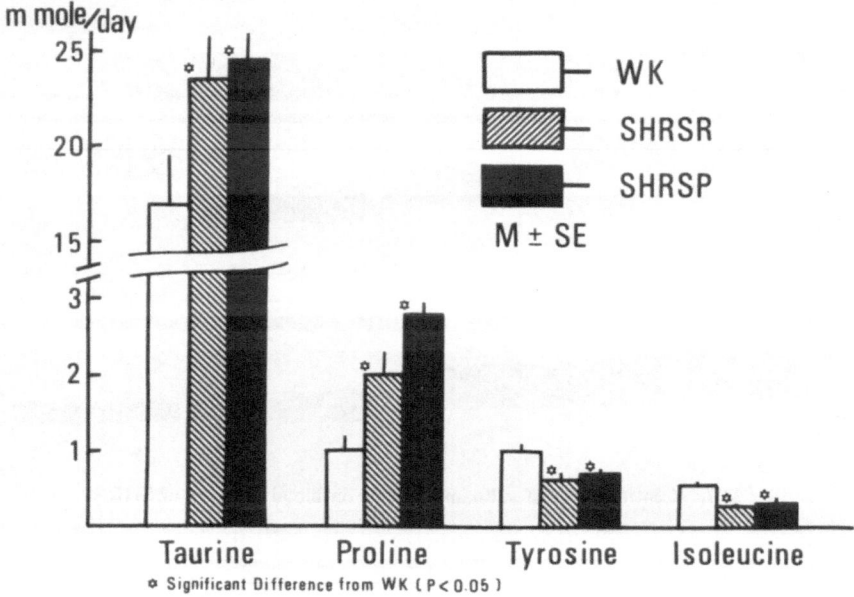

Fig. 27.8. Excretion of amino acids in 24-hr urine collected from SHRSP, SHRSR, and Wistar-Kyoto (WK) rats.

POSSIBLE ROLE OF TAURINE IN HYPERTENSION AND ATHEROSCLEROTIC DISEASES

Taurine, which was reduced in the liver and serum of SHRSP, attenuated the development of severe hypertension in this model when administered perorally. Taurine levels in the brain were increased in these treated SHRSP, and the direct intraventricular injection of taurine and sulfoamino acids decreased blood pressure. These findings suggest that taurine, whose levels might be insufficient in SHRSP due to the reduction in serum and liver taurine content, serves as a neuromodulator to affect central blood pressure regulation. The blood pressure effect of taurine observed in patients with essential hypertension further suggests that a similar mechanism might be involved in hypertension in man (Tsunoo et al., 1968). Dietary prevention of severe hypertension and stroke in these models may partly be ascribed to the effect of sulfoamino acids on central blood pressure regulation.

On the other hand, these hypertensive models quickly develop hypercholestrolemia, as well as arterial fat deposition, within a few weeks when fed on a hypercholesterolemic diet (Yamori et al., 1977a,b). Since taurine administration attenuated the development of severe reactive hypercholesterolemia and since the turnover of cholesterol (the decay of radioactive cholesterol administered intravenously) was delayed in these models, the reduction of taurine in the liver (which conjugates with bile acids and is important in cholesterol excretion) may be the cause of hypercholesterolemia and, therefore, atherogenesis in these models (Yamori et al., 1979b). Consequently, in SHRSP that are predisposed to hypertension, as well as hypercholesterolemia, the decrease in taurine level in the liver and serum may be involved in the pathogenesis of both severe hypertension and reactive hypercholesterolemia. These findings give us a new insight into the possible common pathogenesis of hypertension and atherosclerosis.

SUMMARY

Taurine, a putative neurotransmitter, is present in brain in relatively large amounts. No pathophysiological role for this amino acid is known. Our studies on experimental animal models for hypertension and stroke have suggested a possible role for taurine in the central regulation of blood pressure, as indicated by the following findings:

1. Stroke-prone SHR, established by the selection from spontaneously hypertensive rats (SHR), develop severe hypertension, with over 90% of the population dying of stroke. Our experimental studies showed that a high protein diet, especially one based on fish protein, decreased the incidence of stroke with a concomitant attenuation of the severe hypertension. Further analysis of dietary constituents indicated that the more effective diets contained greater amounts of the sulfoamino acids.

2. Stroke-prone SHR were fed on a normal diet supplemented with various sulfoamino acids. Methionine and taurine, which cross the blood-brain barrier relatively easily, effectively decreased blood pressure.

3. Long-term administration of taurine (3% in drinking water) increased taurine levels in the brain of stroke-prone SHR and significantly decreased their blood pressure.

4. The intraventricular or intracisternal injection of taurine reduced blood pressure in stroke-prone SHR and normotensive rats.

5. The endogenous content of taurine in the liver and the serum of stroke-prone SHR was markedly decreased compared to that in normotensive Wistar-Kyoto (WK) rats, from which SHR were derived. Taurine levels in tissues of SHR were intermediate between stroke-prone SHR and WK rats.

6. Urinary amino acid analysis showed that urinary taurine excretion was significantly increased in stroke-prone SHR compared with WK rats, indicating that the loss of taurine into the urine may be responsible for the reduced serum taurine. Loss of serum taurine may possibly be related to the severe hypertension in this model.

REFERENCES

Chesney, R. W.; Jax, D. K.; Scriver, C. R.; and Mohyuddin, F. Taurine transport in mammalian kidney. In *Taurine and Neurological Disorders,* A. Barbeau and R. J. Huxtable. eds. Raven Press, New York (1978), pp. 73–93.

Cutler, R. W. P.; and Coull, B. M. Amino acid transport in the brain. In *Taurine and Neurological Disorders,* A. Barbeau and R. J. Huxtable. eds. Raven Press, New York (1978), pp. 95–107.

Davison, A. N.; and Kaczmarek, L. K. Taurine—A possible neurotransmitter? *Nature,* 234, 107–108 (1971).

Folkow, B.; Hallbäck, M.; Lundgren, Y.; Sivertsson, R.; and Weiss, L. Importance of adaptive changes in vascular design for establishment of primary hypertension, studied in man and in spontaneously hypertensive rats. *Circulat. Res.,* 32 and 33 (Suppl. 1), 2–16 (1973).

Glowinski, J.; and Iversen, L. Regional studies of catecholamines in the rat brain: I. The disposition of [³H]-norepinephrine, [³H]-dopamine and [³H]-dopa in various regions of the brain. *J. Neurochem.,* 13, 655–669 (1966).

Huxtable, R.; and Bressler, R. Taurine concentrations in congestive heart failure. *Science,* 184, 1187–1188 (1974).

Jacobsen, J. G.; and Smith, L. H., Jr. Biochemistry and physiology of taurine and taurine derivatives. *Physiol. Rev.,* 48, 424–511 (1968).

Kaczmarek, L. K.; and Davison, A. N. Uptake and release of taurine from rat brain slices. *J. Neurochem.,* 19, 2355–2362 (1972).

Lombardini, J. B. Regional and subcellular studies on taurine in the rat central nervous system. In *Taurine,* R. Huxtable and A. Barbeau, eds. Raven Press, New York (1976), pp. 311–326.

Nagaoka, A.; and Lovenberg, W. Regional changes in the activities of aminergic biosynthetic enzymes in the brains of hypertensive rats. *Eur. J. Pharmac.,* 43, 297–306 (1977).

Nara, Y.; Yamori, Y.; and Lovenberg, W. Effect of dietary taurine on blood pressure in spontaneously hypertensive rats. *Biochem. Pharmacol.,* 27, 2689–2692 (1978).

Oja, S. S.; and Lähdesmäki, P. Is taurine an inhibitory neurotransmitter? *Med. Biol.,* 52, 138–143 (1974).

Okamoto, K.; Yamori, Y.; and Nagaoka, A. Establishment of the stroke-prone spontaneously hypertensive rats (SHR). *Circulat. Res.,* 34 and 35 (Suppl. 1), 143–153 (1974).

Oldendorf, W. H. Brain uptake of radiolabeled amino acids, amines, and hexoses after arterial injection. *Am. J. Physiol.,* 221, 1629–1639 (1971).

Spaeth, D. G.; and Schneider, D. L. Turnover of taurine in rat tissues. *J. Nutr.,* 104, 179–186 (1974).

Tsunoo, Sh.; Horisaka, K.; and Yamaquchi, A. Ω-amino sulfonic acid, especially on the pharmacology and clinical application of taurine. *J. Schowa, Medical Assoc.,* 28, 301–316 (1968) (in Japanese).

Yamori, Y. Neural and non-neural mechanisms in spontaneous hypertension. *Clin. Sci. Mol. Med.,* 51, 431s–434s (1976).

Yamori, Y.; and Okamoto, K. Spontaneous hypertension in the rats; a model of human "essential" hypertension. *Proc. 80th Congr. Germ. Soc. Intern. Med.,* 168–170 (1974).

Yamori, Y.; Nagaoka, A.; and Okamoto, K. Importance of genetic factors in hypertensive cerebrovascular lesions; an evidence obtained by successive selective breeding of stroke-prone and -resistant SHR. *Japan. Circulat. J.,* 38, 1095–1100 (1974a).

Yamori, Y.; Tomimoto, K.; Ooshima, A.; Hazama, F.; and Okamoto, K. Developmental course of hypertension in the SHR—Substrains susceptible to hypertensive cerebrovascular lesions. *Japan. Heart J.,* 15, 209–210 (1974b).

Yamori, Y.; Horie, R.; Sato, M.; Akiguchi, I.; Ohtaka, M.; Nara, Y.; and Fukase, M. New Models of SHR for studies on stroke and atherogenesis. *Clin. Exp. Pharmacol. Physiol.,* 199–203 (1976).

Yamori, Y.; Horie, R.; Akiguchi, I.; Nara, Y.; Ohtaka, M.; and Fukase, M. Pathogenic mechanisms and prevention of stroke in stroke-prone spontaneously hypertensive rats. In *Hypertension and Brain Mechanism,* W. de Jong, A. P. Provoost, and A. P. Shapiro, eds. Progress in Brain Research 47 (1977a), pp. 217–234.

Yamori, Y.; Horie, R.; Ohtaka, M.; Nara, Y.; Ohta, K.; Okamoto, K.; Handa, H.; and Fukase, M. Pathogenic approach to the prophylaxis of stroke and atherogenesis in SHR. In *Spontaneous Hypertension: Its Pathogenesis and Complications.* DHEW Publication No. (NIH) 77-1179 (1977b), pp. 269–278.

Yamori, Y.; Horie, R.; Ikeda, K.; Nara, Y.; and Lovenberg, W. Prophylactic effect of dietary protein on stroke and its mechanisms. In *Prophylactic Approach to Hypertensive Diseases,* Y. Yamori, W. Lovenberg, and E. Freis, eds. Raven Press, New York, pp. 497–504 (1979a).

Yamori, Y.; Iritani, N.; Nara, Y.; Horie, R.; Ohtaka, M.; and Ooshima, A. Hypertension and atherogenesis in animal models—A new aspect on the relationship. *Japan. Heart J.,* 20 (Suppl.), 760–761 (1979b).

Copyright © 1981, Spectrum Publications, Inc.
The Effects of Taurine on Excitable Tissues

CHAPTER 28

Factors that Modify the Tissue Concentration or Metabolism of Taurine

Steven I. Baskin
Colin M. Finney

Although the presence of taurine in biological tissues has been known for almost a century and a half, the functional significance(s) of this compound and its immediate precursors has not, as yet, been clearly delineated. Proposed roles for taurine in retina, muscle, and heart as well as other tissues are described in other chapters of this volume.

A number of factors that alter taurine levels and/or metabolism may, in many experimental protocols, interfere with obtaining meaningful taurine values. Alternately, analysis of methodological, pharmacological, and physiological factors that alter basal taurine levels could provide insight into the function(s) that taurine and its precursors may serve in biological systems. These types of taurine perturbation are especially relevant in light of the inability of classical dietary manipulations to alter taurine tissue concentrations (Sturman, 1973).

METHODOLOGICAL FACTORS

Of major importance is the necessity of adequately characterizing the compound and/or compounds subjected to manipulation or analysis. In an effort to increase the sensitivity and to avoid the skin irritability of the dinitroflorobenzene derivative assay we were using (Baskin et al., 1976), we developed an assay employing precolumn derivatization of samples with dabsyl chloride prior to high pressure liquid chromatography (Krusz et al., 1978a). What may be more important than the particular details of this assay is what was uncovered using this particular method of analysis. A small peak occurring on the tail of the taurine peak from the pineal gland of the rat had the same retention time as purified homotaurine. It is our suggestion that homotaurine may occur in biological samples. If in fact this is the case, it may have been possible that homotaurine and taurine, being so similar chemically, cochromatographed, or that homotaurine was masked by the larger concentration of taurine in separations carried out previously. Thus, some of the neurotransmitter effects ascribed to taurine may be due to homotaurine, in that homotaurine may be a neurotransmitter, and taurine a neuromodulator, in the same CNS system. In iontophoresis experiments, homotaurine appears to have greater ability to produce hyperpolarization than taurine (Curtis and Watkins, 1963), while Adembri et al. (1974) have shown that homotaurine may be more potent than taurine in preventing cobalt or hyperbaric oxygen-induced convulsions. In this respect is is interesting that homocysteic acid (a possible homotaurine precursor) has strong excitatory properties itself (Curtis and Watkins, 1963). More experiments are needed to verify this suggestion.

Similarly, the true biological significance(s) of taurine becomes somwhat shaded since it is known that hypotaurine (the immediate precursor of taurine) is converted to taurine in certain assay procedures; indeed, the photo oxidation of hypotaurine to taurine by UV irradiation has been recorded (Ricci et al., 1978). The conversion of hypotaurine to taurine nonenzymatically has been described by a number of investigators, including Cavallini et al. (1976). Using the "dabsyl assay," we found that the conversion of hypotaurine to taurine was dependent on the hypotaurine concentration above 40 μm of hypotaurine. Such a conversion could produce an artificially high taurine and low hypotaurine.

A number of investigations have implicated hypotaurine as having a physiological role itself and not functioning simply as a precursor of taurine (Kochakian, 1976). A major question to be resolved is whether the diverse physiological actions attributed to the analogues of taurine are artifacts of their structural relatedness, or whether taurine and its analogues form a family of compounds in which slightly different molecules have different physiological functions.

We are currently using a method involving the reaction of taurine and its analogues with orthophthaldehyde (OPA). We still observe a small conversion of hypotaurine to taurine with the OPA method.

PHARMACOLOGICAL FACTORS

A fundamental means to gain insight into the function(s) of taurine in biological systems would be to examine the effect of drugs on taurine metabolism/ concentration. Table 28.1 illustrates some of the agents that will affect or be affected by taurine.

These and related drugs can be classified in part into different groups related to the pharmacological interaction between taurine and the particular drug.

Drugs Whose Metabolism Is Affected by Taurine

Taurine has a well-documented history of conjugation with bile acids. Following the report that phenylacetic acid was conjugated with taurine (James et al., 1972), a number of compounds have been reported to be conjugated with taurine, including arylacetic acids (Idle et al., 1978), fenclofenac (Jordan and Rance, 1974), and cholic acid (Vessey, 1978). R. T. Williams and his coworkers (Idle et al., 1978) have found that conjugation with taurine is species dependent. The ferret, bushbaby, and the dog are known to produce taurine conjugates, while in other species, taurine conjugation plays a relatively minor or unknown role (Idle et al., 1978). The extent to which taurine conjugates are found in man and its role in the detoxification of xenobiotics have not been completely established at this time.

Convulsive Agents Affected by Taurine

Barbiturates, phenytoin, strychnine, amantadine, carbamazepine,

Table 28.1.   Drugs that will affect taurine or are affected by taurine

Aspirin (Rylance and Myhal, 1971)	3-Acetylpyridine (Rhode et al., 1978)
Chlorpromazine (Piha et al., 1962)	Amantadine (Gaut and Nauss, 1976)
Alcohol (Turner and Brum, 1964)	Ionophores X537A and A23187 (Pasantes-Morales et al., 1976)
Morphine (Kuriyama and Yoneda, 1978)	
Strychnine (Bonaventure et al., 1974)	Isoniazid (Johnston et al., 1967)
Phenytoin (this chapter)	Cardiac glycosides (Minnich et al., 1976)
Propranolol (Baskin et al., 1979)	Arylacetic acids (James et al., 1972)
Verapamil (Pasantes-Morales et al., 1976)	Carbamazepine (Baskin, unpublished)
Fenclofenac (Jordan and Rance, 1974)	Barbiturates (Iwata et al., 1978b)
Colchicine (Kostos and Kocsis, 1961)	Phenylbutazone (Rylance and Myhal, 1971)
Reserpine (Iwata et al., 1976)	Cholic acid (Spaeth and Schneider, 1976)

chlorpromazine, and perhaps alcohol are all agents that have been shown to have convulsant potential. For a number of these compounds, there appears to be a relationship between an induced loss of taurine from the tissue and the toxicity of the compound. For example, amantadine, in large doses, has been shown to cause convulsions, possibly through inhibition of taurine uptake (Gaut and Nauss, 1976). Indeed, the taurine loss observed with chlorpromazine is in the cerebellum (Piha et al., 1962), an area of the central nervous system thought to be associated with convulsant actions of the drug. Later studies of synaptosome models of neurones indicated that chlorpromazine did not directly affect taurine influx or efflux (Lähdesmäki et al., 1975).

In the case of alcohol, the loss of taurine appears to be associated with alcohol withdrawal. Urine taurine concentrations were found to be elevated during alcoholism and to drop below control values during withdrawal (Turner and Brum, 1964). Other studies have shown that taurine may counteract ethanol's depressant actions (Iida and Hikichi, 1976). On this basis, taurine has been used in the treatment of alcoholism (Ikeda, 1977), though this is still quite controversial.

Barbiturates seem to engender taurine responses distinct from alcohol. Iwata and his colleagues, in a recent study (1978b), have traced an increase in cerebellar taurine during habituation of rats to barbital sodium. Upon withdrawal, the cerebellar taurine concentrations returned to control values, but cerebral cortex values increased. Within 10 days, brain levels had returned to compatible levels with the controls for both the cerebellum and cerebral cortex.

In order to examine the effects of a pharmacological agent that would produce anticonvulsant effects at therapeutic concentrations and convulsant effects at high doses, we have studied the ability of phenytoin to alter taurine concentrations in regions of the brain. We found that phenytoin produced a dose-dependent increase in cerebellar, brain stem, and cerebral taurine concentrations (Fig. 28.1). However, as the concentration of phenytoin is increased toward toxicity, there is a dramatic reduction in taurine concentrations in these same areas. As with the lower doses, where the effect was to increase taurine in the cerebellum, the highest dose brought about the greatest decrease in the cerebellum. The effect of time-response curves of action (anticonvulsant) and the ability to increase taurine concentration in the brains of mice were compared and found to be consistent with each other. In summary, the ability of phenytoin to increase taurine concentrations in the brain supports the hypothesis that this compound and possibly others, such as carbamazepine, barbiturates, and ethanol, exert their therapeutic effects by increasing the proposed "endogenous anticonvulsant" taurine.

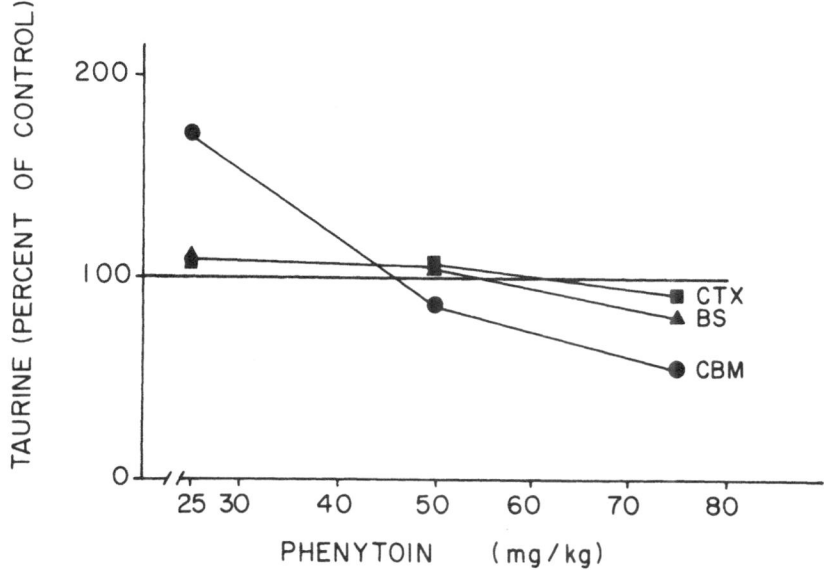

Fig. 28.1. Effect of phenytoin on taurine concentrations in rat brains. Key: CTX cerebral cortex; BS, brainstem; CBM, cerebellum. In this experiment, 25, 50, and 75 mg/kg phenytoin in physiological saline were injected, intraperitoneally 1 hr prior to sacrifice, into Swiss-Webster mice (30–40 g). The cerebellum, brainstem, and cerebrum were rapidly separated and homogenized with sea sand, 10 ml of picric acid. The homogenate was centrifuged at 1,000 rpm for 15 min, and 5 ml of the supernatant was then passed over a column containing two separate ion-exchange resins, the top layer composed of Dowex AG2-X8, the bottom of Dowex 50W-X8 resin. The combined column was then washed twice. A 0.4-ml aliquot of each sample was taken, and NaOH, dimethyl sulfoxide, and dinitrofluorobenzene were added. These solutions were mixed on a vortex for 30 sec. At the end of the hour, HCl and distilled water were added. Each solution was then extracted with water-saturated ethyl acetate and shaken mechanically for 10 min. The extracted aqueous sample and taurine standards were then read at 350 nm. Absorption was compared against a blank and taurine standards containing 1, 2, and 4 mcg/ml. (Baskin et al., 1976.)

Drugs That Are Affected by Their Ability to Interact with Calcium

Propranolol, verapamil, cardiac glycosides, the ionophores X537A and A23187, and catecholamines (including fusidic acid), as well as calcium itself, all have an effect on calcium movements (Baskin et al., 1980). Although some of the drugs named have other effects, they all affect calcium. All of these agents have also been claimed to affect taurine metabolism. For instance, Pasantes-Morales et al., (1976) have shown a marked loss of retinal tuarine when exposed to the ionophores X537A and A23187. This loss is dependent on increased calcium turnover. The use of ouabain and verapamil confirmed the role of calcium in taurine release from the retina. Although many agents may modify taurine concentrations by modifying calcium levels or fluxes, the significance of this in the normal animal is uncertain.

Other Drugs Affecting Tissue Taurine Concentrations

Rylance and Myhal (1971) tested the influence of various anti-inflammatory drugs on the excretion of taurine. Although several had no effect, both aspirin and phenylbutazone lowered urinary taurine. As serum taurine was found to rise, the reduced taurine clearance was hypothesized to involve decreased cellular permeability in the kidney.

Kuriyama and Yoneda (1978) reported the effect of morphine on taurine in the rat central nervous system. Acute administration of morphine had no effect on taurine levels, but chronic morphine administration induced increases in taurine concentration in parts of the spinal cord. Morphine has been demonstrated to have many actions on many systems, so it is unclear if any significance can be attributed to this taurine increase.

Kostos and Kocsis (1961) reported that colchicine increased taurine excretion in rats but did not reduce tissue taurine levels. Increased taurine excretion has also been noted upon isoniazid administration (Johnston et al., 1967).

PHYSIOLOGICAL FACTORS

It is obvious that taurine concentrations in tissues are affected by various physiological factors. In the paragraphs below, we discuss only a few, both because most of the physiological factors that control taurine concentrations are not known and because these factors are ones that we have been involved with ourselves.

Nutrition

Various investigators have reported that taurine concentrations are remarkedly resistant to dietary manipulation (Awapara, 1956; Sturman, 1973). Metabolic pools of taurine in most tissues seem to be dominated by large, slowly exchanging pools, with a minor pool of quickly exchanging taurine. This large inert fraction of taurine may reflect an important role for taurine or alternatively, a close coupling of dietary intake, transport, and biosynthesis.

Age

Studies on the effects of age on tissue taurine concentrations have been reported for several species. Taurine was reported to be four to five times higher during fetal life than during maturity in human and monkey brains (Sturman and Gaull, 1975). Similar findings of higher concentrations of taurine during gestation have been reported for tissues of rats (Oja et al., 1968), mice (Lajtha and Toth, 1973), and chickens (Levi and Morisi, 1971). Linear reductions of taurine concentrations with age have been reported for monkey brains, until weaning, after which levels remain essentially the same till maturity (Sturman and Gaull, 1975).

Our own work with taurine concentrations in pineal glands of both rats and monkeys has revealed that taurine concentrations remain at a reasonably constant level throughout maturity (Krusz et al., 1978a,b); however, in very old rats, taurine concentrations decreased significantly.

Diurnal Rhythms

Although relatively poorly documented as yet, it appears likely that taurine concentrations may undergo diurnal rhythms in many tissues. Nir et al. (1973) reported that the pineal taurine content of immature and mature rats was significantly higher during light periods (13:00–14:00) than during dark periods (01:00–02:00), and that continuous light increased pineal taurine content while continuous dark depressed it. In most cases the change, though statistically significant, was not particularly dramatic, the difference between light- and dark-cycle taurine concentrations varying less than 10%. Grosso et al. (1978) have more recently reported a circadian rhythm of greater magnitude in the rat pineal.

Iwata et al. (1978a) documented a much larger change, on the order of 60% in the light-dark cycle of rat brain taurine concentrations. Taurine was maximal during the period 01:00–02:00 P.M., with lowest concentrations occurring at 14:00. Adrenalectomy resulted in phase-shifting the taurine maximum to 06:00–10:00. This same group of investigators mention in another report (Matsuda et al., 1978) that they have unpublished evidence for a circadian rhythm in the rat heart.

Pasantes-Morales and her co-workers (1973, 1976) have shown that taurine concentrations increase in chicken retinas maintained in continuous darkness compared to those animals maintained in a natural light-dark cycle. Although this finding may reflect a direct influence on retinal taurine levels by light, the possibility of a short-term (diurnal) change mediated by light has not been investigated.

In a study of human volunteers, we have found evidence for a diurnal rhythm in the taurine concentration of blood platelets. Taurine concentrations were found to be maximal in the morning (08:00), with concentrations dropping to a minimum from 12:00 to 16:00, then rising to close to the maximum during the night (Fig. 28.2). Averaged values may not represent the most appropriate statistical description, as individuals display rhythms that may be related to their typical light-dark cycle, which in humans is usually not controllable to the extent that it is for laboratory animals.

Iwata et al. (1978a) concluded that the taurine rhythm in the brain was not related to any rhythm in the blood because of the inability of taurine to cross the blood-brain barrier. Whether the blood taurine rhythm is related to any pineal taurine rhythm is unknown at the present, as well as which other tissue taurine concentrations might drive or be driven by this rhythm.

Species

Numerous examples abound in the literature of contradictory results of analyses of concentrations of taurine between species, or of taurine's effects between species. As examples we cite our own finding of higher taurine concentrations in the ventricles versus auricles of rats, beef, pigs, guinea pigs, and rabbits, while the reverse (higher concentration in auricles than in ventricles)

Fig. 28.2. Durnal (24 hr) rhythm of taurine concentrations in blood platelets of four volunteers; each data point equals mean of two replicate analyses. *Sampling:* One 5-ml vacutainer, containing 7.2 mg EDTA, of blood drawn from each volunteer at each time point. Sample centrifuged at 400 g for 10 min, and 0.5 ml of the resulting platelet-rich plasma recentrifuged at 15,000 g for 15 min. After decanting the supernatant, the platelet pellet was suspended in 250 μliter H_2O and frozen for analysis. *Extraction:* Samples freeze-thawed three times and 5 μliter toluene added to disrupt the membranes. Membranes removed by centrifugation at 15,00 g for 10 min. A sample volume of 25 μliter of supernatant was mixed with 25 μliter 5% TCA to percipitate proteins, and 25 μliter H_2O was added and sample centrifuged at 15,000 g for 10 min to remove proteins. Ethyl ether, 150 μliter was mixed with supernatant to remove TCA, and centrifuged at 15,000 g for 5 min. The ether was decanted and sample frozen for analysis. *Analysis:* Quantification of amino acids was determined using a Waters HPLC employing a fatty acid analysis column run with 500 mg SDS/5 ml acetic acid/1000 ml H_2O as a solvent. Detection was by post column derivitization with O-phthalaldehyde (1 ml of 650 mg OPA/200 μliter 2-mercaptoethanol/1000 ml H_2O in 100 ml of 30 g borate/1,000 ml H_2O, pH adjusted to 10.4 with KOH). This was metered into the column effluent line and fluorescent derivatives detected with a Waters Model 420 E.C. Fluorescence Detector. Calibration was by standard additions to samples prior to the extraction procedure. Samples were analyzed in duplicate.

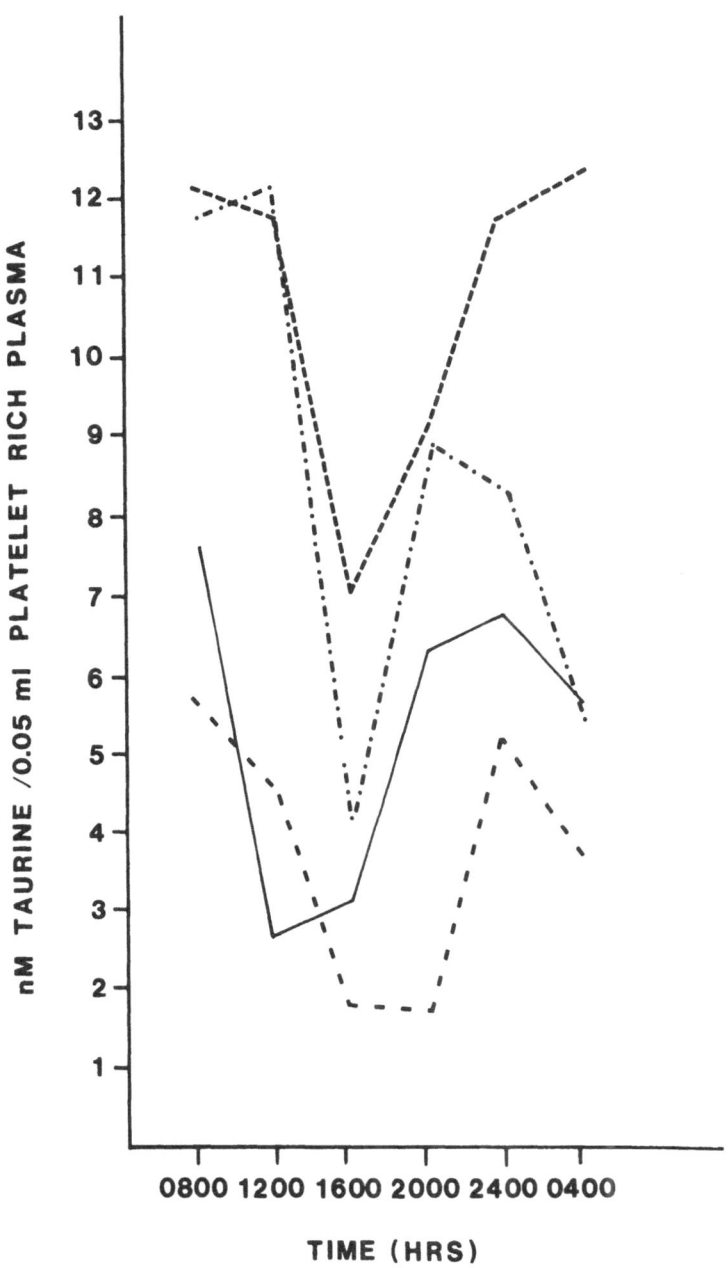

was true of mice and dogs (Kocsis et al., 1976); and the finding by Dietrich and Diacono (1971) that taurine antagonizes low calcium depression of guinea pig heart contractility while potentiating the same phenomenon in rats. Idle et al. (1978), in a comparison of taurine conjugation of arylacetic acids in rats versus ferrets, reemphasized species differences for another physiological process. Comparison of data between species must be viewed with caution until the experimental paradigms are well characterized.

Endocrine Status

Taurine's relationship with endocrine processes is not well defined. Kochakian (1974) observed that after castration, taurine concentrations are increased in mouse kidney, while hypotaurine is reduced approximately 75%. Treatment with testosterone returned the values to close to control values.

Matsuda et al. (1978) reported that glucocorticoids increased the taurine content of cardiac tissue in rats, in some cases up to 50%. Other steroids, such as testosterone and 17β-estradiol, were not effective in altering taurine levels.

Taurine has recently been observed to increase prolactin secretion, while not affecting LH secretion. A taurine precursor, cysteic acid, had just the opposite effect (Scheibel et al., 1978, 1979).

We have recently shown that taurine concentrations in human blood platelets are correlated with the thyroid status of the individual (Baskin et al., 1979). Hypothyroid patients had significantly lower taurine levels than controls, while hyperthyroid patients' taurine concentrations were significantly higher. Those hyperthyroid patients undergoing treatment with antithyroid drugs and/or propranolol had taurine levels similar to the controls (Fig. 28.3).

Osmotic Pressure

Although not usually considered by terrestrial biologists, an osmoregulatory role has long been regarded by marine biologists as a prime physiological role for taurine. Taurine concentrations have been demonstrated to increase substantially under increasing salinity regimens in mollusks (Potts, 1958; Lynch and Wood, 1966), crustaceans (Shaw, 1958), echinoderms (Jeuniaux et al., 1962), and chordates (Fugelli and Zachariassen, 1976). Taurine concentrations can be expected to increase in cells under salinity stress, and this has been used recently to enhance synthesis of taurine in lobster nerves (Finney, 1978).

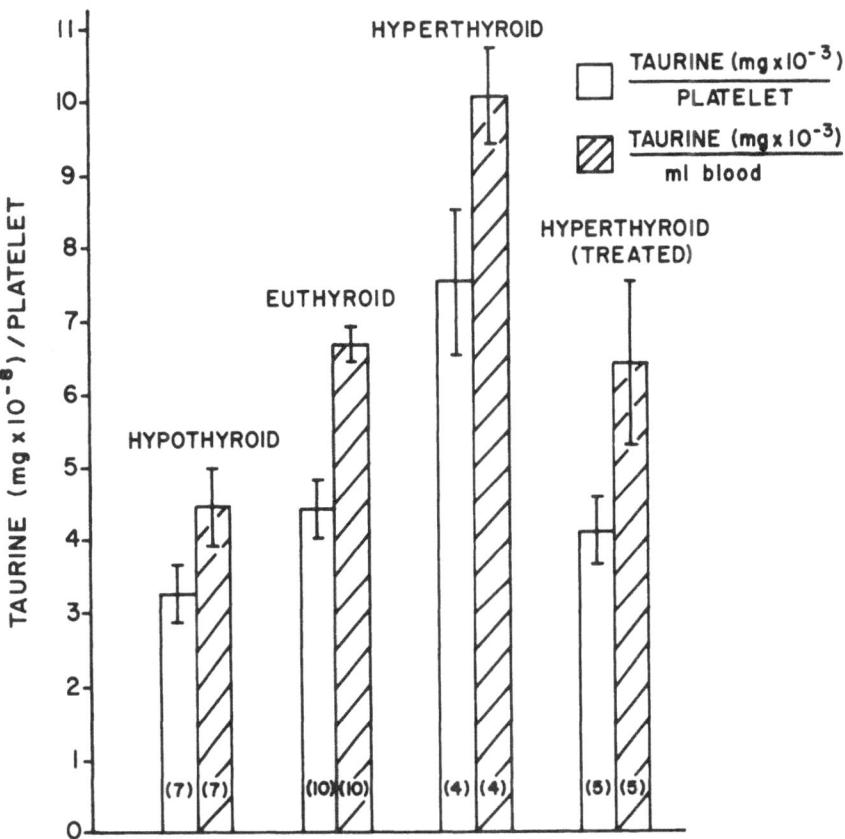

Fig. 28.3.   Taurine concentration in blood platelets in various thyroid states. Approximately 15 ml of blood were collected by venipuncture with a 20-gauge needle, into siliconized Vacutainer® tubes with 1.5 ml of 3.2% sodium citrate. The samples were inverted four times and then centrifuged at 225 *g* for 5 min. The plasma volume was measured, and a platelet count was taken with a dilution of 0.1 ml of plasma with 4.0 of 0.1 N saline, using a hemocytometer. EDTA, 0.075 ml (0.077M), was added to each ml of plasma to facilitate collection and prevent aggregation of platelets. Plasma was cooled on ice for 5 min and centrifuged at 1,500 *g* for 20 min at room temperature. The plasma was discarded, and the platelet button was resuspended in 2 ml of distilled water. The sample was then freeze-thawed three times to ensure complete lysis of platelet membranes. As a precaution, one drop of toluene was then added to disrupt any remaining platelet membranes. The sample was centrifuged at 5,400 *g* for 20 min at ambient temperature to separate the platelet membrane. The measured supernatant was assayed for taurine by the procedure described in the legend to Fig. 28.1.

Stress

Stress induced by surgical trauma has been reported to increase the intracellular concentration of taurine (Vinnars et al., 1975) and to increase taurine excretion between the 1st and 9th post operative days (Turner and Brum, 1964). As corticosteroids increased in the latter study, the increased taurine excretion was attributed to the inhibition of muscle protein synthesis.

Iwata et al. (1976) have shown that stress induced by physical restraint resulted in a twofold increase in urinary excretion, although tissue levels remained similar to control values.

ACKNOWLEDGMENT

This work was supported in part by NIH grants CA22170 and AG00003.

REFERENCES

Adembri, G.; Bartolini, A.; Bartolini, R.; Giotti, A.; and L. Zilletti. Anticonvulsive action of homotaurine and taurine. *Brit. J. Pharmacol.*, 52, 439 (1974).

Awapara, J. The taurine concentration of organs from fed and fasted rats. *J. Biol. Chem.*, 281, 571–576 (1956).

Baskin, S. I.; Cohen E.; and Kocsis, J. Taurine changes in visual tissues with age. In *Taurine*, R. Huxtable and A. Barbeau, eds. Raven Press, New York, pp. 201–207 (1976).

Baskin, S.; Klekotka, S.; Kendrick Z.; and Bartuska, D. Correlation of platelet taurine levels with thyroid function. *J. Endo. Invest.*, 2, 245–249 (1979).

Baskin, S. I.; Zaydon, P. T.; Orr, P. L.; and Katz, T. C. The pharmacokinetics of taurine in bovine Purkinje fibers. *Circ. Res.*, 47, 763–769 (1980).

Bonaventure, N.; Wioland, N.; and Mandel, P. Antagonists of the putative inhibitory transmitter effects of taurine and GABA in the retina. *Brain Res.*, 80, 281–289 (1974).

Cavallini, D.; Scandurra, R.; Dupre, S.; Federici, G.; Santoro, L.; Ricci, G.; and Bara, D. Alternative pathways of taurine biosynthesis. In *Taurine*, R. Huxtable and A. Barbeau, eds. Raven Press, New York, pp. 59–66 (1976).

Curtis, D.; and Watkins, J. Acidic amino acids with strong excitatory actions on mammalian neurons. *J. Physiol.* (London), 166, 1–14 (1963).

Dietrich, J.; and Diacono, J. Comparison between ouabain and taurine effects on isolated rat and guinea pig hearts in low calcium medium. *Life Sci.*, 10, 499–507 (1971).

Finney, C. Isotopic labelling of taurine: Implications for its synthesis in selected tissues of *Homarus americanus. Comp. Biochem. Physiol.*, 61B, 409–413 (1978).

Fugelli, K.; and Zachariassen, K. The distribution of taurine, gamma aminobutyric acid and inorganic ions between plasma and erythrocytes in flounder *(Platichtys flesus)* at different plasma osmolalities. *Comp. Biochem. Physiol.*, 55A, 173–177 (1976).

Gaut, Z.; and Nauss, C. Uptake of taurine by human platelets: A possible model for brain. In *Taurine*, R. Huxtable and A. Barbeau, eds. Raven Press, New York, pp. 91–98 (1976).

Grosso, D.; Bressler, R.; and Benson, B. Circadian rhythm and uptake of taurine by rat pineal gland. *Life Sci.*, 22, 1789–1798 (1978).

Idle, J.; Millburn, P.; and Williams, R. Taurine conjugates as metabolites of arylacetic acids in the ferret. *Xenobiotica*, 8(4), 253–264 (1978).

Iida, S.; and Hikichi, M. Effect of taurine on ethanol induced sleeping time in mice. *J. Studies on Alcohol*, 37(1), 19–26 (1976).

Ikeda, H. Effects of taurine on alcohol withdrawal. *Lancet*, Sept. 3, 1977, 509 (1977).

Iwata, H.; Baba, A.; and Yoneda, Y. Some factors affecting the concentration of tissue or urinary taurine. In *Taurine*, R. Huxtable and A. Barbeau, eds. Raven Press, New York, pp. 85–90 (1976).

Iwata, H.; Matsuda, T.; Yamagami, S.; Tsukamoto, T.; and Baba, A. Circadian periodicity of taurine content of rat brain. *Brain Res.*, 143, 383–386 (1978a).

Iwata, H.; Matsuda, T.; Yamagami, S.; Hirata, Y.; and Baba, A. Changes of taurine content in the brain tissue of barbiturate dependent rats. *Biochem. Pharm.*, 27, 1955–1959 (1978b).

James, M.; Smith, R.; Williams, R.; and Reidenberg, M. The conjugation of phenylacetic acid in man, sub-human primates and some non-primate species. *Proc. Roy. Soc.* (London), B182, 25 (1972).

Jeuniaux, C.; Bricteax-Gregoire, S.; and Florkin, M. Regulation osmotique intracellulaire chez *Asterias rubens* L.: Role du glycolle et de la taurine. *Cahiers Biol. Marine*, 3, 107–113 (1962).

Johnston, F.; Donald, E.; and Mercer, N. Effect of isoniazid on the urinary excretion of taurine by men. *J. Nutr.*, 93, 310–316 (1967).

Jordan, B.; and Rance, M. Taurine conjugation of fenclofenac in the dog. *Pharmacology*, 26, 359–361 (1974).

Kochakian, C. The free amino acids of the mouse kidney: Effect of castration and androgen. *Alabama J. Med. Sci.*, 11(4), 333–339 (1974).

Kochakian, C. Influence of testosterone on the concentration of hypotaurine and taurine in the reproductive tract of the male guinea pig, rat, and mouse. In *Taurine*, R. Huxtable and A. Barbeau, eds. Raven Press, New York, pp. 327–334 (1976).

Kocsis, J. J.; Kostos, V. J.; and Baskin, S. I. Taurine levels in the heart tissues of various species. In *Taurine*, R. Huxtable and A. Barbeau, eds. Raven Press, New York, pp. 145–153 (1976).

Kostos, V.; and Kocsis, J. Effect of colchicine on taurine excretion. *Proc. Soc. Exptl. Biol. Med.*, 106, 659–660 (1961).

Krusz, J. C.; Kendrick, Z. V.; and Baskin, S. I. Taurine content in the pineal of the aging monkey. In *Aging and the Non-Human Primate*, D. Bowden, ed. Raven Press, New York pp. 106–111 (1978a).

Krusz, J.; Dix R.; and Baskin, S. I. Factors that affect uptake and endogenous content of taurine in the pineal. *Fed. Proc.*, 37, 907 (1978b).

Kuriyama, K.; and Yoneda, Y. Morphine induced alterations of γ-aminobutyric acid and taurine contents and L-glutamate decarboxylase activity in rat spinal cord and thalmus: Possible correlates with analgesic action of morphine. *Brain Res.*, 148, 163–179 (1978).

Lähdesmäki, P.; Pasula, M.; and Oja, S. Effect of electrical stimulation and chlorpromazine on the uptake and release of taurine, γ-aminobutyric acid and glutamic acid in mouse brain synaptosomes. *J. Neurochem.*, 25, 675–680 (1975).

Lajtha, A.; and Toth, J. Perinatal changes in the free amino acid pool of the brain in mice. *Brain Res.*, 55, 238–241 (1973).

Levi, G.; and Morisi, G. Free amino acids and related compounds in chick brain during development. *Brain Res.*, 26, 131–140 (1971).

Lynch, M.; and Wood, L. Effects of environmental salinity on free amino acids of *Crassostrea virginica*. *Gmelin. Comp. Biochem. Physiol.*, 19, 783–790 (1966).

Matsuda, T.; Yamagami, S.; Mizui, T.; Baba, A.; and Iwata, H. Increase of cardiac taurine by glucocorticoids. *Biochem. Pharmacol.*, 27, 2973–2975 (1978).

Minnich, J.; Cruz, M.; and Franklin, G. Ouabain-induced seizures: Decreased free amino acids in cerebellum. *Trans. Amer. Soc. Neurochem.*, 7, 121 (abstract) (1976).

Nir, L.; Briel, G.; Dames, W.; and Neuhoff, V. Rat pineal free amino acids diurnal rhythm and effect of light. *Arch. Int. Physiol. Biochem.*, 81, 617–627 (1973).

Oja, A.; Uusitalo, A.; Vahvelainen, M.; and Piha, S. Changes in cerebral and hepatic amino acids in rat and guinea pig during development. *Brain Res.*, 11, 655–661 (1968).

Pasantes-Morales, H.; Klethi, J.; Urban P.; and Mandel, P. Influence of light and dark on the free amino acid pattern of the developing chick retina. *Brain Res.*, 57, 59–65 (1973).

Pasantes-Morales, H.; Salceda R.; and Lopez-Colome, A. The role of taurine in tetina: Factors affecting its release. In *Taurine*, R. Huxtable and A. Barbeau, eds. Raven Press, New York, pp. 191–200 (1976).

Piha, R.; Oja, S.; and Uusitalo, J. The effect of chlorpromazine on free amino acids in the rat brain. *Annales Medicinae Experimentalis et Biologiae Fenniae*, 40 (Suppl. 6), 1–27 (1962).

Potts, W. The inorganic and amino acid composition of some lamellibranch muscles. *J. Exp. Biol.*, 35, 749–764 (1958).

Rhode, B.; Rea, M.; and McBride, W. The effect of 3-acetylpyridine on the levels of taurine in different regions of the brain. *Brain Res.*, 156, 202–205 (1978).

Ricci, G.; Dupre, S.; Federici, G.; Spoto, G.; Matarese, R. M.; and Cavallini, D. Oxidation of hypotaurine to taurine by ultraviolet radiation. *Physiol. Chem. Phys.*, 10(5), 435–441 (1978).

Rylance, H.; and Myhal, D. Taurine excretion and the influence of drugs. *Clin. Chim. Acta.*, 35, 159–164 (1971).

Scheibel, J.; Elsasser, T.; and Ondo, J. Taurine and pituitary hormone secretion. *Fed. Proc.*, 37(3), 437 (abstract) (1978).

Scheibel, J.; Elsasser, T.; and Ondo, J. Effects of taurine and related metabolites on the secretion of LH and prolactin. *Fed. Proc.*, 38(3), 927 (abstract) (1979).

Shaw, J. Osmoregulation in the muscle fibers of *Carcinus maenas*. *J. Exp. Biol.*, 35, 920–929 (1958).

Spaeth, D. G.; and Schneider, D. D. Taurine metabolism: Effects of diet and bile salt metabolism. In: *Taurine*, R. Huxtable and A. Barbeau, eds. Raven Press, New York, pp. 35–44 (1976).

Sturman, J. A. Taurine pool sizes in the rat: Effects of vitamin B_6 deficiency and high taurine diet. *J. Nutr.*, 103, 1566–1580 (1973).

Sturman, J. A.; and Gaull, G. E. Taurine in the brain liver of the developing human and monkey. *J. Neurochem.*, 25, 831–835 (1975).

Turner, F.; and Brum, V. The urinary excretion of free taurine in acute and chronic disease following surgical trauma, and in patients with acute alcoholism. *J. Surg. Res.*, 4(9), 423–431 (1964).

Vessey, D. The biochemical basis for conjugation of bile acids with either glycine or taurine. *Biochem. J.*, 174, 621–626 (1978).

Vinnars, E.; Bergstrom, J.; and Furst, P. Influence of the postoperative state on the intracellular free amino acids in human muscle tissue. *Annals of Surgery*, 182(6), 665–671 (1975).

Copyright ©1981, Spectrum Publications, Inc.
The Effects of Taurine on Excitable Tissues

CHAPTER 29

Taurine and Myocardial Ischemia

J. B. Lombardini
M. F. Crass III

INTRODUCTION

It is now well recognized that taurine is a constituent of cardiac muscle, comprising over 50% of the free amino acid pool of this tissue (Jacobsen and Smith, 1968). However, the function of taurine in heart is not known. The initial stimulus to investigate the role of taurine in cardiac tissue was the pioneering work of Read and Welty (1963; Welty and Read, 1963). These workers reported that intravenous administration of taurine to dogs resulted in the disappearance of arrhythmias induced by large doses of epinephrine and digoxin. Supporting data for the role of taurine as an antiarrhythmic agent have been reported by Novelli et al. (1969) and Guidotti and Giotti (1970). However, it was the observations of Chazov et al. (1974) that fortified the potential role of taurine as an antiarrhythmic agent. Taurine was shown to "normalize" the electrocardiographic abnormalities produced by toxic doses of strophanthin-K in isolated perfused guinea pig hearts and in dogs *in vivo*. While the above results have stimulated much thought, they are not without criticism. First, the interpretation of the effects of taurine on the abnormal electrocardiogram has been challenged by Hinton et al. (1975), who reported that taurine was ineffective in counteracting ventricular arrhythmias induced by deslanoside in the cat. Second, the premise that it is actually isethionic acid, a postulated anionic metabolite of taurine, that is the active component in protecting cardiac muscle from toxic doses of cardiac drugs by intracellular retention of K^+ ions has been disputed. The recent work of Fellman et al. (1978) and that of Cavallini et al. (1978) have definitively ascertained that taurine is not a precursor for the formation of isethionic acid in heart.

Peterson et al. (1973) noted for the first time that the levels of taurine were altered in diseased cardiac muscle. It was observed in dogs with surgically induced right ventricular congestive failure that the content of taurine increased threefold while the content in the left ventricle remained constant. These studies were expanded by Huxtable and Bressler (1974a,b), who analyzed postmortem tissue samples of patients who died of left ventricular congestive failure. They observed that left ventricular tissue had an elevated taurine content (twofold) compared to similar tissue obtained from patients who died of noncardiac related events. In addition, cardiac muscle taurine content was also elevated in hypertension in both humans and rats. Thus, the antithetical observations reported by us (Crass and Lombardini, 1977, 1978) concerning the levels of taurine in ischemic or otherwise oxygen-deficient cardiac tissue have been intriguing. We reported a significant loss of taurine from cardiac muscle in two animal models, namely, ligation of a major coronary vessle in the dog and anoxic perfusion of the isolated rat heart.

While myocardial ischemia is known to produce numerous biochemical and morphological changes (Opie, 1972; Idell-Wenger and Neely, 1977; Hoffstein et al., 1975), our observations are the first to report loss of taurine in ischemic cardiac tissue.

The present communication is designed to elaborate on and augment our previous studies on the loss of taurine in the ischemic dog heart and the oxygen-deficient rat heart. Furthermore, this presentation also addresses the effects of cardiac necrosis resulting from high doses of sympathomimetic agents (isoproterenol and methoxamine) on the levels of taurine and the other amino acids in heart, liver, and blood of the rat. Finally, preliminary clinical results from patients who have had documented myocardial infarctions are discussed with reference to blood taurine levels. An important question to be answered is whether the loss of taurine from the ischemic heart is specific or a general phenomenon with concomitant loss of other small molecular weight compounds such as other amino acids.

MATERIALS AND METHODS

Coronary Artery Occlusion

Adult mongrel dogs of either sex, weighing 15–35 kg, were anesthetized with sodium pentobarbital (30 mg · kg^{-1}). A tracheostomy was performed, and a Harvard positive pressure respirator was used to ventilate the animals with room air. A left thoracotomy exposed the pericardium, which was then incised and sutured to the chest wall, forming a cradle for the heart. The circumflex branch of the left main coronary artery was isolated and was either ligated (ischemic) or left patent (control). Heart rate was monitored at 10-min intervals throughout the 4 hr experimental period. The electrocardiogram (Lead II) was obtained from standard limb leads. Arterial pressures were determined by placing a catheter into the descending aorta via the left femoral artery. Ventricular peak systolic and end diastolic pressures were obtained by placement of a catheter in the left ventricle via the right common carotid artery. Blood gases and pH were maintained within physiologic range. Pressures were recorded with Statham P23Db transducers connected to a Model 79 Grass recorder.

Perfusion Techniques

Hearts from 250–300 g, ether-anesthetized male Sprague-Dawley rats were quickly excised and mounted on a heart perfusion apparatus (Morgan et al., 1965; Neely et al., 1967; Crass et al., 1971). Coronary perfusion pressure was maintained constant at 82 mm Hg. One group of hearts, designated "initial," was perfusion-washed for 5 min in a nonrecirculated system with Krebs-Henseleit bicarbonate buffer containing 5.5 mM glucose. This group was subdivided into two groups and perfused with the above buffer, which was continuously equilibrated with either 95% O_2–5% CO_2 (oxygenated controls) or 95% N_2–5% CO_2 (oxygen-deficient hearts). A second group of rat hearts, designated "residual," was subjected to an additional 30 min of perfusion in a recirculated system with either oxygenated or oxygen-deficient medium. Mean coronary flow rates (ml · min^{-1} · g wet wt.$^{-1}$ ± SE) were 10.9 ± 0.8 and 16.8 ± 1.0 for oxygenated controls and oxygen-deficient hearts, respectively. Mean heart rates (beats · min^{-1} ± SE) were 286 ± 24 for oxygenated hearts and 130 ± 38 for oxygen-deficient hearts.

Biochemical Analyses

Specific areas of dog cardiac muscle were biopsied with a 15-mm (ID) stainless steel boring tube attached to a 3/8 in. electric hand drill. Tissue sections were immediately frozen in liquid nitrogen, after which they were cut into designated regions with a small circular sawblade attached to an electric Dremel tool. Weighed tissue samples were homogenized for 20 sec in 3 volumes of 2% perchloric acid in a Virtis Model 45 homogenizer at half speed. The homogenate was then centrifuged for 10 min at 12,000 g. The pH of the clear deproteinized extract was adjusted to pH 7.0 and then analyzed for taurine content by an enzymatic double isotope derivative assay developed in the authors' laboratory (Lombardini, 1975). In a second procedure, the pH of the extract was adjusted to pH 2.4 and analyzed on a Beckman 121 automatic amino acid analyzer.

Rat ventricles were excised while being perfused and freeze-clamped with Wollenberger tongs precooled in liquid nitrogen. The ventricles were then weighed and analyzed for taurine as above. The perfusion medium was sampled at 10 min intervals and analyzed for taurine content.

Male Sprague-Dawley rats weighing 200–235 g were injected subcutaneously with either DL-isoproterenol (Sigma Chemical Co.) (80 mg · kg^{-1}) or methoxamine (gift from Burroughs Wellcome Co.) (20 mg · kg^{-1}). After 7 hr the animals were anesthetized with ether and 5–7 ml of blood were withdrawn from the inferior vena cava and placed in tubes containing EDTA. The blood was deproteinized with an equal volume of 5% perchloric acid and centrifuged for 10 min at 12,000 g. Ventricular tissue and liver were rapidly removed, rinsed in cold saline, and blotted to remove excess moisture. The tissues were then weighed and homogenized in 2 volumes of 2% perchloric acid with a Polytron homogenizer. Taurine content was determined by the amino acid analyzer procedure. Contents of the individual amino acids in cardiac tissue were determined by utilizing PA-28 (for acid and neutral amino acids) and PA-35 (for basic amino acids) resins and sodium citrate buffers, as outlined in the Beckman manual for the Model 121 amino acid analyzer.

Dry/wet tissue weight ratios were calculated for heart and liver in all experiments in order to express the taurine or amino acid content on a dry weight basis.

RESULTS

Effects of Coronary Artery Occlusion on Taurine Levels in Cardiac Tissue (Dog)

Distribution of Taurine in Normal Cardiac Tissue

In order to assess the effects of ischemia on cardiac tissue levels of taurine, the regional distribution of taurine content in the normal myocardium of the dog was initially measured (Table 29.1). The highest levels of taurine were found in the atria, specifically in the right atrium. The content of taurine in right and left ventricles was less than that of atrial tissue. An increasing outer-to-inner transmural gradient of taurine content was observed in the posterolateral wall of the left ventricle; the subendocardial region contained substantially more taurine than the midventricular region ($p < 0.01$) or the

Table 29.1.   Effects of ligation of the circumflex branch of the left coronary artery on levels of taurine in various regions of the dog heart[a,b]

Region	Control	Ischemia (μmol · g dry wt.$^{-1}$)	Decrease	Percent
Right atrium	96.8 ± 3.5 (5)	71.3 ± 7.2 (4)[c]	25.5	26
Left atrium	71.3 ± 7.2 (5)	54.7 ± 2.7 (4)[c]	16.6	25
Right ventricle				
Subepicardium	46.1 ± 5.5 (5)	—	—	—
Midventricle	47.0 (2)	—	—	—
Subendocardium	48.2 ± 3.3 (5)	—	—	—
Left ventricle				
Posterolateral wall	55.6 ± 3.0 (5)	29.5 ± 2.1 (5)[d]	26.1	47
Subepicardium	45.0 ± 3.8 (5)	25.3 ± 2.6 (5)[d]	19.7	44
Midventricle	53.7 ± 3.1 (5)	27.7 ± 2.6 (5)[d]	26.0	48
Subendocardium	68.0 ± 2.1 (5)	35.6 ± 4.4 (5)[d]	32.4	48

[a]Each value represents the mean ± SE. Numbers of animals in each series is indicated in parentheses.

[b]Data from Crass and Lombardini (1977).

[c]Equals $p < 0.02$ relative to control values.

[d]Equals $p < 0.001$ relative to control values.

subepicardial region ($p < 0.001$). No transmural gradient was observed in the right ventricle.

Distribution of Taurine in Ischemic Cardiac Tissue

Taurine content of all sampled regions of the dog heart decreased after circumflex artery ligation (Table 29.1). The effect of ischemia was greatest in the left ventricle, where taurine levels decreased from 55.6 ± 3.0 to 29.5 ± 2.1 μmol · g dry wt.$^{-1}$. Analyses of transmural sections of the ischemic left ventricle revealed that the increasing outer-to-inner gradient previously observed in control tissue had disappeared; that is, no significant differences in taurine content between the subepicardial, midventricular, and subendocardial sections were noted. All three subregions of the left ventricle lost significant quantities of taurine (44% or more of their preligation content in each layer). Loss of taurine was greatest in the subendocardial layer.

Some taurine loss was also observed in "nonischemic" areas. Both right and left atria decreased in taurine content.

Effects of Anoxic Perfusion on Taurine Levels in Cardiac Muscle (Rat)

Hearts were perfused in a nonrecirculating system for 5 min with oxygenated or nonoxygenated buffer ("initial") in order to washout the coronary vascular bed and thus establish control values for ventricular taurine content (Crass and Lombardini, 1978). The values for taurine in the normoxic and oxygen-deficient ventricles were 154.4 ± 2.6 and 146.4 ± 3.7 μmol · g dry wt.$^{-1}$, respectively ($p > 0.10$; Table 29.2). When hearts were perfused in a recirculating system with oxygenated medium for an additional 30 min ("residual"), the taurine content did not change (159.6 ± 2.7; $p > 0.1$). However, when oxygen-deficient medium was used as the perfusate, the ventricular taurine content was significantly reduced (132.2 ± 3.6; $p < 0.02$).

Taurine concentrations were also determined in the perfusate at 10 min intervals during the 30 min "residual" recirculating period (Fig. 29.1). In the oxygenated controls, a small quantity of taurine (3.9 ± 0.2 μmol · g dry wt.$^{-1}$) was observed in the 10 min sample, but no further changes were observed during the remaining 20 min of perfusion. On the contrary, in the oxygen-deficient hearts (95% nitrogen–5% carbon dioxide), a 3.3-fold increase in taurine concentration was observed after 10 min (13.0 ± 1.6) and was increased further (5-fold) after 30 min (19.4 ± 2.6).

Effects of Sympathomimetic Agents on Taurine Levels in Cardiac Muscle and Blood (Rat)

In the next series of experiments, cardiotoxic doses of DL-isoproterenol (ISO) and methoxamine were administered to male rats. The results, shown in Fig. 29.2, indicated that necrosis-inducing amounts of ISO caused a 29% reduction of ventricular taurine content (control = 102.5 ± 2.4; ISO = 72.4

Table 29.2. Effects of anoxic perfusion on the taurine content of rat ventricles[a,b]

	Taurine (μmol · g dry wt.$^{-1}$)	
	Initial	Residual
Oxygen (95%) – CO_2 (5%)	154.4 ± 2.6 (8)	159.6 ± 2.7 (11)
Nitrogen (95%) – CO_2 (5%)	146.4 ± 3.7 (4)	132.2 ± 3.6 (9)

[a]Each value represents the mean ± SE. Number of animals per group is in parentheses.
[b]Reproduced from Crass and Lombardini (1978), with the permission of the Editor, *Proceedings of the Society for Experimental Biology and Medicine.*

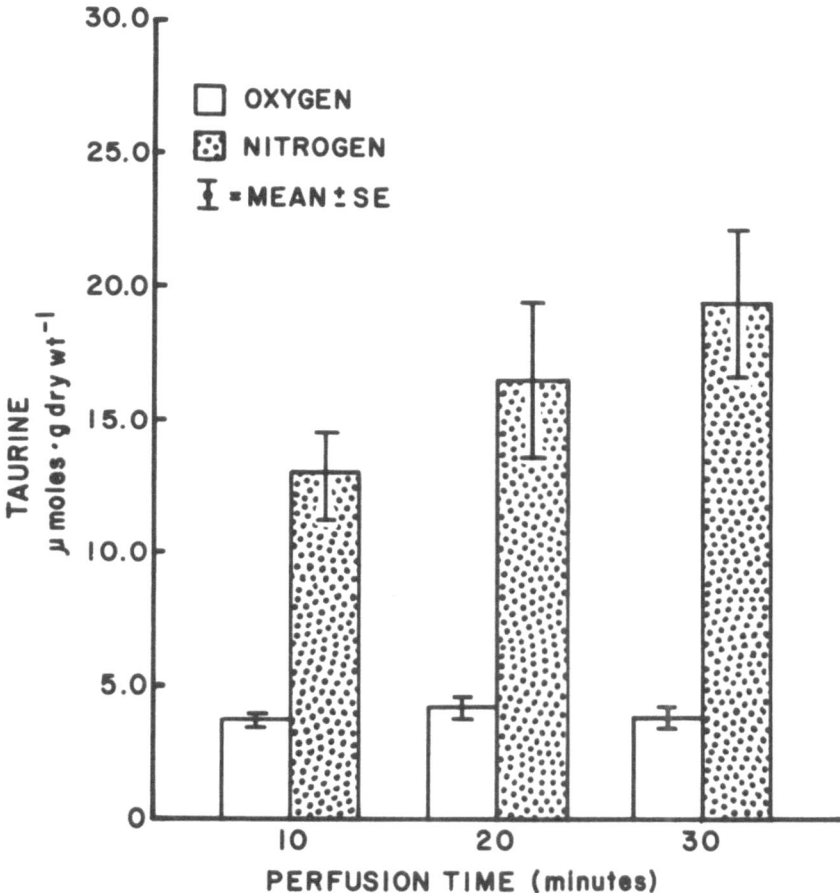

Fig. 29.1. Release of tissue taurine into perfusates of the isolated rat heart. Hearts were per-fused with Krebs-Henseleit bicarbonate buffer containing 5.5 mM glucose equilibrated with either 95% N_2–5% CO_2 or 95% O_2–5% CO_2. Each bar represents the mean cumulative taurine value in recirculated perfusates of six hearts ± SE.

± 3.8 μmol · g dry wt.$^{-1}$), while liver content did not change (control = 13.7 ± 4.2; ISO = 14.4 ± 2.2). Moreover, taurine levels in blood increased from 0.263 ± 0.021 to 0.479 ± 0.041, an 82% increase. In order to determine whether the loss of taurine in cardiac tissue was specific for taurine, the total amino acid pattern was analyzed (Fig. 29.3). Ventricular tissue levels of alanine, aspartate, glutamate, threonine, and glutamine plus asparagine

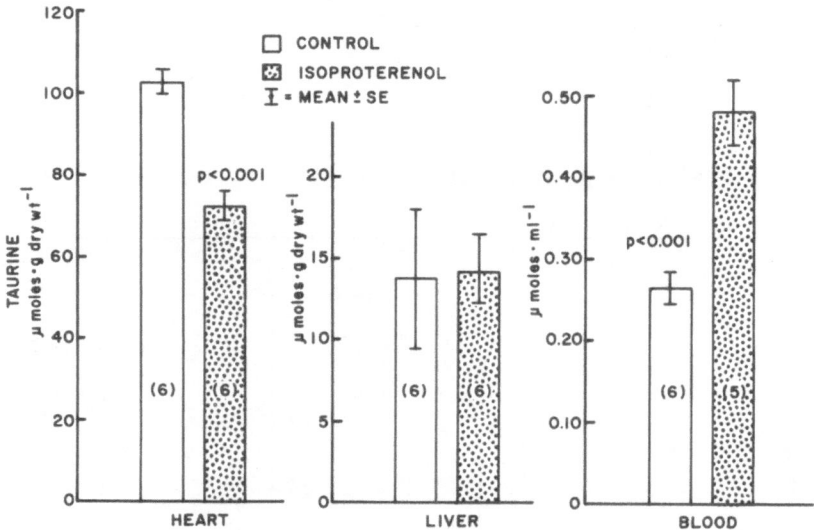

Fig. 29.2. Effect of DL-isoproterenol on taurine levels in rat ventricles, liver, and blood. Animals were killed 7 hr after administration of 80 mg · kg^{-1} of DL-isoproterenol or saline. Number of animals per group is shown in parentheses.

(combined) were the only amino acids, other than taurine, to significantly change after 7 hr of IO treatment. Alanine and asparate increased by 37% (ala: control = 12.6 ± 0.3; ISO = 17.3 ± 0.8) and 41% (asp: control = 8.1 ± 0.4; ISO = 11.4 ± 0.3), respectively. Conversely, glutamine plus asparagine decreased by 37% after ISO administration, a loss of 17.4 μmol (46.5 ± 5.0 to 29.1 ± 2.5). Minor quantitative losses were also observed for glutamate and threonine.

A summation (Table 29.3) of all ninhydrin-positive substances (NPS) in the rat myocardium as measured by the amino acid analyzer technique indicated that taurine (102.5 ± 2.4 μmol · g dry wt.$^{-1}$) comprised 34.5% of the total NPS (297.0 ± 7.9 μmol), or 41.8% of all amino acids (245.4 ± 3.3 μmol) of control animals. Seven hours after the administration of ISO, the taurine content decreased to 29.0% of the total NPS and 36.5% of all amino acids. The only other amino acids that quantitatively contribute to the decrease in total amino acid content are glutamine plus asparagine (Fig. 29.3).

Methoxamine, an α-adrenergic agonist, also reduced ventricular tissue taurine content (107.1 ± 5.9 to 70.5 ± 8.9; Fig. 29.4), a loss comparable to that seen with ISO. However, the blood levels of taurine increased by a factor of 2.5 (0.318 ± 0.023 to 0.829 ± 0.105), a significantly greater increase than was observed after administration of ISO ($p < 0.001$).

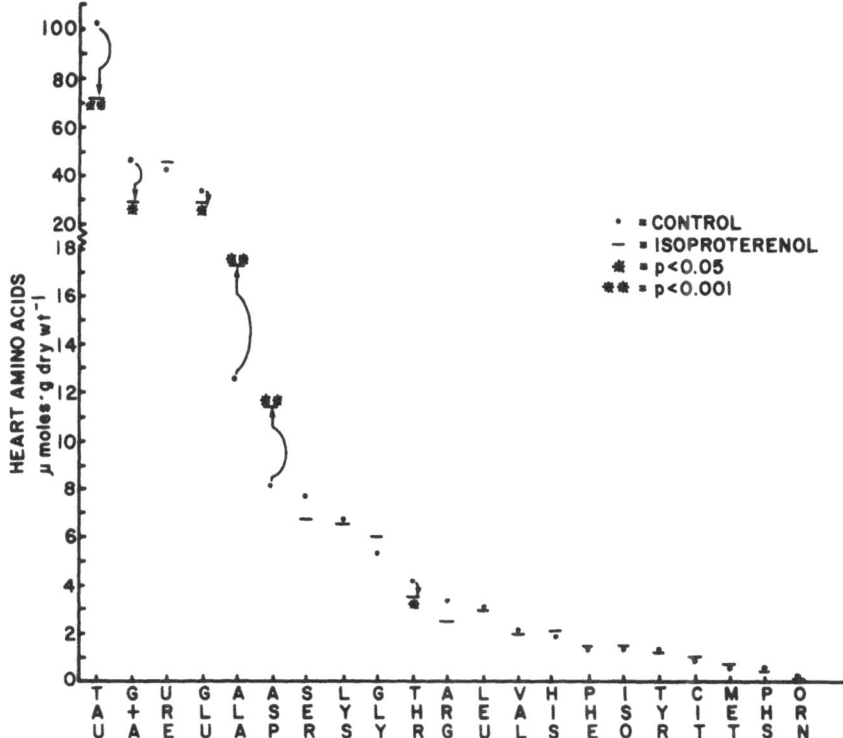

Fig. 29.3.   Amino acid pattern of rat ventricles 7 hr after administration of DL-isoproterenol (80 mg · kg⁻¹). Data represent mean values for five hearts. The control animals are represented by dots (.) and the DL-isoproterenol-treated animals by bars (−). Standard abbreviations for the amino acids are used except for glutamine plus asparagine (G + A) and phosphoserine (PHS). Urea is abbreviated as URE.

Effects of Acute Myocardial Infarction on Taurine Levels in Blood (Human)

Preliminary data on blood taurine content of patients who had a myocardial infarction suggested that taurine concentrations were significantly elevated when compared to levels of patients admitted for other illnesses (unpublished data of J. B. Lombardini and M. W. Cooper). Thus far, five patients with diagnosed acute myocardial infarctions (elevated MB isoenzyme band of creatine phosphokinase) have been monitored for abnormally elevated blood taurine levels. These patients had increased blood taurine levels ranging from 0.279 to 0.571 μmol · ml whole blood⁻¹ (mean = 0.361 $\pm$ 0.054), a statistically significant ($p < 0.01$) increase over control blood taurine levels (0.226 $\pm$ 0.009). A sixth patient had an increase in isoenzyme Band

Table 29.3.   Summation of ventricular tissue content of ninhydrin-positive substances (NPS)[a]

	Control		DL-Isoproterenol	
	μmol · g dry wt.$^{-1}$	Percent taurine	μmol · g dry wt.$^{-1}$	Percent taurine
Taurine	102.5 ± 2.4	—	72.4 ± 3.8	—
Amino acids	142.9 ± 3.7	—	126.0 ± 6.3	—
Other ninhydrin-positive substances[b]	51.6 ± 6.4	—	51.5 ± 4.2	—
Total amino acids plus taurine	245.4 ± 3.3	41.8	198.4 ± 7.2	36.5
Total ninhydrin-positive substances	297.0 ± 7.9	34.5	249.9 ± 7.4	29.0

[a]Each value represents the results from five animals (mean ± SE).
[b]Includes urea, phosphoethanolamine, and ethanolamine.

1 of lactate dehydrogenase and was diagnosed as having had a myocardial infarction. However, the MB isoenzyme band of creatine phosphokinase did not appear, and his blood taurine concentration was only 0.176.

DISCUSSION

The results of the present experimental animal studies utilizing three different models for ischemic injury—coronary artery ligation in dogs, perfusion of rat hearts in an anoxic medium, and high doses of sympathomimetic agents in rats—clearly indicated that taurine was released in significant quantities from the injured myocardium. Moreover, preliminary data from clinical studies involving myocardial infarction in man also suggested, by virtue of increased blood taurine, that taurine is released from the damaged human myocardium.

It has been reported (Kocsis et al., 1976) that there are significant differences between species in the taurine content of cardiac tissue. For example, frog heart (*Rana pipiens*) contained 5.0 ± 0.6 μmol of taurine per gram fresh tissue (mean ± SD); hamster heart contained 15.8 ± 2.4 μmol per gram; and mouse heart, 39.4 ± 7.1 μmol per gram. Kocsis et al. (1976) evaluated nine mammalian species and noted that the taurine contents in ventricular (V) tissue of the rat, guinea pig, rabbit, beef, pig, and cat were higher than those of

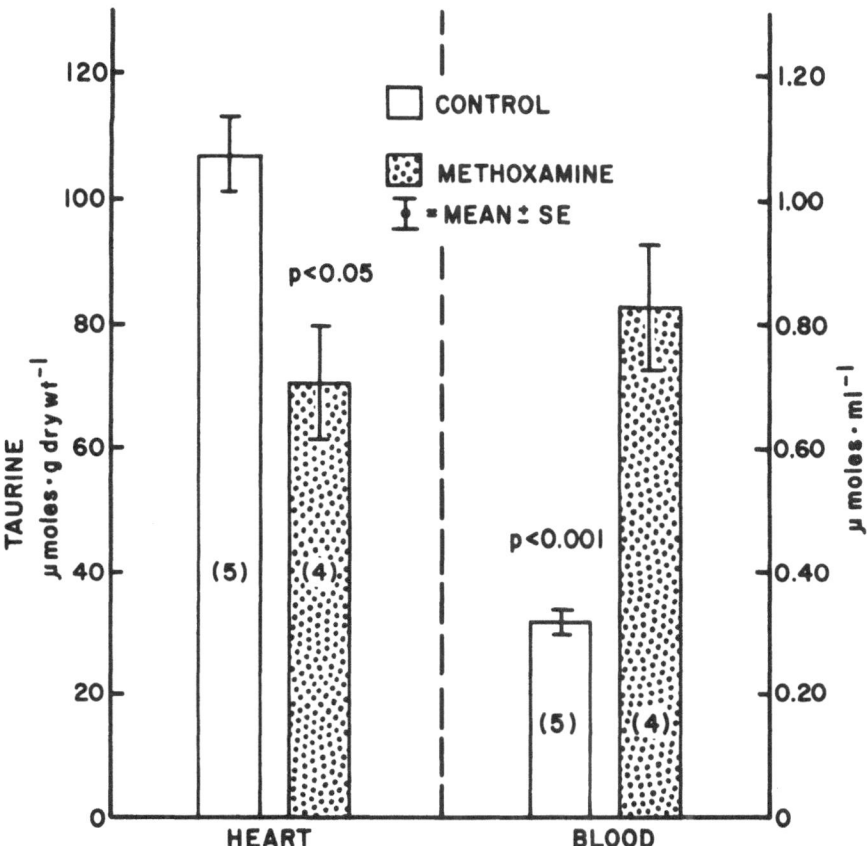

Fig. 29.4. Effect of methoxamine on taurine levels in rat ventricles and blood. Animals were killed 7 hr after administration of 20 mg · kg⁻¹ of methoxamine. Control animals received 0.9% NaCl. The number of animals in each group is shown in parentheses.

the auricles (A). Interestingly, a reversed ratio was demonstrated in dog and mouse cardiac tissues.

Kocsis et al. (1976) reported no differences between the left and right ventricles or left and right auricles in the dog. However, in observations from our laboratory (Crass and Lombardini, 1977), a significant difference ($p < 0.02$) between the right (96.8 ± 3.5 μmol · g dry wt.⁻¹) and the left atrium (71.3 ± 7.2) of the dog was noted.

Canine ventricles exhibited no regional distribution of taurine when samples where analyzed from different areas, including the base and apex (Kocsis et al., 1976). However, transmural differences in the left ventricle were noted by both groups of investigators (Kocsis et al., 1976; Crass and Lombardini, 1977). As shown in Table 29.1, a taurine concentration gradient was observed in the left ventricle, with the content lowest in the subepicardial region (45.0 ± 3.8 μmol · g dry wt.$^{-1}$), higher in the midventricular region (53.7 ± 3.1), and greatest in the subendocardial region (68.0 ± 2.1) ($p < 0.001$). No transmural gradient was observed in the right ventricle.

The communications of Kocsis et al. (1976) and Crass and Lombardini (1977) concerning the regional heterogeneity of taurine in the dog heart have stimulated thought as to the physiologic function of taurine in the myocardium. For example, it has been reported that Purkinje fibers of cat and beef hearts contain 2½–3 times more taurine than left ventricular myocardial tissue, suggesting that taurine may function in the conduction of the ventricular action potential (Kocsis et al., 1976). In the ischemic left ventricle, the net loss of taurine was found to be greatest in the subendocardial region (Table 29.1), an area anatomically rich in Purkinje fibers (Cardwell and Abramson, 1931) and most susceptable to ischemic damage (Moir, 1972). The above observations suggested that conduction disturbances resulting in electrical abnormalities might be attributed to the loss of tissue taurine due to ischemic injury in the endocardial region. Metabolically, this region undergoes more profound changes in metabolites such as pyruvate, lactate, ATP (Griggs et al., 1972), and glycogen (Crass and Sterrett, 1975) than the outer regions in experimental ischemia. The decrease in right and left atrial taurine cannot be readily explained. Because blood supply is not compromised to these regions, an alternative explanation other than severe hypoxia or cell death must be invoked. Hemodynamic disturbances that involve the entire myocardium occur during the sequelae of acute myocardial infarction. Changes in specific cardiodynamic parameters—such as cardiac output, heart rate, blood pressure, and stroke volume—could produce a generalized cardiac "stress," which could conceivably result in alteration of taurine levels. Experiments are currently underway to determine if these nonischemic changes are of a permanent or transient nature.

Guidotti et al. (1971) reported that isolated guinea pig auricles lost significant quantities of taurine when incubated in Tyrode solution. Experiments in our laboratory have demonstrated that perfusion of rat hearts with anoxic, taurine-free medium resulted in a significant loss of taurine from the ventricles. In our studies, release of taurine was time dependent only when anoxic perfusion conditions were employed. Although a small quantity of perfusate taurine (3.88 μmol · g dry wt.$^{-1}$) was observed after 10 min with an oxygenated medium, the "loss" was within analytical error. These results thus demonstrate that whole heart hypoxia, under conditions in which cell death was highly unlikely, also resulted in taurine loss.

It is well known that large doses (80 mg · kg⁻¹) of the β-agonist DL-iso-proterenol produce myocardial necrosis similar to infarction in experimental animal models (Rona et al., 1959; Kutsuna, 1972; Rona et al., 1973). It has also been suggested that this synthetic catecholamine induces myocardial ischemia by lowering coronary perfusion pressure (Handforth, 1962; Daniell et al., 1967). Due to the ischemia-generated consequences of DL-isopro-terenol treatment and because it has been recently communicated by Chubb and Huxtable (1978a,b) that DL-isoproterenol modifies taurine concentra-tions in the rat heart, we decided to examine the effects of high doses of DL-isoproterenol on the total amino acid spectrum in rat cardiac tissue.

Administration of 80 mg · kg⁻¹ of DL-isoproterenol to rats produced a 29% decrease in the taurine content of cardiac tissue but no change in the tau-rine levels of liver. In addition to depleted taurine levels, it was also demon-strated that the levels of glutamine plus asparagine (these two amino acids elute as one peak under the conditions employed with the amino acid ana-lyzer), glutamate, and threonine decreased significantly when ventricular tis-sue was analyzed 7 hr after DL-isoproterenol administration. On the con-trary, levels of alanine and aspartate increased by 37% and 41%, respec-tively. The levels of all other amino acids did not change. Thus, specificity of release of certain amino acids, particularly taurine, was noted. The data on the analyses of the total amino acid spectrum suggested that taurine loss from the damaged myocardium is not entirely an outcome of cell death, since there is not a comprehensive loss of all amino acids.

It was interesting to note that opposite effects on cardiac taurine levels can be produced by isoproterenol, depending upon dosage and regimen. Our results indicate that high doses of DL-isoproterenol (80 mg · kg⁻¹) cause car-diac tissue to release large quantities of taurine. On the other hand, the in-vestigations of Chubb and Huxtable (1978a,b) have demonstrated that chronic daily doses (5 mg · kg⁻¹ s.c.) of isoproterenol, which are known to produce cardiac hypertrophy (Stanton et al., 1969), increased the content of taurine in heart when expressed on a per heart basis. These workers reported that levels of taurine after isoproterenol treatment increased from 30 to 58 μmol · heart⁻¹. However, when expressed on a per gram wet weight basis, an apparent increase was also observed (33.6 ± 1.0 to 42.7 ± 7.0 μmol), al-though this change was not statistically significant. Furthermore, in the per-fused rat heart (Langendorff technique), Huxtable and Chubb (1976, 1977; Chubb and Huxtable, 1978c) reported that DL-isoproterenol stimulated the rate of taurine influx.

It has been demonstrated that methoxamine (an α-agonist and vasocon-strictor), which is known to exhibit pharmacological effects on the cardio-vascular system that are opposite to DL-isoproterenol, also reduced the taurine content of the myocardium. The magnitude of the decrease in taurine content after methoxamine administration was similar to that observed after DL-isoproterenol treatment; that is, a loss of approximately 30%.

The increase in blood levels of taurine after administration of either DL-isoproterenol or methoxamine is most likely a reflection of taurine loss from cardiac tissue. However, methoxamine increased blood levels of taurine to a greater extent. This result is perhaps indirectly due to the vasoconstricting properties of methoxamine, which are known to lower renal blood flow and consequently limit the production of urine. Thus, the excretion of taurine in the urine, a major route of taurine elimination (Chesney et al., 1978), may be inhibited in methoxamine-treated animals.

Perhaps one of the most significant observations from our studies is the prodigious quantity of taurine that was released from the occluded or chemically damaged heart. The taurine content of the dog left ventricle decreased by 26.1 μmol · g dry wt.$^{-1}$ (47%) after ligation of the circumflex branch of the left main coronary artery (Table 29.1); the oxygen-deficient perfused rat heart ventricles lost approximately 20 μmol of taurine (10–15%; Table 29.2); and toxic doses of the sympathomimetic agents isoproterenol (Fig. 29.2) and methoxamine (Fig. 29.4), provoked taurine loses of 30 (29%) and 36 (34%) μmol, respectively.

A second important observation is that only specific amino acids—glutamine plus asparagine, glutamate, threonine, and taurine—are released from cardiac tissue after DL-isoproterenol administration. The specificity of the release indicates that DL-isoproterenol, which is known to produce ischemia and cardiac necrosis in the catecholamine model, does not disrupt myocardial cells in such a way that all metabolites are randomly expelled. Perhaps a conservation mechanism for most of the amino acids, but not for taurine, is available to the cell, thus maintaining sufficient concentrations of substrates and enabling protein synthesis to more rapidly recover after the initial insult.

A third observation worthy of note is that taurine levels increase in the blood after the administration of DL-isoproterenol. Thus, damage to the myocardium can be monitored indirectly. This observation in the rat animal model was especially significant vis-à-vis our preliminary studies on patients who have had myocardial infarctions. In the latter, it also appears that the concentrations of taurine in blood increase after hypoxic insult. However, among the more important questions remaining is whether taurine levels rise proportionally to the extent of cardiac damage. A greater number of patients in this series and careful scrutiny of the clinical data on these patients will be necessary before this query can be properly answered.

Two further questions can be posed: (1) What is the function of taurine in cardiac tissue? and (2) By what mechanism does taurine perform this function? Much speculation has suggested that taurine may have a "protective" or membrane-stabilizing effect(s) against arrhythmia-inducing factors, such as drugs, trauma, and ischemia. Attempts have been made to explain the antiarrhythmic properties of taurine by the observations that taurine produces a hyperpolarization of the membrane potential and a reduction in the time course of the muscle action potential via acceleration of the repolarization phase (Gruener et al., 1975; Gruener and Bryant, 1975; Gruener et al., 1976). However, these phenomena could be secondary to a taurine-induced selective increase in membrane permeability to potassium and chloride ions but not to sodium ions (Gruener and Bryant, 1975). Moreover, the role of taurine in affecting calcium fluxes in cardiac tissue must be considered. Dolara et al. (1973, 1976, 1978) have reported that taurine exerts an ameliorating effect on the decrease in contractile force in guinea pig hearts perfused with a calcium-free Tyrode solution. These investigators also reported that the calcium content is increased in taurine-treated hearts and proposed that taurine might modify calcium turnover by: (1) a direct calcium-taurine interaction, (2) by forming a neutral carrier system for calcium, or (3) by forming a complex involving taurine, calcium, and a cellular macromolecule. On the other hand, it has been reported by Entman et al. (1977) that taurine has no effect on calcium transport, binding, or release, or on ATPase activity in sarcoplasmic reticulum isolated from dog cardiac muscle.

SUMMARY

The present status of knowledge of the distribution of taurine in the myocardium and the effects of ischemia are reviewed. The content of taurine in heart tissue varies in different animal species. Furthermore, it has been demonstrated that a regional distribution of taurine exists within the hearts of a given animal species. Specifically, atria contain higher levels of taurine than the ventricles in the dog. Also, a gradient exists in the left ventricular wall of the myocardium, the subendocardium having a greater taurine content then the midventricle or subepicardium.

The effects of ischemia on taurine levels were noted in three different types of animal models: ligation of the left circumflex artery in the dog, perfusion of rat hearts with buffer deficient in oxygen, and administration of toxic doses of sympathomimetic agents. In each model, it was demonstrated that the myocardium lost significant quantities of taurine. Analysis of individual and total amino acid contents in rat ventricles after administration of DL-isoproterenol revealed that loss was relatively specific for taurine. Glutamine plus asparagine, glutamate, and threonine levels of heart tissue were decreased also in isoproterenol-treated animals, while alanine and aspartate increased significantly.

Preliminary results also indicated that blood taurine levels increased in patients with clinically diagnosed myocardial infarction.

ACKNOWLEDGMENT

This study was supported in part by a grant from the Texas Affiliate of the American Heart Association.

The skillful technical assistance of Mrs. S. Paulette Decker is gratefully acknowledged. We wish to thank Mr. Harvey O. Olney, Department of Biochemistry, Texas Tech University Health Sciences Center, for his time and effort in performing the amino acid analyses. We also wish to express our appreciation to the technical and nursing staffs of the Pathology Clinical Laboratory and the Coronary Care Unit of the Texas Tech University Health Sciences Center Hospital for their cooperation in drawing the patient blood samples.

REFERENCES

Cardwell, J. C.; and Abramson, D. I. The atrioventricular conduction system of the beef heart. *Am. J. Anat.,* 49, 167–192 (1931).

Cavallini, D.; Duprè, S.; Federici, G.; Solinas, S.; Ricci, G.; Antonucci, A.; Spoto, G.; and Matarese, M. Isethionic acid as a taurine co-metabolite. In *Taurine and Neurological Disorders,* A. Barbeau and R. J. Huxtable, eds. Raven Press, New York (1978), pp. 29–34.

Chazov, E. I.; Malchikova, L. S.; Lipina, N. V.; Asafov, G. B.; and Smirnov, V. N. Taurine and electrical activity of the heart. *Circ. Res.,* 34 and 35 (Suppl. III), 11–21 (1974).

Chesney, R. W.; Jax, D. K.; Scriver, C. R.; and Mohyuddin, F. Taurine transport in mammalian kidney. In *Taurine and Neurological Disorders,* A. Barbeau and R. J. Huxtable, eds. Raven Press, New York (1978), pp. 73–93.

Chubb, J.; and Huxtable, R. J. Transport and biosynthesis of taurine in the stressed heart. In *Taurine and Neurological Disorders,* E. Barbeau and R. J. Huxtable, eds. Raven Press, New York (1978a), pp. 161–178.

Chubb, J.; and Huxtable, R. The effects of isoproterenol on taurine concentration in the rat heart. *Eur. J. Pharmacol.* 48:357–367 (1978b).

Chubb, J. and Huxtable, R.: Isoproterenol-stimulated taurine influx in the perfused rat heart. *Eur. J. Pharmacol.,* 48, 369–376 (1978c).

Crass, M. F., III; and Lombardini, J. B. Loss of cardiac muscle taurine after acute left ventricular ischemia. *Life Sciences,* 21, 951–958 (1977).

Crass, M. F., III; and Lombardini, J. B. Release of tissue taurine from the oxygen-deficient perfused rat heart. *Proc. Soc. Exp. Biol. Med.,* 157, 486–488 (1978).

Crass, M. F., III; McCaskill, E. S.; Shipp, J. C.; and Murthy, V. K. Metabolism of endogenous lipids in cardiac muscle: Effect of pressure development. *Am. J. Physiol.,* 220, 428–435 (1971).

Crass, M. F., III; and Sterrett, P. R. Distribution of glycogen and lipids in the ischemic canine left ventricle: Biochemical and light and electron microscopic correlates. In *Recent Advances in Studies on Cardiac Structure and Metabolism,* P. Roy and G. Rona, eds. University Park Press, Baltimore (1975), Vol. 10, pp. 251–263.

Daniell, H. B.; Bagwell, E. E.; and Walton, R. P. Limitation of myocardial function by reduced coronary blood flow during isoproterenol action. *Circ. Res.,* 21, 85–98 (1967).

Dolara, P.; Agresti, A.; Giotti, A.; and Pasquini, G. Effect of taurine on calcium kinetics of guinea-pig heart. *Eur. J. Pharmacol.,* 24, 352–358 (1973).

Dolara, P.; Agresti, A.; Giotti, A.; and Sorace, E. The effect of taurine on calcium exchange of sarcoplasmic reticulum of guinea-pig heart studied by means of dialysis kinetics. *Can. J. Physiol. Pharmacol.,* 54, 529–533 (1976).

Dolara, P.; Ledda, F.; Mugelli, A.; Mantelli, L.; Zilletti, L.; Franconi, F.; and Giotti, A. Effect of taurine on calcium, inotropism, and electrical activity of the heart. In *Taurine and Neurological Disorders,* A. Barbeau and R. J. Huxtable, eds. Raven Press, New York (1978), pp. 151–159.

Entman, M. L.; Bornet, E. P.; and Bressler, R. The effect of taurine on cardiac sarcoplasmic reticulum. *Life Sciences,* 21, 543–550 (1977).

Fellman, J. H.; Roth, E. S.; and Fugita, T. S. Taurine is not metabolized to isethionate in mammalian tissue. In *Taurine and Neurological Disorders,* A. Barbeau and R. J. Huxtable, eds. Raven Press, New York (1978), pp. 19–24.

Griggs, D. M., Jr.; Tchokoev, V. V.; and Chen, C. C. Transmural differences in ventricular tissue substrate levels due to coronary constriction. *Am. J. Physiol.,* 222, 705–709 (1972).

Gruener, R.; and Bryant, H. J. Excitability modulation by taurine: Action on axon membrane permeabilities. *J. Pharmac. Exp. Ther.,* 194, 514–521 (1975).

Gruener, R.; Markovitz, D.; Huxtable, R.; and Bressler, R. Excitability modulation by taurine: Transmembrane measurements of neuromuscular transmission. *J. Neurol. Sci.,* 24, 351–360 (1975).

Gruener, R.; Bryant, H.; Markovitz, D.; Huxtable, R.; and Bressler, R. Ionic actions of taurine on nerve and muscle membranes: Electrophysiologic studies. In *Taurine,* R. Huxtable and A. Barbeau, eds. Raven Press, New York (1976), pp. 225–242.

Guidotti, A.; and Giotti, A. Taurina e sistema cardiovascolare. *Recent. Progr. Med.* (Rome), 49, 61–97 (1970).

Guidotti, A.; Badiani, G.; and Giotti, A. Potentiation by taurine of inotropic effect of strophanthin-K on guinea-pig isolated auricles. *Pharmacol. Res. Comm.,* 3, 29–38 (1971).

Handforth, C. P. Isoproterenol-induced myocardial infarction in animals. *Arch Path.,* 73, 161–165 (1962).

Hinton, J. R.; Souza, J. D.; and Gillis, R. A. Deleterious effects of taurine in cats with digitalis-induced arrhythmias. *Eur. J. Pharmacol.,* 33, 383–387 (1975).

Foffstein, S.; Gennaro, D.; Weissmann, G.; Hirsch, J.; Streuli, F.; and Fox, A. C. Cytochemical localization of lysosomal enzyme activity in normal and ischemic dog myocardium. *Am. J. Pathol.,* 79, 193–206 (1975).

Huxtable, R.; and Bressler, R. Taurine concentrations in congestive heart failure. *Science,* 184, 1187–1188 (1974a).

Huxtable, R.; and Bressler, R. Elevation of taurine in human congestive heart failure. *Life Sciences,* 14, 1353–1359 (1974b).

Huxtable, R.; and Chubb, J. Taurine and isoproterenol toxicity. *Proc. West. Pharmacol. Soc.,* 19, 316–319 (1976).

Huxtable, R.; and Chubb, J. Adrenergic stimulation of taurine transport by the heart. *Science,* 198, 409–411 (1977).

Idell-Wenger, J. A.; and Neely, J. R. Effects of ischemia on myocardial fatty acid oxidation. In *Pathophysiology and Therapeutics of Myocardial Ischemia,* A. M. Lefer, G. J. Kelliher, and M. J. Rovetto, eds. Spectrum Publications, New York (1977), pp. 227–238.

Jacobsen, J. G.; and Smith, L. H., Jr. Biochemistry and physiology of taurine and taurine derivatives. *Physiol. Rev.,* 48, 424–511 (1968).

Kocsis, J. J.; Kostos, V. J.; and Baskin, S. I. Taurine levels in the heart tissues of various species. In *Taurine,* R. Huxtable and A. Barbeau, eds. Raven Press, New York (1976), pp. 145–153.

Kutsuna, F. Electron microscopic studies on isoproterenol-induced myocardial lesions in rats. *Japan. Heart J.,* 13, 168–175 (1972).

Lombardini, J. B. An enzymatic derivative double isotope assay for measuring tissue levels of taurine. *J. Pharmacol. Exp. Ther.,* 193, 301–308 (1975).

Moir, T. W. Subendocardial distribution of coronary blood flow and the effect of antianginal drugs. *Circ. Res.,* 30, 621–627 (1972).

Morgan, H. E.; Neely, J. R.; Wood, R. E.; Liebecq, C.; Liebermeister, H.; and Park, C. R. Factors affecting glucose transport in heart muscle and erythrocytes. *Fed. Proc.,* 24, 1040–1045 (1965).

Neely, J. R.; Liebermeister, H.; Battersby, E. J.; and Morgan, H. E. Effect of pressure development on oxygen consumption by isolated rat heart. *Am. J. Physiol.,* 212, 804–814 (1967).

Novelli, G. P.; Ariano, M.; and Francini, R. Un nuovo medicamento per la prevenzione delle aritmie: la taurina. *Minerva Anest.,* 35, 1241–1250 (1969).

Opie, L. H. Metabolic response during impending myocardial infarction: I. Relevance of studies of glucose and fatty acid metabolism in animals. *Circulation,* 45, 483–490 (1972).

Peterson, M. B.; Mead, R. J.; and Welty, J. D. Free amino acids in congestive heart failure. *J. Mol. Cell. Cardiol.,* 5, 139–147 (1973).

Read, W. O.; and Welty, J. D. Effect of taurine on epinephrine and digoxin induced irregularities of the dog heart. *J. Pharmacol. Exp. Ther.,* 139, 283–289 (1963).

Rona, G.; Chappel, C. I.; Balazs, T.; and Gaudry, R. An infarct-like myocardial lesion and other toxic manifestations produced by isoproterenol in the rat. *Arch. Path.,* 67, 443–455 (1959).

Rona, G.; Boutet, M.; Huttner, I.; and Peters, H. Pathogenesis of isoproterenol-induced myocardial alterations: Functional and morphological correlates. In *Recent Advances in Studies on Cardiac Structure and Metabolism,* N. S. Dhalla, ed. University Park Press, Baltimore (1973), Vol. 3, pp. 507–525.

Stanton, H. C.; Brenner, G.; and Mayfield, E. D., Jr. Studies on isoproterenol-induced cardiomegaly in rats. *Am. Heart J.,* 77, 72–80 (1969).

Welty, J. D.; and Read, W. O. Studies on the function of taurine in the heart. *Proc. S.D. Acad. Sci.,* 42, 157–163 (1963).

Discussion

DR. JOHN SILVERIO (Wyeth Laboratories): Has any work been done suggesting that breast milk may protect the premature baby against retrolental fibroplasia?

DR. DAVID RASSIN (Institute for Basic Research in Mental Retardation, New York): As far as I know, there has been no demonstrated morphologic or biochemical detriment to the newborn infant receiving a taurine-deficient diet, although there is some indication that changes occur in the bile acids. We still don't know if this has a negative influence on retrolental fibroplasia.

DR. LOUISE JOHNSON (Graduate Hospital, University of Pennsylvania): Dr. Rassin, what kind of taurine supplement are you using in your studies? Also, why have none of the commercial formulas added taurine if it is possible to feed taurine to infants?

DR. DAVID RASSIN (Institute for Basic Research in Mental Retardation, New York): The major reason it is not being added to the formulas in the United States at the present time is because of FDA regulations. The taurine supplement studies are being carried out in Finland. We're giving exactly the same amount of taurine as human milk-fed children would receive.

DR. WALTER LOVENBERG (National Institutes of Health): Dr. Rassin, do you see any physiological or clinical changes in the infants that receive taurine?

DR. DAVID RASSIN (Institute for Basic Research in Mental Retardation, New York): The taurine supplement studies only began two years ago. If you consider both the number of infants required for these studies and the time involved, it will probably be six or seven years before the desired information will be available.

DR. CLAIRE LATHERS (Medical College of Pennsylvania): Dr. Baskin, have you looked at the changes in taurine content of various animal epileptic models?

DR. STEVEN BASKIN (Medical College of Pennsylvania): We looked at audiogenic seizures in Flora O'Grady (Bronx, N.Y.) strain mice and found that taurine levels decreased. Dr. Barbeau has looked at ouabain-induced convulsions and also observed a decrease in taurine levels.

DR. CLAIRE LATHERS (Medical College of Pennsylvania): Are there any experiments that show whether exogenous taurine is an anticonvulsant agent?

DR. STEVEN BASKIN (Medical College of Pennsylvania): Taurine doesn't readily penetrate the blood-brain barrier. Only in temporal lobe epilepsy, in which there may be a breakdown of the blood-brain barrier, do you seem to see an effect of taurine. When the blood-brain barrier is in its natural state, the effect of taurine seems to be greatly reduced. Thus, a more lipophilic or less acidic analogue of taurine may be preferred as an antiepileptic agent. Perhaps Dilantin is such an agent.

DR. BARRY LOMBARDINI (Texas Tech University): Dr. Tamas Frigyesi and I have looked at a series of cats with a penicillin epileptogenic focus. We find no difference in the taurine levels of a variety of regions between the penicillin focus site and the con-ftralateral site. Moreover, in high doses we find that taurine itself is epileptogenic.

DR. RYAN HUXTABLE (University of Arizona): In our audiogenic epileptic model, taurine is strongly anticonvulsant and has a very long-lasting effect. However, there is no difference in taurine levels between these and non-seizure control animals. Therefore, although taurine is antiepileptic, it is not due to replacement therapy. In this regard, some clinical trials have been carried out with taurine in a number of countries and the results are mixed. In about half of the cases, the patients seem to be helped. We have been treating one patient in Tucson with taurine, and the results are favorable. Although this patient was receiving other medication at the same time, during the period of taurine administration, he was remarkably seizure free even though he had normally been seizuring several times a day.

DR. S. S. OJA (Tampere University of Tampere): To my knowledge, most of the clinical taurine trials have been carried out on patients who have been very unresponsive to other medication, suggesting that they don't represent the normal epileptic population but rather are the worst cases.

DR. CHRISTINA VANDERWENDE (Rutgers University): We've looked at the effect of taurine on morphine-induced seizures in mice. One observes two different LD50s, one where the animals die through seizure activity, and the other where they die from a depressant effect on the CNS. We examined a dose of morphine that would cause the animal to die through seizures. However, pretreatment with taurine reduced the number of seizures and protected the animal from death due to seizure activity. On the other hand taurine had no effect on animals treated with a dose of morphine that caused death by CNS depression. Thus, the taurine effect seems to relate specifically to the seizure-induced mechanism.

DR. JAMES KOCSIS (Jefferson Medical College): Dr. VanderWende, is the dose that you give for producing morphine seizures higher than the dose that kills them by depression?

DR. CHRISTINA VANDERWENDE (Rutgers University): Yes. It is both a dose-dependent phenomenon and a time-dependent function. You get both CNS depression and seizure activity at the same time, but this is further complicated by postictal depression. Most likely, death due to seizure activity is related to the postictal depressive stage.

DR. DAVID RASSIN (Institute for Basic Research in Mental Retardation, New York): Dr. Lombardini, does the plasma taurine content of patients suffering from a myocardial infarct remain elevated for a long time after the infarct?

DR. BARRY LOMBARDINI (Texas Tech University): Our data is still preliminary. However, the rise in taurine levels appears to parallel the increase in CPK isoenzyme activity. It increases about 12 to 18 hours after the infarct and falls about 30 hours later.

DR. NARANJAN DHALLA (University of Manitoba): Dr. Lombardini, is there any relationship between tissue taurine depletion and myocardial cell damage in any of your models?

DR. BARRY LOMBARDINI (Texas Tech University): We haven't carried out any pathological studies on the extent of myocardial damage. The point that we were trying to make is that taurine is specifically being leached out of the heart. If ischemia or hypoxia were leading to a general necrosis and cell death, one would expect a loss of all of the amino acids rather than just four of them. There is clearly a selective action.

DR. GENE LENTINI (Philadelphia College of Osteopathy): Is there a correlation between taurine and methionine levels?

DR. BARRY LOMBARDINI (Texas Tech University): We have not examined that question directly; however, in the ischemic myocardium the two amino acids do not appear to be related.

DR. GENE LENTINI (Philadelphia College of Osteopathy): Dr. Lombardini, is there a relationship between the loss of the four amino acids during ischemia and their normal membrane permeabilities?

DR. BARRY LOMBARDINI (Texas Tech University): I really don't know.

Index